# THE HELIUM LIQUIDS

# The Helium Liquids

Proceedings of the Fifteenth Scottish Universities Summer School in Physics, 1974

A NATO Advanced Study Institute

*Edited by*

JONATHAN G. M. ARMITAGE

*Department of Physics*
*University of St Andrews*
*Scotland*

and

IAN E. FARQUHAR

*Department of Theoretical Physics*
*University of St Andrews*
*Scotland*

1975

ACADEMIC PRESS

LONDON · NEW YORK · SAN FRANCISCO

A Subsidiary of Harcourt Brace Jovanovich, Publishers

ACADEMIC PRESS INC. (LONDON) LTD.
24/28 Oval Road
London NW1

*United States Edition published by*
ACADEMIC PRESS INC.
111 Fifth Avenue
New York, New York 10003





Library of Congress Catalog Card Number: 75–19616
ISBN: 0–12–062 550–4

Printed in Great Britain by
Page Bros (Norwich) Ltd, Norwich

# List of Participants

## ORGANIZING COMMITTEE

ALLEN, PROFESSOR J. F. *University of St Andrews, Scotland (Director).*
ARMITAGE, MR J. G. M. *University of St Andrews, Scotland (Secretary and Joint Editor).*
McCALL, MR D. M. *University of St Andrews, Scotland (Treasurer).*
FARQUHAR, DR I. E. *University of St Andrews, Scotland (Joint Editor).*
BARKLA, MR. H. M. *University of St Andrews, Scotland (Social Secretary).*
HARPER, PROFESSOR P. G. *Heriot-Watt University, Scotland.*
CLARK, DR R. C. *University of Aberdeen, Scotland.*
HORNE, MISS K. M. *University of St Andrews, Scotland (Steward).*
MILLER, MR D. B. *University of St Andrews, Scotland (Steward).*

## GOVERNING COMMITTEE

DINGLE, PROFESSOR R. B. *University of St Andrews, Scotland (Chairman).*
FINLAYSON, DR D. M. *University of St Andrews, Scotland (Secretary-Treasurer).*

## LECTURERS

ANDERSON, DR P. W. *Cavendish Laboratory, Cambridge, England; also Bell Laboratories, Murray Hill, New Jersey 07974, U.S.A.*
BAYM, PROFESSOR G. *University of Illinois, Urbana, Illinois, 61801, U.S.A.*
BRINKMAN, DR W. F. *Bell Laboratories, Holmdel, New Jersey 07733, U.S.A. also H. C. Ørsted Institute, Copenhagen, Denmark.*
CAMPBELL, DR L. J. *Los Alamos Scientific Laboratory, University of California, Los Alamos, New Mexico, 87544, U.S.A.*
CHALLIS, PROFESSOR L. J. *Department of Physics, University of Nottingham, England.*
CHESTER, PROFESSOR G. V. *Laboratory of Atomic and Solid State Physics, Cornell University, Ithaca, N.Y. 14850, U.S.A.*
EGGINGTON, DR ANN *Department of Theoretical Physics, Oxford University, Oxford, England.*
PUTTERMAN, DR S. *Physics Department, University of California, Los Angeles, 90024, U.S.A.*
RICKAYZEN, PROFESSOR G. *Physics Laboratory, The University, Canterbury, Kent, England.*
RON, PROFESSOR A. *Physics Department, Technion-Israel Institute of Technology, Haifa, Israel.*
WHEATLEY, PROFESSOR J. C. *Department of Physics, University of California, San Diego, La Jolla, California, 92037, U.S.A.*

ALNAIMI, MR A. E. *University College of North Wales, Wales.*
ATAD, MR E. *Tel-Aviv University, Israel.*
AZIZ, MR S. M. *University of St Andrews, Scotland.*
BAROCCHI, PROFESSOR F. *Laboratorio di Elettronica Quantistica, Universita di Firenze, Italy.*
BARSOUMIAN, MR K. H. *American University of Beirut, Lebanon.*
BISHOP, DR R. F. *The University, Manchester, England.*
BUCHAN, MR G. *University of Aberdeen, Scotland.*
CASTAING, DR B. *Groupe de Physique des Solides de l'Ecole Normale Supérieure (CNRS) France.*
CHOU, MR C. S. *University of St Andrews, Scotland.*
COLCLOUGH, DR A. R. *National Physical Laboratory, England.*
COLE, DR M. W. *University of Washington, U.S.A.*
COLINET, MRS E. *Université Libre de Bruxelles, Belgium.*
CRARY, MR S. B. *University of Washington, U.S.A.*
CROSS, MR M. C. *University of Cambridge, England.*
DE HAAN, DR M. J. *Université Libre de Bruxelles, Belgium.*
DE HAAS, DR W. *Kamerlingh Onnes Laboratorium, The Netherlands.*
DELHEY, MR P. P. J. *Reactor Centrum Nederland, The Netherlands.*
DE MICHELIS, DR C. G. *University of Milan, Italy.*
DE VOOGT, DR W. J. P. *Kamerlingh Onnes Laboratorium, The Netherlands.*
FATOUROS, DR P. P. *Ohio State University, U.S.A.*
FERNANDEZ, DR J. F. *Instituto Venezolano de Investigaciones Científicas, Venezuela.*
FLOUQUET, DR J. *Laboratoire de Physique des Solides, (CNRS) France.*
GILLESSEN, DR P. *Justus Liebig Universität, Germany.*
GOMERSALL, MR I. R. *University of Bristol, England.*
GRAHAM, PROFESSOR R. *Universität Stuttgart, Germany.*
GREENSPOON, MR S. *University of Waterloo, Canada.*
HAAVASOJA, MR T. *Helsinki University of Technology, Finland.*
HEBRAL, DR B. *CRTBT (CNRS), Grenoble, France.*
IJSSELSTEIN, DR R. R. *Kamerlingh Onnes Laboratorium, The Netherlands.*
IKUSHIMA, PROFESSOR A. *University of Tokyo, Japan.*
ITOH, DR N. *University of Cambridge, England.*
JOSE, MR J. V. *Brown University, U.S.A.*
JUNKER, MR W. R. *Brown University, U.S.A.*
KARATCHENTZEFF, MR M. *Université de Paris VI, France.*
KESSEL, DR W. *Physikalisch-Technische Bundesanstalt, Germany.*
KIEWIET, MR C. W. *The University, Manchester, England.*
KING, MISS C. A. *University of London, England.*
KOOPS, DR W. *Kamerlingh Onnes Laboratorium, The Netherlands.*
KORCZYNSKYJ, MR Y. *University of East Anglia, England.*
LEE, MR J. F. *Columbia University, U.S.A.*
LEROUX, MR A. *CRTBT (CNRS), Grenoble, France.*
LOPONEN, MR M. *Helsinki University of Technology, Finland.*
MAIN, MR P. C. *The University, Manchester, England.*

MERCAIDE, MR J. *University of St Andrews, Scotland.*
MEYUHAS, MR R. *Tel-Aviv University, Israel.*
MILLER, MR D. B. *University of St Andrews, Scotland.*
MITCHELL, MR H. B. *University of Sussex, England.*
NAGURNEY, MR L. S. *Brown University, U.S.A.*
NAMAIZAWA, DR H. *University of Saarland, Germany.*
PEARLMAN, PROFESSOR N. *Purdue University, U.S.A.*
PENMAN, MR D. A. *The University, Manchester, England.*
PENNINGS, DR N. H. *Kamerlingh Onnes Laboratorium, The Netherlands.*
SALEMINK, DR H. *Katholieke Universiteit, The Netherlands.*
SCHUMACHER, DR G. *CRTBT (CNRS), Grenoble, France.*
SCHWARTZ, MR M. *Tel-Aviv University, Israel.*
SERVADIO, DR S. *Dipartimento di Scienze Fisiche, Ancona, Italy.*
SHAHZAMANIAN, MR M. A. *University of Sussex, England.*
STRAYER, DR M. *The University, Manchester, England.*
SLUCKIN, MR T. *University of Nottingham, England.*
TAKAGI, DR S. *University of Sussex, England.*
TARTAGLIA, DR A. *Instituto de Fisica Sperimentale del Politecnico, Turin, Italy.*
TITUS, DR W. *Carleton College, Minnesota, U.S.A.*
TOFT, MR M. *University of St Andrews, Scotland.*
TOSATTI, DR E. *Universität Stuttgart, Germany.*
VAN DER BOOG, DR A. G. M. *Kamerlingh Onnes Laboratorium, The Netherlands.*
VAN DIJK DR J. D. P. *Kamerlingh Onnes Laboratorium, The Netherlands.*
VAN SPRONSEN, DR E. *Kamerlingh Onnes Laboratorium, The Netherlands.*
VAROQUAUX, DR E. *Institut d'Electronique Fundamentale (CNRS), Orsay, France.*
VOSS, MR M. *Germany.*
WALLENBORN, DR J. *Université Libre de Bruxelles, Belgium.*
WOYNAROVICH, MR F. *University of Bristol, England.*
ZIFF, MR R. *Rockefeller University, U.S.A.*

## VISITOR

DOBBS, PROFESSOR E. R. *University of London, England.*

# Director's Preface

The Fifteenth Scottish Universities Summer School in Physics was held in the University of St Andrews from 21 July to 10 August 1974.

The Helium Liquids are subjects of intense study at the present time partly because the theory of the superfluid state of Helium-4 still poses considerable problems, but also and even more importantly, because of the recent discoveries and theoretical interpretations of superfluid states of Helium-3.

We were pleased to be able to secure a body of lecturers who were not only famous for their contributions to the subject but also well-known for their outstanding ability to lecture and expound the subject. The result was a very successful school attended by a large group of dedicated and enthusiastic students. Its success is reflected in these Proceedings which I believe will prove a significant contribution to the literature on the superfluids.

The School was sponsored by and had the support of the eight Scottish Universities and the Scientific Affairs Division of NATO.

The participants, the lecturers and their families were housed in University residences, mostly in Hamilton Hall, and we were greatly indebted to the Warden of Hamilton, Miss Anne Wright and her staff for the excellent organization, catering and other arrangements. The lectures and seminars were held in the Physics Building of the University.

Finally, it is a pleasure to acknowledge the great contributions made by all of those who helped to organize and run the School. Jonathan Armitage was the Secretary; he bore most of the load and has also been one of the co-editors. The other co-editor was Ian Farquhar. Donald McCall looked after the money very efficiently. Hugh Barkla organized the social side, ably assisted by his wife Stella who arranged the ladies' programme. The two stewards were research students Karen Horne and Danny Miller; they were always on hand when problems arose, and acted as couriers on all of the tours.

The devoted and efficient work of all of these was the main reason for the School's undoubted success.

J. F. Allen
School of Physical Sciences
University of St Andrews

## Editors' Note

The articles in this volume have been prepared by the authors and are based closely on the lectures presented at the Summer School. Editing has been minimal, directed towards correcting errors and achieving some degree of uniformity in notation throughout the volume.

We are grateful to Norma Pacholek, Tricia Summerfield and Linda McLean for secretarial assistance in preparing the final version of the manuscripts.

J. G. M. Armitage
I. E. Farquhar

# Contents

# 1. INTRODUCTORY LECTURES ON LIQUID HELIUM FOUR AND THREE

G. V. Chester

*Laboratory of Atomic and Solid State Physics, Cornell University, Clark Hall, Ithica, New York 14850, USA.*

## Introduction

In these lectures I will talk about six topics:

I Liquid Helium Isotopes as Liquids.
II Excitations in Liquid $He^4$ and $He^3$.
III Bose Condensation.
IV Hydrodynamics of Liquid $He^4$.
V Helium in Films and Pores.
VI Recent Developments in Liquid $He^3$.

I have chosen these topics first, because I believe they cover many of the most important areas of theory and experiment and second, because in each of them significant research is being carried out.

In four hours of lectures I cannot talk at length on these topics. My approach will be to try to present a concise summary of the background of the subject and then to point to unsolved problems which are significant research topics. The treatment will therefore be highly condensed. To help the reader I have tried to provide sufficient references so that she or he can penetrate the literature of the subject.

One further point deserves emphasis. There is no coherent body of theory for liquid $He^4$. This makes it very difficult to present an introduction to the subject which is not rather fragmented. I have tried to compensate for this by providing some connection between the topics I shall discuss. On the other hand, Fermi liquid theory does provide a coherent theoretical framework in which to discuss $He^3$. Likewise, extensions of the BCS theory of superconductivity theory to the new superfluid phases of $He^3$ also provides the necessary framework for that topic. This is the main reason that I have largely limited my discussion of topics in $He^3$ to Section VI.

References 1–18 will help the student who seeks an introduction to some aspect of the behaviour of liquid $He^4$ and $He^3$.

## I. Liquid $He^4$ and Liquid $He^3$–As Liquids

Under this title I want to consider the liquid helium isotopes as *liquids*. This is not often done—one usually concentrates on the more exotic properties of the superfluid phase of $He^4$, or the very low temperature properties of $He^3$. The reason for returning to this aspect of the helium isotopes is that there has been in the last three or four years a dramatic advance in our understanding of *dense classical fluids*. Furthermore, in the last two years it has been shown that many of the ideas used in understanding classical fluids are applicable to the helium isotopes, which are of course, moderately dense fluids dominated by quantum effects.

The basic features we would like to be able to understand in any fluids system are: (1) thermodynamic properties such as the equation of state and specific heat, (2) transport properties such as diffusion and heat transport, (3) phase transformations such as condensation and crystallization and (4) the space and time correlations in the fluid. In this brief review I shall limit myself to thermodynamic properties and spatial correlations.

First, in dense classical fluids a comparatively simple and accurate physical picture has emerged from detailed computer simulations† of these systems. Unfortunately no review article is available on this material. Key references are (19–22). I must emphasize that while we have a physical picture that provides insight and allows us to make predictions we do not have an analytical theory. This however is perhaps not as important as it might seem, as long as we understand the physical phenomena. The classical fluids that have been extensively studied by simulation techniques are fluids of simple spherical atoms such as liquid argon. Second, there is strong evidence[(23–25)] from very recent computer simulations on boson systems that the same physical picture is applicable to liquid $He^4$.

Finally, there is indirect[(25)] evidence that with some modification we will be able to understand some of the basic fluid properties of $He^3$ in the same

† A computer simulation of a classical dense fluid can be performed either by solving the equation of motion for $N$ (usually several hundred) particles interacting via a realistic two body potential (such as the Lennard-Jones potential) or by a Monte Carlo evaluation of the thermodynamic properties of the system. The first method can be used to study both equilibrium and nonequilibrium properties; the second is confined to equilibrium properties. Both these methods are used in the references referred to in this section. A simulation can provide very detailed *microscopic* information about the fluid.

way. I have therefore divided this section into the following subsections.
(a) Review of the theory of dense classical fluids,
(b) The hard sphere model of $He^4$,
(c) The hard sphere model of $He^3$.

### (a) *A Review of the Theory of Dense Classical Fluids*

All the fluids I have mentioned consist of spherical atoms whose interaction is very strongly repulsive at short distances and rather weakly attractive at larger distances. It turns out that the heuristic Lennard-Jones potential:

$$V(r) = 4\varepsilon[(\sigma/r)^{12} - (\sigma/r)^6]$$

provides a basis for a good qualitative (or even semi-quantitative) description of the properties of simple classical and quantum fluids. It is not completely satisfactory but nevertheless very good. The attractive part of the potential is strong enough to condense all these fluids—even the quantum fluids condense at low temperatures. The classical fluids such as argon, xenon, krypton all obey rather accurately a law of corresponding states(26) which says that if $\varepsilon$ and $\sigma$ are used to scale energies and lengths then all the physical properties of these substances are universal functions of the reduced temperature and the reduced density. This means that once we understand one such system we have understood them all. Thus the equation of state for all of them has the universal form

$$p_c^* = p_c^*(T^*, V^*)$$

where $p^*$ is the reduced pressure $(p\sigma^3/\varepsilon)$ and is a universal function of $T^*$, the reduced temperature $(kT/\varepsilon)$, and $V^*$, the reduced volume $(V/N\sigma^3)$. For the isotopes of helium quantum effects are very important and these enter the equation of state and other physical properties via the so called De Boer parameter(26) $\Lambda^* = \hbar/\sigma\sqrt{(m\varepsilon)}$. This parameter varies from substance to substance and from isotope to isotope. The equation of state for a quantum system thus has the form:

$$p_q^* = p_q^*(T^*, V^*, \Lambda^*)$$

and is thus not a universal function due to the $\Lambda^*$ dependence. It will also depend on the statistics obeyed by the particles.

All these fluids crystallize under pressure. The 'classical' fluids crystallize at very low pressures at low temperature. Quantum fluids require a finite pressure even at absolute zero for them to crystallize. It turns out that pressure is not the best variable to understand crystallization—a much better one is density—we shall return to this later.

The basic physical picture we now have of these classical fluids is that to a very good approximation the repulsive part of the potential can be replaced by a hard sphere potential and the attractive part can be treated as a perturbation whose main effect is to provide a mean attractive potential in which the hard spheres move. In essence this picture tells us that the details of the potential are not important and that one can understand most of the physical properties of dense classical fluids in terms of a fluid of hard spheres moving in a mean field potential. This model really goes back to Van der Waals, as has recently been emphasized by Alder.(19) It was first put forward in the modern literature by Longuet-Higgins and Widom(27) who showed that it gave an excellent account of the behaviour of liquid argon at the triple point. The mean field attractive potential does not play an important role in determining the microscope motion of the particles in the fluid. If it did we could not logically refer to it as a *mean field*. Thus the potential energy of interaction of the fluid can be replaced by

$$V \rightarrow V_{HS} + U(\rho)$$

to first approximation, where $V_{HS}$ is the hard sphere interaction and $U(\rho)$ is the mean field of the attractive part of the potential.

Three points should be made at this stage: first, the model can be considerably refined and thus made more accurate. One can divide the original potential into repulsive and attractive parts, find the best way of representing the repulsive part by a hard sphere model and then treat the remaining attractive portion by perturbation theory.(22) The details of these procedures need not concern us because they do not add to the physical picture—they simply provide better ways of computing numbers using it.

Second, if the physical picture is correct, then it follows that the spatial correlations between the particles must be essentially the same as those in a hard sphere fluid. We show in Figs. 1, 2 and 3 the structure factor for a Lennard-Jones model liquid argon, krypton, and that of a hard sphere system at different densities. The agreement is excellent. (The figures also show that the structure factor for Lennard-Jones liquid argon agrees very well with the best experimental data on argon.)

Third, how can we understand why this simple picture is so accurate? What must be happening is that the random particle motions are averaging out the weaker parts of the potential leaving only the very repulsive part unchanged—i.e. effectively reducing it to a hard sphere potential. This averaging process will clearly be more effective the denser the fluid and the faster the particles move—i.e. the hotter the fluid. One would not expect such effective averaging in a low density fluid and this is confirmed by computer simulations. There is also evidence(28) that the averaging process is more effective in a disordered fluid than in a crystal.

Perhaps one of the most striking predictions of this model is that any

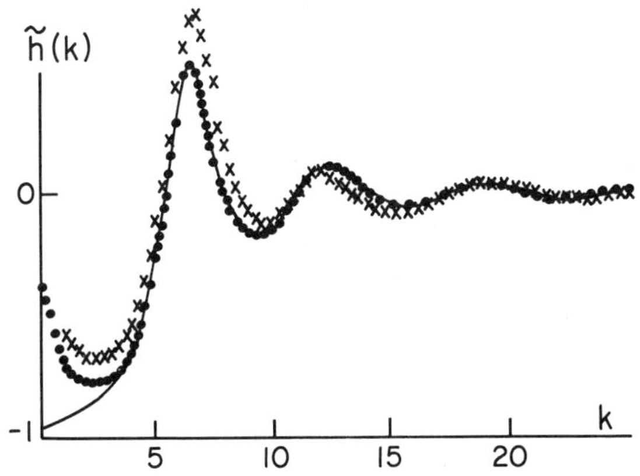


FIG. 1. $\tilde{h}(k)$ as a function of $k$. Dots: Results from molecular dynamics for $T = 1{\cdot}326$, $\rho = 0{\cdot}5246$. Solid line: hard-sphere model, with the hard-sphere diameter $a = 1{\cdot}0$ and the hard-sphere density $\hat{\rho} = 0{\cdot}57$. Crosses: x-ray rxperiment on argon at the same density and $T = 1{\cdot}28$. $\tilde{h}(k) = S(k) - 1$.

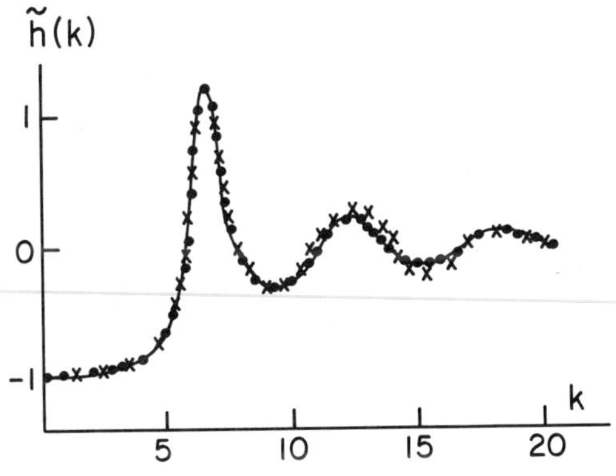


FIG. 2. $\tilde{h}(k)$ as a function of $k$. Dots: Results from molecular dynamics for $T = 1{\cdot}827$, $\rho = 0{\cdot}75$. Solid line: hard-sphere model, with $a = 1{\cdot}03$, $\hat{\rho} = 0{\cdot}817$. Crosses: neutron epxeriment on krypton. $\rho = 0{\cdot}77$, $T = 0{\cdot}77$. $\tilde{h}(k) = S(k) - 1$.

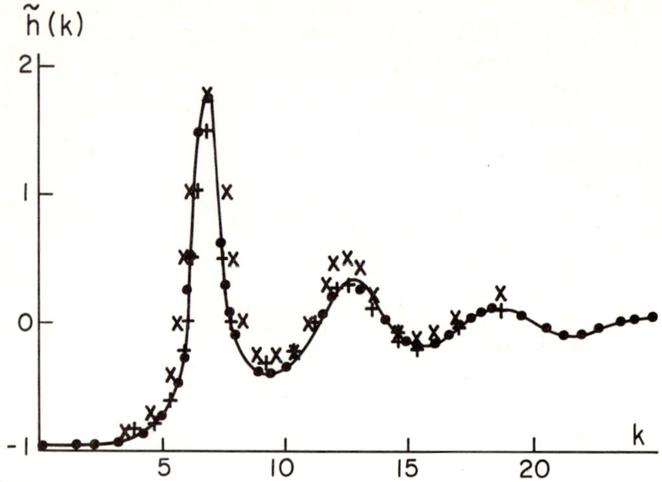


FIG. 3. $\tilde{h}(k)$ as a function of $k$. Dots: Results from molecular dynamics for $T = 0{\cdot}723$, $\rho = 0{\cdot}844$. Solid line: hard-sphere model with $a = 1{\cdot}026$, $\hat{\rho} = 0{\cdot}91$. +: neutron experiment on argon at the same density and $T = 0{\cdot}7$. X: x-ray experiment on argon at the same density and $T = 0{\cdot}7$. $\tilde{h}(k) = S(k) - 1$.

classical system with a strongly repulsive potential at short distances should crystallize at approximately the same reduced density, and that this reduced density should be equal to that at which the hard sphere system crystallizes. This prediction has been confirmed by computer experiments.

## (b) *The Hard Sphere Model of He*$^4$

In our discussion of quantum fluids we shall limit our considerations to absolute zero, i.e. structure and properties of the ground state.

The recent paper by Kalos, Levesque and Verlet[(25)] has shown that the picture we have outlined for classical fluids is also applicable to liquid $He^4$ as far as its basic structural properties are concerned i.e. equation of state and spatial correlations. The paper also suggests that it will also be applicable to liquid $He^3$—basically because Fermi statistics play a small role in determining the equation of state and static particle correlations. (They do of course play a vital role in specific heat and transport properties.)

What these authors showed†—by exact computer simulations of the Bose

† The method used by these authors amounts to an exact solution of Schrödinger's equation for the ground state of a boson system. However the equation is not solved by any of the "usual" numerical techniques but by a Monte Carlo method. This method can be applied because the ground state wave function can be shown to be of one sign and so can be taken to be always positive. The wave function itself therefore has a probability interpretation and it is this that allows the integration of the equation by Monte Carlo techniques. The method is not directly applicable to excited states whose wave function must of course contain nodes, for the same reason it is also not applicable to Fermi systems.

quantum system—was that the equation of state, static correlation function $g(r)$, and the crystallization density of a Lennard-Jones model of helium could all be accurately calculated from a perturbation scheme in which a hard sphere model was the first approximation and the attractive part of the Lennard-Jones potential was included as a perturbative correction. *This is the same approach as was used in classical fluids.* The most striking confirmation of the approach is shown in Fig. 4 where we compare the structure function $S(k)$ for a quantum hard sphere system with that of liquid $He^4$. Finally the crystallization phenomena of the hard sphere model and the Lennard-Jones model can be understood in the following way. The hard sphere boson system crystallizes at absolute zero wheh $n\sigma_{HS}^3 = 0{\cdot}25$; $\sigma_{HS}$ is the hard sphere radius. The Lennard-Jones system crystallizes when $n\sigma^3 = 0{\cdot}43$, $\sigma =$ Lennard-Jones parameter. At first sight there is no connection between these two. conditions. However, if we use the following argument then these results become closely related. We ask what length should be used to characterize the Lennard-Jones potential in the quantum system. Clearly we should look for the length that is equivalent to $\sigma_{HS}$. The simplest way to answer the question is to look at the two body scattering problem. Then we discover that $\sigma_{HS}$ is the *s*-wave scattering length at zero energy. We therefore ask what is the equivalent length for the Lennard-Jones potential and find that it is given by $\tilde{\sigma} = 0{\cdot}84\sigma$.† If we use this length to characterize the Lennard-Jones potential we find that $n\tilde{\sigma}^3 = 0{\cdot}26$ which is remarkable. I have little doubt that any more elaborate scheme would produce the same kind of agreement with the hard sphere system.

It is worthwhile to summarize some of the results of the computer simulations on $He^4$. We have already displayed the results for the structure factor $S(k)$ (Fig. 4). The small binding energy, about $-7$ K per particle, is the difference between a large potential energy term ($-19$ K per particle) and a large kinetic energy (12 K) per particle). The theoretical crystallization density is given by $n\sigma^3 = 0{\cdot}45$; the experimental data yields $n\sigma^3 = 0{\cdot}43$. From the results I think it is reasonable to conclude that we can now conduct accurate simulations of $He^4$ at absolute zero.

### (c) *The Hard Sphere Model of $He^3$*

For liquid $He^4$ we could use the exact computer results for boson hard spheres as a basis for our discussion. The method used to solve Schrödinger's equation is not directly applicable to Fermi systems and so we do not have similar results available. Nevertheless I think that the following points are worth making.

† This is of course the scattering length for the repulsive part of the potential. This is defined to be that part of the potential which produces a repulsive force.

(i) Both $He^3$ and $He^4$ are composed of atoms which interact via essentially identical forces. We have seen that the basic structure of the wave function for the boson system is completely dominated by the hard core of the potential—this is strikingly revealed by Fig. 4.

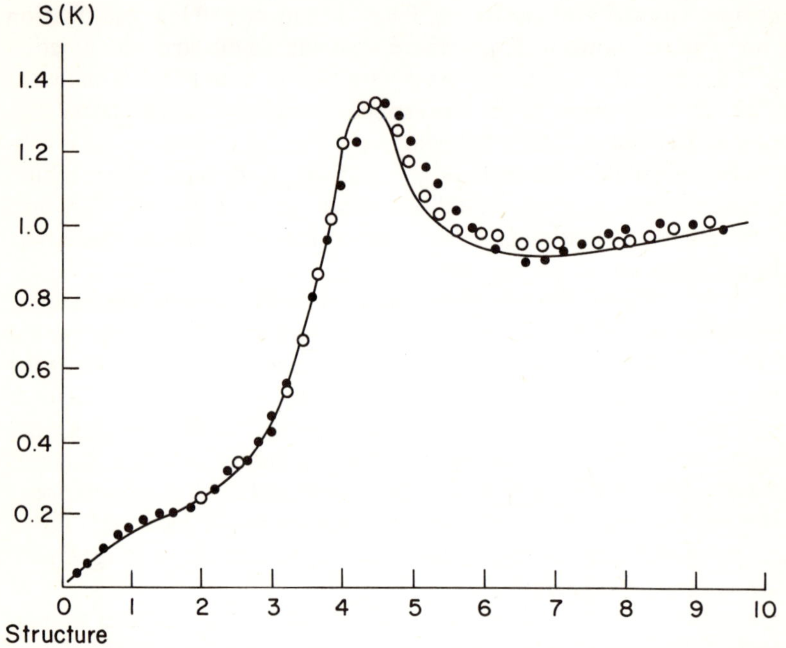


FIG. 4. The structure factor $S(k)$ at normal density for liquid helium. Solid line: experiment. Full circles: "exact" hard spheres results. Open circles: experimental data adjusted to ensure that the radial distribution function obtained by Fourier transform of $S(k)$ is zero at short distances.

(ii) The basic question we have to answer is how much change will be introduced into the wave function by the Fermi statistics obeyed by the $He^3$ atoms? We have some fairly convincing evidence from variational calculations[(23)] that very little extra structure will be introduced. In essence what was found in the calculation was that the full antisymmetry of the trial wave function could be very well approximated if only two particle permutations were taken into account in the calculation of the physical properties. This is indeed the simplest approximation to a fully antisymmetric state—apart from the obvious zeroth approximation of ignoring the antisymmetry entirely. There is probably a simple physical explanation of this—namely the very strong repulsive forces already force a complex nodal structure on the wave function—the antisymmetry requirement does very little to alter this structure.

(iii) However we must in conclusion point out that the variational calculations on which we based our argument are not nearly so successful for $He^3$ as for $He^4$. For $He^4$ the energy discrepancy between the variational results[(23)] and exact result is about 15%. The corresponding figure for $He^3$ is about 30%. Thus the argument we have presented may turn out to be erroneous. I believe that this will not be the case and that the basic structural properties, binding energy and crystallization density for $He^3$ will soon be well understood on the basis of a hard sphere model with the attractive part of the potential handled by perturbation theory.

## II. EXCITATIONS IN LIQUID $He^4$ AND $He^3$

In the first lecture I reviewed the simplest structural properties of liquid $He^4$ and $He^3$. This review was limited to the ground state. As a natural extension of these considerations I shall review what is known about the elementary excitations in these systems. This of course means that we are looking at the simplest excited states of these fluids rather than the ground state.

I shall start by reviewing what is known about excitations in liquid $He^4$ and I will divide the topic as follows:

(a) The phonon–roton spectrum, basic data and theory,
(b) Problems connected with the phonon–roton spectrum,
(c) The excitation spectrum at higher energies and momenta.

### (a) *The Phonon–Roton Spectrum*

This spectrum is now well established experimentally[(29, 17)] and much of the spectrum has been studied as a function of both density and temperature.

The basic data at vapour pressure is shown in Fig. 5. The width of the roton minimum is shown in Fig. 6, the pressure dependence of the same quantity is shown in Fig. 7. The excitations belonging to this spectrum are good elementary excitations up to about 1·5 K above which temperature they acquire an appreciable width (see Fig. 6) showing that they are becoming shorter lived. Above about 1·8 K the excitations near the roton minimum are so broad that they are no longer *independent* excitations of the system. On the other hand very long wave length (hydrodynamic) modes of both first and second sound propagate over an even wider range of temperature. These excitations account extremely accurately for the specific heat and normal fluid density up to 1·8 K—*provided we allow for their temperature dependence*[(30)] (see also (b) below). Most of the measurements of the spectrum have been made by neutron scattering. For long wavelengths (small $k$) ultrasonic and specific heat measurements have been used to explore this part of the spectrum. It is important to realise that due to the very large density of states in the neighbourhood of the roton minimum, these excitations dominate the thermodynamic and hydrodynamic properties over much of the temperature range. It is easy to show that for a spectrum with the shape shown in Fig. 5 no excitations can decay into excitations of lower energy and momenta.[31, 32] The spectrum is thus a stable one; the conservation of energy and momentum prevents decay for a spectrum of this shape. However beyond the point marked (A) on the graph, the spectrum does become unstable and excitations of that energy and above can decay into two rotons. This point was emphasized

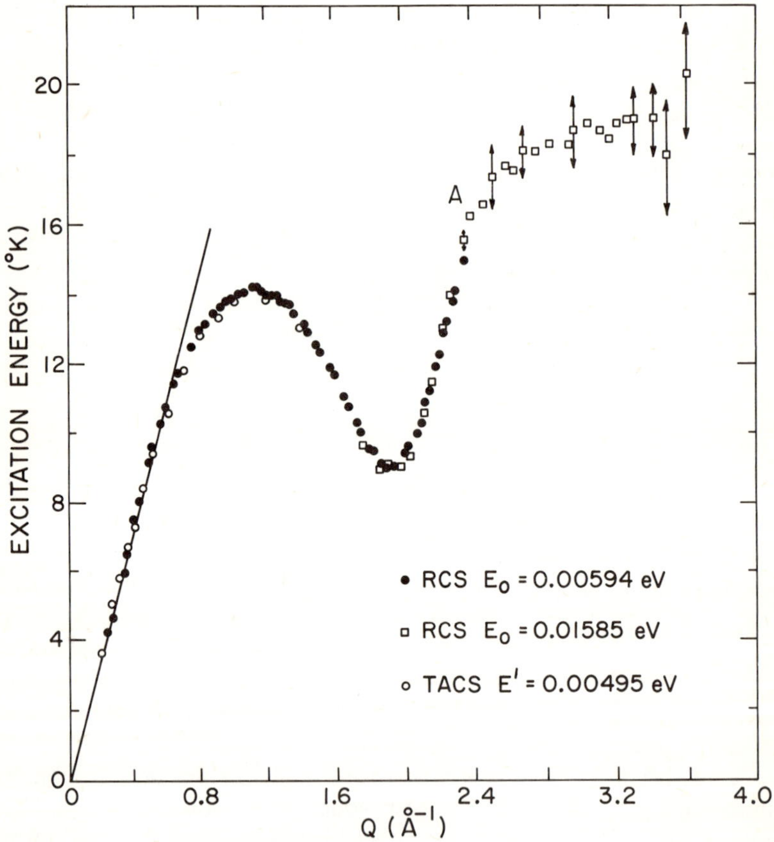


FIG. 5. The experimental results for the energies of the one-phonon excitations at 1·1 K.

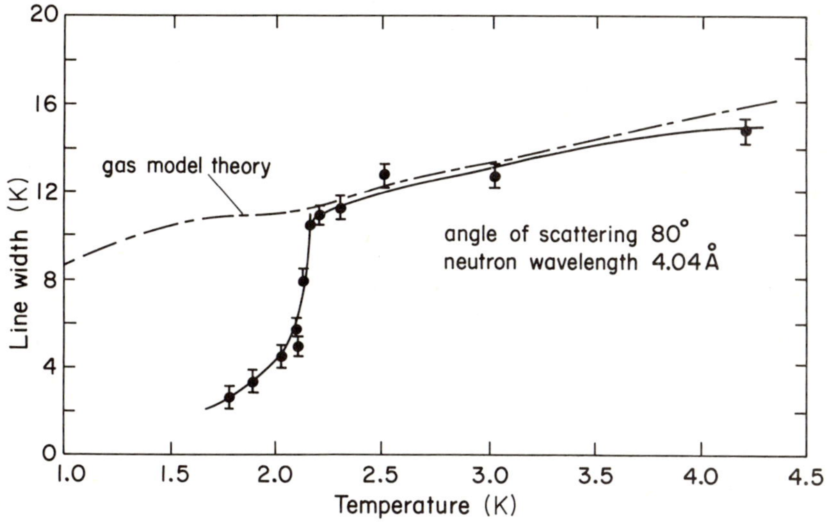


FIG. 6. The line width of neutrons scattered from liquid $^4$He as a function of temperature.

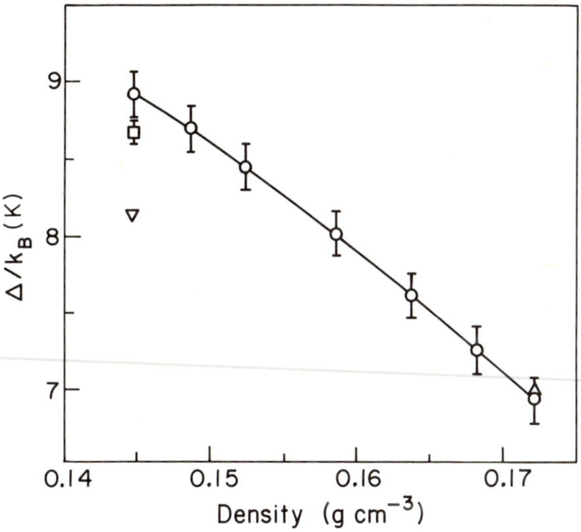


FIG. 7. The dependence of the energy gap $\Delta$ on the density: ○ from entropy; △, □, ▽, different measurements from neutron data.

by Pitaevskii[(33)] who developed a theoretical model which allowed for this and which led to a flattening of the spectrum beyond the point (A) and finally an effective termination of the spectrum. Unfortunately the experimental data does not confirm the details of Pitaevskii's model.

Finally, recent experimental work[(34)] suggests that the spectrum may be slightly concave at very small $k$. This would lead to an unstable portion of the spectrum.

At long wavelengths the spectrum corresponds to hydrodynamic modes and the slope of the curve is equal to the measured sound velocity.[(29)] It is remarkable how the linear portion of the spectrum persists up to about 0·8 Å$^{-1}$. It has been suggested[(35)] that most of this linear region corresponds to zero-sound density waves. (See the subsection on excitations in $He^3$). But no convincing experimental or theoretical evidence has been found in support of the conjecture.

Many suggestions have been made to explain the nature of the roton minimum. Landau originally suggested some kind of vortex motion—an idea later taken up by Feynman[(36)]—we shall return to these suggestions in section (b). I want to emphasize that there is no real understanding of the physical motions that underly this part of the spectrum.

Finally we note that the spectrum is very sensitive to pressure changes. Both the velocity of sound—which controls the small $k$ part of the spectrum—and the roton minimum change markedly as the pressure is increased (see Fig. 7 and Ref. 37).

### (b) *Problems Connected With the Phonon–Roton Spectrum*

First, we have no adequate theory of the spectrum apart from the hydrodynamic sound wave region. For many years it was believed that the Feynman–Cohen variational theory was at least a semiquantitative theory. Recent work[(38)] has shown that this theory does not describe at all well the behaviour of the roton minimum (and the neighbouring parts of the spectrum) as a function of pressure. We believe that the agreement between theory and experiment at vapour pressure is somewhat fortuitous. This is disappointing for we now know that we will have to substantially improve the Feynman–Cohen theory in order to have even a semi-quantitative theory. The results of the theory are compared with experiment in Fig. 8. What does this theory say about the nature of a roton? Unfortunately, very little. The trial wave function has the form:

$$\psi_T = \psi_0 \sum_j \exp(i\mathbf{k}_j \cdot \mathbf{r}_j)(1 - i \sum_l g_{jl})$$

where $g_{jl} = A\mathbf{k} \cdot \mathbf{r}_{jl}/|\mathbf{r}_{jl}|^3$. The function $g_{jl}$ was chosen on the basis of the

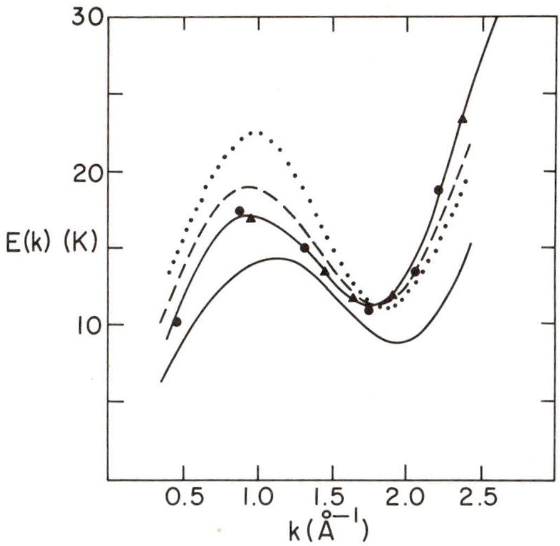


FIG. 8. The FC spectrum, as determined in this work, for three densities: solid line, $\rho = \rho_0$; dashed line, $\rho = 1{\cdot}1\,\rho_0$; dots, $= 1{\cdot}2\,\rho_0$. The raw computer calculations are shown only for $\rho = \rho_0$. The lower solid curve is the experimental dispersion curve.

model calculation for the backflow round a single impurity in liquid helium. If one "sits" on the $j$th atom and looks at the motions of the other atoms then the function $g_{jl}$ tells us that superimposed on the random zero point motion there is a long range dipole flow field around particle $j$. This led Feynman and Cohen to suggest that a roton corresponds to a very small vortex ring because such a vortex motion has a dipole flow field associated with it at large distances. Unfortunately this is not a very convincing argument because other vortex motions have dipole flows at large distances (e.g. a spherical vortex). Moreover one would expect the circulation of the flow to be quantized, and if this is so then how does this kind of flow develop smoothly into high frequency "phonons" as the wavelength gets longer? We see no convincing reason to believe that vortex motion is necessarily associated with a roton. The roton remains a mystery and so does a very large part of the rest of the spectrum.

Second, there is much debate about and measurement of the spectrum at very small $k$. It is possible that for very small $k$ the spectrum is concave upwards and is therefore unstable.[34] Apart from the stability, the details of the spectrum at very long wavelengths, while interesting, do not seem to me to be particularly significant.

Third, what causes the spectrum to end? If we look at the production cross section for excitations with different momenta (Fig. 9) we see a very striking

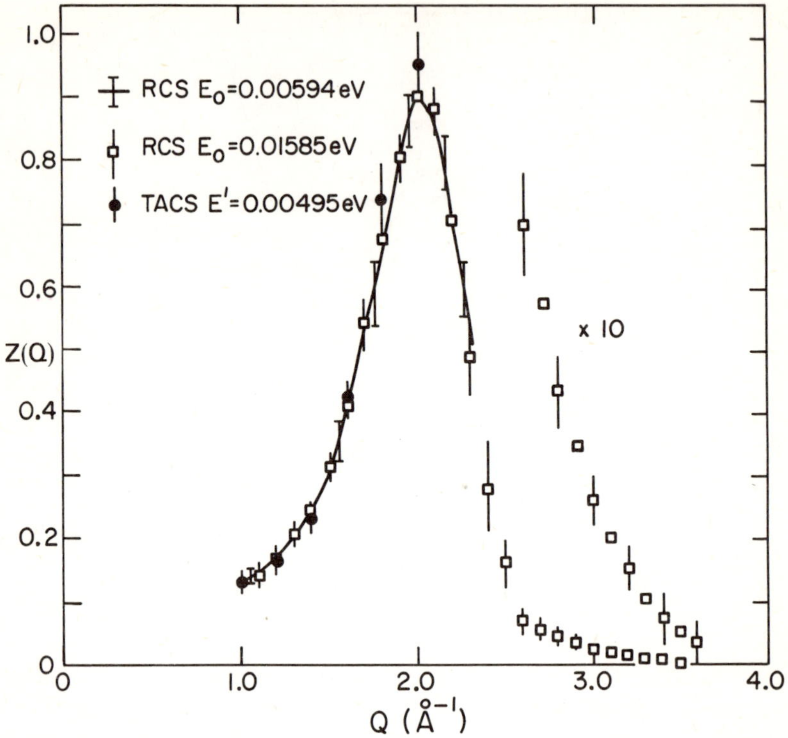


FIG. 9. The contribution $Z(Q)$ of the one-phonon excitations to the total scattering $S(Q)$ at 1·1 K. The solid line through the origin shows the theoretical small $Q$ limit at 0 K.

peak at the roton minimum and almost no height at all in the region where the spectrum levels and becomes very broad. This in itself tells us that it is very difficult for a neutron to produce these kind of excitations. (They are also of sufficiently higher energy that they are only very weakly excited thermally.) As we have already remembered, Pitaevskii suggested that when the spectrum reaches a height of twice the roton minimum it is bound to become completely unstable against decay into two rotons of appropriate momenta. This certainly will lead to a substantial broadening of the spectrum but it is not obvious that it must lead to its termination.

Finally, an old mystery still remains. It was shown many years ago[30] that if one simply substituted into the standard formulae for the specific heat and normal fluid density the measured *temperature dependent* spectrum then excellent agreement with the experimental values for these quantities is found up to 1·8 K. This is considerably beyond the point at which the width of the rotons begins to grow significantly. The important point here is that the spectrum is measured by means of neutron scattering and is thus related to "poles" of the fourier transform $[S(\mathbf{k}, \omega)]$ of the space time $G(\mathbf{r}, t)$ correlation function. It is a remarkable fact that the poles of this function account for the measured thermodynamic properties of the system over a wide range of temperature. These latter properties are related to the behaviour of the single particle Green's function—which in principle is a very different quantity from $S(\mathbf{k}, \omega)$. This problem has been partly unravelled at very low temperature. It has been shown[39] that the phonon poles of the single particle Green's function account for the $T^3$ specific and that[40] the same poles account for the $T^4$ dependence of $\rho_n$. The mystery still remains at higher temperatures, particularly in the region where the spectrum is considerably broadened.

(c) *Excitations at Higher Energies and Momenta*

A few years ago[41] the Chalk River group discovered what might be called a second branch of the excitation spectrum. It is shown in Fig. 10 as a curve with a broad band of shading across it. The shaded area shows in an approximate fashion the width of the excitations in this branch. In Fig. 11 we show

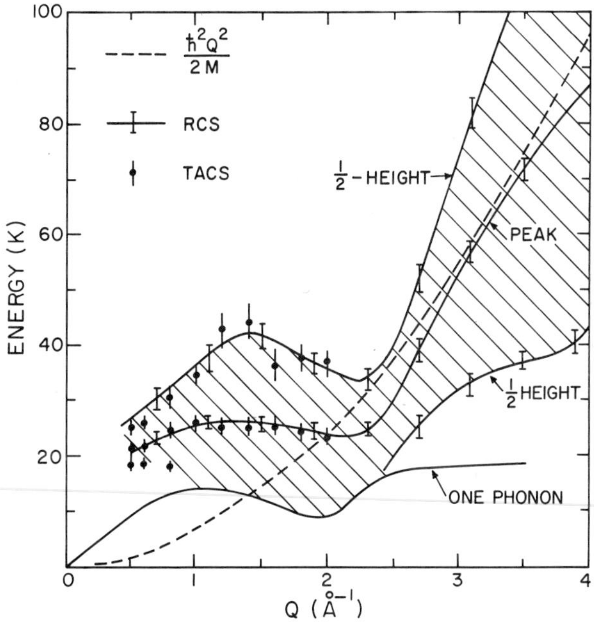


FIG. 10. The energy/wave-vector dependence of the scattering at 1·1 K. Shown are the one-phonon dispersion curve, the upper and lower energies corresponding to half the peak intensity, and the mean energy, of the multiphonon peak. The results were obtained with both types of spectrometer.

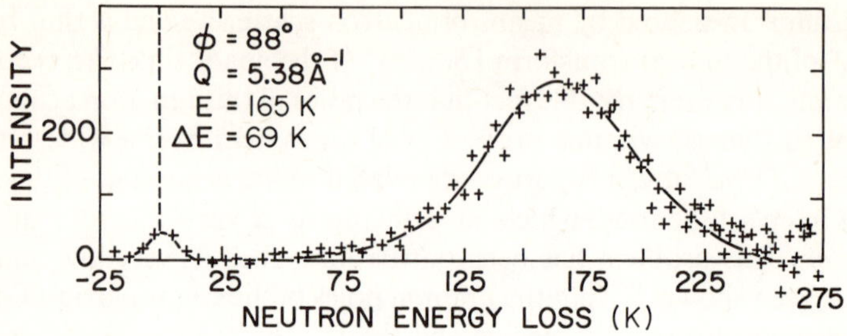


FIG. 11. High energy inelastic neutron scattering with energy transfer up to 275 K. The small elastic peak represents scattering from the apparatus.

the actual neutron data for a momentum transfer of 5·38 $\text{Å}^{-1}$. The broad group of neutrons centered on an energy of 165 K corresponds to one particular excitation on the first curve. In Fig. 12 we show the cross section for production of these excitations. It is very small at low momenta, remains rather small at momenta around the roton minimum and then climbs to high values for the larger momenta.

Before attempting to interpret this branch of the spectrum in any detail several general points should be made. First the higher energy part of the spectrum, which rather accurately follows the curve $\varepsilon(k) = h^2k^2/2m$, has also been found in solid helium.[(42)] This observation strongly suggests that this branch corresponds to the short time evolution of free particle motion. This motion is initiated by the neutron transferring a large amount (50–100 K) of energy to a single particle which then propagates freely for a very short period of time. The linewidth $\Delta E$ suggests that this time interval is of the order of

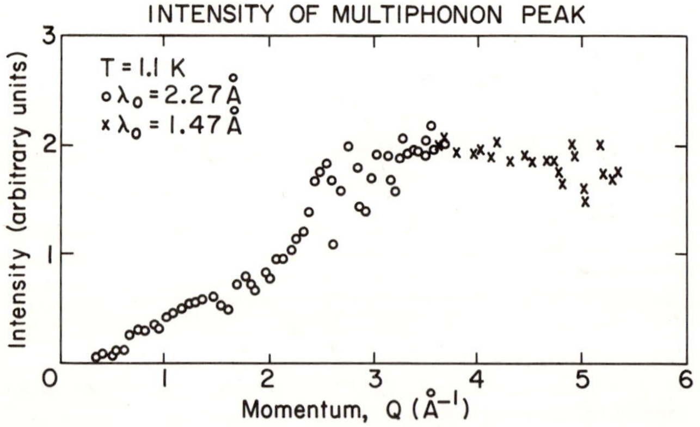


FIG. 12. The intensity of the "single particle" peak as a function of momentum transfer.

$\sim 10^{-13}$ sec.† Second, the part of this spectrum which is approximately flat and lies below $2\,\text{Å}^{-1}$ may not correspond at all to any genuine elementary excitation. Notice that it is very broad and not very easily excited. It has been suggested that it corresponds to two "phonon" processes.[17,43] Such processes can lead to a peak in the neutron scattering in liquid helium, in a crystalline phase they usually do not do so. The other suggestion that has been made is that it corresponds to a two roton bound state—at low momenta.[44] There is good evidence from the Raman scattering process in which two rotons are produced simultaneously[45] that they can form a bound pair. It is therefore at least plausible that a gound pair excitation might exist near this range of momenta values. The broad width associated with this part of the spectrum could arise from the very short lifetime of the bound state.

Thus the main portion of this spectrum has a very simple interpretation—it is free particle motion. The energy independent part is the subject of more speculation. However the width of the free particle part is also mysterious. In Fig. 13a and 13b we show it for two temperatures as a function of momentum transfer. The oscillations are very intriguing as is the fact that the width is qualitatively the same below and above the lambda transition. Thus whatever mechanisms are causing the decay of the free particle motion are qualitatively the same above and below the lambda point. No satisfactory explanation of the oscillations has yet been put forward.

I should emphasize that the motion really is *free* particle motion because the mass of the particle is the bare mass and not an effective mass. The high energy of the particle means that all the "dressing" effects have been stripped off and the effective mass reduced to the bare mass.

Finally we should mention that this type of neutron scattering has been used as a means of estimating the number of particles with zero-momentum in the fluid. We shall return to this topic in detail in lecture III when we discuss the Bose-condensate fraction.

*Excitations in $He^3$*. The excitations that are known to exist in $He^3$ are really much better understood than those I have been describing in $He^4$. There are two well established collective excitations, first sound and zero sound.[46] The excitations that dominate thermal and transport properties at low temperatures are the quasi-particle excitations first proposed by Landau.[46]

First sound excitations in $He^3$ are of course the same hydrodynamic mode as first sound in $He^4$ .They have a different velocity because the compressibilities of the fluids are different. Zero sound was predicted to exist by Landau on the basis of Fermi liquid theory. It is a density wave at high frequencies ($\omega\tau \gg 1$) and the system is not in local equilibrium (which would require

† The time taken by a helium atom with average kinetic energy to traverse one interparticle distance is also about $10^{-13}$ sec.

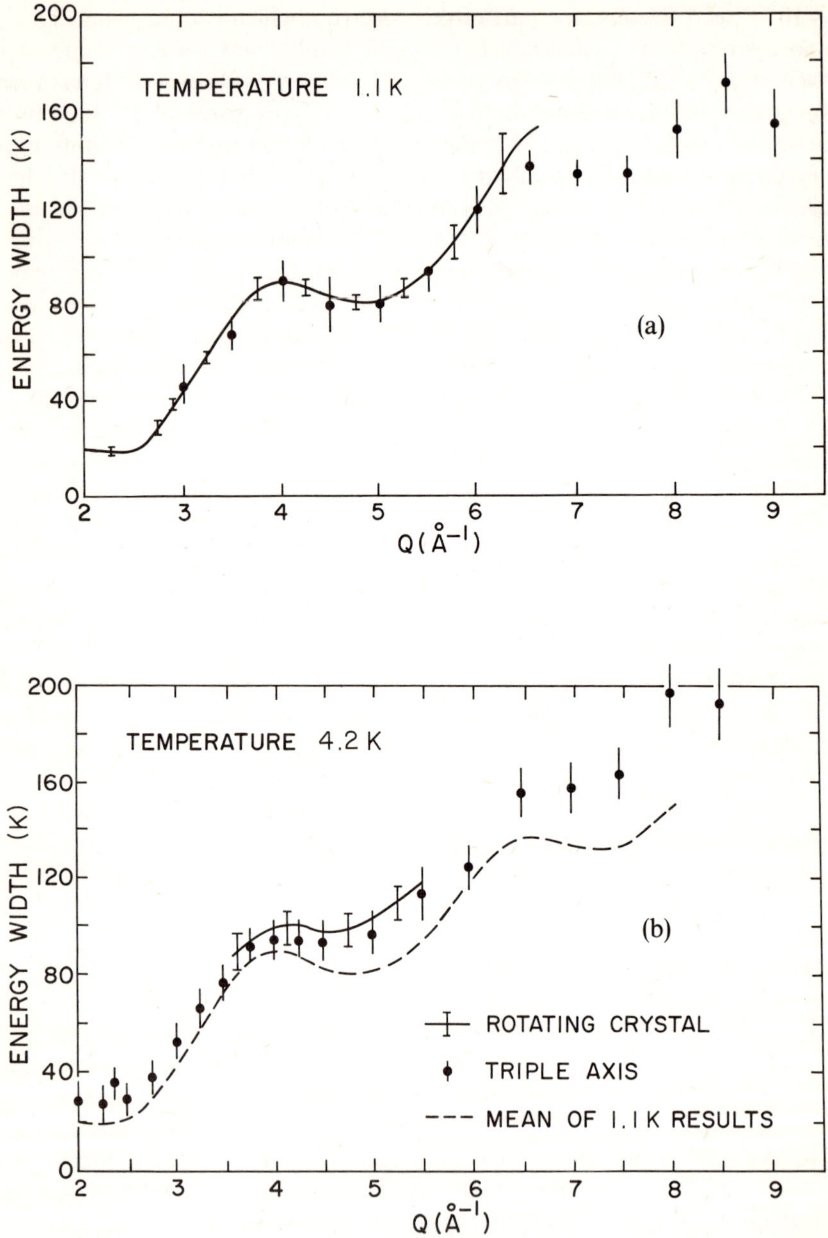


FIG. 13. The width of the peak of the scattering at large $Q$ as determined by the half-height at 1·1 K and 4·2 K.

($\omega\tau \ll 1$)). The velocity of propagation cannot be predicted numerically from Fermi liquid theory without assumptions. Experimentally it is 6 m/sec greater than that of first sound.(47) The frequency and temperature dependence of the attenuation can however be predicted. The prediction is that the attenuation coefficient should vary as $T^2$ and be independent of the frequency; this has been very well confirmed by measurement,(47) and this confirmation can be regarded as one of the basic tests of Landau theory. As we have mentioned it has often been speculated that the long linear region of the $He^4$ spectrum corresponds to zero sound—but no evidence is available to confirm that conjecture.

The Landau–Fermi liquid theory is built upon the concept of quasi-particle excitations. These are basically defined in a self-consistent fashion by the theory, because the excitation energy of a quasi-particle is functionally dependent upon the distribution of the other quasi-particles. This makes the Landau theory rather more subtle than other theories of excitations. It is far from a single particle picture. This means that relating one physical property to another is generally not possible in this theory. Nevertheless the measurements of the low temperature properties of liquid $He^3$ are all consistent with the predictions of Fermi liquid theory, and hence of the quasi-particle concept. There are three measurements to which I would like to draw your attention. First, the existence and frequency dependence of the attenuation of zero sound is, as I have already mentioned, a striking confirmation of the theory. Second, a rather special spin resonance experiment has been performed(48) and this too completely confirmed the predictions of the theory—which were for the first time precise numerical predictions. Finally, recent work at Sussex University(49) has shown that one can see long mean free path effects from the quasi-particles at very low temperatures. In essence these measurements show that the quasi-particles form a "Knudsen gas" of excitations at very low temperatures—which also constitutes a striking verification of the quasi-particle concept.

As this was being written an excellent review article covering most aspects of excitations in $He^4$ appeared in preprint form by L. Reatto.(18)

## III. Bose Condensation in Liquid $He^4$

There seem to me to be two interesting questions concerning Bose condensation: (a) does the phenomenon occur in liquid $He^4$ (b) what is the connection between Bose condensation and superfluidity?

We can always define a single particle momentum distribution for a system—whether the particles interact or not. In a non-interacting system the momentum carried by a particle remains with it until it interacts with the boundary. In an interacting system the momentum of any one particle is continually fluctuating due to collisions with other particles. Nevertheless we can still ask, in a statistical sense, for the mean number of particles with momentum **p**. (In any classical system it is known that this mean number, $n(\mathbf{p})$, is always a Maxwellian distribution[50] $\propto \exp(-\mathbf{p}^2/2mkT)$, no matter how strong the interactions or what equilibrium state the system is in.) Unfortunately we know very little about $n(\mathbf{p})$ for quantum systems. The ideal Bose gas has a distribution function $n(\mathbf{p})$ with a striking anomaly:[50] namely at all temperatures below $T_0$

$$n(\mathbf{p}) = N_0(T/T_0)\,\delta(\mathbf{p}) + f(\mathbf{p})$$

where $kT_0 \propto \hbar^2\rho^{2/3}/m$, $N_0(T/T_0) = N[1 - (T/T_0)^{3/2}]$. The temperature $T_0$ is the so called Bose condensation temperature of the ideal Bose gas. Clearly, $N_0(T/T_0)$ is the mean number of particles with zero momentum, while $f(\mathbf{p})$ is the mean number of particles with momentum $p > 0$; $f(\mathbf{p}) = \{\exp[\beta\varepsilon(\mathbf{p})] - 1\}^{-1}$. The existence of the term proportional to $\delta(\mathbf{p})$ is the phenomenon of Bose condensation—a finite fraction of all the particles have zero momentum. (Here I am assuming a description using periodic boundary conditions, consequently $\mathbf{p} = 0$ is a momentum eigenvalue.) This phenomenon is of course limited to Bose systems—it is quite impossible in Fermi systems. (In BCS theory there is the analogous phenomenon of condensation into a *pair state.*[51])

### (a) *Does Bose–Einstein Condensation occur in Liquid Helium*?

This question can be examined from both theoretical and experimental points of view. First, we look at the theoretical evidence. Several features of these calculations are worth mentioning. First, both variational and exact calculations[25] on hard sphere systems give the same results, namely that at $T = 0$, the number of particles with zero momentum is $10\% \pm 1\%$. Second, variational calculations on the Lennard-Jones system also give the same result.[23] Third, the number in the zero momentum state is a very sensitive function of the density. The results we have quoted are all for zero pressure. At a density just below crystallization the number has dropped to $2\% \pm 1\%$.[23] Thus a 20% change in density produces a five fold reduction in the "condensate fraction." Finally exact computations on a model hard-sphere plus square well system also show that the condensate fraction is close to 10%.

The results for this system are particularly important as they tell us that the condensate fraction is not sensitive to the potential. Summarizing all the available evidence from theoretical calculations tells us that the condensate fraction is close to 10% at zero pressure and temperature.

The experimental data is more difficult to assess because it is at present impossible to directly measure the momentum distribution of a strongly interacting quantum system. There is no difficulty in principle—rather no one has been able to think of how to measure $n(\mathbf{p})$ directly.

Several years ago it was suggested[52] that neutron scattering at very high energies from the fluid could yield data that would lead to an estimate of $n_0$. The basic idea has been discussed in lecture II. A very high energy neutron can communicate a large amount of energy to a single particle which will then propagate for a short period of time as a free particle before collision processes cause the excitation to decay. Thus if $S(\mathbf{k}, \omega)$ is the scattering cross section for momentum transfer $\mathbf{k}$ and energy transfer $\omega$, then for large $\mathbf{k}$ and $\omega$ we can easily show[52] that

$$S(\mathbf{k}, \omega) = \sum_{\mathbf{p}} n_{\mathbf{p}}\, \delta\!\left(\omega - k^2/2m - \frac{\mathbf{k}\cdot\mathbf{p}}{m}\right);$$

the last term in the delta function represents the Doppler shift due to the initial momentum of the particle. This expression is not exact because it ignores final state interactions and these will lead to a broadening of the delta function. We will return to this point in a moment. If we accept this expression for $S(\mathbf{k}, \omega)$ then we see that if there is a condensate, i.e. $n_0 \propto N$ then

$$S(\mathbf{k}, \omega) = n_0\, \delta(\omega - k^2/2m) + f(\mathbf{k}, \omega).$$

This says that there should be a sharp spike, superimposed on a broad background. The height of this peak is a direct measure of $n_0$. Unfortunately this simple formula for $S(\mathbf{k}, \omega)$ has to be modified due to final state interactions and as a consequence the $\delta$ function will be broadened. The most one can reasonably hope for is that the neutron scattering cross section will be narrower and have a higher central peak when a condensate is present. Thus in liquid helium we might hope for a qualitative change in the scattering between say 1 K and above the lambda point where presumably no condensate is present. In Fig. 14[53] we make such a comparison and there is indeed a narrowing of the central portion of the scattering.[53] However we still have to decide whether this is really due to a condensate or whether it might be due to some other effect. Second, can we extract from the data a reliable numerical estimate of the magnitude of $n_0$? A lot of work has been done on this topic[54] and the current estimate is that $n_0$ is approximately 3 %. This result is in complete disagreement with the results of the computer experiments to which we already have referred. Two resolutions of this conflict are possible. The interpretation of the experimental data may be wrong or the results of the computer simulation may be wrong At the present time I believe that there must be something wrong with the interpretation of the data. In support of the computer results I would point out that (i) the exact and variational

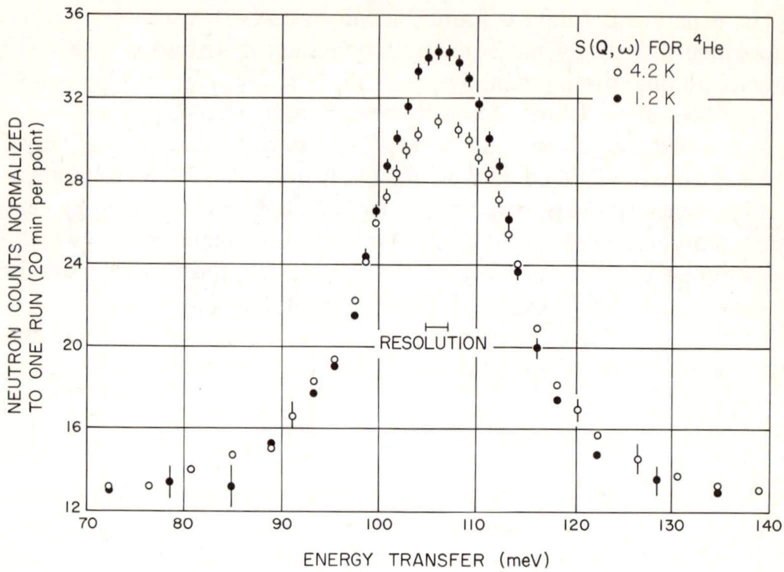


FIG. 14. $S(\vec{Q}, \omega)$ for $^4$He at 4·2 and 1·2 K.

results are the same, (ii) the result of 10% seems to be independent of the detailed form of the interaction. Thus if these results are wrong—a considerable number of very different numerical computations must be wrong. This seems to me to be much less likely than that the method of analysing the experimental data has some undiscovered flaw or uncontrolled approximation.

(b) *What is the connection between Bose Condensation and Superfluidity*?

One point should be made at once; as far as I am aware no general connection has been generally established between the two phenomena. We are therefore dealing with speculations, approximate calculations or partial results. Although there has been some discussion of this topic in the literature no comprehensive review of the speculations is available.

Some preliminaries should be dealt with first. The definition we gave of $n_0$ can be generalized to very general physical situations in a Bose system. First in equilibrium when the system is uniform we know that the single particle density matrix[55] $\sigma_1(\mathbf{x}, \mathbf{x}')$ is a function of $\mathbf{x}$–$\mathbf{x}'$ alone. This implies that its eigenfunctions defined by the equation

$$\int \sigma_1(\mathbf{x} - \mathbf{x}')\psi_\lambda(\mathbf{x}')\,d^3\mathbf{x}' = c_\lambda\psi_\lambda(\mathbf{x})$$

are plane waves $\psi_\lambda \to \psi_{\mathbf{k}} = \exp(i\mathbf{k}\cdot\mathbf{x})$ and this means that the eigenvalues $c_\lambda \to \bar{n}_{\mathbf{k}}$ where $\bar{n}_{\mathbf{k}}$ is the mean number of particles with momentum $\mathbf{k}$. We say that Bose–Einstein condensation has taken place in the single particle state $\psi_{\mathbf{k}}$ if the corresponding $\bar{n}_{\mathbf{k}}$ is proportional to the total number of particles $N$.[(55)] The single particle density matrix can, of course, be defined at any temperature and consequently the above definition is applicable at any temperature. This definition can be easily extended to nonequilibrium situations in which the system is inhomogeneous and $\sigma(\mathbf{x}, \mathbf{x}')$ is *not* a function of $\mathbf{x}$–$\mathbf{x}'$ alone. We then no longer know the form of the eigenfunctions but nevertheless the eigenvalue equation

$$\int \sigma_1(\mathbf{x}, \mathbf{x}')\psi_\lambda(\mathbf{x}')\,\mathrm{d}^3\mathbf{x}' = \bar{n}_\lambda \psi_\lambda(\mathbf{x}')$$

is still valid and since $\sigma_1(\mathbf{x}, \mathbf{x}')$ is hermitian the eigenvalues $\bar{n}_\lambda$ are real. They have the interpretation that $\bar{n}_\lambda$ is the mean number of particles with single particle properties corresponding to $\psi_\lambda(\mathbf{x})$; and $\Sigma_\lambda \bar{n}_\lambda = N$. We again say that Bose–Einstein condensation has taken place into the single particle state $\psi_\lambda$ if the corresponding $\bar{n}_\lambda$ is proportional to $N$.

What has Bose condensation defined in this way to do with superfluid helium? At present we cannot answer that question rigorously. There are however several plausible remarks which strongly suggest that superfluidity can only occur if there is a Bose condensate in the system. Whether or not the converse is true is a question to which we shall return later.

The connection between Bose condensation and superfluidity is probably as follows.[(56)] First if *one* $\bar{n}_\lambda$ is of the order $N$, then this state clearly stands out in the decomposition of $\sigma_1(\mathbf{x}, \mathbf{x}')$ into its eigenfunctions. We can always write

$$\sigma_1(\mathbf{x}, \mathbf{x}') = \sum_\lambda \bar{n}_\lambda \psi_\lambda^*(\mathbf{x})\psi_\lambda(\mathbf{x}'),$$

and it obviously makes sense to single out of this sum the term $\bar{n}_\lambda \psi_\lambda^* \psi_\lambda$ for which $\bar{n}_\lambda$ is proportional to $N$;

$$\sigma_1(\mathbf{x}, \mathbf{x}') = n_\lambda \psi_\lambda^*(\mathbf{x})\psi_\lambda(\mathbf{x}') + \sum_{\nu \neq \lambda} \psi_\nu^*(\mathbf{x})\psi_\nu(\mathbf{x}').$$

In general $\psi_\lambda$ will be a complex function $\exp[iS(\mathbf{x})]R(\mathbf{x})$, with $S$ and $R$ real functions. We identify the superfluid velocity as being equal to $(\hbar/m)\nabla S$ and this definition immediately leads to $\mathbf{v}_s$ being everywhere curl free—as we expect. It then also easily follows† that the superfluid circulation $\int_c \mathbf{v}_s \cdot \mathrm{d}\mathbf{l}$ will be quantized in units of $h/m$. Thus, in summary we can say that the notion of Bose–Einstein condensation leads us at once to the possibility of defining the

† This requires that the density matrix be single valued. If we assume that the ground state is single valued then it follows that all other states—and hence the density matrix—must be.

superfluid velocity field in a general and possibly rigorous fashion. It does not so obviously lead us to see how to break up the total density into superfluid and normal components. It also does not lead immediately to an equation of motion for $\mathbf{v}_s$. These arguments are usually believed to be sufficiently compelling that it is a generally held belief that superfluidity will occur if the system has a Bose–Einstein condensate.† Very strong circumstantial evidence to support this contention comes from the highly successful BCS theory of superconductivity, the central feature of which is the macroscopic occupation of a two-particle state of the two-particle density matrix—the existence, in other words, of Cooper-pairs.

The definition we have given of Bose–Einstein condensation is probably the simplest that can be given. However, it turns out to be much more convenient for many purposes to work with a formalism in which the total number of particles is not fixed and the conjugate variable, namely the phase, becomes well defined. Details of this approach will be found in references.[57, 58]

It is much more difficult to define what one means by superfluidity in general terms and then convert this to a precise mathematical prescription for the superfluid density $\rho_s$. Two quite different approaches are possible. First, superfluid systems respond in a fundamentally different way to external forces. Thus the linear response of a superfluid system is different from that of a normal system. For example, if an ion in liquid helium has an external field applied to it then at very low temperatures it will simply accelerate through the fluid—without experiencing any resistance. Ultimately this acceleration will be stopped by the production of vorticity but there is a range of velocities for which free acceleration occurs. In a normal fluid the response to a field would be a steady drift in which the external field is balanced by the resistive force of the fluid. Thus superfluidity can be characterized by the response of the system to arbitrary weak forces and indeed one can in this way arrive at a mathematical definition of the superfluid (or normal fluid) density.[59] Alternatively the response of a fluid system to the slow rotation of a noncylindrical container can be calculated. In particular, the angular momentum after a long time can be calculated and an effective moment of inertia $I_{\text{eff}}$ can then be defined by the equation

$$L = I_{\text{eff}}\,\omega$$

where $\omega$ is the speed of rotation of the vessel and $L$ is the angular momentum of the fluid formula for $I_{\text{eff}}$. In a normal fluid $I_{\text{eff}}$ is equal to the classical moment of inertia ($I_c$) of the fluid rotating as a rigid body. In a superfluid

† The exceptional case of the ideal Bose gas must of course be excluded as it has a zero Landau critical velocity.

$I_{\text{eff}}$ is less than $I_c$ because the fluid cannot be brought into "full" rotation by the motion of the boundary—only a fraction of the fluid need respond to the motion of the walls. The normal fluid density $\rho_n$ can be defined by the equation

$$I_{\text{eff}} = \frac{\rho_n}{\rho} I_c$$

where $\rho$ is the average density of the fluid. This approach(59) finally leads to an expression for $\rho_n$ and hence $\rho_s = (\rho - \rho_n)$ in terms of the transverse current–current response function and it can be (provided assumptions are made about the nature of the excited states) shown that this expression for $\rho_n$ is identical to that first derived by Landau.(60)

This approach of using the response of the system to rotation has been used to establish a connection between Bose-condensation and superfluidity. Pasquale and Tarbet(61) showed that a system with a condensate would respond under rotation in a different manner to one without a condensate. A more general proof appears in the paper by Kohn and Sherrington.(62) These authors showed the presence of a Bose condensate was a necessary and sufficient condition for non-classical rotation. However these discussions by no means attack the general problem of the connection between a Bose condensate and the appearance of superfluidity. They only establish the connection between condensation and the response to rotation. Most importantly they do not discuss the stability of flow states—which is the most obvious manifestation of superfluidity.

The second approach to the discussion of superfluidity is to try to base the definition on the occurrence of metastable flow states. Such states are of course the most obvious manifestation of superfluidity. Persistent currents in helium "persist" for tens of hours(63) without measureable decay. In this approach one has to be able to determine whether such states exist for a given system and to what "degree" they are metastable. I believe this is a very much more difficult question than those I have just discussed. Even more difficult questions arise when one attempts to understand whether there is any connection between the two approaches.

The only general discussions of persistent flow states are due to Landau(31) which we mentioned in the lecture on excitations and Langer and Fisher.(64) The latter approach will be discussed in lecture IV. Landau showed by a simple argument based on conservation of energy and momentum that a flow state could not decay into excitations if the spectrum were such that $[\varepsilon(\mathbf{p})/\mathbf{p}]_{\text{min}}$ had a finite value. This is of course not a complete discussion because there may be other "excitations" into which the flow could decay. Moreover the argument does not rely directly on the presence of Bose condensation in the system. It does rely directly on the shape of the spectrum for small momenta (and elsewhere). Indeed the well known Landau formula for the critical velocity†

$$v_c = \left[\frac{\varepsilon(\mathbf{p})}{\mathbf{p}}\right]_{\min}$$

† The argument is as follows. Initially we suppose that the liquid contains no excitations. If the body moving through it creates a certain number of excitations, say $n_{\mathbf{p}}$ with momentum $\mathbf{p}$, then the increment in energy of the fluid is

$$\delta E_F = \sum_{\mathbf{p}} n_{\mathbf{p}} \varepsilon(\mathbf{p}) \tag{1}$$

and the increment in momentum is

$$\delta P_F = \sum_{\mathbf{p}} n_{\mathbf{p}} \mathbf{p}. \tag{2}$$

On the other hand the change in energy of the body is

$$\delta E_B = \mathbf{p}_f^2/2M - p_i^2/2M, \tag{3}$$

where $\mathbf{p}_f$ and $\mathbf{p}_i$ are the final and initial momenta of the body. If we set

$$\mathbf{p}_f - \mathbf{p}_i = \delta \mathbf{P}_B \tag{4}$$

where $\delta \mathbf{P}_B$ is the change in momenta of the body, we find

$$\delta E_B = \mathbf{v}_i \cdot \delta \mathbf{P}_B + (\delta \mathbf{P}_B)^2/2M \tag{5}$$

where $\mathbf{v}_i$ is the initial velocity of the body. Now from (1) we find

$$\delta E_F \geqslant \sum_{\mathbf{p}} n_{\mathbf{p}} |\mathbf{p}| \cdot \left|\frac{\varepsilon(\mathbf{p})}{\mathbf{p}}\right|_{\min}, \tag{6}$$

where $|\varepsilon(\mathbf{p})/\mathbf{p}|_{\min}$ is the minimum value that this ratio can take on for any value of $\mathbf{p}$. We assumed that all the $n_{\mathbf{p}} \geqslant 0$; we only create excitations. From (2)

$$|\delta \mathbf{P}_F| \leqslant \sum_{\mathbf{p}} n_{\mathbf{p}} |\mathbf{p}|, \tag{7}$$

and combining (6) and (7) we have

$$\delta E_F \geqslant |\delta \mathbf{P}_F| \left|\frac{\varepsilon(\mathbf{p})}{\mathbf{p}}\right|_{\min}. \tag{8}$$

We now assume that in the transition, in which the excitations are created in the fluid, energy and momentum are conserved. This implies that

$$\delta E_F = -\delta E_B \tag{9a}$$

$$\delta P_F = -\delta \mathbf{P}_B. \tag{9b}$$

If we substitute (9) into (3) and use (8) then we have at once

$$\mathbf{v}_i \cdot \delta \mathbf{P}_F - (\delta \mathbf{P}_F)^2/2M \geqslant |\delta \mathbf{P}_F| \left|\frac{\varepsilon(\mathbf{p})}{\mathbf{p}}\right|_{\min}. \tag{10}$$

For a massive body we can neglect the term $(\delta \mathbf{P}_F)^2/2M$; (10) then implies that

$$|v_i| \geqslant \left|\frac{\varepsilon(p)}{\mathbf{p}}\right|_{\min}. \tag{11}$$

If we do not neglect the term $(\delta \mathbf{P}_f)^2/2M$ then the condition becomes

$$|\mathbf{v}_i + \mathbf{v}_f|/2 \geqslant \left|\frac{\varepsilon(\mathbf{p})}{\mathbf{p}}\right|_{\min}. \tag{12}$$

where $\varepsilon(\mathbf{p})$ is the energy of an excitation with momentum $\mathbf{p}$ would produce zero for the critical speed if $\varepsilon(\mathbf{p})$ behaved like $\mathbf{p}^2$ for small $\mathbf{p}$. We see now a rough connection between the response approach (and the formula for $\rho_n$) and the Landau argument for a finite critical velocity. If $\rho_n/\rho$ is to be less than unity and $v_c$ is to be finite, then we apparently must have an absence of low lying modes of excitation in the fluid.

This is however an oversimplified picture. We know from superconductivity theory that one can have superconductivity with low lying excitations. These are the so-called "gapless" superconductors. A very similar situation can arise in liquid $He^4$ if we have $He^3$ impurities present for then we have a low lying impurity excitation branch with $\varepsilon(k) \sim k^2/2m^*$. This can be described phenomenologically by saying that we have a finite fraction of normal fluid present at absolute zero. However from a microscopic point of view the rotational response function will still be non-classical because only the normal fluid can be brought into saturation.

It is generally believed that the absence of low lying single particle modes in $He^4$ is due to the presence of a Bose condensate.[65] A detailed diagrammatic analysis of the spectrum of an interacting Bose system supports this belief. One is thus tempted to say that Bose condensation in an *interacting* system is sufficient to ensure stable superflow. Is the converse true: is Bose condensation necessary for superfluidity? There is one example of an interacting Bose system that has no condensate but nevertheless in which there are no low lying modes. (The lowest modes behave like phonons $\varepsilon(\mathbf{p}) \sim |\mathbf{p}|$.) This is the one dimensional Lieb[66] model of bosons interacting via repulsive delta functions. The system has a very large population in the zero momentum state—but does not have a condensate. This at once tells us that bose condensation is not a necessary condition for the absence of low lying modes. Unfortunately we do not know whether the model exhibits any form of superfluidity.

Two final remarks: Applications of the Bogoliubov inequality can be made[67] to decide whether two (or nearly two) dimensional Bose systems can exhibit Bose condensation. One finds that the inequality tells us nothing about physically realizable systems and so has nothing to say about our basic discussion. Finally, recent discussions[68] about the occurrence of Bose condensation in states with crystalline order add little of value to what we have already said. We do not really know if these states would exhibit superflow or not.

## IV. SUPERFLUID HYDRODYNAMICS

In this lecture I shall talk about the following topics:

(a) Ideal two fluid hydrodynamics,
(b) Two fluid hydrodynamics with vorticity,
(c) The stability of superflow and creation of vorticity.

These three topics cover most of the important elements of superflow in bulk $He^4$. In the next lecture I shall say a little about superflow in pores and films. I have chosen these three topics because they illustrate the variety of flow regimes that are observed in superfluid $He^4$. These range all the way from highly stable superflow at low speeds, through metastable superflow states, states with weak macroscopic vorticity to turbulent flow states in which the superfluid is believed to be in a turbulent state.

### (a) *Ideal two fluid hydrodynamics*

These equations were first derived by Landau.[(31)] The fluid is described by two velocity fields $\mathbf{v}_n$ and $\mathbf{v}_s$ together with scalar density and temperature fields. The appearance of two velocity fields is unique to superfluid systems; $\mathbf{v}_n$ is often called the velocity of the normal component, $\mathbf{v}_s$ that of the superfluid component. A more physical way of understanding these fields can be got from the excitation picture of the fluid. The mean speed of the fluid in the absence of excitations is $v_s$; the mass current is thus $\rho\mathbf{v}_s$. If excitations are introduced (thermally) they can be made to transport momentum independently of the motion of the background fluid. A simple kinetic argument shows that this momentum must be (to lowest order) proportional to $\mathbf{v}_n - \mathbf{v}_s$ where $\mathbf{v}_n$ is the mean *group* velocity of the excitations. The multiplying factor which converts $\mathbf{v}_n - \mathbf{v}_s$ into the momentum carried by the excitations is called the normal fluid density $\rho_n$. Thus the total mass current $j$ is given by

$$\mathbf{j} = \rho\mathbf{v}_s + \rho_n(\mathbf{v}_n - \mathbf{v}_s) \quad (1a)$$

$$= (\rho - \rho_n)\mathbf{v}_s + \rho_n\mathbf{v}_n \quad (1b)$$

$$= \rho_s\mathbf{v}_s + \rho_n\mathbf{v}_n. \quad (1c)$$

I have gone through this discussion rather quickly to bring out that $\rho_s$ and $\rho_n$ are really not such fundamental quantities as $\rho\mathbf{v}_s$ and $\rho_n(\mathbf{v}_n - \mathbf{v}_s)$. Moreover, (1a) is really the most physical expression for the mass current—because the two terms on the right hand side have a very direct physical interpretation. Finally, $\rho_s$ appears as a derived quantity; it is merely defined as $\rho - \rho_n$.

What is the underlying assumption that allows us to derive (in some way) two fluid hydrodynamics? After all we must be making some very fundamental assumption that distinguishes superfluid helium from other fluids—for if we did not then we would necessarily arrive at the hydrodynamic equations of a normal fluid. The fundamental assumption is simply that the full specification of the motion of the fluid requires two velocity fields $\mathbf{v}_s$ and

$\mathbf{v}_n$. In the excitation picture we are asserting that the mean group velocity ($\mathbf{v}_n$) of the excitations can be prescribed independently of the motion of the fluid as whole—which is prescribed by $\mathbf{v}_s$. (As we pointed out in the last lecture we believe that the underlying microscopic reason for the appearance of two velocity fields is that there is a macroscopic occupation of a single eigenstate of the single particle density matrix. This idea leads naturally to the idea of a new flow field $\mathbf{v}_s$ given by the gradient of the phase of the macroscopically occupied state.)

This basic assumption has to be made in either of the two standard methods of deriving two fluid hydrodynamics.[69, 70] The first of these methods is purely phenomenological in approach. In outline, one proceeds as follows. A form for the energy density of the fluid is assumed. This is then combined with the basic conservation equations for mass, momentum, and energy and an equation of motion for $\mathbf{v}_s$. This latter equation is *assumed* to have the form

$$\frac{\partial \mathbf{v}_s}{\partial t} + \nabla(\phi + \mathbf{v}_s^2/2) = 0$$

where $\phi$ is some scalar function still to be determined. This equation is of course consistent with the basic assumption that $\nabla \wedge \mathbf{v}_s = 0$ everywhere. The form for the energy density $E$ is such that

$$\mathrm{d}E = T\mathrm{d}S + \mu \mathrm{d}\rho + (\mathbf{v}_n - \mathbf{v}_s)\cdot \mathrm{d}\mathbf{p}$$

where $E$ and $\mathbf{p}$ are the energy and momentum density of the fluid in a coordinate system moving with velocity $\mathbf{v}_s$. Clearly these two equations contain the basic assumptions; namely that $\nabla \wedge \mathbf{v}_s = 0$ and that two independent velocity fields are needed to specify the energy density. It is a fairly straightforward matter to derive the equation of motion for $\mathbf{v}_n$ and $\mathbf{v}_s$ from this approach. One finds that $\phi = \mu$, the chemical potential of the fluid. The equation of motion for the total momentum, $\mathbf{p}$, takes the form

$$\frac{\partial \mathbf{p}}{\partial t} + \nabla \Pi = 0$$

$$\Pi = \rho_n \mathbf{v}_n \mathbf{v}_n + \rho_s \mathbf{v}_s \mathbf{v}_s + PI$$

where $P$ is the pressure, $\Pi$ is the stress tensor and $I$ the unit tensor.

This derivation can be generalized in two ways: one can allow for dissipative processes,[71] and we can also change the derivation radically by allowing for vorticity in the superfluid. Allowing for dissipation processes is straightforward—but the equations become quite complicated. However the equation of motion for $\mathbf{v}_s$ is still assumed to have the form

$$\frac{\partial \mathbf{v}_s}{\partial t} + \nabla(\mathbf{v}_s^2/2 + h) = 0$$

where $h$ is a new scalar function. This choice is perfectly consistent with dissipation. A clear phenomenological derivation will be found in reference.[(69)] Allowing for the presence of vorticity in the superfluid is a more subtle matter to which we will return in Section (c).

It is important to realize that the two assumptions that $\nabla \wedge \mathbf{v}_s = 0$ and that two velocity fields are required to specify the flow of the fluid are the basic assumptions that are needed to determine the form of the two fluid equations—whether dissipation is allowed for or not.[(71)]

The second approach[(70)] is based on the excitation picture of $He^4$. One writes down a Boltzmann equation for the excitations in the usual way. However in deriving hydrodynamics from it one assumes that the local equilibrium distribution has the form

$$\exp\{-\beta[\varepsilon(\mathbf{p}) - \mathbf{p}\cdot(\mathbf{v}_n - \mathbf{v}_s)]\}.$$

The superfluid velocity field has crept into the analysis by the assumption that the excitations are propagating in a moving medium and that a Galilean transformation can always be found that will take one from the lab frame to a local frame in the fluid in which the medium is at rest. This assumption has two parts: first that $\mathbf{v}_s$ can be prescribed independently of $\mathbf{v}_n$ and second that a Galilean transformation will bring the fluid to rest. The first assumption is of course the one we have already emphasized underlies two fluid hydrodynamics and the second is equivalent to saying that $\mathbf{v}_s$ is curl free everywhere. For if there were rotation in a fluid then a change to a rotating frame would be necessary to bring the fluid to rest.[(32)] From the Boltzmann equation for the excitations and the assumption of curl free motion one can derive the standard two fluid equations.

These equations are enormously rich in that they describe with considerable accuracy a wide range of phenomena in helium. I would like to emphasize the following points.

(i) The equation of motion for $\mathbf{v}_s$ in its most general form is:

$$\frac{\partial \mathbf{v}_s}{\partial t} = -\nabla(\mathbf{v}_s^2/2 + h).$$

Since the right hande side is the gradient of a scalar function this enables us to derive the usual vorticity conservation laws for the superfluid vorticity $\omega_s = \nabla \wedge \mathbf{v}_s$, namely that the vorticity in the superfluid is invariable in time. Since we have already assumed that it is always zero this should not surprise us; it is a consequence of that assumption.

(ii) The equation for $\mathbf{v}_s$ is first order in spatial derivatives of $\mathbf{v}_s$, hence only one boundary condition can be applied. Since we must have no mass transport through a surface we must have $\mathbf{v}_{sn} = 0$ at any surface. The parallel component of $\mathbf{v}_s$ is thus quite free at the surface.

(iii) The equation for $\mathbf{v}_n$ is second order in spatial derivatives and this means that boundary conditions for $\mathbf{v}_n$ perpendicular and parallel to a surface must be applied to determine any solution of the equation. Since there is a viscosity term $\eta_n \nabla^2 \mathbf{v}_n$ in the equation it is assumed that $\mathbf{v}_n$ does not slip at a boundary.[(69)]

In this discussion I have tried to emphasize the underlying assumptions in these two methods of deriving two fluid hydrodynamics. It is important to realize that neither has anything directly to say about the creation of vorticity in the superfluid or about the quantization of circulation.

A fundamental question remains. Can these equations be derived from a microscopic theory of liquid helium? As far as I am aware no general derivation has been given. But several limited derivations have appeared in the literature. The Bogoliubov theory of the weakly coupled Bose gas[(72)] can be used to give a simple form of the hydrodynamics—essentially limited to uniform flow states and low temperatures where phonons are excited. More general derivations limited to homogeneous situations—describable by a free energy have been presented by Bogoliubov[(73)] and Hohenberg and Martin.[(74)] However as Martin[(75)] has stressed no general derivation which deals with genuinely inhomogeneous situations has been given. Recently a substantial advance has been published by Josephson[(76, 77)] who has derived, under certain assumptions, the basic equation for the superfluid velocity field from a microscopic theory. Three basic assumptions are made in this derivation: (i) the system has a Bose condensate, (ii) all quantities vary slowly in space and time and (iii) the only possible excitations in the system are phonons. Finally the derivation is limited to absolute zero. I must emphasize that as far as we know the fundamental equation for $\mathbf{v}_s$

$$\partial \mathbf{v}_s / \partial t = -\nabla(\mu + \tfrac{1}{2}\mathbf{v}_s^2)$$

has only been derived when $\mathbf{v}_s$ and $\mu$ vary slowly in space and time. We therefore have no reason to believe that it can be applied on spatial scales of the size of a vortex core—i.e. a few interparticle spacings.

### (b) *Two fluid hydrodynamics with vorticity*

There are a number[(6, 78)] of very well known experiments which demonstrate conclusively that superfluid helium can support vorticity in the superfluid and that the circulation associated with that vorticity is quantized in units of $h/m$, where $m$ is the mass of the helium atom. Neither of these phenomena are contained in the equations we have discussed in this lecture. In lecture III we pointed out the very simple origin of the quantization of circulation for $\mathbf{v}_s$—it was simply linked to the single valuedness of the states of the system.[79] But that discussion sheds no light at all on how vorticity might arise in the fluid. We shall postpone discussion of this matter to Section (c)—for the moment we concentrate on the derivation of hydrodynamic equations with vorticity present. Two methods have been used.[80, 81]

Before proceeding to discuss these derivations there is an important point to be made. In these equations when we speak of vorticity in the superfluid we mean macroscopic vorticity *not* microscopic vorticity. Thus we still assert that $\nabla \wedge \mathbf{v}_s = 0$ everywhere on a *microscopic* scale except in small regions—which are the cores of vortices. When we calculate the circulation around some small but macroscopic circuit we will always find a non-zero result if there is a finite density of vortices in the fluid. Hence, on a macroscopic scale the superfluid will appear to be non curl free. We should therefore distinguish in our equations between the microscopic velocity field $\mathbf{v}_n$ and a macroscopic field $\tilde{\mathbf{v}}_s$. We have not done so in this section.

In the first method, due to Bekarevitch and Kahlatnikov,[80] one simply generalizes the form of the energy density to allow for the appearance of a term due to vorticity in the superfluid. The change in the energy density now becomes

$$\mathrm{d}E = T\,\mathrm{d}S + \mu \mathrm{d}\rho + (\mathbf{v}_n - \mathbf{v}_s)\cdot \mathrm{d}\mathbf{p} + \lambda\,\mathrm{d}\omega$$

where the last term $\lambda\,\mathrm{d}\omega$ represents the contribution due to vorticity $\boldsymbol{\omega}$. If the same methods are applied as before then a new set of two fluid equations emerge. For example the equation of motion for $\mathbf{v}_s$ now reads

$$\frac{\partial \mathbf{v}_s}{\partial t} = -\nabla(\mathbf{v}_s^2/2 + \mu) + \mathbf{f}$$

where $\mathbf{f}$ is given by

$$\mathbf{f} = \beta\frac{\rho_s\rho_n}{2\rho}\hat{\omega} \wedge [\boldsymbol{\omega} \wedge (\mathbf{v}_s - \mathbf{v}_n)] + \beta'\frac{\rho_s\rho_n}{2\rho}\boldsymbol{\omega} \wedge (\mathbf{v}_n - \mathbf{v}_s)$$

$$-\beta\frac{\rho_s\rho_n}{2\rho}\nu\hat{\omega} \wedge (\boldsymbol{\omega}\cdot\nabla)\boldsymbol{\omega} + \rho_s\nu(1 - \beta'\rho_n/2\rho)(\boldsymbol{\omega}\cdot\nabla)\hat{\omega}$$

and $\hat{\omega}$ is the unit vector in the direction of $\boldsymbol{\omega}$. The coefficients $\beta$, $\beta'$, $\nu$ are phenomenological coefficients that cannot be determined from this approach. They can be expressed in terms of the detailed properties of a vortex line. The new terms are obviously dependent on $\nabla \wedge \mathbf{v}_s$ and vanish if $\mathbf{v}_s$ is curl free. Similar terms emerge in the equation for $\mathbf{v}_n$; as far as I am aware no derivations of these equations based on the Boltzmann equation have been made. The second approach[81] does however use the excitation picture. It considers in some detail how the motion of the excitations will be affected by the presence of vortices in the superfluid. This of course depends on using some model of the interaction of excitations with the vortex core and with the velocity field surrounding a vortex. When this is carried through the resulting two fluid equations have the same form as those derived by the purely phenomenological approach.[80] Now however the coefficients in the equation can be related to the underlying microscopic model of excitations interacting with vortices. There is good experimental evidence[81] that these equations give a good description of superfluid helium with vorticity present.

### (c) *The stability of superflow and creation of vorticity*

In our discussion up to this point we have avoided the question of how vorticity might be created in the fluid. Two possibilities present themselves. A quantum transition could take place from a situation with no vorticity to one with non-zero vorticity. This transition might be brought about by external forces acting on the system or alternatively it might come about as a fluctuation phenomena in which a high energy metastable flow state "decays" to a lower energy state with vorticity present. On the other hand *thermal* fluctuations might also cause a metastable flow state to decay to one of lower energy with vorticity present. The first kind of transition phenomena still remains very speculative and I do not know of any direct experimental evidence that supports the idea. However, there is a large body of evidence[8] that supports the basic idea of thermal fluctuations providing the mechanism of transition between states with different "amounts" of vorticity. If the circulation associated with vortex motion were not quantized then thermal fluctuation could easily cause a metastable flow state to decay into others of lower energy. The energy of the flow would rapidly be degraded into the energy of vortices and we would not observe long lived superfluid flow states. Thus quantization of circulation is vital if we are to stabilise superflow against decay into vorticity. Once we accept this then we see that there will in effect be a finite energy barrier between one flow state and the next. For example an ion which is being freely accelerated through superfluid helium cannot create arbitrarily small vortex rings because there is a minimum circulation ($h/m$) associated with a vortex ring. The thermal fluctuations have to be large enough to carry the system over the finite energy barrier. This idea has been worked out in some detail by Fisher and Langer[64] and many of the results agree rather well with very careful measurements on the decay of superfluid flow states by Reppy.[8] Since this topic will be discussed in detail by Dr L J Campbell in these proceedings I will merely outline the argument.

We assume that there is an activation energy for the creation of vorticity $E_a(\mathbf{v}_s, T)$ that is independent of the size and shape of the system. That is, we are assuming that the decay of the metastable flow state is an intrinsic phenomenon. Thus in a container of cross section $A$ and length $L$ (i.e. periodically extended in the last direction) the rate $R$ at which fluctuations of energy $E_a$ will occur will have the form

$$R = ALv_0 \exp[-E_a/kT]$$

where $v_0$ is the fundamental frequency of events per unit volume. Next we argue that the basic process of fluctuation will take the system from one metastable flow state to the next state separated by one unit of circulation. Hence

$$\Delta\mathbf{v}_s = -(h/m)L$$

is the change in velocity per fluctuation event. If we multiply this by $R$ we then get

$$\frac{\mathrm{d}v_s}{\mathrm{d}t} = R\Delta\mathbf{v}_s = -h/m\, Av_0 \exp[-E_a/kT]$$

which is a nonlinear differential equation for $\mathbf{v}_s$; it expresses the rate of change of $\mathbf{v}_s$ due to activation processes in which vorticity is created. How does this equation define a critical velocity? The simple operational argument is that when $|\mathrm{d}\mathbf{v}_s/\mathrm{d}t|$ becomes undetectably small with any particular experimental apparatus then $\mathbf{v}_s$ will have fallen below the critical $\mathbf{v}_{sc}$. If we call this value $\Phi_0$ then

$$\Phi_0 = \frac{h}{m} Av_0 \exp[-E_a(\mathbf{v}_{sc})/kT],$$

again a non linear equation—this time for $\mathbf{v}_{sc}$. Two developments of these ideas are particularly interesting. First, we argue that only in the neighbourhood of $\mathbf{v}_{sc}$ will we be able to detect $\mathrm{d}\mathbf{v}_s/\mathrm{d}t$. This says that the exponential varies so rapidly that if $\mathbf{v}_s$ is varied very much from $\mathbf{v}_{sc}$, then $E_a$ changes too much and we either see no decay or no metastable flow. If this is accepted then we can expand $E_a$ in the form $E_a^0 - \alpha(\mathbf{v}_s - \mathbf{v}_{sc})$ which leads to an equation of the form

$$\frac{\mathrm{d}\mathbf{v}^s}{\mathrm{d}t} \propto \exp[\alpha(\mathbf{v}_s - \mathbf{v}_{sc})/kT].$$

For long times this has the solution

$$\mathbf{v}_s \simeq \mathbf{v}_{sc} - \frac{kT}{\alpha} \ln(at).$$

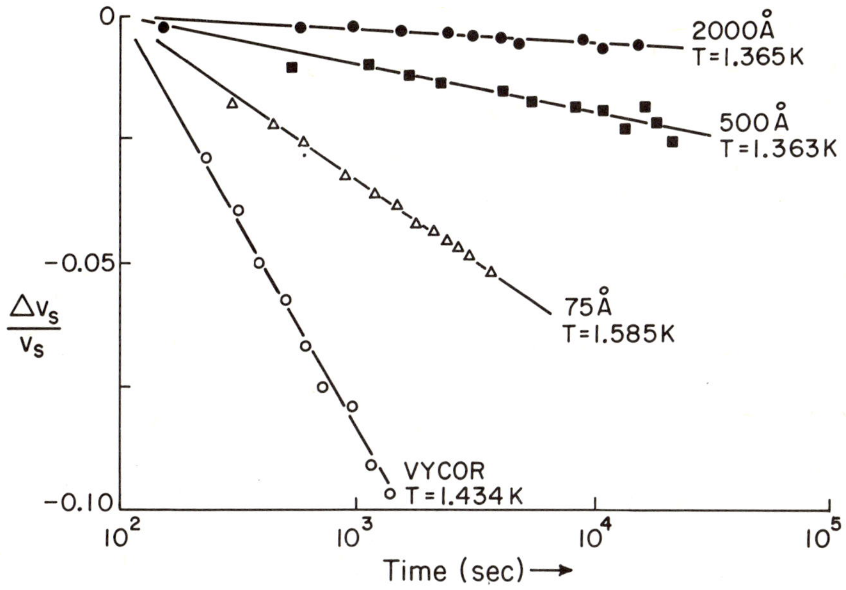


FIG. 15. The fractional change in superfluid velocity during the decay process is shown for persistent currents in a number of geometries as a function of time.

Thus $\mathbf{v}_s$ should decay logarithmically at long times—a prediction which has (near $T_\lambda$) been accurately verified; see Fig. 15. Second, a simple dimensional argument can be made which shows that in three dimensions

$$E_a \propto \frac{\rho_s}{\mathbf{v}_s}\left(\frac{h}{m}\right)^3,$$

$h/m$ appears as the unit of circulation. This will immediately give $\mathbf{v}_{sc} \propto \rho_s$ in qualitative agreement with experiment. In two dimensions the same argument leads to,

$$E_a \propto \rho_s d(h/m)^2$$

where $d$ is the size in the third dimension. This form for the activation energy predicts no metastable flow states at all above some well defined temperature.

This theory and its experimental verification mark an important development in the theory of superfluidity. For the first time we have a physical picture of how a flow state can decay in helium and this picture is well confirmed by experiments. Mechanisms of decay at very low temperatures are still very speculative.

## V. HELIUM IN FILMS AND PORES

At first sight it seems strange to study liquid $He^4$ in thin films and porous media such as Vycor glass or compressed powders. However, the flow of helium in these systems is one of the most striking manifestations of superfluidity.

(a) The temperature at which superflow sets in is a function of the thickness of the helium film or pore size.[9]

(b) One can pass from a non-superfluid film to a superfluid film by adding more material. In other words these systems allow one to pass to the superfluid state in an entirely different way as compared with bulk helium.

(c) When superfluid helium flows in these systems only a fraction of the fluid participates in the flow. The superfluid density is reduced ($\sim 30\%$ often) below its value in the bulk fluid.[82]

(d) There is substantial and convincing evidence that the superfluid density is finite at the temperature at which superflow first sets in.[83] Thus at temperatures above the onset temperature there is in some sense (to be explained) a finite density of superfluid.

All of these striking phenomena suggest that by studying these systems we may be able to gain some real insights into the phenomenon of superfluidity; this is the most important reason for studying helium in films and pores.

Since this topic will receive quite extensive treatment by Dr Ann Eggington and Dr L. J. Campbell in these proceedings I will limit my lecture to two aspects of the field. First I want to emphasize the complexity of the field and hence the difficulty of obtaining a definitive understanding of many of the phenomena. Second I want to point to several possible misconceptions which frequently arise in this area.

I will begin with a simplified description of how a thin film of helium is thought to be built up. If helium atoms are allowed to condense on a substrate then as more and more material is added a first layer of helium forms at a fairly high density. This is because of the strong attractive forces between the substrate and the helium atoms. It thus becomes energetically favourable to pack the helium atoms closer together than in the bulk fluid. The extra binding energy to the substrate more than compensates the loss in binding energy between the helium atoms due to the increase in density of the first layer. As the layer builds up it can go through a variety of phases. It may be mobile and gas-like at low densities, become a fluid at intermediate density and probably a solid at the highest densities.[84] Much of what goes on in the first layer is greatly influenced by the substrate. It is often stated[85] that the first layer is a solid because it appears to be subject to a pressure somewhat larger than the pressure at which bulk helium solidifies. This does not however mean that the helium atoms are localized due to their mutual interactions. They may well be localized on the substrate at quite low density due to binding to energetically favourable sites.[84] Preliminary calculations that we have carried out in a two-dimensional Lennard–Jones system show that the solidification density is higher than one might expect from the same model in three dimensions.

If more helium is now added it is believed to form a second layer on top of the first very tightly bound layer. Its final density is less than that of the first layer but is probably still somewhat higher than that of the bulk fluid at zero pressure. As still further material is added a portion of the film becomes superfluid. One thus passes through the superfluid transition by increasing one particular dimension of the system. If we could understand this phenomena I believe we would have a much deeper insight into the nature of superfluidity.

In my introduction to this lecture I mentioned the complexity of this field of experiment and theory. I want to enlarge on this topic a little so that you will be more fully aware of some of the basic reasons for this complexity. The most obvious reason is the presence of the substrate. It must clearly influence in a variety of ways the behaviour of helium in films and pores. The substrate should be characterized as carefully as possible and as many experiments as possible should be conducted on identical substrates to allow careful comparison of the results.

The second reason is that we are dealing with nearly two-dimensional systems about which we do not have a large body of either data or theory to use as a guide to understanding the behaviour of helium in these situations. On the theoretical side there is a strong tendency to take over the data on three-dimensional systems and scale the results in some way to interpret two-dimensional data. This is I believe very dangerous—until verified to be correct.

Let us now turn to some of the possible misconceptions which from time to time seem to plague this field.

### (a) *The Existence of Order in Two-Dimensional Systems*

A very considerable amount of theoretical work on the existence of various kinds of order in two-dimensional systems has been carried out in the last five years.[86a, b] Out of this body of work two results *appear* to have some relevance to the study of thin helium films. First it has been possible to show[67] rigorously and under very general conditions that Bose condensation cannot occur in an infinitely extended two-dimensional or nearly two-dimensional system. A similar[87] result has been proved which shows that crystalline order cannot exist under the same conditions. Both these results have no relevance to the actual experimental situation because the experiments are conducted on systems which are not infinitely extended in two dimensions. They are conducted on systems whose lateral dimensions in no case even approaches 1 cm. If the same rigorous mathematics is applied to this situation one simply cannot conclude[87] that either crystalline order or Bose condensations must be absent. *Finite* two-dimensional systems can perfectly well show these kinds of order. Indeed two-dimensional systems have to be almost astronomical in extent before the order disappears. In addition to this argument we have already emphasized that we do not actually know whether Bose condensation is necessary for superfluidity.

A good example which illustrates the general point is that one can "photograph" the two-dimensional flux lattice that occurs in type two superconductors. In an infinitely extended specimen such two-dimensional order cannot occur. In a finite system it can as the experiments show.

### (b) *The Concept of an Order Parameter and the Coherence Length*

There is a basic concept concerning the behaviour of helium in thin films which should be brought out. This is the concept of the *coherence length* for the *order parameter* in the system. In many systems one can readily define a physical quantity that describes the order in the system. For example, the magnetic moment of a ferromagnet, or the spatial order in a crystal which can be defined by the spatial order of the atoms. These quantities are often referred to as order parameters. Such parameters may either be uniform for a homogeneous system or they may vary in space if the system is inhomogeneous. Thus we might think of the magnetization as being a function of position in space $M(\mathbf{r})$. The coherence length for an order parameter may be defined as the shortest distance over which the order parameter can vary appreciably. Thus if the system were forced into a situation in which the order parameter had to vary appreciably over a distance shorter than the coherence length then the order would break down locally and the order parameter would vanish in a local region. These ideas have been greatly developed and exploited in the theory of superconductivity where there is a highly developed order parameter theory—the Landau–Ginsberg theory. This theory was originally proposed on a phenomenological basis by Landau[88]—it can be derived from the BCS theory of superconductivity under very well controlled approximations.[89] In this theory (in its simplest form) the order parameter is a complex scalar $\psi(\mathbf{r}, t) = \exp[iS(\mathbf{r}, t)]R(\mathbf{r}, t)$. It has a natural length scale associated with it—the coherence length $\xi(T)$, which is a function of temperature. For a clean superconductor near $T_c$, $\xi = \xi_0(1 - T/T_c)^{-\frac{1}{2}}$ where $T_c$ is the superconducting transition temperature and $\xi_0 = \hbar v_F/\pi\Delta$ and numerically is of the order of $10^4$ Å. If a superconducting material is forced into a situation where the order parameter has to vary locally over distances short compared with $\xi$ then the order will vanish locally and $|\psi|$ will vanish in that region and the material will locally become "normal".

Unfortunately in helium we do not have such a coherent body of theory to work with and we are on much less certain ground. It is possible to define[(11)] the order parameter in helium; it is proportional to the condensed state wave function we defined in lecture III. However we do not know how to derive an equation for its amplitude—corresponding to the Landau–Ginsburg equation. There are theories based on the Hartree approximation to a Bose system interacting via repulsive delta function potentials which do lead to a set of equations very similar to Landau–Ginsberg theory. This is the so called Gross–Pitaevskii theory—which is certainly useful as a rough qualitative guide to what might be happening in helium. However the theory in its simple form makes no distinction between the superfluid density and total density and one is not strictly dealing with an order parameter theory. Moreover the model is so idealized that it is difficult to see how to relate the parameters of the model in any realistic way to those of liquid helium. In other words although it is possible to define an order parameter in a satisfactory way we do not know how to derive an equation for its amplitude. (The equation of motion for its phase can be derived in a simplified form and is often referred to as the Josephson equation.[(11)] It has the form:

$$\frac{\mathrm{d}\phi(\mathbf{r})}{\mathrm{d}t} = -\frac{h}{m}\mu(\mathbf{r})$$

where $\mu(\mathbf{r})$ is the local chemical potential.) The best we can do is to try to use the analogy with superconductivity as a guide.

Let us suppose that we can define an order parameter $\psi(\mathbf{r})$ for helium. In a stationary situation it will be independent of time and will be a function that depends only on spatial variables. Let us further suppose that there is a coherence length to be associated with $\psi(\mathbf{r})$ in the sense I have just explained. There is indirect evidence that this length is of the order of a few angstroms or at most one or two interparticle spacings at low temperatures.† Suppose

† The size of the coherence length is usually estimated from arguments about this size of the core of a vortex in helium.[(6, 78)] The argument runs as follows. Since a vortex has vorticity associated with it we expect the order parameter to vanish in the region in which the vorticity is distributed. This is because in any region in which the flow is not curl free there can be no superflow (which at the microscopic level must be always curl free; see lecture IV) because the superfluid velocity is the gradient of the phase of the order parameter. For there to be no superflow we must thus have the amplitude of the order parameter zero. The best estimates of the region in which the vorticity is concentrated in a vortex are of the order of one or two interparticle distances, at low temperatures. Over this region the order parameter must vanish and we therefore conclude that this is a rough estimate of the coherence length.

then we return to our very thin film of liquid helium. Then if the thickness of the film is of the order of the coherence length $\xi$ (or less) and if the boundary condition on $\psi$ is that it must vanish at the boundary of the helium then we will expect it to vanish everywhere in the fluid. This is because we have demanded that $\psi$ vary more rapidly than it is allowed to do and consequently the order will collapse. There are two key ideas in this argument: the size of the coherence length and the boundary conditions that the order parameter must obey.†

### (c) *The Reduction of the Superfluid Density*

In bulk helium the superfluid density rapidly approaches the total density, less than 1% is normal fluid at 1 K. In a thin film the superfluid density is reduced far below this modest reduction and may typically be reduced by 20–30% of its bulk value. We thus have another phenomena to understand—namely even though superfluid order has set in—the fraction of fluid participatibg in it is reduced considerably below its bulk value. The main point I want to make is that I do not think that it is adequate to argue that this is entirely the result of the reduction in the magnitude of the order parameter due to the severe "bending" it has to undergo at each surface of the film. The basic argument against this is that superflow has fundamentally to do with the "free" slippage of fluid past a rough boundary. One cannot completely understand superflow by looking at smooth boundaries. Any fluid can flow without friction past a smooth plane surface. This argument suggests that the superfluid density will depend to some extent on the nature of the substrate. I give, in the appendix, a simple variational argument that leads to the following suggestive expression:‡

$$\tilde{\rho}_s \leqslant \left[ \int \frac{\mathrm{d}x}{\tilde{\rho}(x)} \right]^{-1}$$

Here $\tilde{\rho}_s$ is the mean superfluid density in the film. The functiou $\tilde{\rho}(x)$ is a mean fluid density. If $z$ is normal to the substrate then,

† The boundary condition on $\psi$ in liquid helium is to be contrasted with the situation in a superconductor. At the surface of a superconductor the approximate boundary condition is that $\partial\psi/\partial n = 0$. Thus for a thin film we find that $\psi$ = constant is the solution and there is not reduction or collapse of the order parameter. If a superconducting film is in contact with a normal metal then the boundary condition is that $\psi = 0$. This leads to a drastic reduction in the magnitude of $\psi$ and hence the depression of the superconducting transition temperature. To have a situation analogous to a helium film with $\psi = 0$ on both boundaries we would have to have a thin film sandwiched between two pieces of normal metal, a situation that is not normally considered.

‡ A similar expression has been derived by Leggett[(90)] using a rather different argument.

$$\tilde{\rho}(x) = L^{-1}d^{-1}\int_0^L\int_0^d \rho(x, y, z)\,dz dy.$$

In other words $\tilde{\rho}(x)$ is an average of the local density $\rho(x, y, z)$ across the thickness of the film and across the lateral dimension perpendicular to the direction of flow. Finally the integration over $[\tilde{\rho}]^{-1}$ is in the dimensionless variable $x/L$. This formula is suggestive because the smaller $\tilde{\rho}(x)$ is in any region of space the larger will the integral of its inverse become and hence the smaller the upper bound to $\rho_s$ becomes. We may generally expect $\tilde{\rho}(x)$ to be small in regions where the wave function must vanish. The rougher the substrate the more nodal surfaces willbe introduced into the wave function and hence $\tilde{\rho}(x)$ will have more holes in it. I should emphasise that the above expression is exact, but may of course by a very crude upper bound. If the substrate is smooth $\tilde{\rho} = \rho$ and we merely recover the uninteresting result that $\rho_s \leqslant \rho$, as we expect.

I believe that this formula does support the contention that the reduction in superfluid density must be due to the way in which the wave function has to adjust to the substrate as the particles flow over it. This does not mean that it is entirely unconnected with the way the order parameter has to bend over to meet the boundary.

To exploit the above formula one must know $\tilde{\rho}_s(x)$. This in turn requires calculations of the exact ground state of helium in the film on a rough substrate. We hope to be able to accomplish this within the next year.

### (d) *The Concept of Layers in Thin Films*

Much of the experimental and theoretical literature on thin films and porous media makes great use of the concept of a more or less definite layered structure near the substrate. I believe this idea should be treated with caution—except for the first tightly bound layer of helium atoms. There is strong evidence for a layered structure from exact computer simulations of bosons in a narrow channel.[(91)] Whether these results will transfer to the film we do not know. Perhaps the most important result these computations have revealed is that one does not have partial layers formed in narrow channels—this may have important implications for films.

### (e) *The Destruction of Superflow by Thermal Fluctuations in Thin Films*

In lecture IV I discussed the thermal fluctuation mechanism for the destruction of superflow and at the end of that section I pointed out that the creation of vortex lines was a likely process in two dimensions. It turns out that special care must be taken in two dimensions because the energy depends logarithmically on the extent of the system, as indeed does the entropy of the vortex line. We thus arrive at the following expression for the free energy.[92]

$$F = E_a - TS = \left(\frac{\pi \hbar^2 \rho}{m} - 2kT\right) \log (R/a),$$

where $R$ is the lateral extension of the system, $a$ the radius of the vortex core, and $\rho$ the superfluid density per unit area. At high temperatures the entropy term, $(-2kT \log R/a)$, dominates and there will be no barrier for the production of pairs of vortex lines. At low temperatures the energy term dominates and there will be a barrier to inhibit the production process. This expression for $F$ leads to a natural definition of the onset temperature; namely that temperature near $F = 0$.[93] This at once gives

$$kT_0 = \frac{\pi \hbar^2 \rho}{2m}.$$

If we convert $\rho$ in this formula to the usual mean superfluid density: $\rho = \langle \rho_s \rangle db$, where $d$ is the thickness in atomic layers and $b$ is the mean interatomic spacing, then we find

$$\frac{\langle \rho_s \rangle d}{T_0} \approx 15 \times 10^{21} \text{ cm}^{-2} \text{ K}^{-1}.$$

There are very striking features of this simple argument; namely that it suggests that there is a finite $\langle \rho_s \rangle$ at the onset temperature $T_0$. There are several experiments to support this point of view.[93] The form of the above result is very well confirmed by the data which we have just quoted, and even the numerical value of the constant is almost correct: the experimental values fall in the range $15 \pm 2 \times 10^{21}$ cm$^{-2}$ K$^{-1}$.

This brings us to an important idea. Namely that one can have "superfluid" order present but the system may not be able to support any form of the superfluid *flow*. We are asserting that the order parameter may be finite throughout the system but because there is no free energy barrier to the formation of vortices no metastable flow states are possible.

## VI. Recent Developments in $He^3$

In the last two years there have been some very significant experimental discoveries at very low temperatures in liquid helium three.[94, 95, 96] It is known that in the range 1–3 millidegrees $He^3$ undergoes two phase transformations to two new superfluid phases. The phase diagram at zero magnetic field is shown in Fig. 16, where the new phases are labelled A and B—the

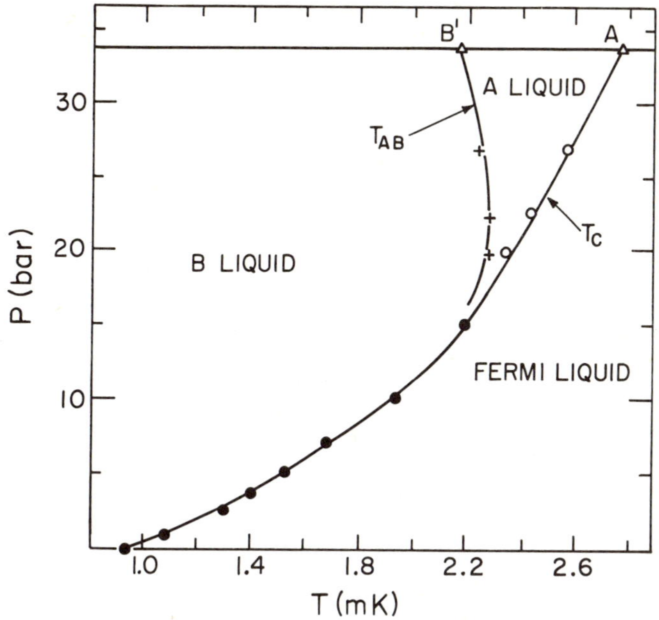


FIG. 16. The phase diagram of liquid $^3$He, data at 32 mT: open circles, $T_c$; crosses, $T_{AB}$; crosses in circles, "average" of $T_c$ and $T_{AB}$. Triangles, melting-curve data.

standard nomenclature We therefore have an example of a neutral (uncharged) Fermi system which is a superfluid. Since Drs J. Wheatley, P. W. Anderson and W. Brinkham will be describing in detail the experimental and theoretical work in this area I will try to complement their lectures by giving (a) a brief history of the area, (b) a qualitative comparison with $He^4$ and superconductors.

(a) *A Brief History*

The first indications that some kind of transition was taking place in late $He^3$ just below three millidegrees came in late 1971. At that time Douglas Osheroff, a graduate student at Cornell, noticed a small kink in the pressure curve of a Pomeranchuk cell as the temperature in the cell was rising. This observation, which at first was interpreted as possibly being due to a magnetic phase transformation in solid helium[94], came many years after the first theoretical speculations[97] that a superfluid might exist.

Theoretical speculations began almost immediately after the initial successes of the BCS theory of superconductivity. The basic theoretical idea has always been that a suitable attractive interaction between quasi-particles in $He^3$ could lead to the formation of a BCS pair state and hence to superfluidity. One basic complication is that the very strong repulsive forces between helium atoms at short distances will certainly lead to similar repulsions between the quasi-particle excitations—hence formation of pair-states with orbital angular momentum $L = 0$ is energetically unfavourable. Hence we have to look for a generalization of the standard form of BCS theory before we can start to make any sound theoretical predictions. Much of the earlier theoretical work was concerned with trying to estimate the transition temperature for different models.[98, 99, 100] Here the almost insurmountable difficulty is that one does not know enough about the interactions between quasi-particles to make reliable predictions. There is in addition considerable difficulty in formulating BCS theory for a system in which the short range interactions are so strong. As the experimental search progressed[101, 102] the theoretical transition temperature fell lower and lower, always keeping ahead of the experiments When no transition appeared down to about 2–3 mK both theoretical and experimental interest faded and comparatively little work was done looking for the transition for a number of years. Hindsight now tells us that it would have been worthwhile to search for the transition in the same temperature region but at elevated pressures. Indeed we could have reasonably argued many years ago that many of the parameters and physical properties of helium are very sensitive functions of pressure (density).[10] Consequently, the absence of a transition at zero pressure is absolutely no indication of what might be present at higher pressures. As I have already mentioned the basic discovery was made in 1971 about six years after interest in the transition had begun to wane. Although the initial discovery was thought to be evidence for a phase transformation in the solid nuclear magnetic resonance measurements[95] soon provided convincing evidence that the transformation was in the liquid phase. These were rapidly followed by sound attenuation measurements,[103] specific heat measurements,[104] and measurements on the damping of a virbrating wire.[105] None of these definitely established that the new phases were superfluid. The first direct evidence came from the fourth sound measurements made at La Jolla[106] and Cornell.[107] Fourth sound is a density wave in a superfluid medium confined to small pores so that the normal fluid is completely clamped by its viscosity to the surface of the pores. Since it can only propagate in a superfluid medium, we now have convincing evidence that the new phases are superfluid. Dr. Wheatley will be reviewing recent experimental work in this area.

(b) *Speculative Comparisons with Helium Four and Superconductors*

We now know of a new superfluid which is a neutral Fermi system. Our previous examples of superfluids were $He^4$, a neutral Bose system, and superconductors which are charged Fermi systems. The discovery of superfluidity in $He^3$ will thus provide us with a novel superfluid system.

As compared with superconductors, three very important differences are obvious. The fluid is neutral, it is not coupled to a lattice or immobile impurities, and finally the BCS pairing is not in the orbital $l = 0$ state. The absence of charge means that the usual electromagnetic probes cannot be used on the system—with the exception of magnetic experiments which interact with the nuclear spins of the system. In a superconductor many of the "fluid" properties of the electron gas are masked or entirely inhibited by the scattering on impurities. For example, it is for these reasons that one cannot observe second sound. In $He^3$ we have every expectation that under suitable conditions we will be able to observe the full range of fluid phenomena that are seen in $He^4$. Since we are dealing with a Fermi system in a BCS state the observation of these phenomena will show that they are basic superfluid phenomena and are in no way special to boson systems. The phenomenological derivation of the two fluid equations which was outlined in lecture IV already points to some extent to the same conclusion. That derivation in no way rests on ideas closely connected with the Boson nature of the system. However, the basic phenomenological assumption of that theory is not general enough to encompass the more general BCS pairing state that is needed to describe $He^3$. There is an underlying assumption in the derivation of two fluid hydrodynamics presented in lecture IV; namely that the fluid is isotropic, and of course there are no spin degrees of freedom. The BCS state describing $He^3$ is anisotropic and in addition the spin degrees of freedom will appear in the hydrodynamics.

In reference 108 you find an outline of some of the basic features of the new phase that will have to be taken into account in constructing a hydrodynamic description. On the assumption that pairing takes place in the $l = 1$ state, 9 complex numbers are needed to specify the order parameter. This is to be compared with one complex number for $He^4$. A simplified version of this general theory still requires six real variables (one amplitude and five angles). Thus the supercurrent has a much more complex structure than in $He^4$. In the same reference the boundary conditions on the order parameter and thus on the currnets are considered. Since the underlying description is that of an anisotropic superfluid (somewhat analogous to a liquid crystal) the boundary conditions play a much more important role than in $He^4$. Finally it is important to realize that the coherence length is very much larger in $He^3$ than in $He^4$.

In lecture V we discussed the concept of coherence length and pointed out that at low temperatures in $He^4$ the coherence length is estimated to be of the order of a few angstroms. In $He^3$ we can as a rough guide use the conventional BCS formula[109] $\hbar V_F/\pi\Delta$ to estimate the coherence length, where $\Delta$ is the energy gap. If we substitute for $\Delta$ a number of the order of $kT_c$ (which is a reasonable guess) then we find that the coherence length is of the order of $10^3$ Å. This difference will make some of the more detailed superfluid properties of $He^3$ very different from those of $He^4$. The most obvious example is that we must expect the detailed structure and size of vortex cores to be very different. A vortex core in $He^4$ is still basically a mystery.[6] We have every reason to belive that it is only a few angstroms in extent—but what goes on inside is a mystery. The main reason for this is that because the size of the core is so small in many phenomena it does not play a major role and hence experiments to explore its structure are very difficult to perform. If the scale of vortex cores in $He^3$ is of the order of the coherence length then we should much more readily be able to design experiments that will test detailed models of their structure. For example, if we can create vortex rings then the energy of the rings will be much more dependent on the core structure than in $He^4$.[78] Likewise the observation of vortex waves[7] should again be very sensitive to the core structure. In summary in $He^3$ there is reasonable hope that we may learn something about the detailed structure of vortices. The major experimental difficulty may be that it will probably be very difficult to do any experiments at all on vortices at a few millidegrees Kelvin.

The large coherence length should also have considerable influence on the behaviour of $He^3$ in thin films and pores. Here, though, we will have to be careful because of the importance of boundary conditions on the anisotropic superfluid.[108]

## Appendix

### A VARIATIONAL ESTIMATE OF $\rho_s$

The basis of the argument is as follows. We know that if superflow takes place at very low speeds then the ground state energy $E_0$ of the system is increased by an amount $\frac{1}{2}\rho_s\tilde{\mathbf{v}}_s^2$, where $\tilde{\mathbf{v}}_s$ is the superfluid velocity and $\rho_s$ the superfluid density. If we construct a trial wave function which allows the fluid to flow and furthermore constrain that flow so that its mean value (averaged over some suitable region) is equal to $\tilde{\mathbf{v}}_s$ then we can at once say that

$$E_0 + \tfrac{1}{2}\rho_s\tilde{\mathbf{v}}_s^2 \leqslant \frac{\langle\psi_T(\tilde{\mathbf{v}}_s|H|\psi_T(\tilde{\mathbf{v}}_s)\rangle}{\langle\psi_T(\tilde{\mathbf{v}}_s)|\psi_T(\tilde{\mathbf{v}}_s)\rangle} \tag{1}$$

where $\psi_T$ is the trial function, in which the fluid flows with mean speed $\tilde{\mathbf{v}}_s$. We now choose

$$\psi_T(\tilde{\mathbf{v}}_s) = e^{iS}\,\psi_0$$

where $\psi_0$ *is the exact ground state of the system.* Then the RHS of (1) becomes

$$E_0 + \frac{\hbar^2}{2m}\int \left|\sum_i \boldsymbol{\nabla}_i S\right|^2 \psi_0^2\, d\tau^N. \tag{2}$$

This allows us to cancel $E_0$ from both sides of (1). If we now choose $S = \sum_i S(\mathbf{r}_i)$ then we can combine (1) and (2) to get

$$\tfrac{1}{2}\rho_s \tilde{\mathbf{v}}_s^2 \leqslant \tfrac{1}{2}\frac{m}{V}\int \mathbf{v}_s^2(\mathbf{r}_i) n_0(\mathbf{r}_i)\, d^3\mathbf{r}_i \tag{3}$$

where $\mathbf{v}_s(\mathbf{r}_i)$ is identified as the local superfluid velocity

$$\mathbf{v}_s(\mathbf{r}_i) = -\frac{\hbar}{m}\boldsymbol{\nabla} S(\mathbf{r}_i) \tag{4}$$

and $n_0(\mathbf{r}_i)$ is the local number density for the ground state of the system.

Clearly we shall do best by choosing $\mathbf{v}_s(\mathbf{r}_i)$ (or equivalentally $S(\mathbf{r}_i)$) so that the RHS of (3) is as small as possible subject to the constraint that the mean value of $\mathbf{v}_s$ is equal to $\tilde{\mathbf{v}}_s$. We set up this constraint by demanding that

$$\int_V \frac{\mathbf{v}_s(\mathbf{r})\, d^3\mathbf{r}}{V} = \tilde{\mathbf{v}}_s. \tag{5}$$

To make the calculation more concrete we take $V = L^2 d$, where $L$ is the lateral dimension of a film and $d$ its thickness. We further simplify the calculation by assuming that $S(\mathbf{r}_i) = S(x_i)$; hence $\mathbf{v}_s(\mathbf{r}_i) = \mathbf{v}_s(x_i)$ where $x$ is one of the lateral directions. Then (5) becomes

$$\int_0^L \frac{\mathbf{v}_s(x)\, dx}{L} = \tilde{\mathbf{v}}_s \tag{6}$$

and the flow energy becomes

$$\tfrac{1}{2}\frac{m}{L^2 d}\int_0^d\int_0^L\int_0^L \mathbf{v}_s^2(x)\, n_0(x, y, z)\, dx dy dz. \tag{7}$$

If we set

$$\tilde{n}(x) = \int_0^L\int_0^d \frac{n_0(y, z)}{Ld}\, dy dz \tag{8}$$

then the flow energy becomes

$$\tfrac{1}{2}\frac{m}{L}\int_0^L \mathbf{v}_s^2(x)\tilde{n}(x)\, dx. \tag{9}$$

We can now easily complete our calculation, we vary (9) with respect to $\mathbf{v}_s(x)$ subject to the constraint given by (6). This leads to

$$\tfrac{1}{2}\rho_s\tilde{\mathbf{v}}_s^2 \leqslant \tfrac{1}{2}\tilde{\mathbf{v}}_s^2 m\left[\frac{1}{L}\int_0^L \frac{\mathrm{d}x}{\tilde{n}(x)}\right]^{-1} \tag{10}$$

or

$$\rho_s \leqslant m\left[\int_0^1 \frac{\mathrm{d}x'}{\tilde{n}(x)}\right]^{-1} \tag{11}$$

where now $x'$ is measured in units of $L$; $x' = x/L$, or finally

$$\rho_s \leqslant \left[\int_0^1 \frac{\mathrm{d}x'}{\tilde{\rho}(x')}\right]^{-1} \tag{12}$$

where

$$\tilde{\rho}(x') = m\tilde{n}(x').$$

It is interesting to note that this result has also been derived by Leggett by a rather different argument.[90]

## References


1. J. Wilks. "The Properties of Liquid and Solid Helium", Clarendon Press, Oxford (1967).
2. D. F. Brewer (Ed.). "Quantum Fluids", North Holland, Amsterdam (1966).
3. R. D. Parks (Ed.). "Superconductivity Vols I and II", Marcel Dekker, New York (1969).
4. I. M. Khalatnikov. "An Introduction to the Theory of Superfluidity", Benjamin, New York (1965).
5. G. Careri (Ed.). "Proceedings of the International School of Physics, Enrico Fermi, Course XXI", Academic Press, New York and London, 1963).
6. W. F. Vinen. *In* "Progress in Low Temperature Physics" (C. J. Gorter, Ed.), North Holland, Amsterdam (1961), Vol. III, p. 1, Review article on vortices in $He^4$
7. H. E. Hall. *Adv. Phys.* **9**, 89 (1960). Review article on vortices in $He^4$.
8. J. S. Langer and J. D. Reppy. *In* "Progress in Low Temperature Physics" (C. J. Gorter, Ed.), North Holland, Amsterdam (1970), Vol. VI, p. 1. Review article on intrinsic dissipation in supercurrents.
9. K. R. Atkins and I. Rudnick. *In* "Progress in Low Temperature Physics" (C. J. Gorter, Ed.), North Holland, Amsterdam (1970), Vol. VI, p. 37. Review article on third sound in thin films.
10. J. C. Wheatley. *In* "Progress in Low Temperature Physics" (C. J. Gorter, Ed.), North Holland, Amsterdam (1970), Vol. VI, p. 77. Review article on the low temperature properties of $He^3$.
11. P. W. Anderson. *Rev. Mod. Phys.* **38**, 298 (1966). Review article on flow and phase in $He^4$.
12. W. F. Vinen. *Rep. Prog. Phys.* **31**, 61 (1968). Macroscopic quantum phenomena.

13. V. P. Peshkov. *In* "Progress in Low Temperature Physics" (C. J. Gorter, Ed.), North Holland, Amsterdam (1964), Vol. IV, p. 1. Article on critical velocities in $He^4$.
14. P. W. Anderson. *In* "Progress in Low Temperature Physics" (C. J. Gorter, Ed.), North Holland, Amsterdam (1967), Vol. V, p. 1. Article on the Josephson effect in superfluids and superconductors.
15. R. de Bruyn Ouboter, K. W. Taconis and W. M. van Alphen. *In* "Progress in Low Temperature Physics" (C. J. Gorter, Ed.), North Holland, Amsterdam (1967), Vol. V, p. 44. Article on dissipation of superflow.
16. E. L. Andronikashvili and Yu. G. Mamaladze. *In* "Progress in Low Temperature Physics" (C. J. Gorter, Ed.), North Holland, Amsterdam (1967), Vol. V, p. 79. Article on rotating $He^4$.
17. A. D. B. Woods and R. A. Cowley, *Rep. Prog. Phys.* **36**, 1135 (1973).
18. L. Reatto, "Excitations in Liquid Helium", preprint from University of Milan (to be published in *Nuovo Cimento*).
19. B. J. Alder. *In* "Physics of Simple Liquids" (H. N. V. Temperley, J. S. Rowlinson and G. S. Rushbrooke, Eds), North Holland, Amsterdam.
20. A. Rahman. *Phys. Rev.* **A136**, 405 (1964).
21. L. Verlet. *Phys. Rev.* **159**, 98 (1967).
22. J. J. Weis and L. Verlet. *Phys. Rev.* **A5**, 939 (1972).
23. D. Schiff and L. Verlet. *Phys. Rev.* **160**, 208 (1967).
24. J. P. Hansen, D. Levesque, and D. Schiff. *Phys. Rev.* **A3**, 776 (1971).
25. M. H. Kalos, D. Levesque, and L. Verlet. *Phys. Rev.* **A9**, 2178 (1974).
26. J. de Boer. "Proceedings of the International School of Physics, Enrico Fermi, Course XXI„ (G. Careri, Ed.), Academic Press, New York and London (1963), p. 1.
27. H. C. Longuet-Higgins, and B. Widom. *Mol. Phys.* **8**, 549 (1964).
28. J. J. Weis, Orsay preprint (to be published).
29. H. Palevsky, K. Otnes, and K. E. Larsson. *Phys. Rev.* **112**, 11 (1958); J. L. Yarnell, G. P. Arnold, P. J. Bendt and E. C. Kerr. *Phys. Rev.* **113**, 1379 (1959).
30. P. J. Bendt, R. D. Cowan, and J. L. Yarnell. *Phys. Rev.* **113**, 1386 (1959).
31. L. D. Landau. *J. Phys. USSR* **5**, 71 (1941); and **11**, 91 (1947).
32. G. V. Chester. *In* "Proceedings of the International School of Physics, Enrico Fermi, Course XXI" (G. Careri, Ed.), Academic Press, New York and London (1963), p. 51.
33. L. P. Pitaevskii. *JETP* **9**, (1959), and **12**, 155 (1961).
34. N. E. Phillips, C. G. Waterfield, and J. K. Hoffer. *Phys. Rev. Lett.* **25**, 1260 (1970); S. G. Eckstein, Y. Eckstein, J. B. Ketterson, and J. H. Vignos. *In* "Physical Acoustics" Vol. VI (W. P. Mason and R. M. Thursten, Eds), Academic Press, New York and London (1970).
35. D. Pines. *In* "Quantum Fluids" (D. Brewer, Ed.), North Holland, Amsterdam (1966), p. 257.
36. R. P. Feynman. *Phys. Rev.* **94**, 262 (1954); R. P. Feynman and M. Cohen. *Phys. Rev.* **102**, 1189 (1956).
37. J. Wilks. "The Properties of Liquid and Solid Helium", Clarendon Press, Oxford (1967), Ch. 5.
38. T. C. Padmore and G. V. Chester. *Phys. Rev.* **A9**, 1725 (1974).
39. W. Götze and H. Wagner, *Physica* **31**, 475 (1965).
40. K. Kehr. *Physica* **33**, 620 (1967).
41. R. A. Cowley and A. D. B. Woods. *Phys. Rev. Lett.* **21**, 787 (1968).

42. E. B. Osgood, V. J. Minkiewicz, T. A. Kitchens and G. Shirane. *Phys. Rev.* **A5**, 1537 (1972).
43. R. A. Cowley and A. D. B. Woods. *Can. J. Phys.* **49**, 177 (1971).
44. J. Ruvalds and A. Zawadowski. *Phys. Rev. Lett.* **25**, 333 (1970); A. Zawadowski, J. Ruvalds and J. Solana. *Phys. Rev.* **A5**, 399 (1972).
45. T. J. Greytak, R. Woerner, J. Yan and R. Benjamin. *Phys. Rev. Lett.* **25**, 1547 (1970).
46. A. A. Abrikosov and I. M. Khalatnikov. *Rep. Prog. in Phys.* **22**, 329 (1959).
47. J. Wilks. "The Properties of Liquid and Solid Helium", Clarendon Press, Oxford (1967), Ch. 18.
48. L. R. Corruccini, D. D. Osheroff, D. M. Lee and R. C. Richardson, *Phys. Rev. Lett.* **27**, 650 (1971).
49. D. S. Betts, D. F. Brewer and R. S Hamilton. *J. Low Temp. Phys.* **14**, 331 (1974).
50. L. D. Landau and E. M. Lifshitz. "Statistical Physics", Pergamon Press.
51. W. F. Vinen. *In* "Superconductivity", (R. D. Parks, Ed.), Marcel Dekker, New York (1969), Vol. II, p. 1167.
52. P. C. Hohenberg and P. M. Platzman. *Phys. Rev.* **152**, 198 (1966).
53. O. K. Harling. *Phys. Rev. Lett.* **24**, 1046 (1970). H. A. Mook, R. Scherm and M. K. Wilkinson. *Phys. Rev.* **A6**, 2268 (1972).
54. L. J. Rodriguez and H. A. Gersch. *Phys. Rev.* **A9**, 2085 (1974).
55. O. Penrose and L. Onsager. *Phys. Rev.* **104**, 576 (1956).
56. O. Penrose. *In* "The Fifth Low Temperature Conference on Low Temperature Physics and Chemistry, 1957" (I. R. Dillinger, Ed.), Madison, Wisconsin (1958).
57. W. F. Vinen. *In* "Quantum Fluids" (D. F. Brewer, Ed.), North Holland, Amsterdam (1966), p. 74.
58. P. W. Anderson. *In* "Quantum Fluids" (D. F. Brewer, Ed.), North Holland, Amsterdam (1966), p. 146.
59. D. Pines. *In* "Low Temperature Physics, LT9" (Daunt, Edwards, Milford and Yaqub, Eds), Plenum Press, New York (1965), p. 61.
60. See reference (59).
61. F. de Pasquale and E. Tabet. *Ann. Phys.* **33**, 381 (1965).
62. W. Kohn and D. Sherrington. *Rev. Mod. Phys.* **42**, 1 (1970).
63. J. R. Clow and J. D. Reppy. *Phys. Rev. Lett.* **16**, 887 (1966).
64. J. S. Langer and M. E. Fisher. *Phys. Rev. Lett.* **19**, 560 (1967).
65. N. N. Bogoliubov. *J. Phys. U.S.S.R.* **11**, 23 (1947); J. Gavoret and P. Nozieres. *Ann. Phys.* (*N.Y.*) **28**, 349 (1964).
66. E. H. Lieb and W. Liniger. *Phys. Rev.* **130**, 1605 (1963).
67. G. V. Chester, M. E. Fisher and N. D. Mermin. *Phys. Rev.* **185**, 760 (1969).
68. G. V. Chester. *Phys. Rev.* **A2**, 256 (1970).
69. I. M. Khalatnikov. "An Introduction to the Theory of Superfluidity", Benjamin, New York (1965), p. 55 ff.
70. *IbId.* p. 113 ff.
71. *Ibid.* p. 63 ff.
72. G. Rickayzen. These proceedings, p. 53 ff.
73. N. N. Bogoliubov. "On the hydrodynamics of a superfluid", Dubna preprint in Russian (1963).
74. P. C. Hohenberg and P. C. Martin, *Ann. Phys.* (*N.Y.*), **34**, 291 (1965).
75. P. C. Martin. *In* "Quantum Fluids" (D. F. Brewer, Ed.), North Holland, Amsterdam (1966), p. 230 ff.
76. B. D. Josephson. *Ann. Phys.* **70**, 229 (1972).

77. M. Droz, Thesis Universite de Geneve (1973).
78. G. W. Rayfield and F. Reif. *Phys. Rev.* **136A**, 1194 (1964).
79. G. V. Chester, R. Metz and L. Reatto. *Phys. Rev.* **175**, 275 (1968).
80. I. L. Bekarevich and I. M. Khalatnikov. *JETP*, **13**, 643 (1961).
81. H. E. Hall. *In* "Proceedings of the International School of Physics, Enrico Fermi, *Course XXI*" (G. Careri, Ed.), Academic Press, New York and London (1963), p. 326.
82. F. D. M. Pobell, H. W. Chan, L. R. Corruccini, R. P. Henkel, S. W. Schwenterly and J. D. Reppy. *Phys. Rev. Lett.* **28**, 542 (1972).
83. I. Rudnick and J. C. Fraser. *J. Low Temp. Phys.* **3**, 225 (1970).
84. R. L. Elgin and D. L. Goodstein. *Phys. Rev.* **A9**, 2657 (1974).
85. J. Wilks. "The Properties of Solid and Liquid Helium", Clarendon Press, Oxford (1967), Ch. 14.
86a. P. C. Hohenberg. *Phys. Rev.* **158**, 383 (1967).
86b. N. D. Mermin. *Phys. Rev.* **176**, 250 (1968).
87. G. V. Chester. *In* "1968 Boulder Lectures in Physics", Vol. XI-B (W. Britten, Ed.), Gordon and Breach (1968), p. 253.
88. V. L. Ginzberg and L. D. Landau. Translation of original paper *in* "Men of Physics", L. D. Landau, Vol. I (D. ter Haar, Ed.), Pergamon Press, 138.
89. N. R. Werthamer. "Superconductivity", Vol. II (R. D. Parks, Ed.), Marcel Dekker, New York (1969). p. 321.
90. A. J. Leggett. *Phys. Rev. Lett.* **25**, 1543 (1970).
91. M. L. Kalos, K. Liu and G. V. Chester. To be published in *Phys. Rev.*
92. J. M. Kosterlitz and D. J. Thouless. *J. Phys.* C: *Solid State Physics* **5**, L124 (1972).
93. E. S. Sabisky and C. H. Anderson. *Phys. Rev. Lett.* **30**, 1122 (1973).
94. D. D. Osheroff, R. C. Richardson and D. M. Lee. *Phys. Rev. Lett.* **28**, 885 (1972).
95. D. D. Osheroff, W. J. Gully, R. C. Richardson, and D. M. Lee. *Phys. Rev. Lett.* **29**, 920 (1972).
96. W. P. Halperin, R. A. Buhrman, and R. C. Richardson. *Bull. Am. Phys. Soc.* **18**, 662 (1973).
97. L. N. Cooper, R. L. Mills, and A. M. Sessler. *Phys. Rev.* **114**, 1377 (1959).
98. V. J. Emery and A. M. Sessler. *Phys. Rev.* **119**, 43 (1960).
99. P. W. Anderson and P. Morel. *Phys. Rev.* **123**, 1911 (1961).
100. R. Balian and N. R. Werthamer. *Phys. Rev.* **131**, 1553 (1963).
101. J. C. Wheatley. *In* "Quantum Fluids" (D. F. Brewer, Ed.), North Holland, Amsterdam (1966), p. 183.
102. E. B. Osgood and J. M. Goodkind. *Phys. Rev. Lett.* **18**, 894 (1967).
103. D. T. Lawson, W. J. Gully, S. Goldstein, R. C. Richardson, and D. M. Lee. *Phys. Rev. Lett.* **30**, 541 (1973). D. N. Paulson, R. T. Johnson, and J. C. Wheatley. *Phys. Rev. Lett.* **30**, 829 (1973).
104. R. A. Webb, T. J. Greytak, R. T. Johnson, and J. C. Wheatley. *Phys. Rev. Lett.* **30**, 210 (1973).
105. T. A. Alvesalo, Yu. D. Anufriyev, H. K. Collan, O. V. Lounasmaa, and P. Wennerström. *Phys. Rev. Lett.* **30**, 962 (1973).
106. H. Kozima, D. N. Paulson and J. C. Wheatley. *Phys. Rev. Lett.* **32**, 141 (1974).
107. A. W. Yanof, E. Smith, D. M. Lee, R. C. Richardson, and J. D. Reppy. *Bull. Am. Phys. Soc.* **19**, 435 (1974). ref AK5.
108. V. Ambegaokar, P. G. de Gennes and D. Rainer. *Phys. Rev.* **A9**, 2676 (1974).
109. A. L. Fetter and P. C. Hohenberg. *In* "Superconductivity" (R. D. Parks, Ed.), Marcel Dekker, New York (1969), Vol. 2, p. 817.

## ACKNOWLEDGEMENTS

The publishers are grateful for permission to reproduce the following figures:

Figs. 1–3 from Verlet, L. *Phys. Rev.* **165**, 205 (1968).
Figs. 5, 9, 10, 13a, 13b from Cowley, R. A. and Woods, A. *Canadian J. Phys.* **49**, 177 (1971).
Figs. 6, 7 from Wilks, J. "The Properties of Liquid and Solid Helium" (Clarendon Press, Oxford, 1967).
Fig. 8 from Padmore, T. C. and Chester, G. V. *Phys. Rev.* **A9**, 1725 (1974).
Figs. 11, 12 from Chester, G. V. Lectures in Theoretical Physics, X1 B. (Britten W. ed.) (Gordon and Breach, 1968).
Fig. 14 from Mook, H. A., Scherm, R. and Wilkinson, M. K. *Phys. Rev.* **A6**, 2268 (1972).
Fig. 15 from Reppy, J. and Langer, J. S. Prog. in Low Temp. Phys. VI (Gorter C. J. ed.) (North Holland, 1970).
Fig. 16 from Ahonen, A., Haikala, M. T., Krusius, M. and Lounasmaa, O. *Phys. Rev. Letts.*, **33**, 628 (1974).

# 2. MANY-BODY THEORY OF SUPERFLUID $He^4$

G. Rickayzen

*Physics Laboratory, The University, Canterbury, Kent, England*

## I. Introduction

There is no microscopic theory of superfluid $He^4$. There are many mathematical models which appear to provide insight into the behaviour of the superfluid but there is no theory which provides quantitative predictions that agree with observation. Even some of the most widely held assumptions of the theory, such as the idea of a condensation in a zero momentum state to which we shall come shortly, cannot be said to be proved beyond reasonable doubt. This means that the application to superfluid $He^4$ of all I have to say, even though most is conventional theory, should be regarded with some scepticism.

The goals of the theory are to explain the thermodynamic and transport properties of liquid $He^4$. Among the thermodynamic properties are the existence of a phase transition of the liquid at a temperature of 2·17 K at saturated vapour pressure, and the anomalous behaviour of the specific heat $C_p$ as a function of temperature as shown in Fig. 1, the shape of this curve giving rise to the transition being called a $\lambda$-transition. The specific heat and other thermodynamic properties depend on the excitation spectrum which is observed directly in inelastic neutron scattering experiments. The observed dispersion curve[(1)] of excitation energy $\varepsilon(\mathbf{k})$ versus wave vector $\mathbf{k}$ is shown in Fig. 2. The main features are that for small wave vectors $\varepsilon(\mathbf{k})$ is proportional to $k$ and that there is a local minimum in the spectrum at $k \approx 2\,\text{Å}^{-1}$ at an energy $\Delta$ given by $\Delta/k_B = 8{\cdot}67 \pm 0{\cdot}04$ K. For reasons that will become clear later the excitations at small wave vector are called phonons while those near the minimum are called rotons.

The flow properties of liquid $He^4$ will be discussed in some detail by Dr Putterman in these Proceedings so I shall not elaborate here. The distinct

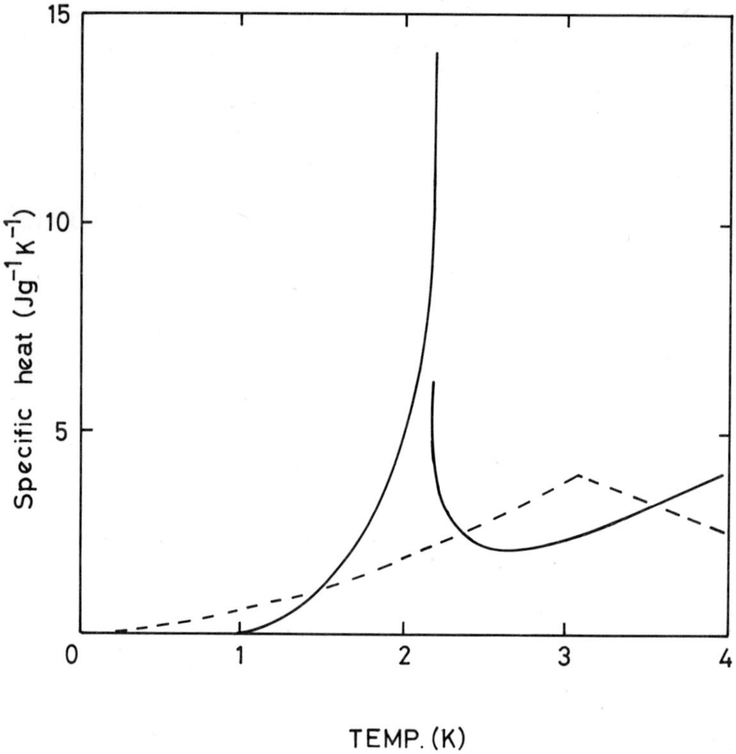


FIG. 1. Specific heat of liquid $He^4$. Broken line shows the specific heat of an ideal Bose–Einstein gas having the same density as liquid $He^4$ (after ref. 11).

feature is that whereas liquid $He^4$ has normal viscous properties above the transition temperature, below that temperature it can flow through narrow capillaries with no measurable viscosity—hence the name superfluid. The chief goal of the theory must be the explanation of this property. It seems to be similar to the property of superconductivity, the ability of electrons in a superconductor to carry an electric current without resistance and there are analogies between the two underlying theories. Indeed it has proved fruitful to develop the analogy between the theories. It had been argued over a long period, particularly by F. London,[(2)] that both are essentially quantum phenomena and the observation of various macroscopic quantum phenomena in superconductors has conclusively shown this to be true in that case. The observation of the corresponding phenomena (cf. Sections IX and X) in superfluid helium is not quite as convincing but there is sufficient evidence to confirm that superfluidity is a quantum phenomenon. We shall take this for granted.

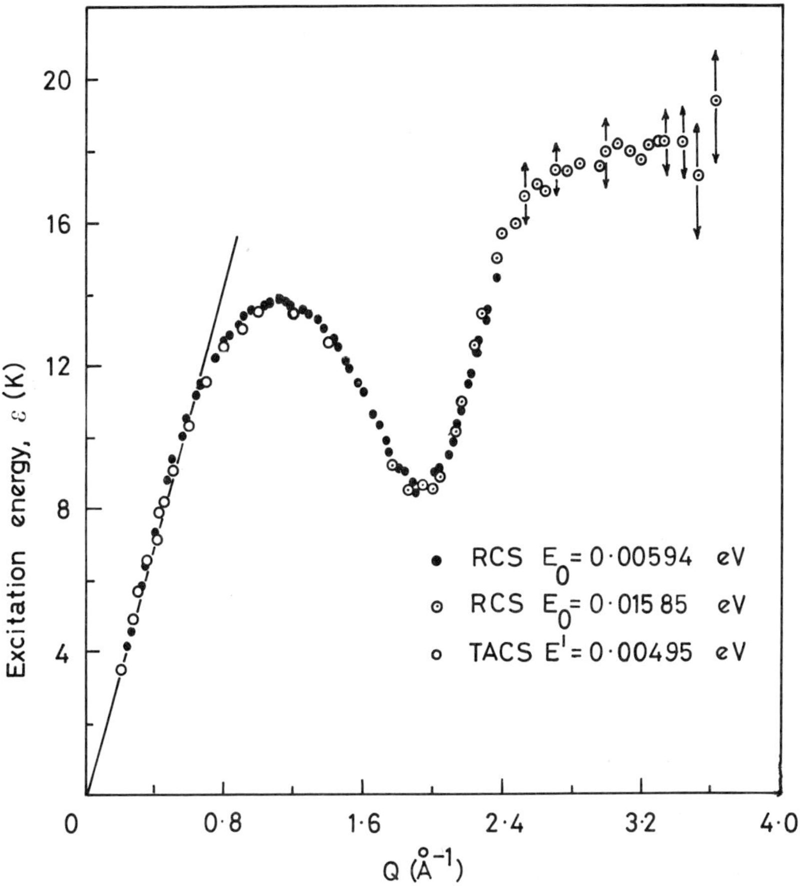


FIG. 2. The dispersion curve for superfluid He$^4$ for the elementary excitations at $T = 1{\cdot}1$ K which is reasonably close to the curve for $T = 0$ (after ref. 1).

Feynman[3] has based a general argument for the observed excitation spectrum at long wavelengths on the bose statistics of the atoms. Any electrically neutral condensed system can be shown on the basis of classical macroscopic arguments to carry sound waves. When quantised, these waves become phonons like the phonons of solids and have a dispersion equation

$$\varepsilon(\mathbf{q}) = \hbar c_s q.$$

Hence, liquid He$^4$ must have a phonon branch to its spectrum at long wavelengths. The unusual feature is that it possesses no other low energy long wavelength excitations, such as free particle excitations. (As the argument in Section IV shows this helps to explain superfluidity.) Feynman's argument for the *absence* of such excitations is based, in very condensed form, on the following considerations:

(i) the ground state of the system will, because of the Bose statistics, be described by a wave function $\Phi(\mathbf{r}_1, \mathbf{r}_2 \ldots \mathbf{r}_N)$ which is symmetric in the co-ordinates, has no nodes, and can be chosen to be positive everywhere;

(ii) the system possesses phonons which are long wavelength excitations in which the density varies;

(iii) any other excitation must have a wave-function $\Psi$ orthogonal to the ground state and so will be positive in some regions, negative in others;

(iv) $\Psi$ must also be orthogonal to the phonon states and so will involve no (or very little) change in density;

(v) $\Psi$ must vary slowly in space, otherwise there is a large contribution to the kinetic energy of the particles;

(vi) any large scale motion of the atoms which does not change the density simply permutes the particles and so is contained in the ground state.

It is thus impossible to find a wave function $\Psi$ which satisfies the restrictions (iii), (iv) and (v). (The reader is advised to look at Feynman's instructive attempts to construct such a wave function.)

In fact, Feynman has argued that the wave function for excitations even out to the roton part of tue spectrum will be close to

$$\Psi_{\mathbf{k}} = \sum_i \exp(i\mathbf{k}\,.\,\mathbf{r}_i)\,\Phi(\mathbf{r}_1 \ldots \mathbf{r}_N).$$

If the hamiltonian is of the form

$$H = \sum -\frac{\hbar^2}{2m}\nabla_i^2 + V$$

the excitation energy for this state is

$$\varepsilon(\mathbf{k}) = \frac{\hbar^2 k^2}{2mS(k)}$$

where

$$S(\mathbf{k}) = (\Phi, \sum_i \exp(-i\mathbf{k}\,.\,\mathbf{r}_i) \sum_j \exp(i\mathbf{k}\,.\,\mathbf{r}_j)\Phi),$$

and is the Fourier transform of the density–density correlation function of the liquid or static structure factor. It is measured in x-ray scattering and neutron scattering experiments. The measured values of $S(\mathbf{k})$ lead to an energy dispersion curve which is satisfactory for small $k$ but is too large by a factor of about 2 at the roton part of the spectrum. Feynman and Cohen[(4)] have used an improved wave function which attempts to account for the backflow of the bulk of the liquid as a single atom moves and so obtained better values for $\varepsilon(\mathbf{k})$. This semi-empirical approach, useful as it is, does not derive all quantities from a basic theory but it can show that observations are consistent with each other and with some general approach.

If quantum theory is important, the statistics of the particles play a significant role. Now at the low temperatures involved and in problems of flow, the electronic states of the atoms are not excited and electrons are certainly not stripped from the atoms. We can, therefore, take the helium atoms to be the immutable constituents of the fluid. Since $He^4$ atoms obey Bose–Einstein statistics, we have to consider a fluid of Bose–Einstein particles. One concept that is paramount in present thinking on superfluid $He^4$ is that of a condensation and this stems from the Bose–Einstein condensation of a gas of independent bosons. We, therefore, discuss this property in the next section.

## II. Bose–Einstein Condensation

Consider a grand canonical ensemble of independent free Bose–Einstein particles of mass $m$ at a chemical potential $\mu$. The different states of the particles have momentum $\mathbf{p}$ and energy $p^2/2m$. The number of particles in the state $\mathbf{p}$ is given by the Bose distribution function

$$n_{\mathbf{p}} = \frac{1}{\exp(\beta\varepsilon_{\mathbf{p}}) - 1} \tag{1}$$

where

$$\varepsilon_{\mathbf{p}} = p^2/m - \mu, \tag{2}$$

$$\beta = (k_B T)^{-1}. \tag{3}$$

Hence, the mean total number of particles present is

$$N = \sum_{\mathbf{p}} n_{\mathbf{p}} = \sum_{\mathbf{p}} \frac{1}{\exp(\beta\varepsilon_{\mathbf{p}}) - 1}. \tag{4}$$

If the particles are, as usual, thought to be in a box of side $L$ with periodic boundary conditions, the allowed values of $p$ are $(n_1 h/L, n_2 h/L, n_3 h/L)$ where $n_1$, $n_2$, $n_3$ are integers. The number of values in a small but macroscopic volume $d^3p$ of momentum space is, therefore, $(L/h)^3$. Hence, provided $|\mu|$ is larger than the separation of the levels ($\sim \hbar^2/2mL^2$) the sum can be replaced by an integral and results in

$$N = \left(\frac{L}{h}\right)^3 \int \frac{d^3p}{\exp(\beta\varepsilon_{\mathbf{p}}) - 1}. \tag{5}$$

If the integration over solid angle is performed and the variable of integration changed from $p$ to $x = \beta\varepsilon_{\mathbf{p}}$, we have

$$\frac{N}{L^3} = 2\pi\left(\frac{1}{h}\right)^3\left(\frac{2m}{\beta}\right)^{3/2} \int_{-\beta\mu}^{\infty} \frac{(x + \beta\mu)^{\frac{1}{2}}\,\mathrm{d}x}{\mathrm{e}^x - 1}. \tag{6}$$

For a given density and temperature this equation determines $\mu$. When $T$ is large, $\beta$ is small and the equation has a solution with $\mu$ large and negative. As $T$ decreases $\mu$ increases (decreases in magnitude) until it vanishes at a temperature $T_c$ given by

$$\begin{aligned}\frac{N}{L^3} &= 2\pi\left(\frac{2m}{\beta_c h^2}\right)^{3/2} \int_0^{\infty} \frac{x^{\frac{1}{2}}\,\mathrm{d}x}{\mathrm{e}^x - 1} \\ &\approx \left(\frac{2\pi m k_B T_c}{h^2}\right)^{3/2} 2{\cdot}612.\end{aligned} \tag{7}$$

Below the temperature $T_c$, it is impossible to satisfy (6) with any real value of $\mu$. Negative (and zero) values of $\mu$ clearly make the right hand side too small. Positive values make the right hand side diverge. However, the equation we should be solving for $\mu$ is equation (4). This can be satisfied by having an infinitesimal value of $\mu \sim N^{-1}$. Then we have to separate the term with $\mathbf{p} = 0$ from the sum. Since $\mu \ll \hbar^2/2mL^2$, $\mu$ can be put equal to zero in the terms for which $\mathbf{p} \neq 0$ and these terms can be replaced by an integral. Thus

$$\begin{aligned}N &= n_0 + \sum_{\mathbf{p}\neq 0} \frac{1}{\exp(\beta\varepsilon_{\mathbf{p}}) - 1} \\ &\approx \frac{1}{-\beta\mu} + L^3\left(\frac{2\pi m k_B T}{h^2}\right)^{3/2} 2{\cdot}612.\end{aligned} \tag{8}$$

Hence, there is a macroscopic number of particles in the state $\mathbf{p} = 0$,

$$n_0 = -\frac{1}{\beta\mu} = N - L^3\left(\frac{2\pi m k_B T}{h^2}\right)^{3/2} 2{\cdot}612. \tag{9}$$

This phenomenon whereby a macroscopic number of bosons is found in a single quantum state is referred to as a Bose–Einstein condensation. At the absolute zero,

$$n_0 = N$$

and all particles condense into the $\mathbf{p} = 0$ state.

There is, therefore, a transition of the free boson gas at the temperature $T_c$. If one uses the mass of a helium atom and the density of liquid He$^4$, $T_c$ is 3·1 K. Despite the remote resemblance of real liquid He$^4$ to a gas of independent bosons, this value of $T_c$ is sufficiently close to that of the transition temperature of the liquid to make one surmise that the two are connected. This connection has been assumed in most models of superfluid helium, and many start with the free boson gas as the zeroth approximation.

It must be emphasized, however, that a free boson gas is not a superfluid. If the particles are given a velocity relative to the walls of a container, they are scattered by the walls and gradually reduced to rest. There is, therefore, no flow without dissipation.

It should be emphasized that the discussion given above of the free boson gas below its transition temperature is not really valid. For the use of the grand canonical ensemble assumes that the fluctuations in the number of particles $\Delta N$, defined by

$$(\Delta N)^2 = \langle N^2 \rangle - \langle N \rangle^2,$$

where the angular brackets denote an emsemble average, is much less than $\langle N \rangle$. When the free boson gas condenses, this condition is not satisfied. It is not difficult to show that $\Delta N \sim N$ and the condition fails. It is however, obvious that the ground state of $N$ independent bosons has all particles in the lowest energy single particle state and there is a condensation.

The model of independent bosons must clearly be improved by taking into account the interactions between particles. To do this, basically two different approaches have been used. The first uses the variational principle of quantum mechanics and depends on a suitable choice for a trial wavefunction which contains the physically necessary features and is of a form such that the multidimensional integrals which result can be performed, at least, on a computer. The second method uses many-body theory, treating the interaction between the bosons as a perturbation. It is mostly this latter method which I shall describe in these lectures. Since it exploits the technique of second quantization, I have added an Appendix which describes this.

## III. Bogoliubov's Model[5]

Consider a gas of weakly interacting bosons and let us look for its ground state and lowest lying excitations. Assuming a potential $V(r)$ between any pair of atoms separated by a distance $r$, we can write the hamiltonian for $N$ bosons in the co-ordinate representation as

$$H = \sum_{i=1}^{N} -\frac{\hbar^2}{2m}\nabla_i^2 + \tfrac{1}{2}\sum_{i\neq j} V(\mathbf{r}_i - \mathbf{r}_j).$$

In terms of creation and annihilation operators, it is written

$$H = \frac{\hbar^2}{2m}\sum_{\mathbf{k}} k^2 a_{\mathbf{k}}^\dagger a_{\mathbf{k}} + \tfrac{1}{2}\sum_{\mathbf{q},\mathbf{k},\mathbf{k}'} V(q)\, a_{\mathbf{k}}^\dagger a_{\mathbf{k}'+\mathbf{q}}^\dagger a_{k'} a_{\mathbf{k}+\mathbf{q}} \tag{10}$$

where

$$[a_{\mathbf{k}}, a_{\mathbf{k}'}^\dagger] = \delta_{\mathbf{k}\mathbf{k}'}, \qquad [a_{\mathbf{k}}, a_{\mathbf{k}'}] = 0. \tag{11}$$

If the interaction $V$ is weak the expectation is that the energy and wave function do not differ radically from those of the non-interacting system, and that even in the perturbed system most of the particles are in the zero momentum state, $\mathbf{k} = 0$ ($\mathbf{p} = \hbar\mathbf{k}$). Hence, for the ground state $|0\rangle$,

$$a_0^\dagger a_0|0\rangle \approx N_0, \qquad a_0 a_0^\dagger|0\rangle \approx N_0, \tag{12}$$

where $N_0$ is the large number of bosons in the state $\mathbf{k} = 0$. In comparison with $N_0$, the righthand side of the commutator

$$a_0 a_0^\dagger - a_0^\dagger a_0 = 1$$

is negligible and we can use

$$[a_0, a_0^\dagger] = 0.$$

The error we make in doing this is presumably $O(1/N_0)$. However, $a_0$, $a_0^\dagger$ now commute with all operators in the problem and can be taken to be $c$-numbers. In view of (12) we take

$$a_0 = \sqrt{N_0}\,e^{i\phi}, \qquad a_0^\dagger = \sqrt{N_0}\,e^{-i\phi} \tag{13}$$

where $\phi$ is an arbitrary constant phase. Since it can be eliminated from the problem by a canonical transformation we take it to be zero.

If the interaction is weak, it is also to be expected that scattering in and out of states with $\mathbf{k} \neq 0$ is weak. We, therefore, retain in the hamiltonian only terms up to the second order in the operators $a_{\mathbf{k}}$, $a_{\mathbf{k}}^\dagger$ ($\mathbf{k} \neq 0$). This approximation can be checked for consistency at the end of the calculation.

With these approximations, the hamiltonian becomes

$$H = \sum_{\mathbf{k}\neq 0}\left[\frac{\hbar^2 k^2}{2m} + N_0 V(0) + N_0 V(k)\right] a_{\mathbf{k}}^\dagger a_{\mathbf{k}} + \sum_{\mathbf{k}\neq 0} N_0 V(k)(a_{\mathbf{k}}^\dagger a_{-\mathbf{k}}^\dagger + a_{-\mathbf{k}} a_{\mathbf{k}}) + \tfrac{1}{2}V(0)N_0^2 \tag{14}$$

The first term contains self-energy corrections due to interactions of particles in states $\mathbf{k}$ with the particles in the condensed state. The second term allows for the creation and destruction of pairs of excitations from and into the condensed state. The final term is the interaction energy of the condensed particles. In this term we have, in finding the excitation energies of the particles, to be careful to note the difference between $N_0$ and the total number of particles and to take account of the fluctuations in the number of particles in the condensed state. In fact, $N_0$ is strictly given by

$$N_0 = N - \sum_{\mathbf{k} \neq 0} a_{\mathbf{k}}^\dagger a_{\mathbf{k}} \tag{15}$$

and the hamiltonian can be written

$$H = \tfrac{1}{2}V(0)N^2 + \sum_{\mathbf{k} \neq 0} \left[\frac{\hbar^2 k^2}{2m} + NV(k)\right] a_{\mathbf{k}}^\dagger a_{\mathbf{k}} + \sum_{\mathbf{k} \neq 0} NV(k)(a_{\mathbf{k}}^\dagger a_{-\mathbf{k}}^\dagger + a_{-\mathbf{k}} a_{\mathbf{k}}). \tag{16}$$

The hamiltonian is quadratic in creation and annihilation operators and so can be diagonalized by a linear transformation. In fact, since states **k** interact only with states ($-$**k**), we can diagonalize by introducing the canonical transformations

$$\begin{aligned} a_{\mathbf{k}} &= u_{\mathbf{k}} \alpha_{\mathbf{k}} + v_{\mathbf{k}} \alpha_{-\mathbf{k}}^\dagger \\ a_{-\mathbf{k}}^\dagger &= w_{\mathbf{k}} \alpha_{\mathbf{k}} + t_{\mathbf{k}} \alpha_{-\mathbf{k}}^\dagger . \end{aligned} \tag{17}$$

Without loss of generality $u$, $v$, $w$, $t$ can all be chosen real. Then, to satisfy

$$[\alpha_{\mathbf{k}}, \alpha_{\mathbf{k}'}^\dagger] = \delta_{\mathbf{k}\mathbf{k}'}, \qquad [\alpha_{\mathbf{k}}, \alpha_{\mathbf{k}'}] = 0$$

we require

$$w_{\mathbf{k}} = v_{\mathbf{k}}, \quad t_{\mathbf{k}} = u_{\mathbf{k}} \quad \text{and} \quad u_{\mathbf{k}}^2 - v_{\mathbf{k}}^2 = 1. \tag{18}$$

If $u_{\mathbf{k}}$, $v_{\mathbf{k}}$ are then chosen so that

$$\left[\frac{\hbar^2 k^2}{2m} + NV(k)\right] 2u_{\mathbf{k}} v_{\mathbf{k}} + NV(k)(u_{\mathbf{k}}^2 + v_{\mathbf{k}}^2) = 0. \tag{19}$$

non-diagonal terms like $\alpha_{-\mathbf{k}}^\dagger \alpha_{\mathbf{k}}^\dagger$ and $\alpha_{\mathbf{k}} \alpha_{-\mathbf{k}}$ are removed from the hamiltonian. The result is

$$\begin{aligned} H = \tfrac{1}{2}V(0)N^2 &+ \sum_{\mathbf{k} \neq 0} \left[\left(\frac{\hbar^2 k^2}{2m} + NV(k)\right)(v_{\mathbf{k}}^2) + NV(k) u_{\mathbf{k}} v_{\mathbf{k}}\right] \\ &+ \sum_{\mathbf{k} \neq 0} \left\{\left[\frac{\hbar^2 k^2}{2m} + NV(k)\right](u_{\mathbf{k}}^2 + v_{\mathbf{k}}^2) + 2NV(k) u_{\mathbf{k}} v_{\mathbf{k}}\right\} \alpha_{\mathbf{k}}^\dagger \alpha_{\mathbf{k}}. \end{aligned} \tag{20}$$

The parameters $u_{\mathbf{k}}$ and $v_{\mathbf{k}}$ are obtained from (18) and (19) which yield

$$u_{\mathbf{k}}v_{\mathbf{k}} = -\frac{NV(k)}{2E_k}, \qquad v_{\mathbf{k}}^2 = \left[\frac{\hbar^2k^2}{2m} + NV(k) - E_k\right]\Big/2E_k \tag{21}$$

where

$$E_k = \left\{\left[\frac{\hbar^2k^2}{2m} + NV(k)\right]^2 - N^2V(k)^2\right\}^{\frac{1}{2}} = \left\{\frac{\hbar^2k^2}{2m}\left[\frac{\hbar^2k^2}{2m} + 2NV(k)\right]\right\}^{\frac{1}{2}}. \tag{22}$$

The sign of $E_k$ is not determined by these equations. However, for the excitation energies to be positive it is necessary for $E_k$ to be taken positive. The ground state $|G\rangle$ of the system is then given by the equations

$$\alpha_{\mathbf{k}}|G\rangle = 0 \text{ for every } \mathbf{k} \tag{23}$$

and the ground state energy is

$$E_0 = \tfrac{1}{2}V(0)N^2 - \tfrac{1}{2}\sum_{\mathbf{k}\neq 0}\left\{\left[\frac{\hbar^2k^2}{2m} + NV(k)\right] - E_k\right\}. \tag{24}$$

The excited states of the hamiltonian are the states $\alpha_{\mathbf{k}}^\dagger|G\rangle$ with energies

$$\left[\frac{\hbar^2k^2}{2m} + NV(k)\right](u_{\mathbf{k}}^2 + v_{\mathbf{k}}^2) + 2NV(k)u_{\mathbf{k}}v_{\mathbf{k}} = E_k. \tag{25}$$

Notice that the approximate hamiltonian (20) which can be written

$$H = E_0 + \sum_{\mathbf{k}} E_k\alpha_{\mathbf{k}}^\dagger\alpha_{\mathbf{k}} \tag{26}$$

describes a gas of independent bosons. These bosons, created by the operator $\alpha_{\mathbf{k}}^\dagger$, are not helium atoms. Let us look more closely at these excitations. It is easy to see from (22) that as $k \to 0$, $E_k \to 0$. For small values of $k$, we have

$$E_k = k\left[\frac{\hbar^2}{m}NV(0)\right]^{\frac{1}{2}} \tag{27}$$

provided that $V(0)$ is finite. It is certainly not infinite because the forces between helium atoms are short range. We shall see later on (Section VII) that $V(0)$ is indeed finite. Hence, $E_k \propto k$ as is true for long wavelength excitations in liquid helium. Indeed one can show that for this model

$$E_k = \hbar c_s k \tag{28}$$

where $c_s$ is the velocity of sound in the system. The sound velocity is, in general, given by the formula

$$c_s = \sqrt{\frac{-\Omega^2}{mN}\frac{\partial p}{\partial\Omega}} = \sqrt{\frac{\Omega^2}{mN}\frac{\partial^2E_0}{\partial\Omega^2}}. \tag{29}$$

A careful calculation from this formula[6] leads to

$$c_s = \left[\frac{NV(0)}{m}\right]^{\frac{1}{2}} \tag{30}$$

and so to (28). Notice that these formulae are all valid if $V(0)$ is positive so that the interaction is predominantly repulsive.

Within our approximations it is possible to show that the excited states are density waves at long wavelength. The operator for the number density is $\psi^\dagger(\mathbf{r})\psi(\mathbf{r})$. A Fourier component of this of wave number $\mathbf{k}$ creates a running density wave. Such a component is

$$\rho_{\mathbf{k}} = \sum_{\mathbf{K}} a^\dagger_{\mathbf{K}+\mathbf{k}} a_{\mathbf{K}}. \tag{31}$$

But the largest terms in this sum are those where one of the subscripts refers to the condensed state. Hence

$$\rho_{\mathbf{k}} \approx \sqrt{N}\,(a^\dagger_{\mathbf{k}} + a_{-\mathbf{k}}). \tag{32}$$

But, as $k \to 0$, $u_k/v_k \to -1$, and

$$\alpha^\dagger_{\mathbf{k}} = u_k a^\dagger_{\mathbf{k}} - v_k a_{-\mathbf{k}} \to u_k(a^\dagger_{\mathbf{k}} + a_{-\mathbf{k}}).$$

Hence, at long wavelengths $\alpha^\dagger_{\mathbf{k}}$ behaves like $\rho_{\mathbf{k}}$ and creates density waves, i.e. phonons, as is to be expected.

The mean number of bosons in the $\mathbf{k} = 0$ single particle state when the system is in the ground state can be calculated from (15) for the corresponding operator. We have

$$\begin{aligned}
\overline{N}_0 &= \langle G|N_0|G\rangle \\
&= N - \langle G|\sum_{k\neq 0} (u_k\alpha^\dagger_{\mathbf{k}} + v_k\alpha_{-\mathbf{k}})(u_k\alpha_{\mathbf{k}} + v_k\alpha^\dagger_{-\mathbf{k}})|G\rangle \\
&= N - \sum_{\mathbf{k}\neq 0} v_k^2 \\
&= N - \sum_{\mathbf{k}\neq 0} \frac{(\hbar^2k^2/2m) + NV(k) - E_k}{2E_k}.
\end{aligned} \tag{33}$$

The sum converges and is finite. There is, therefore, a depletion of the number in the condensed state from $N$, the number for the free boson gas. If $V(k)$ varies slowly with $k$ we can replace it by $V(0)$ in the sum to find

$$\overline{N}_0/N = 1 - \frac{1}{3\pi^2\hbar^3}(m^3V(0)^3N\Omega^2)^{\frac{1}{2}}. \tag{34}$$

Note that the depletion is not analytic in $V$ and cannot, therefore, be obtained by straight forward perturbation theory.

It is of some interest to examine the form of the wave function $|G\rangle$ of the ground state. We have seen that $|G\rangle$ satisfies

$$\alpha_{\mathbf{k}}|G\rangle \equiv (u_k a_{\mathbf{k}} - v_k a^{\dagger}_{-\mathbf{k}})|G\rangle = 0 \text{ for every } k. \tag{35}$$

Since the different degrees of freedom are independent and the state **k** interferes only with the state $-\mathbf{k}$ we can write

$$|G\rangle = \prod_{\mathbf{k}}{}' f_{\mathbf{k}}(a^{\dagger}_{\mathbf{k}}, a^{\dagger}_{-\mathbf{k}})|0\rangle \tag{36}$$

where the product is taken over half of **k** space and $f_{\mathbf{k}}$ depends only on $a^{\dagger}_{\mathbf{k}}$ and $a^{\dagger}_{-\mathbf{k}}$. Since

$$[a_{\mathbf{k}}, a^{\dagger}_{\mathbf{k}}] = 1,$$

$a_{\mathbf{k}}$ behaves like $d/da^{\dagger}_{\mathbf{k}}$ when acting on a function of $a^{\dagger}_{\mathbf{k}}$. Hence, from (35) and (36) $f_{\mathbf{k}}$ must satisfy

$$\left(u_k \frac{d}{da^{\dagger}_{\mathbf{k}}} - v_k a^{\dagger}_{-\mathbf{k}}\right) f_{\mathbf{k}} = 0. \tag{37}$$

Also, from (35) with $\alpha_k$ replaced by $\alpha_{-k}$

$$\left(u_k \frac{d}{da^{\dagger}_{-\mathbf{k}}} - v_k a^{\dagger}_{\mathbf{k}}\right) f_{\mathbf{k}} = 0. \tag{38}$$

The solutions of these equations are

$$f_{\mathbf{k}} = \exp\left[v_k a^{\dagger}_{\mathbf{k}} a^{\dagger}_{-\mathbf{k}}/u_k\right].$$

Hence,

$$|G\rangle = \exp\left(\sum_k{}' v_k a^{\dagger}_{\mathbf{k}} a^{\dagger}_{-\mathbf{k}}/u_k\right)|0\rangle$$

$$= \sum_{n=0}^{\infty} \frac{1}{n!} \left(\sum_k{}' v_k a^{\dagger}_{\mathbf{k}} a^{\dagger}_{-\mathbf{k}}/u_k\right)^n |0\rangle. \tag{39}$$

This state is a superposition of independent particle states in which the particular single particle states **k** and $-\mathbf{k}$ are either both occupied or both unoccupied. This is analogous to the ground state of a superconductor in which, according to the theory of BCS,[7] single particle states are occupied in pairs, the paired states being **k**, spin up, with $-\mathbf{k}$, spin down.

Because we have treated the operators $a_0$, $a_0^{\dagger}$ as *c*-numbers the wave function (29) shows only the occupation of the $\mathbf{k} \neq 0$ single particle states. The *n*th term in the sum is a state with $2n$ bosons in the $\mathbf{k} \neq 0$ states. The remaining $(N - 2n)$ bosons are in the state $\mathbf{k} = 0$. Hence the *n*th term in the sum should be written

$$\frac{1}{n![(N-2n)!]^{\frac{1}{2}}} a_0^{\dagger N-2n} \left(\sum_{\mathbf{k}}{}' v_k a^{\dagger}_{\mathbf{k}} a^{\dagger}_{-\mathbf{k}}/u_k\right)^n |0\rangle$$

where the factor $[(N - 2n)!]^{-\frac{1}{2}}$ is introduced to leave the normalization unchanged. (Our procedure for finding $|G\rangle$ from (35) to (39) leaves the dependence on $a_0^\dagger$ undetermined and we are now using the known number of particles present to rectify the balance.) Hence

$$|G\rangle = \sum_{n=0}^{N/2} \frac{1}{n![(N - 2n)!]^{\frac{1}{2}}} a_0^{\dagger N-2n} \sum_{\mathbf{k}}{}' (v_{\mathbf{k}} a_{\mathbf{k}}^\dagger a_{-\mathbf{k}}^\dagger / u_{\mathbf{k}})^n |0\rangle \tag{40}$$

where, within the approximations we have used, we limit the sum over $n$ to ensure no more than $N$ particles are present.

In configuration space, the wave function $\Psi(\mathbf{r}_1, \mathbf{r}_2 \ldots \mathbf{r}_N)$ is given by (see Appendix)

$$\Psi(\mathbf{r}_1, \mathbf{r}_2 \ldots \mathbf{r}_N) = \langle 0 | \psi(\mathbf{r}_N) \ldots \psi(\mathbf{r}_1) | G\rangle.$$

With $|G\rangle$ given by (40) one finds

$$\Psi(\mathbf{r}_1, \mathbf{r}_2 \ldots \mathbf{r}_N) = \Omega^{-N/2} \sum_n [(N - 2n)!]^{\frac{1}{2}} \sum P f(r_{12}) f(r_{34}) \ldots f(r_{(n-1),n})$$

where

$$f(r) = \sum_{\mathbf{k}} e^{i\mathbf{k}\cdot\mathbf{r}} (v_k/u_k), \qquad r_{ij} = |\mathbf{r}_i - \mathbf{r}_j|, \tag{41}$$

and $\Sigma P$ indicates the sum over all different ways of pairing the $N$ coordinates. The wave function in configuration space, therefore, also shows that the many-body wave function is a coherent superposition of wave functions in which the bosons are paired.

We have looked only at the ground state and low-lying excitations of the model. It is possible, in the same spirit, to look at the thermodynamic properties of the model. This analysis has been carried out by Lee and Yang[(8)] who showed that the model has a phase transition with a discontinuity in the specific heat. We shall not repeat the discussion here because it will appear as a special case of the more general discussion we give in Section VI.

## IV. States of Flow and Landau's Criterion

We have seen that a gas of independent bosons is not superfluid—but what of the weakly interacting gas. It might be thought that since the model hamiltonian (26) describes independent bosons, the weakly interacting gas is not a superfluid. This would be, however, to mistake the meaning of the model hamiltonian which describes the behaviour of a few excitations created out of the ground state. If the system is put into a state of flow, a macroscopic change takes place and the treatment of the complete hamiltonian (10) must be considered *ab initio*.

Let us suppose that the gas is in its ground state and that suddenly all particles are given an extra velocity $\mathbf{V}$ (relative to the container) so that the gas, as a whole, moves with this velocity. The question is now whether, if the system moves under its own forces, it will come to rest. If the particles are all given an extra velocity $\mathbf{V}$, those in the condensed state $\mathbf{k} = 0$ will be transferred to the state $\mathbf{k}_0 = m\mathbf{V}/\hbar$ and this is the state which is macroscopically occupied. We have, therefore, to find the eigenstates and energies of the hamiltonian for this new situation. Thus we put

$$a_{\mathbf{k}_0} = \sqrt{N_0}, \qquad a^{\dagger}_{\mathbf{k}_0} = \sqrt{N_0} \tag{42}$$

where $N_0$ is a macroscopic number and deal with the hamiltonian in a similar way to that of the last section. The result of this procedure can be found from an argument of Landau[9] which applies quite generally to homogeneous systems and, therefore, to this model as well. We, therefore, present the argument in its general form, based simply on the Gallilean invariance of the system.

The system possesses a ground state $E_0$ with zero momentum and microscopic excitations with momentum $\mathbf{p}$ and excitation energy $\varepsilon(\mathbf{p})$. Now let us suppose that we give the system as a whole a velocity $\mathbf{V}$. As all particles will have the same extra velocity their interaction energy is unchanged and the increase in energy is due to the change of kinetic energy. An initial state of the system with momentum $\mathbf{P}$ and energy $E$ will, when set in motion, have momentum $\mathbf{P} + Nm\mathbf{V}$ and energy

$$E + \frac{(\mathbf{P} + Nm\mathbf{V})^2}{2Nm} - \frac{P^2}{2Nm} = E + \mathbf{V} \cdot \mathbf{P} + \tfrac{1}{2}Nm\mathbf{V}^2 \tag{43}$$

where the second and third terms on the left hand side represent the change in the kinetic energy of the centre of mass motion. The new energy of the ground state ($\mathbf{P} = 0$) will be

$$E_0 + \tfrac{1}{2}Nm\mathbf{V}^2 \tag{44}$$

while that of the ground state with an excitation ($\mathbf{P} = \mathbf{p}$) will be

$$E_0 + \varepsilon(\mathbf{p}) + \mathbf{V} \cdot \mathbf{p} + \tfrac{1}{2}Nm\,\mathbf{V}^2.$$

The new excitation energy will, therefore, be

$$\varepsilon(\mathbf{p}) + \mathbf{V} \cdot \mathbf{p}. \tag{45}$$

If this energy is positive it will be impossible to reduce the system to rest by random scattering because such scattering cannot increase the energy of the system. Hence, the system will flow without dissipation if

$$\varepsilon(\mathbf{p}) + \mathbf{V} \cdot \mathbf{p} > 0$$

for all excitations. Since the left hand side is least when **V** and **p** are in opposite directions, there will be no dissipation if

$$\varepsilon(p)/p > V. \tag{46}$$

Since the long wavelength excitations in Bogoliubov's model have $\varepsilon(p)$ equal to $c_s p$, they will not reduce the flow until $V$ is equal to the velocity of sound.

It is often said that if the inequality (46) is not obeyed the fluid is not a superfluid. This is not quite correct. All it shows is that scattering can create some excitations and so reduce the total flow but not that there will be

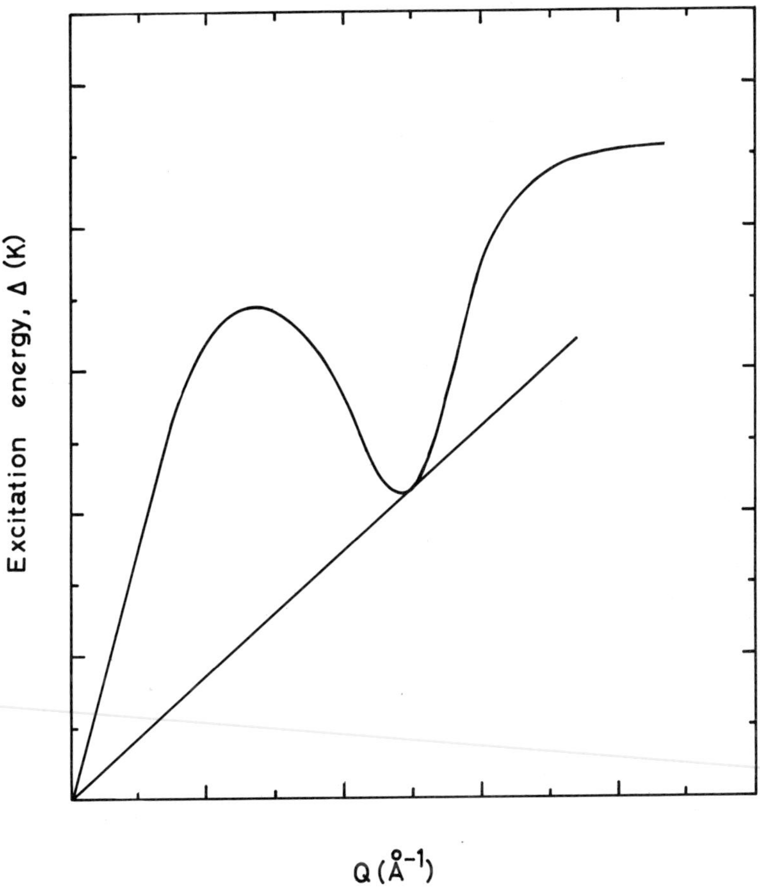


FIG. 3. The dispersion curve for superfluid He$^4$ with the tangent from the origin to the roton spectrum shown.

sufficient to reduce the current to zero. As the case of gapless superconductivity[10] (where $\varepsilon(p)/p$ is zero) shows, Landau's criterion is a sufficient condition for superfluidity but not a necessary one. As long as it is satisfied all the atoms flow with the velocity $\mathbf{V}$ at absolute zero. Hence the number of superfluid atoms $N_s$ is equal to $N$ at the absolute zero.

Since Landau's argument is quite general it can be applied to real liquid helium. If the argument is correct the excitation spectrum of superfluid helium should provide us with a lower bound to the critical velocity $v_c$, that is, the greatest velocity with which the liquid will flow through a narrow channel. The long wavelength excitations of liquid helium are phonons so that if these are to be responsible for viscosity the critical velocity must be greater than the velocity of sound $\sim 240\ \text{ms}^{-1}$. This is several orders of magnitude greater than the observed critical velocities which in thin films are of the order of 20 to 30 cm $\text{s}^{-1}$. Hence, one looks for other excitations for which $\varepsilon(p)/p$ might be less. In terms of the dispersion curve, $\varepsilon(p)/p$ for any excitation is the gradient of the line joining any point of the curve to the origin. From Fig. 3, it can be seen that this is a minimum for rotons, that is excitations near the local minimum of the curve. For these, the critical velocity is greater than $58\ \text{ms}^{-1}$. This is still well above the observed velocity.[11] This discrepancy between the observed and theoretical critical velocity has stimulated a search for other excitations that could explain the results. One possibility is a vortex to be discussed in Section IX. However, this is a macroscopic excitation and it is not clear that it would be excited sufficiently rapidly by the walls of the capillary to stop the flow. The explanation of the observed critical velocities will be discussed in detail in the other lectures.

## V. Perturbation Theory for Normal Bose Systems

Although Bogoliubov's model displays some of the properties of liquid helium, a phonon spectrum, superfluidity, and a phase transition, it is a long way quantitatively from a description of liquid helium. For example, in the model the fraction of bosons condensed into the $\mathbf{k} = 0$ state is close to unity whereas, in liquid $\text{He}^4$, recent estimates[12] suggest that it is well below 10%. (It cannot yet be said even to have been shown unequivocally to be non-zero). It is, therefore, necessary to find methods for dealing with a boson system in which only a small proportion of the particles has condensed. The method we shall describe is that of perturbation theory. This begins with the expansion of the sum-over-states in terms of the interaction. Subsequent summation of an infinite number of terms of the resulting series can lead to results which are valid even when the interaction is not small.

We shall begin by discussing the case when the system is not a superfluid (for example, liquid He$^4$ above its transition temperature) and go on in the next section to discuss modifications due to any condensation. We shall follow the method described in the books by Parry[(13)] and by Abrikosov, Gor'kov and Dzyaloshinskii[(14)] where further details can be found. We try to calculate the thermodynamic properties by calculating first the grand partition function $Z$ for a grand canonical ensemble and deriving other quantities from it. In quantum mechanics, $Z$ is given by[(15)]

$$Z = \sum_n \exp\left[-\beta(E_n - \mu N_n)\right] \tag{47}$$

where the subscripts $n$ label the states, $E_n$ and $N_n$ being the energy of and number of particles in the state $n$. The chemical potential is denoted by $\mu$. One can rewrite (47) as

$$Z = \operatorname{Tr} \exp\left[-\beta \mathscr{H}\right] \tag{48}$$

where

$$\mathscr{H} = H - \mu N_{\text{op}}. \tag{49}$$

$H$ is the hamiltonian and $N_{\text{op}}$ is the operator for the number of bosons. The operator $\mathscr{H}$ is then split into two parts

$$\mathscr{H} = \mathscr{H}_0 + V \tag{50}$$

where $\mathscr{H}_0$ is an operator (usually describing independent bosons) all of whose eigenvalues and eigenvectors are known. $Z$ is then expanded as a power series in $V$.

If we put

$$U(\beta) = \exp(-\beta \mathscr{H}), \tag{51}$$

then

$$\frac{dU}{d\beta} = -\mathscr{H}U = -(\mathscr{H}_0 + V)U(\beta). \tag{52}$$

The operator $U(\beta)$ is the unique solution of this equation with

$$U(0) = 1. \tag{53}$$

To obtain the series in $V$ we write

$$U(\beta) = \exp(-\beta \mathscr{H}_0)S(\beta). \tag{54}$$

If $V$ is zero then $S(\beta)$ is unity. We look for a series solution of the equation for $S(\beta)$. If expression (54) is substituted in equation (52), there results the equation

$$\frac{dS}{d\beta} = -\hat{V}(\beta)S(\beta) \tag{55}$$

where

$$\hat{V}(\beta) = \exp\beta\mathscr{H}_0\, V \exp(-\beta\mathscr{H}_0). \tag{56}$$

This operator is in a Heisenberg representation but with imaginary "time" $\beta$. Equation (51) can be solved by iteration using as starting value for $S(\beta)$, its value when $V$ is zero. Then

$$S(\beta) = 1 + \sum_{n=1}^{\infty} \int_0^{\beta} \hat{V}(\tau_n)\,d\tau_n \int_0^{\tau_n} \hat{V}(\tau_{n-1})\,d\tau_{n-1} \ldots \int_0^{\tau_2} \hat{V}(\tau_1)\,d\tau_1. \tag{57}$$

This is the required power series and when substituted in (54), (51) and (48) yields a power series for $Z$.

The symmetry of the integrand in (57) can be exploited to write the $n$th term in a slightly different way. If the $\hat{V}$'s were not operators but $c$-numbers their order in the integrand would not matter. Any change of variables which permuted the variables $\tau_1, \tau_2 \ldots \tau_n$ would not change the integral nor the integrand. It would change the limits on the integrals and so cover a different region of the $\tau$-space. All $n!$ permutations of the variables would cover the space of $\tau$ variables, each variable ranging from 0 to $\beta$ (a hyper-cube of edge $\beta$), just once. Hence the $n$th term in (57) could be written

$$\frac{1}{n!}\int_0^{\beta} d\tau_n \int_0^{\beta} d\tau_{n-1} \ldots \int_0^{\beta} d\tau_1\, \hat{V}(\tau_n) \ldots \hat{V}(\tau_1). \tag{58}$$

However, the $\hat{V}$'s are operators which do not necessarily commute (because $\mathscr{H}_0$ does not commute with $V$) and their order must be preserved in any change of variables. In the expression (57) the $\hat{V}$'s are ordered according to their arguments $\tau$, the least $\tau$ value being on the right. Provided we preserve this order, we can write the $n$th term as in (58). To show that the operators $\hat{V}$ are to be ordered in this way we introduce an ordering operator $P$ which simply tells us to order the operators on which it acts according to their values of $\tau$, the smaller (often called earlier because of the analogy with time) being on the right. Then, we can write

$$S(\beta) = 1 + \sum_{n=1}^{\infty} \frac{1}{n!}\int_0^{\beta} d\tau_n \ldots \int_0^{\beta} d\tau_1\, P[\hat{V}(\tau_n) \ldots \hat{V}(\tau_1)], \tag{59}$$

and

$$Z = \mathrm{Tr}\exp(-\beta\mathscr{H}_0) \sum_{n=0}^{\infty} \frac{1}{n!}\int_0^{\beta} d\tau_n \ldots \int_0^{\beta} d\tau_1\, P[\hat{V}(\tau_n) \ldots \hat{V}(\tau_1)]. \tag{60}$$

It is convenient to write $Z$ in terms of $Z_0$, the partition function, when $V$ is zero and $\rho_0$, the density matrix, when $V$ is zero. Then

$$Z_0 = \text{Tr} \exp(-\beta\mathscr{H}_0) \tag{61}$$

$$\rho_0 = \exp(-\beta\mathscr{H}_0)/Z_0 \tag{62}$$

and

$$Z/Z_0 = \text{Tr}\, \rho_0 \sum_{n=0}^{\infty} \frac{1}{n!} \int_0^\beta d\tau_n \ldots \int_0^\beta d\tau_1\, P[\hat{V}(\tau_n) \ldots \hat{V}(\tau_1)]. \tag{63}$$

It is usually assumed that is is permissible to perform the trace before the integration and one is, therefore, faced with the evaluation of

$$\text{Tr}\,\{\rho_0 P\hat{V}(\tau_n) \ldots \hat{V}(\tau_1)\}. \tag{64}$$

To proceed further it is necessary to particularize $\mathscr{H}_0$ and $V$ further. We shall assume that the interaction between the bosons (helium atoms) is given adequately by two-body forces and that the hamiltonian can be given in the form of (10), We shall take $V$ to be the interaction, that is the second term of this Hamiltonian. Recalling that $\mathscr{H}$ includes the chemical potential, we can write

$$\mathscr{H}_0 = \sum_{\mathbf{k}} \varepsilon_{\mathbf{k}} a_{\mathbf{k}}^\dagger a_{\mathbf{k}}, \qquad \varepsilon_{\mathbf{k}} = \frac{\hbar^2 k^2}{2m} - \mu. \tag{65}$$

Then the trace in (64) can be evaluated using Wick's theorem[16] generalized for finite temperatures. We outline this theorem below.

Since $V$ is a sum of products of creation and annihilation operators we require

$$a_{\mathbf{k}}(\tau) = \exp(\tau\mathscr{H}_0) a_{\mathbf{k}} \exp(-\tau\mathscr{H}_0) \tag{66}$$

which, according to (A21), is

$$a_{\mathbf{k}}(\tau) = a_{\mathbf{k}} \exp(-\tau\varepsilon_{\mathbf{k}}). \tag{67}$$

Similarly,

$$a_{\mathbf{k}}^\dagger(\tau) = a_{\mathbf{k}}^\dagger \exp(\tau\varepsilon_{\mathbf{k}}). \tag{68}$$

Hence the trace of (64) can be written completely and explicitly in terms of the unperturbed creation and annihilation operators. A typical term in (64) is of the form

$$\text{Tr}\,\{\rho_0 P A(\tau_1) B(\tau_2) C(\tau_3) \ldots\} \tag{69}$$

where $A$, $B$, $C$ stand for different creation and annihilation operators. Some of the $\tau$'s may be the same because the operators come from the same $V$. In this case they are ordered as in $V$. Otherwise, they are ordered according to the operator $P$. For (69) to be non-zero there must be equal numbers of creation and annihilation operators in the bracket.

Since $\mathscr{H}_0$ and, hence, $\rho_0$ are diagonal in the number operators in each single particle state $\mathbf{k}$, (69) will be non-zero only if the bracket includes equal numbers of creation and annihilation operators for each state $\mathbf{k}$. If we ignore for the moment the ordering operator $P$, a typical non-zero term to evaluate is

$$\mathrm{Tr}\,\{\rho_0 a^\dagger_{\mathbf{k}_1} a^\dagger_{\mathbf{k}_2} a_{\mathbf{k}_3} a^\dagger_{\mathbf{k}_3} a_{\mathbf{k}_1} a_{\mathbf{k}_2}\}. \tag{70}$$

If we note that the basis states can be taken to be

$$|n_{l_1}, n_{l_2}, n_{l_3} \ldots\rangle = |n_{l_1}\rangle |n_{l_2}\rangle |n_{l_3}\rangle \ldots$$

and that

$$\exp(-\beta\mathscr{H}_0) = \prod_{\mathbf{k}} \exp(-\beta\varepsilon_{\mathbf{k}} a^\dagger_{\mathbf{k}} a_{\mathbf{k}}),$$

because the operators belonging to different degrees of freedom $\mathbf{k}$ commute, we can write (70) (if $\mathbf{k}_1 \neq \mathbf{k}_2$, $\mathbf{k}_2 \neq \mathbf{k}_3$, $\mathbf{k}_3 \neq \mathbf{k}_1$) as

$$\mathrm{Tr}\,\{\rho_0 a^\dagger_{\mathbf{k}_1} a_{\mathbf{k}_1}\}\,\mathrm{Tr}\,\{\rho_0 a^\dagger_{\mathbf{k}_2} a_{\mathbf{k}_2}\}\,\mathrm{Tr}\,\{\rho_0 a_{\mathbf{k}_3} a^\dagger_{\mathbf{k}_3}\} \tag{71}$$

where inside any pair in expression (71) the operators are ordered as in (70). Since the operators $a_{\mathbf{k}}(\tau)$, $a^\dagger_{\mathbf{k}}(\tau)$ differ from $a_{\mathbf{k}}$, $a^\dagger_{\mathbf{k}}$ respectively only by constant factors, the result remains true if each $a_{\mathbf{k}}$ is replaced by an operator $a_{\mathbf{k}}(\tau)$. Similarly, since the equality of (70) and (71) is valid for any ordering provided the pairs are ordered in (70) as in (71) we have

$$\begin{aligned} &\mathrm{Tr}\,\{\rho_0 P a^\dagger_{\mathbf{k}_1}(\tau_1) a^\dagger_{\mathbf{k}_2}(\tau_2) a_{\mathbf{k}_3}(\tau_3) a^\dagger_{\mathbf{k}_3}(\tau_4) a_{\mathbf{k}_1}(\tau_5) a_{\mathbf{k}_2}(\tau_6)\} \\ &= \mathrm{Tr}\,\{\rho_0 P a^\dagger_{\mathbf{k}_1}(\tau_1) a_{\mathbf{k}_1}(\tau_5)\}\,\mathrm{Tr}\,\{\rho_0 P a^\dagger_{\mathbf{k}2}(\tau_2) a_{\mathbf{k}_2}(\tau_6)\}\,\mathrm{Tr}\,\{\rho_0 P a^\dagger_{\mathbf{k}_3}(\tau_3) a_{\mathbf{k}_3}(\tau_4)\} \end{aligned} \tag{72}$$

Wick's theorem extends this to the case when operators referring to a particular state $\mathbf{k}$ are repeated any number of times. The result is (for a full proof see Parry, Section 2.2[13]) that the trace in (69) is given by

$$\begin{aligned} &\mathrm{Tr}\,\{\rho_0 P A(\tau_1) B(\tau_2) C(\tau_3) D(\tau_4) \ldots F(\tau_{n-1}) G(\tau_n)\} \\ &= \sum \mathrm{Tr}\,\{\rho_0 P A(\tau_1) B(\tau_2)\}\,\mathrm{Tr}\,\{\rho_0 P C(\tau_3) D(\tau_4)\} \ldots \mathrm{Tr}\,\{\rho_0 P E(\tau_{n-1}) G(\tau_n)\} \end{aligned} \tag{73}$$

where in each term of the sum the operators on the left hand side are paired and appear in each bracket in the same order as on the left and the sum is over all possible pairings. If no operator $a_{\mathbf{k}}$ or $a^\dagger_{\mathbf{k}}$ is repeated (73) is of the same type as (72) which we proved. For each bracket on the right is non-zero only if $a_{\mathbf{k}}$ and the corresponding $a^\dagger_{\mathbf{k}}$ are paired. Hence, both sides are non-zero only when both $a_{\mathbf{k}}$ and $a^\dagger_{\mathbf{k}}$ appear among the operators and, if there is at most one creation operator for each $\mathbf{k}$, only one term in the sum on the right is non-zero. The remarkable thing is that (73) remains true however often any operator is repeated.

Each term in the expansion is then given in terms of the quantities $\mathrm{Tr}\,\{\rho_0 P a_{\mathbf{k}}(\tau_1) a_{\mathbf{k}}^{\dagger}(\tau_2)\}$, called Green's functions and denoted by $g(\mathbf{k}, \tau_1 - \tau_2)$. Then

$$\begin{aligned} g(\mathbf{k}, \tau_1 - \tau_2) &= \mathrm{Tr}\,\{\rho_0 P a_{\mathbf{k}}(\tau_1) a_{\mathbf{k}}^{\dagger}(\tau_2)\} \\ &= \mathrm{Tr}\,\{\rho_0 a_{\mathbf{k}}(\tau_1) a_{\mathbf{k}}^{\dagger}(\tau_2)\}\theta(\tau_1 - \tau_2) + \mathrm{Tr}\,\{\rho_0 a_{\mathbf{k}}^{\dagger}(\tau_2) a_{\mathbf{k}}(\tau_1)\}\theta(\tau_2 - \tau_1) \\ &= \exp\left[\varepsilon_{\mathbf{k}}(\tau_2 - \tau_1)\right](\mathrm{Tr}\,\{\rho_0(n_{\mathbf{k}} + 1)\}\theta(\tau_1 - \tau_2) + \\ &\quad + \mathrm{Tr}\,\{\rho_0 n_{\mathbf{k}}\}\theta(\tau_2 - \tau_1)] \\ &= \exp\left[\varepsilon_{\mathbf{k}}(\tau_2 - \tau_1)\right] \frac{\exp(\beta\varepsilon_{\mathbf{k}})\theta(\tau_1 - \tau_2) + \theta(\tau_2 - \tau_1)}{\exp(\beta\varepsilon_{\mathbf{k}}) - 1}, \end{aligned} \tag{74}$$

which is a function only of the difference $(\tau_1 - \tau_2)$ as the notation anticipated.

If Wick's theorem is used for each term in the perturbation series of (64), it is a mammoth task to keep count of all the possible pairings. This task is considerably simplified with the use of Feynman diagrams which we illustrate with some simple examples.

The first order terms require the evaluation of

$$\begin{aligned} \mathrm{Tr}\,\{\rho_0 \hat{V}(\tau)\} &= \tfrac{1}{2}\mathrm{Tr}\,\{\rho_0 \sum_{\mathbf{q}, \mathbf{k}, \mathbf{k}'} V(\mathbf{q}) a_{\mathbf{k}}^{\dagger}(\tau) a_{\mathbf{k}'+\mathbf{q}}^{\dagger}(\tau) a_{\mathbf{k}'}(\tau) a_{\mathbf{k}+\mathbf{q}}(\tau)\} \\ &= \tfrac{1}{2} \sum_{\mathbf{q}, \mathbf{k}, \mathbf{k}'} V(\mathbf{q})\left[g(\mathbf{k}, 0^-) g(\mathbf{k}^1, 0^-)\delta_{\mathbf{q}, 0} + g(\mathbf{k}, 0^-) g(\mathbf{k} + \mathbf{q}, 0^-)\right. \\ &\qquad \left. \times\, \delta_{\mathbf{k}'\mathbf{k}}\right] \end{aligned} \tag{75}$$

where

$$g(\mathbf{k}, 0^-) = \lim_{\tau \to 0} g(\mathbf{k}, -\tau)\theta(\tau), \tag{76}$$

and is used to take care of the order of operators at the same $\tau$. As this is always the appropriate limit for $g(\mathbf{k}, 0)$ we shall usually drop the minus sign. Then (75) can be written

$$\mathrm{Tr}\,\{\rho_0 \hat{V}(\tau)\} = \tfrac{1}{2} V(0) \sum_{\mathbf{k}, \mathbf{k}'} g(\mathbf{k}, 0) g(\mathbf{k}', 0) + \tfrac{1}{2} \sum_{\mathbf{q}, \mathbf{k}} V(\mathbf{q}) g(\mathbf{k}, 0) g(\mathbf{k} + \mathbf{q}, 0). \tag{77}$$

This is a sum of terms each of which is a product of a potential and two Green's functions. There are two kinds of terms represented by the two diagrams of Fig. 4. In each diagram the dashed line represents the potential, the solid lines the Green's functions and the vertices labelled by $\tau$, the imaginary times (in this case $\tau$) at which the interactions take place. Each line is labelled by the wave-vector of the corresponding function. For future reference the

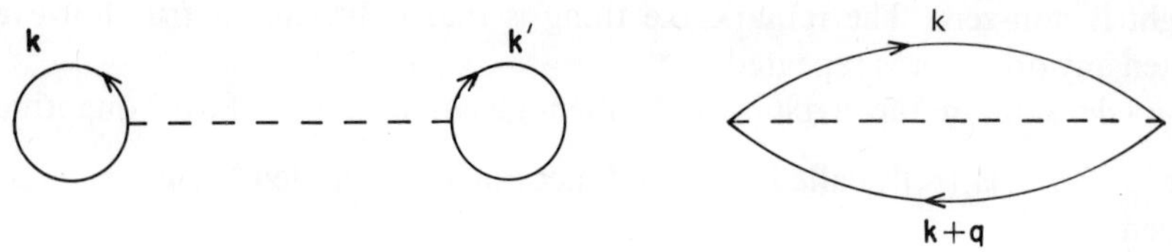


FIG. 4. The diagrams of the first order for Ω.

solid lines carry arrows which are directed from the vertex at which $a_{\mathbf{k}}$ operates to that at which $a_{\mathbf{k}}^{\dagger}$ operates. Thus

$$g(\mathbf{k}, \tau_1 - \tau_2) = \mathrm{Tr}\,\{\rho_0 P a_{\mathbf{k}}(\tau_1) a_{\mathbf{k}}^{\dagger}(\tau_2)\}$$

is represented by a solid line as in Fig. 5.

FIG. 5. The diagram for the Green's function $g(\mathbf{k}, \tau_1 - \tau_2)$.

If we regard the wave vector (or momentum) **k** as propagating in the direction of the arrow, wave vector (momentum) is conserved at every vertex. There is no need to put an arrow in the dashed line because $V(\mathbf{q})$ and $V(-\mathbf{q})$ are identical. The diagrams of Fig. 4(a) and Fig. 4(b) then represent, respectively the products

$$V(0) g(\mathbf{k}, 0) g(\mathbf{k}', 0) \quad \text{and} \quad V(\mathbf{q}) g(\mathbf{k}, 0) g(\mathbf{k} + \mathbf{q}, 0).$$

The result (77) is found by multiplying each of these by $\frac{1}{2}$ and summing over the free wave vectors $\mathbf{k}, \mathbf{k}'$ in the first case and $\mathbf{k}, \mathbf{q}$ in the second.

The relationship between terms in the series and diagrams is more fully brought out in a consideration of the second order term in the series

$$\begin{aligned} &\mathrm{Tr}\,\{\rho_0 \hat{V}(\tau_1) \hat{V}(\tau_2)\} \\ &\quad = \tfrac{1}{4}\mathrm{Tr}\,\{\rho_0 \textstyle\sum V(\mathbf{q}_1) a_{\mathbf{k}_1}^{\dagger}(\tau_1) a_{\mathbf{k}_1'+\mathbf{q}_1}^{\dagger}(\tau_1) a_{\mathbf{k}_1'}(\tau_1) a_{\mathbf{k}_1+\mathbf{q}_1}(\tau_1) \\ &\qquad \times \textstyle\sum V(\mathbf{q}_2) a_{\mathbf{k}_2}^{\dagger}(\tau_2) a_{\mathbf{k}_2'+\mathbf{q}_2}^{\dagger}(\tau_2) a_{\mathbf{k}_2'}(\tau_2) a_{\mathbf{k}_2+\mathbf{q}_2}(\tau_2)\} \end{aligned} \tag{78}$$

Any non-zero term arising from the use of (79) will involve the product of two potentials and four Green's functions. Corresponding diagrams will contain two dashed lines for the potentials; one will carry the labels $\tau_1$ at the vertices at its ends and the other the labels $\tau_2$ at its ends. The vertices are joined by four solid lines one for each Green's function. One Green's function enters and one leaves each vertex. The topologically distinct possible diagrams are shown in Fig. 6. Diagrams (c) and (d) are changed when the labels

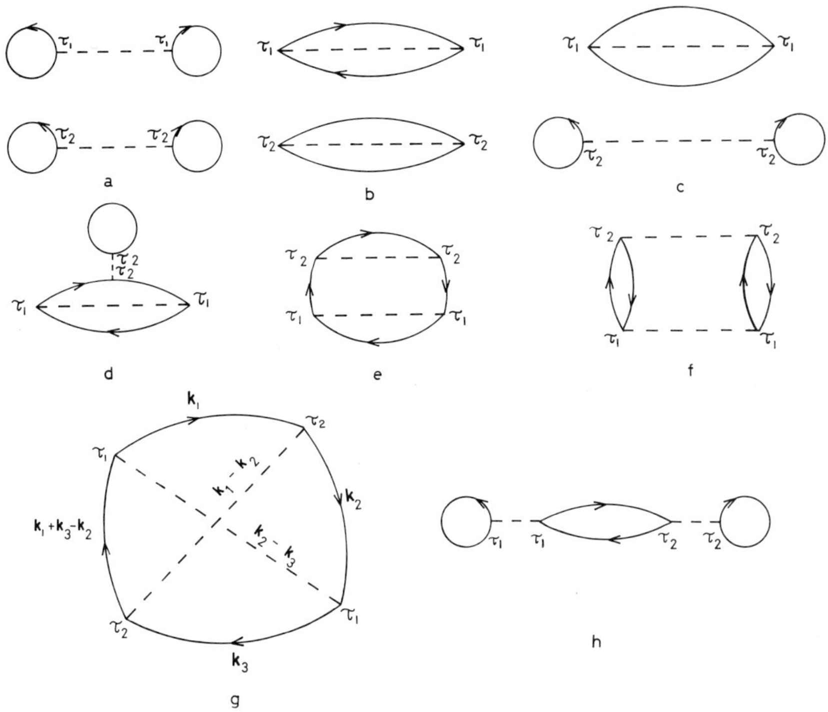


FIG. 6. All topologically different second order diagrams for $\Omega$.

$\tau_1$ and $\tau_2$ are interchanged and the contributions of all distinct possibilities have to be included. Each line has a wave vector associated with it and total wave vector is conserved at each vertex. The diagrams (a) to (c) then have four free momenta which are to be summed over and diagrams (d) to (h) have three free momenta to be summed over. For example, the contribution from Fig. 6(g) to (78) is

$$\frac{1}{4}\sum_{\mathbf{k}_1,\mathbf{k}_2,\mathbf{k}_3} V(\mathbf{k}_1-\mathbf{k}_2)V(\mathbf{k}_2-\mathbf{k}_3)g(\mathbf{k}_1,\tau_1-\tau_2)g(\mathbf{k}_2,\tau_2-\tau_1) \times g(\mathbf{k}_3,\tau_1-\tau_2)g(\mathbf{k}_1+\mathbf{k}_3-\mathbf{k}_2,\tau_2-\tau_1). \tag{79}$$

This contribution will actually arise twice according to the two ways in which the annihilation operators at $\tau_1$ can be paired with the creation operators at $\tau_2$. Hence, the total contribution from diagrams equivalent to Fig. 6(g) to (78) is twice expression (79).

In the same way one relates the contribution of the $n$th order term in the series to diagrams and one can write down the contribution of each $n$th order diagram to the series. In fact, each $n$th order diagram will contain $n$

dashed lines corresponding to the potentials, each with the same labels $\tau_i$ at the vertices at its two ends. Arrowed full lines join the vertices in pairs in such a way that one full line enters and leaves each vertex. All topologically distinct diagrams that can be drawn according to this prescription make a contribution to the $n$th order term of the series. To each line a wave-vector is associated in such a way that wave vector is conserved. Then the contribution of each topologically distinct diagram of order $n$ comprises the following factors:

(i) $\frac{1}{2}V(\mathbf{q})$ for each dashed line with wave-vector $\mathbf{q}$;

(ii) $g(\mathbf{k}, \tau_1 - \tau_2)$ for each solid line with wave-vector $\mathbf{k}$ and going from vertex labelled $\tau_1$ to vertex labelled $\tau_2$;

(iii) $(n!)^{-1}$ for the overall factor in the expansion;

(iv) a numerical factor for the number of times a particular diagram arises from the expansion of (73);

To obtain the $n$th term in the series, this must be summed over all free wave-vectors and integrated over the $n$ different labels $\tau$ from 0 to $\beta$. The factor which requires care in its determination is the numerical factor (iv).

This result can be usefully simplified if one takes note of the difference between connected and disconnected diagrams. A disconnected diagram can be divided into two parts by a line which does not cut any line of the diagram. A connected diagram cannot. In Fig. 6, (a), (b), (c) are disconnected and (d) to (h) are connected. Let us call the contribution of a connected diagram $C_i$. Then the contribution of a disconnected diagram, because the labels of the disconnected parts are independent, is apart from a numerical factor, the product of the contributions $C_1, C_2, \ldots$ of the separate parts. Since there are terms in the series with disconnected diagrams containing each connected part any number of times, the series will be of the form

$$\sum_{l_1 l_2 l_3 \ldots} a_{l_1 l_2 l_3 \ldots} C_1^{l_1} C_2^{l_2} C_3^{l_3} \ldots$$

where every $C_i$ and every integral value of $l_i$ occurs in the sum. Some combinatorial algebra shows that (see ref. (13), Section 2.5 or ref. (17), p. 386)

$$a_{l_1 l_2 l_3} \ldots = \prod_{l_i} \frac{1}{l_i!}$$

and that the sum is

$$Z/Z_0 = \exp(\Sigma C_i) \tag{80}$$

where $\Sigma C_i$ is the sum over connected diagrams. Since $Z$ is related to the thermodynamic potential $\Omega$ by

$$\Omega = -kT \ln Z,$$

we have

$$\Omega = \Omega_0 - \beta^{-1} \Sigma C_i \tag{81}$$

where the sum is over all connected diagrams.

The reduction to connected diagrams has the further advantage that the different terms in the sum have the same dependence on volume. This dependence can be determined from the facts that $V(\mathbf{q})$ is inversely proportional to volume and, in a normal system, every sum of wave-vector is proportional to volume since

$$\sum_{\mathbf{k}} \equiv \frac{\text{vol}}{(2\pi)^3} \int \mathrm{d}^3\mathrm{k}.$$

Now a connected diagram of order $n$ will have $n$ dashed lines and so $n$ potentials $V(\mathbf{q})$ and a contribution proportional to $(\text{vol})^{-n}$. However, it will contain $n + 1$ free wave-vectors, one for each dashed line and one for *one* Green's function (conservation of wave-vector determines the remainder). Hence there are $(n + 1)$ sums which lead to a factor $(\text{vol})^{n+1}$. Therefore,

$$C_i \propto \text{vol}, \qquad \text{for every } i.$$

This ensures that $\Omega$ is proportional to volume as it should be. Note that a disconnected diagram with $r$ connected parts will be proportional to $(\text{vol})^r$. Hence the original expansion mixes up different volume dependences in the same order.

The rules can be further simplified with the introduction of the Fourier transforms of the Green's functions. Now the Green's function $g(\mathbf{k}, \tau)$ is required only for $-\beta < \tau \leqslant \beta$ and in this interval it is given by (74). It, therefore, has the property that if $0 < \tau \leqslant \beta$, so that $-\beta < (\tau - \beta) \leqslant 0$, then

$$g(\mathbf{k}, \tau - \beta) = \frac{\exp[-\varepsilon_{\mathbf{k}}(\tau - \beta)]}{\exp(\beta\varepsilon_{\mathbf{k}}) - 1} = g(\mathbf{k}, \tau) \tag{82}$$

Hence, in the interval $-\beta < \tau \leqslant \beta$, $g(\mathbf{k}, \tau)$ can be expanded in a Fourier series in terms of $\exp(\mathrm{i}\omega_p\tau)$ where $\omega_p$ is $2\pi p/\beta$ and $p$ is an integer. Hence we can write

$$g(\mathbf{k}, \tau) = -\beta^{-1} \sum_{\omega_p} g(\mathbf{k}, \omega_p) \exp(-\mathrm{i}\omega_p\tau) \tag{83}$$

where,

$$\begin{aligned} g(\mathbf{k}, \mathrm{i}\omega_p) &= -\int_0^\beta \mathrm{d}\tau\, g(\mathbf{k}, \tau) \exp(\mathrm{i}\omega_p\tau) \\ &= (\mathrm{i}\omega_p - \varepsilon_{\mathbf{k}})^{-1} \end{aligned} \tag{84}$$

the last equation being obtained by explicit integration. The variables $\omega_p$ have the dimensions of energy so we shall often refer to them as energies.

If the Fourier series (83) is used for every Green's function, the integrations over $\tau$ can be performed explicitly. In Fig. 7, we show the part of a diagram

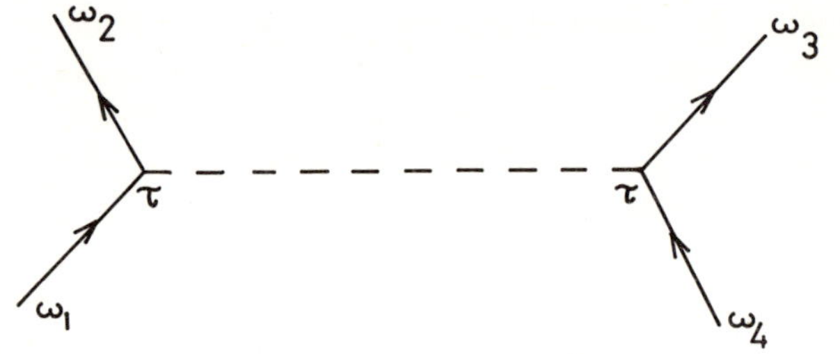


FIG. 7. An interaction with the energies of the boson lines indicated.

containing one dashed line and the boson lines leading in and out of the vertices labelled $\tau$ at its ends. For this diagram the $\tau$ integral

$$\int_0^\beta d\tau \exp\left[i(\omega_1 + \omega_3 - \omega_2 - \omega_4)\tau\right] = \beta\delta_{\omega_1+\omega_3,\,\omega_2+\omega_4}.$$

The delta function ensures that there is conservation of energy at each vertex. This means that we can assign an energy to each line of the diagram subject to this conservation and use the same rules as before except that $-\beta^{-1}(i\omega_p - \varepsilon_{\mathbf{k}})^{-1}$ is the contribution of each full line, $\frac{1}{2}\beta V(\mathbf{q})$ the contribution of each dashed line and one ultimately sums over all free energy variables as well as wave-vectors. In ensuring conservation of energy of each vertex it is convenient to imagine that the dashed line carries an energy, for example, that in Fig. 7 carries the energy $\omega_1 - \omega_2$.

As an example we consider the contribution of Fig. 6(g) to $\Omega$. We show it again in Fig. 8 with the energy labels attached. The contribution is

$$-2\beta^{-3} \sum_{\omega_1,\omega_2,\omega_3} \sum_{\mathbf{k}_1,\mathbf{k}_2,\mathbf{k}_3} \tfrac{1}{2}V(\mathbf{k}_1 - \mathbf{k}_2)\tfrac{1}{2}V(\mathbf{k}_2 - \mathbf{k}_3)g(\mathbf{k}_1, i\omega_1)g(\mathbf{k}_2, i\omega_2)g(\mathbf{k}_3, i\omega_3)$$
$$\times\, g(\mathbf{k}_1 + \mathbf{k}_3 - \mathbf{k}_2, i\overline{\omega_1 + \omega_3 - \omega_2}),$$

where the origin of the first factor 2 was discussed after equation (79). This provides us with the rules for evaluating the contribution of any diagram to $\Omega$. In any particular problem we have to decide which diagrams are likely to make the most significant contributions and sum these.

One particular and commonly used sum is that which leads to the full or renormalized Green's function for a boson. Consider a boson line which

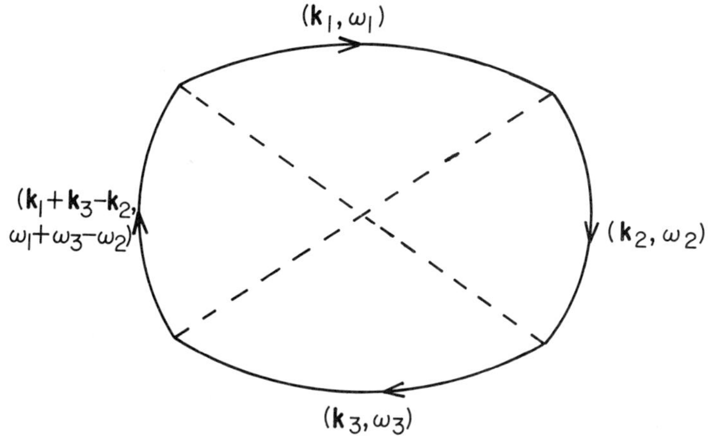


FIG. 8. A second order diagram with the momenta **k** and energies $\omega_p$ of the lines indicated.

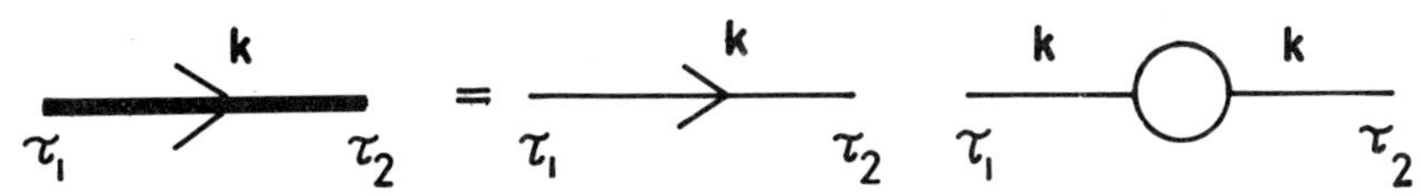


FIG. 9. The full Green's function, $G$, (represented by the thick line) related to $g$ (the thin lines) and the self-energy, the circle.

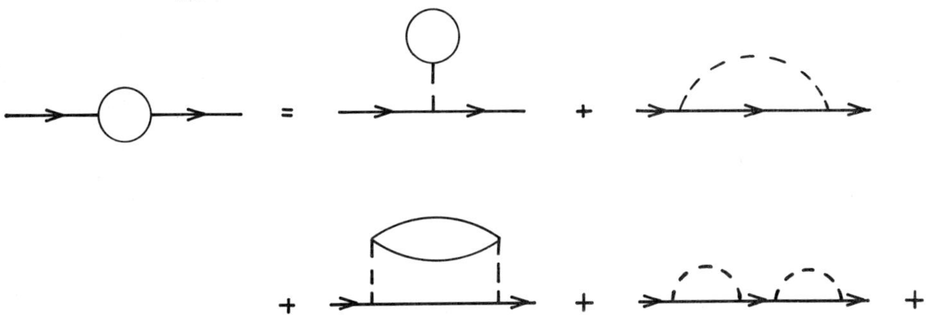

FIG. 10. Some diagrams for the self-energy.

runs between two vertices labelled $\tau_1$ and $\tau_2$ and carries a wave-vector **k**. If its direction is from $\tau_1$ to $\tau_2$ it contributes the Green's function $g(\mathbf{k}, \tau_1 - \tau_2)$. Now one can insert in this line all possibilities allowed by the rules which do not affect any other part of the diagram. This part can, therefore, be considered separately. Let us sum all possibilities of this kind as indicated in Fig. 9 where the full line indicates the complete sum and the circle indicates the sum of all insertions. Some contributions to this are indicated in Fig. 10.

Any $\tau$ integrations and sums over wave-vector in the insertions are performed. Then, because wave-vector is conserved everywhere and all quantities depend on differences of $\tau$'s ,the sum is a function only of $\mathbf{k}$ and $\tau_1 - \tau_2$; it is denoted by $G(\mathbf{k}, \tau_1 - \tau_2)$ and is called the full Green's function. It can be shown[12, 13] that it is actually given by

$$G(\mathbf{k}, \tau_1 - \tau_2) = \mathrm{Tr}\,\{\rho P \hat{a}_{\mathbf{k}}(\tau_1)\hat{a}^{\dagger}_{\mathbf{k}}(\tau_2)\} \tag{85}$$

where

$$\begin{aligned} \rho &= \exp(-\beta\mathscr{H})/Z \\ \hat{a}_{\mathbf{k}}(\tau) &= \exp(\tau\mathscr{H})a_{\mathbf{k}}\exp(-\tau\mathscr{H}) \\ \hat{a}^{\dagger}_{\mathbf{k}}(\tau) &= \exp(\tau\mathscr{H})a^{\dagger}_{\mathbf{k}}\exp(-\tau\mathscr{H}), \end{aligned} \tag{86}$$

so that $G$ is related to $\mathscr{H}$ in the same way as $g$ is related to $\mathscr{H}_0$, the reason for its name. We can also introduce the Fourier series for $G(\mathbf{k}, \tau)$ by

$$G(\mathbf{k}, \tau) = -\beta^{-1}\sum_{\omega_p} G(\mathbf{k}, \mathrm{i}\omega_p)\exp(-\mathrm{i}\omega_p\tau)$$

and this is related to $g(\mathbf{k}, \mathrm{i}\omega_p)$ through the same diagrams, where now the boson lines with free ends (often called the entering and leaving lines) carry the wave-vector $\mathbf{k}$ and energy $\omega_p$.

The contribution of the circle in Fig. 9 is often called the self-energy and can be usefully sub-divided. Any self-energy part which cannot be sub-divided into two by a line which cuts exactly one boson (solid) line is called a proper self-energy part. The self-energy contributions of the first three diagrams on the right of Fig. 10 are proper self-energy parts. The fourth is not. Let us call the sum of proper self-energy parts $\Sigma(\mathbf{k}, \mathrm{i}\omega_p)$ because of the conservation rules it can depend only on $\mathbf{k}$ and $\omega_p$) and denote it by a filled circle in a diagram. Then the complete self-energy part is related to $\Sigma$ as indicated in Fig. 11, which indicates that every proper self-energy part will appear in the total self-energy and any number of times. Explicitly, Fig. 11 represents the equation

$$\begin{aligned} G(\mathbf{k}_1, \mathrm{i}\omega_p) - g(\mathbf{k}, \mathrm{i}\omega_p) &= g(\mathbf{k}, \mathrm{i}\omega_p)\Sigma(\mathbf{k}, \mathrm{i}\omega_p)g(\mathbf{k}, \mathrm{i}\omega_p) + g\Sigma g\Sigma g \\ &\quad + g\Sigma g\Sigma g\Sigma g + \ldots = g(\mathbf{k}, \mathrm{i}\omega_p)\Sigma(\mathbf{k}, \mathrm{i}\omega_p)G(\mathbf{k}, \mathrm{i}\omega_p). \end{aligned} \tag{87}$$

Hence,

$$G(\mathbf{k}, \mathrm{i}\omega_p) = [g(\mathbf{k}, \mathrm{i}\omega_p)^{-1} - \Sigma(\mathbf{k}, \mathrm{i}\omega_p)]^{-1}.$$

FIG. 11. The total self-energy (hollow circle) related to the proper self-energy (filled circles).

If $\Sigma(\mathbf{k}, i\omega_p)$ is independent of $\omega_p$,

$$G(\mathbf{k}, i\omega_p) = [i\omega_p - \varepsilon_{\mathbf{k}} - \Sigma(\mathbf{k})]^{-1},$$

which shows that in this case it simply corrects the energy $\varepsilon_{\mathbf{k}}$. This is the reason for its name. In general, the effect of $\Sigma(\mathbf{k}, i\omega_p)$ is more subtle.

The full Green's function is of interest in itself as well as a step towards calculating $\Omega$. In fact, once $G(\mathbf{k}, i\omega_p)$ is known $\Omega$ can be calculated directly from it without considering the contribution of $G$ to the diagrams for $\Omega$. Thus the study of $\Omega$ can be replaced by the study of $G$. $G$ is also of interest for its relation to the excitation energies of the system. To see this, write (85) in terms of the exact (although unknown) eigenstates $|m\rangle$ and energies $E_m$ of the many-body system. Then

$$G(\mathbf{k}, \tau_1 - \tau_2) = Z^{-1} \sum_{m,n} [\exp(-\beta E_m)\theta(\tau_1 - \tau_2) + \exp(-\beta E_n)\theta(\tau_2 - \tau_1)] \times |\langle m|a_{\mathbf{k}}|n\rangle|^2 \exp[(E_m - E_n)(\tau_1 - \tau_2)]$$

The Fourier transform of this is

$$G(\mathbf{k}, i\omega_p) = Z^{-1} \sum_{m,n} \frac{[\exp(-\beta E_m) - \exp(-\beta E_n)]|\langle m|a_{\mathbf{k}}|n\rangle|^2}{\beta(-i\omega_p + E_m - E_n)}. \tag{88}$$

If we write formally,

$$G(\mathbf{k}, \omega) = Z^{-1} \sum_{m,n} \frac{[\exp(-\beta E_m) - \exp(-\beta E_n)]|\langle m|a_{\mathbf{k}}|n\rangle|^2}{\beta(-\omega + E_m - E_n)}, \tag{89}$$

then $G(\mathbf{k}, \omega)$ is a function of the complex variable $\omega$ which is analytic everywhere except where $\omega = (E_m - E_n)$, that is, on the real axis, which behaves like [constant/$\omega$] as $|\omega|$ becomes large, and which is equal to $G(\mathbf{k}, i\omega_p)$ where $\omega = i\omega_p$. There are theorems in complex variable theory[18] which show that such a function is defined uniquely by (88), that is, when $G(\mathbf{k}, i\omega_p)$ is found, $G(\mathbf{k}, \omega)$ is determined uniquely everywhere. Hence, we can regard the determination of $G(\mathbf{k}, i\omega_p)$ as the determination of $G(\mathbf{k}, \omega)$. But $G(\mathbf{k}, \omega)$ is singular only at places on the real axis where there are real energy differences $(E_m - E_n)$. Hence, an exploration of the analytic properties of $G(\mathbf{k}, \omega)$ will often determine at least some of the excitation energies of the system.

It is possible in a similar manner to explore the properties of higher order Green's functions such as

$$G(\mathbf{k}_1, \mathbf{k}_2, \mathbf{k}_3, \mathbf{k}_4; \tau_1, \tau_2, \tau_3, \tau_4) = \mathrm{Tr}\,\{\rho P \hat{a}_{\mathbf{k}_1}(\tau_1)\hat{a}_{\mathbf{k}_2}(\tau_2)\hat{a}^\dagger_{\mathbf{k}_3}(\tau_3)\hat{a}^\dagger_{\mathbf{k}_4}(\tau_4)\}$$

and to give rules for calculating them from diagrams. Unfortunately, we do not have room to expand on this here. The results of inelastic neutron scattering can, however, be related to the second order Green's functions written in this equation.

The rules in this section enable us to calculate, at least in principle, the properties of any normal system of bosons interacting through two-body forces within any desired approximation. We now have to consider whether these rules apply to a condensed system and, if not, how they should be modified.

## VI. Perturbation Theory for a Condensed System

If we try to apply the rules of the preceding section directly to an interacting gas of bosons near absolute zero we encounter a number of difficulties. First, the chemical potential may be such that $\varepsilon_k$ is negative or zero for some $\mathbf{k}$. Then Wick's theorem fails because the series in

$$\mathrm{Tr}\,\{\rho_0 a_{\mathbf{k}}^\dagger a_{\mathbf{k}}\} = \sum_{n_k} n_k \exp(-\beta\varepsilon_k n_k)/[\sum \exp(-\beta\varepsilon_k n_k)]$$

both diverge. This difficulty can, if necessary, be dealt with formally by writing the term in $\mathscr{H}$ involving $\mu$ as

$$-\mu \sum a_{\mathbf{k}}^\dagger a_{\mathbf{k}} = -(\mu - \delta\mu) \sum a_{\mathbf{k}}^\dagger a_{\mathbf{k}} - \delta\mu \sum a_{\mathbf{k}}^\dagger a_{\mathbf{k}},$$

and taking the first of these in $\mathscr{H}_0$ and the second in $V$. If $\delta\mu$ is chosen so that $(\hbar^2k^2/2m - \mu + \delta\mu)$ is positive for all $\mathbf{k}$, individual terms in the expansion can found as before. The extra perturbation $\delta\mu\Sigma a_{\mathbf{k}}^\dagger a_{\mathbf{k}}$ affects only the self-energies of the particles.

A second difficulty arises from the volume dependence of the various terms. In the last section it was stated that each connected diagram makes a contribution to $\Omega$ which is proportional to volume and so to $N$ the total number of particles. This statement was based on the assumption that each term in a sum over $\mathbf{k}$ is of order 1 and that a sum over $\mathbf{k}$ is, therefore, of order $N$. This assumption is false when there is a condensation. For then,

$$\mathrm{Tr}\,\{\rho a_0^\dagger a_0\} \sim N$$

and, therefore,

$$G(0, \tau) \sim N.$$

But there will be a contribution to $\Omega$ from any $n$th order connected diagram each of whose boson lines carries a momentum zero. This contribution will be of order

$$N^{-n} N^{2n} = N^n,$$

where there is a factor $N^{-1}$ for each dashed line and a factor $N$ for each full Green's function. This means that the early terms in the series expansion will grow rapidly as $n$ increases. This argument will fail when $n$ is greater than or of order $N$ but it strongly suggests that the expansion of the last section is a poor starting point for approximating a condensed system.

Because of this effect of the zero momentum states on the perturbation series we should like to separate these states. One way to do this is suggested by Bogoliubov's approach. If there is a condensation then the fluctuation in the number of particles in the zero momentum state is small. Hence in the thermodynamic state, and those it may fluctuate to, $a_0^\dagger a_0 \sim N_0$, the number of condensed particles. Hence, in the commutation relation

$$[a_0, a_0^\dagger] = 1,$$

the relative error in dropping the unity is $(1/N_0)$ and is negligibly small. This means that, as in Bogoliubov's calculation we can treat $a_0$ as a $c$-number and take

$$a_0 = \sqrt{N_0} = a_0^\dagger \tag{90}$$

where for the reasons given in Section III we take $a_0$ to be real. We could include the effect of the small fluctuations referred to by taking

$$a_0 = \sqrt{N_0} + \xi_0, \qquad [\xi_0, \xi_0^\dagger] = 1$$

and treating $\xi_0$ like the operators $a_{\mathbf{k}}$ ($\mathbf{k} \neq 0$) but this does not introduce a significant correction to the thermodynamics. For $N_0$ macroscopic but small, the fluctuations due to $\xi_0$ do become important but this is true only close to $T_c$. Even though the approximation (90) is similar to Bogoliubov's, the treatment of this section is different because it is not assumed that $N_0$ is very close to $N$.

For each value of $N_0$ the calculation to be given below leads to a thermodynamic potential $\Omega$ which depends on $N_0$. We are thus implicitly assuming that $N_0$ is a macroscopic thermodynamic parameter like volume or entropy for each value of which there exists a different thermodynamic state. If isolated, the system will be found to have the value of $N_0$ which minimises $\Omega(N_0)$. Hence, $N_0$ is ultimately to be determined from

$$\frac{\partial \Omega}{\partial N_0} = 0. \tag{91}$$

As in the last section, the chemical potential is determined from

$$\frac{\partial \Omega}{\partial \mu} = -N,$$

although, in practice, the equivalent equation (103) (below) is usually used.

If the substitution (90) is made in the hamiltonian one finds

$$
\begin{aligned}
\mathscr{H} &= -\mu N_0 + \mathscr{H}_0 + V + \tfrac{1}{2}N_0^2 V(0),\\
\mathscr{H}_0 &= \sum{}' \varepsilon_{\mathbf{k}} a_{\mathbf{k}}^{\dagger} a_{\mathbf{k}},\\
V &= V_1 + V_2 + V_3 + V_4,\\
V_1 &= N_0 \sum{}' [V(\mathbf{k}) + V(0)] a_{\mathbf{k}}^{\dagger} a_{\mathbf{k}},\\
V_2 &= \tfrac{1}{2} N_0 \sum{}' V(\mathbf{k})(a_{\mathbf{k}}^{\dagger} a_{-\mathbf{k}}^{\dagger} + a_{-\mathbf{k}} a_{\mathbf{k}}),\\
V_3 &= N_0^{\frac{1}{2}} \sum_{\mathbf{k},\mathbf{q}}{}' V(\mathbf{q})(a_{\mathbf{k}}^{\dagger} a_{\mathbf{q}}^{\dagger} a_{\mathbf{k}+\mathbf{q}} + a_{\mathbf{k}}^{\dagger} a_{\mathbf{q}} a_{\mathbf{k}-\mathbf{q}}),\\
V_4 &= \tfrac{1}{2} \sum_{\mathbf{k},\mathbf{k}',\mathbf{q}}{}' V(\mathbf{q})(a_{\mathbf{k}}^{\dagger} a_{\mathbf{k}'+\mathbf{q}}^{\dagger} a_{\mathbf{k}'} a_{\mathbf{k}+\mathbf{q}}),
\end{aligned} \tag{92}
$$

where the dash on a sum indicates that the term with $\mathbf{k} = 0$ is omitted. As all sums are of this kind we shall omit the dash in future. Now we can perform a perturbation expansion as in the last section with $\mathscr{H}_0$ as the unperturbed part and $V$ as the perturbation. As $\mathscr{H}_0$ is still a hamiltonian for independent bosons, Wick's theorem is still valid and the different terms in the expansion can be analysed in terms of diagrams as in Section V. The only differences arise from the four kinds of interaction terms $V_1$, $V_2$, $V_3$, $V_4$ which give rise to dashed lines with four different kinds of behaviour at the vertices at their ends as illustrated in Fig. 12. Figure 12(a) shows the effect of the first term of $V_1$ and one can see that this is a contribution to the self-energy. Figure 12(b) illustrates the second term of $V_1$ and is likewise a contribution to the self-energy. Figure 12(c) illustrates the first term of $V_2$. The Hermitian conjugate term is represented by a similar diagram with the arrows reversed. Figure 12(d) represents the first term of $V_3$ and its Hermitian conjugate is represented by the same diagram with the arrows reversed. Finally, Fig. 12(e) represents $V_4$ and is similar to the diagrams considered in the last section. The rules for obtaining the contributions of different diagrams are the same as in the preceding section except that the dash line attached to only three boson lines

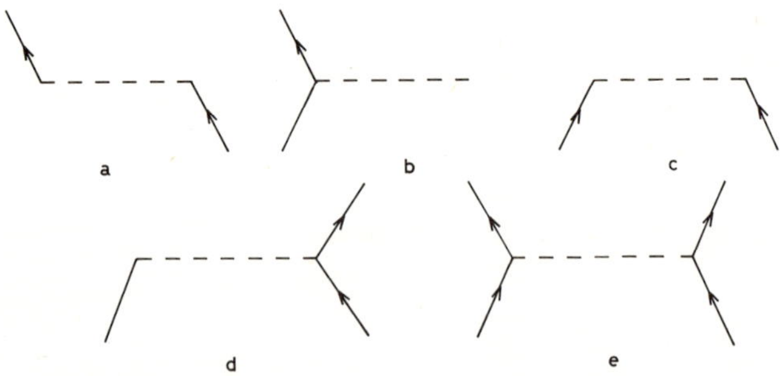


FIG. 12. The representative diagrams of the interaction terms of Bogoliubov's model, (a) and (b) represent $V_1$, (c) $V_2$, (d) $V_3$ and (e) $V_4$.

contributes an extra factor $N_0^{\frac{1}{2}}$ and one attached to only two boson lines has an extra factor $N_0$. These factors replace the effects of the operators $a_0$ and $a_0^\dagger$.

The further analysis also follows that of Section V and, in particular, the partition function and thermodynamic potential can be given in terms of connected diagrams. If we write

$$\begin{aligned} Z &= \operatorname{Tr} \exp(-\beta\mathscr{H}) \\ &= \exp[\beta\mu N_0 - \tfrac{1}{2}\beta V(0)N_0^2] \operatorname{Tr} \exp[-\beta(\mathscr{H}_0 + V)] \\ &= \exp(\beta\mu N_0 - \tfrac{1}{2}\beta V(0)N_0^2)Z' \end{aligned} \tag{93}$$

Then,

$$Z' = Z_0 \exp(\Sigma C_i), \qquad Z_0 = \operatorname{Tr} \exp(-\beta\mathscr{H}_0)$$

where $\Sigma C_i$ is the sum of all connected diagrams. In this case the connected diagrams of first order are shown in Fig. 13. The thermodynamic potential

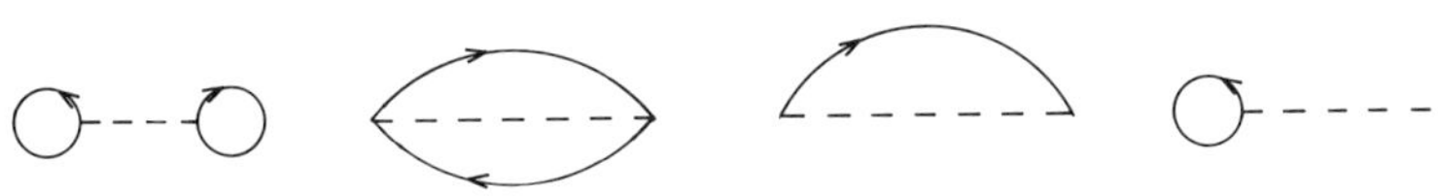

FIG. 13. All first order diagrams of the condensed system.

$\Omega$ is given by

$$\begin{aligned} \Omega &= -kT \ln Z \\ &= \Omega' - \mu N_0 + \tfrac{1}{2}\beta V(0)N_0^2, \end{aligned} \tag{94}$$

$$\Omega' = \Omega_0 - kT\Sigma C_i. \tag{95}$$

When we come to analyse the full Green's function we find some differences from the treatment of Section V. Consider all the insertions we can make in a solid boson line joining two vertices, the line leaving one and entering the other. This can still be illustrated by Fig. 9 where the circle indicates the sum of all insertions. It can be shown that the full thick line still represents the full Green's function $G(\mathbf{k}, \tau_1 - \tau_2)$ where

$$G(\mathbf{k}, \tau_1 - \tau_2) = \operatorname{Tr}\{\rho P\hat{a}(\tau_1)\hat{a}^\dagger(\tau_2)\}$$

where the $\mathscr{H}_0$ and $V$ of this section replace those of Section V. Differences arise when we come to define the proper self-energy part by cutting a diagram with a line so that it cuts just one boson line. For example, consider the contribution to $G$ shown in Fig. 14 which can be cut through single boson lines in a number of places. The part between $\tau_7$ and $\tau_8$ is like a self-energy part discussed in Section V; it has a single boson line entering and a single boson

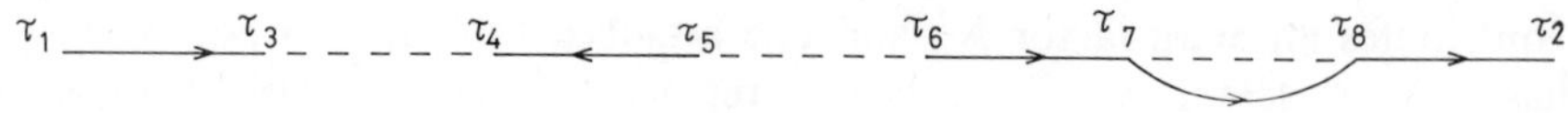


FIG. 14. A representative self-energy of the condensed system.

line leaving. On the other hand the part between $\tau_3$ and $\tau_4$ has two boson lines entering and cannot be included in an integral equation for $G$ of the type considered in Section V. Similarly, the part between $\tau_5$ and $\tau_6$ has two boson lines leaving and is new. Hence, if we wish to use an integral equation of the form (87) we have to treat the whole of the diagram from $\tau_3$ to $\tau_6$ as a proper self-energy part.

An alternative approach is to consider that a boson line leaving one vertex $\tau_1$ and eventually found to leave another $\tau_2$, without being connected in any other way to the remainder of the diagram (for example, the line from $\tau_1$ to $\tau_5$ in Fig. 14) represents a second Green's function to be denoted by $F(\mathbf{k}, \tau_1 - \tau_2)$. We shall illustrate this diagramatically by a thick solid line with two arrows on it directed away from the vertices at its ends. Correspondingly there are three kinds of proper self-energy parts as illustrated in Fig. 15. One

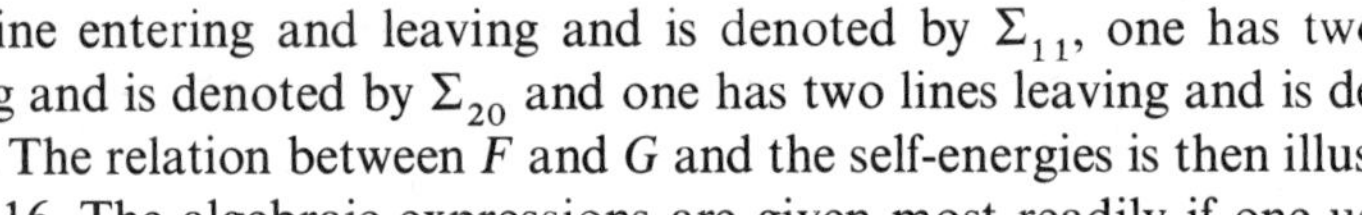


FIG. 15. The three different proper self-energy parts (a) $\Sigma_{11}$, (b) $\Sigma_{20}$, (c) $\Sigma_{02}$.

has a line entering and leaving and is denoted by $\Sigma_{11}$, one has two lines entering and is denoted by $\Sigma_{20}$ and one has two lines leaving and is denoted by $\Sigma_{02}$. The relation between $F$ and $G$ and the self-energies is then illustrated in Fig. 16. The algebraic expressions are given most readily if one uses the

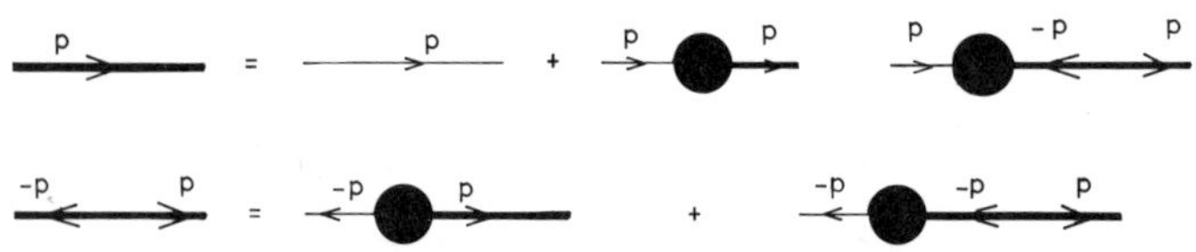


FIG. 16. Diagrammatic representation of the integral equations for $G$ (the thick line with one arrow) and $F$ (the thick line with two arrows). ($p = (\mathbf{k}, \omega_p)$, $-p = (-\mathbf{k}, -\omega_p)$).

Fourier transforms, $G(\mathbf{k}, i\omega_p)$, $F(\mathbf{k}, i\omega_p)$. Since the total wave-vector and energy entering a vertex is zero we have

$$G(\mathbf{k}, i\omega_p) = g(\mathbf{k}, i\omega_p) + g(\mathbf{k}, i\omega_p)\Sigma_{11}(\mathbf{k}, i\omega_p)G(\mathbf{k}, i\omega_p) + g(\mathbf{k}, i\omega_p)\Sigma_{20}(\mathbf{k}, i\omega_p)F(\mathbf{k}, i\omega_p)$$

$$F(\mathbf{k}, i\omega_p) = g(-\mathbf{k}, -i\omega_p)\Sigma_{02}(\mathbf{k}, i\omega_p)G(\mathbf{k}, i\omega_p) + g(-\mathbf{k}, -i\omega_p)\Sigma_{11}(-\mathbf{k}, -i\omega_p)F(\mathbf{k}, i\omega_p),$$

whence

$$G(\mathbf{k}, i\omega_p)^{-1} = g(\mathbf{k}, i\omega_p)^{-1} - \Sigma_{11}(\mathbf{k}, i\omega_p) - \frac{\Sigma_{02}(\mathbf{k}, i\omega_p)\Sigma_{20}(\mathbf{k}, i\omega_p)}{g^{-1}(-\mathbf{k}, -i\omega_p) - \Sigma_{11}(-\mathbf{k}, -i\omega_p)}. \tag{96}$$

Hence the single particle Green's function can be written down once we have expressions for the three self-energies.

As an illustration of the use of perturbation theory we show that the equations of this section lead to the results of Section III for a weakly interacting gas. In this case, we can use the first order results for the self-energies. There are illustrated in Fig. 17, from which we derive

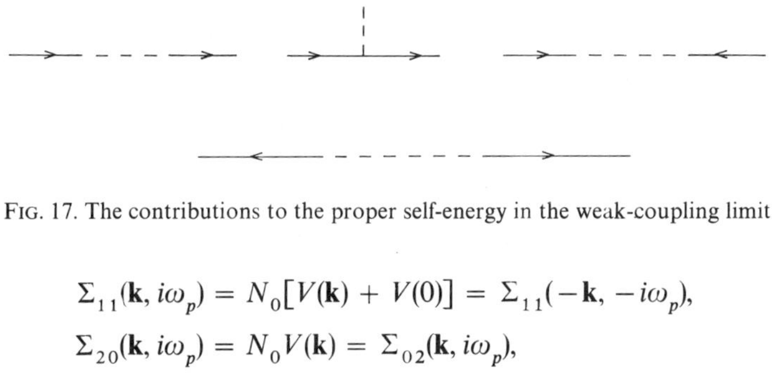

FIG. 17. The contributions to the proper self-energy in the weak-coupling limit.

$$\Sigma_{11}(\mathbf{k}, i\omega_p) = N_0[V(\mathbf{k}) + V(0)] = \Sigma_{11}(-\mathbf{k}, -i\omega_p),$$
$$\Sigma_{20}(\mathbf{k}, i\omega_p) = N_0 V(\mathbf{k}) = \Sigma_{02}(\mathbf{k}, i\omega_p),$$

and

$$G(\mathbf{k}, i\omega_p) = \frac{-i\omega_p - \varepsilon_k - N_0[V(\mathbf{k}) + V(0)]}{\omega_p^2 + \{\varepsilon_k + N_0[V(\mathbf{k}) + V(0)]\}^2 - N_0^2 V(\mathbf{k})^2} \tag{97}$$

We still need to determine $\mu$ (which is contained in $\varepsilon_k$) and $N_0$. For the former it is not very convenient to use the thermodynamic potential. However, it may be shown that (cf. equation (103)) $\mu$ is generally given exactly by

$$\mu = \lim_{k \to 0} [\Sigma_{11}(\mathbf{k}, 0) - \Sigma_{02}(\mathbf{k}, 0)]. \tag{98}$$

In our case this leads to

$$\mu = N_0 V(0).$$

Hence

$$G(\mathbf{k}, \omega) = \frac{-\omega - \hbar^2k^2/2m - N_0V(k)}{-\omega^2 + (\hbar^2k^2/2m)[\hbar^2k^2/2m + 2N_0V(\mathbf{k})]}$$

This has poles at

$$\omega = \left\{\frac{\hbar^2k^2}{2m}\left[\frac{\hbar^2k^2}{2m} + 2N_0V(\mathbf{k})\right]\right\}^{\frac{1}{2}} \tag{99}$$

which are, therefore, excitation energies. These are in agreement with those of Section III provided $N_0$ is the same. One can show that as $T \to 0$, $N_0 \to N$ and one does, therefore, obtain Bogoliubov's results.

## VII. The T-matrix

As an approach to the theory of liquid helium, the treatment given in Section III and at the end of Section VI falls short because the interaction between two helium atoms is never weak. When two atoms come very close the interaction is very strongly repulsive, so strongly that the potential does not have a finite Fourier transform. The potential is believed to be close to[(19)],

$$V(r) = 4E\left[\left(\frac{\beta}{r}\right)^{12} - \left(\frac{\sigma}{r}\right)^{6}\right], \qquad E = 10{\cdot}2\,\mathrm{K}, \quad \sigma = 2{\cdot}556\,\text{Å}, \tag{100}$$

which, because of its singularity at $r = 0$, does not have a transform $V(q)$. Sometimes the repulsion at near distances is represented by an infinitely hard core. This too does not have a transform. Thus for these cases we cannot write down the initial hamiltonian in the form (10) on which the rest of the analysis is based. If the potential is singular two atoms will not approach each other closely and we must include this effect in the calculation.

If the gas is dilute we can expect that the interaction between two atoms will suffer only negligible interference from other atoms. Hence we can try to treat the interaction between any two atoms exactly but to treat the effect of this on the many-body system as weak.

Let us suppose for the moment that the potential between two atoms has a Fourier transform but that it is not weak. For example, we could suppose that the hard core is large, not infinite, and let it become infinite at the end of the calculation. Then we can draw diagrams as before. The first order scattering (Born approximation) of two bosons is represented by the part of a diagram shown in Fig. 18(a). If the scattering is strong, however, we should allow these two particles to interact any number of times. This means that we should add to Fig. 18(a) all parts like Fig. 18(b) where any number

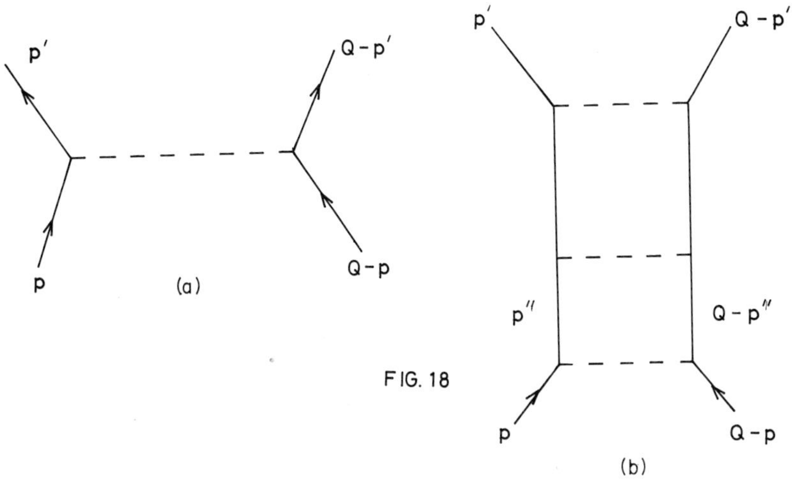


FIG. 18. Diagrams which contribute to the $t$-matrix ($p = (k, \omega)$, $Q = (\mathbf{q}, \Omega)$).

of dashed lines is included. This means that we are replacing the Born approximation to the scattering between two particles by the full scattering amplitude.

It might be thought that one should also include diagrams in which the dashed lines cross, for example Fig. 19(a). However, an analysis of the intermediate states involved in this process shows that it can take place only if

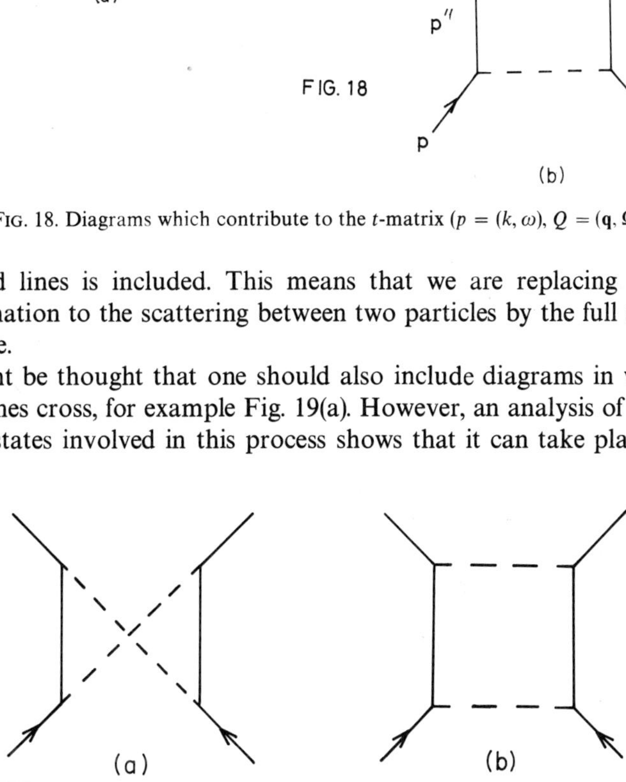


FIG. 19. Two second order contributions to the $t$-matrix.

more than two bosons are involved, at least one in an excited state. Hence, we expect this contribution to be smaller at low densities and low temperatures.

The sum of the contributions of the diagrams of Fig. 18 can be written in terms of the variables shown in the diagram as $t(\mathbf{k}\omega, \mathbf{k}'\omega'; \mathbf{q}\Omega)$ where $t$ is usually called the $t$-matrix. It satisifies the equation

$$t(\mathbf{k}, \omega, \mathbf{k}'\omega'; \mathbf{q}\Omega) = V(\mathbf{k} - \mathbf{k}') + \beta^{-1} \sum_{\omega_1} \int d^3k_1 V(\mathbf{k} - \mathbf{k}_1)$$
$$\times\, g(\mathbf{k}_1, i\omega_1) g(\mathbf{q} - \mathbf{k}_1, i(\Omega - \omega_1)) t(\mathbf{k}_1\omega_1, \mathbf{k}'\omega'; \mathbf{q}\Omega),$$

from which one can deduce that $t$ does not depend on $\omega$ and $\omega'$. In general this equation cannot be solved analytically. For the case, however, when $t$ is required for small $\mathbf{k}, \mathbf{k}', \mathbf{q}, \Omega$ and low density, it can be shown that[14, 20]

$$t(\mathbf{k}, \mathbf{k}'; \mathbf{q}, \Omega) = \frac{4\pi}{m} f(\mathbf{k}, \mathbf{k}') \approx \frac{4\pi}{m} f(0, 0) = \frac{4\pi}{m} f_0 \tag{101}$$

where $f(\mathbf{k}, \mathbf{k}')$ is the exact amplitude for the scattering of two free helium atoms from a state of relative momentum $2\hbar\mathbf{k}$ to one of relative momentum $2\hbar\mathbf{k}'$. Thus the effect of the repeated scattering is to replace the potential by the scattering amplitude; in the equations of Section II as well as (97) and (99), $V(k)$ and $V(0)$ should be replaced by $(4\pi\hbar^2/mL^3) f_0$. For hard spheres of radius $a$, a commonly used model, $f_0 = a$.

This treatment is still essentially a weak-coupling treatment which is valid if $N_0 \approx N$ at the absolute zero. In fact, as we pointed out earlier, limited experimental evidence and machine calculations strongly suggest that $N_0 \ll N$ at the absolute zero and that the simple approximations used here must be improved considerably if they are to be applied to liquid helium. There have been a number of attempts in this direction but none is yet quantitatively successful. A good recent review of such attempts and their achievements is contained in the review article of Woods and Cowley.[1] So far this approach seems to have been most useful in increasing our understanding of the long wavelength part of the spectrum but few, if any, useful results have emerged concerning the roton part of the spectrum.

## VIII. Exact Results

In view of the difficulty of obtaining overall quantitative agreement with experiment, it is useful to have a number of exact results as a check on partial theories and on experiment. These results can be proved using the techniques of perturbation theory just described, but using all terms in the series; some can be proved from general relations. We have only the space to state these results and indicate how they can be proved.

We have seen already that the one-particle Green's function generally has the form in equation (96). From the structure of the diagrams it is not difficult to show that

$$\Sigma_{20}(\mathbf{k}, \omega) = \Sigma_{02}(\mathbf{k}, \omega).$$

Then, if one writes

$$\Sigma_e(\mathbf{k}, \omega) = \tfrac{1}{2}[\Sigma_{11}(\mathbf{k}, \omega) + \Sigma_{11}(-\mathbf{k}, -\omega)]$$

$$\Sigma_0(\mathbf{k}, \omega) = \tfrac{1}{2}[\Sigma_{11}(\mathbf{k}, \omega) - \Sigma_{11}(-\mathbf{k}, -\omega)]$$

one finds

$$G(\mathbf{k}, \omega) = \frac{+\omega + \varepsilon + \Sigma_e - \Sigma_0}{(\omega - \Sigma_0)^2 - (\varepsilon + \Sigma_e)^2 + \Sigma_{20}^2}. \tag{102}$$

This has poles where

$$[\omega - \Sigma_0(\mathbf{k}, \omega)]^2 - [\varepsilon_k + \Sigma_e(\mathbf{k}, \omega)]^2 + \Sigma_{20}(\mathbf{k}, \omega)^2 = 0$$

and for small $\omega$ and $k$ provides the excitation energies $\omega(\mathbf{k})$ of the system. Now we have argued quite generally that for small $\mathbf{k}$, we have only excitations with energies $\hbar ck$. Hence, as $\omega$ and $\mathbf{k}$ tend to zero the equation must be satisfied and

$$-[\mu - \Sigma_e(0, 0)]^2 + \Sigma_{20}(0, 0)^2 = 0;$$

here we have used the fact that $\Sigma_0(0, 0)$ is odd in its arguments and so must be zero, and that $\varepsilon_0 = -\mu$. Hence,

$$\mu = \Sigma_{11}(0, 0) \pm \Sigma_{20}(0, 0).$$

An analysis of the diagrams[12, 13] shows that this must be so and that the negative sign must be taken. Strictly the $\Sigma$'s are not defined for $\mathbf{k} = 0$ so that this equation should be interpreted as

$$\mu = \lim_{\mathbf{k} \to 0} [\Sigma_{11}(\mathbf{k}, 0) - \Sigma_{20}(\mathbf{k}, 0)]. \tag{103}$$

This equation is in practice the simplest to use in order to determine $\mu$.

An expansion of the denominator of $G(\mathbf{k}, \omega)$ for small $\mathbf{k}$ and $\omega$ leads to an exact expression for the velocity of sound $c$ in terms of the self-energies. In fact a diagrammatic analysis for the absolute zero of temperature[21] shows that as $\omega$ and $\mathbf{k}$ tend to zero

$$G(\mathbf{k}, \omega) = \frac{N_0 mc^2}{N\hbar^2(\omega^2 - c^2k^2)}$$

with $c$ the macroscopic sound velocity. This confirms that, if there is a condensation, the only low energy microscopic excitations are phonons.

One can also obtain exact results for higher order Green's functions. For example the dynamic structure factor of the liquid is

$$S(q, \omega) = \int \mathrm{d}t\, \mathrm{e}^{i\omega t}\, \mathrm{Tr}\, \{\rho_q(t)\rho_{-q}(0)\}$$

in which $\rho_{\mathbf{q}}$ is the particle density operator of wave-vector **q**,

$$\alpha_{\mathbf{q}} = \sum_{\mathbf{k}} a^{\dagger}_{\mathbf{k}+\mathbf{q}} a_{\mathbf{k}}$$

and $\rho_q(t)$ is the Heisenberg operator in real time. $S(\mathbf{q}, \omega)$ is a special two-particle Green's function. From the fact that particle number is conserved one can prove[22] the $f$-sum rule

$$\int_{-\infty}^{\infty} \omega S(\mathbf{q}, \omega)\, \mathrm{d}\omega = \hbar q^2/2m.$$

This can provide a useful test for the consistency of experiments or approximate theories.

## IX. States of Flow

We saw in Section VI, that we could ascribe different thermodynamic states to liquid helium corresponding to different values of $N_0$ and that, in any given situation, the system would come to equilibrium in the state with the $N_0$ which makes the thermodynamic potential $\Omega(N_0)$ a minimum. Similarly in the Bogoliubov model we showed in Section IV, that there would exist macroscopic states in which the single particle state $k$ ($\neq 0$) is occupied and that these states would represent states of superfluid flow. This argument can clearly be carried forward in the manner of Section VI so that perturbation theory can be applied to all orders and one is not restricted to weak interactions. Thus one starts with the assumption that

$$a_{\mathbf{k}} = N^{\frac{1}{2}}_{0\mathbf{k}} = a^{\dagger}_{\mathbf{k}}$$

for one specific value of **k** and calculates the thermodynamic potential as a function of $N_{0\mathbf{k}}$. The state of flow in which the superfluid may be found is given by the value of $N_{0\mathbf{k}}$ which makes $\Omega(N_{0\mathbf{k}})$ a minimum. If such a non-zero $N_{0\mathbf{k}}$ can be found, the excitation energies are positive (otherwise $\Omega(N_{0\mathbf{k}})$ cannot be a minimum) and the flow is indeed a superfluid one. Because the equilibrium state is one with only phonons as the low energy excitations, Landau's criterion shows that, at least for small **k**, superfluid flow is possible and that thermodynamic states with non-zero $N_{0\mathbf{k}}$. exist.

In a state of flow, the superfluid current and density can be found as follows. The operator for the mass current density, when it is uniform, is

$$\mathbf{j} = \frac{\hbar}{L^3} \sum_{\mathbf{k}'} \mathbf{k}' a^{\dagger}_{k} a_{k'}, \qquad L^3 = \text{volume}$$

Hence, its expectation value is

$$\mathbf{j} = \frac{\hbar}{L_3} \sum_{\mathbf{k}'} \mathbf{k}' \operatorname{Tr} \{\rho a_{\mathbf{k}'}^\dagger a_{\mathbf{k}'}\}$$

$$= \frac{\hbar}{L^3} \sum_{\mathbf{k}'} \mathbf{k}' G(\mathbf{k}', 0_-).$$

For small values of $\mathbf{k}_s$ (the macroscopically occupied state) this will be proportional to $\mathbf{k}_s$ and we have

$$\bar{\mathbf{j}} = \alpha\hbar\mathbf{k}_s. \tag{104}$$

It is usual to interpret this in the two-fluid model by supposing that the condensation into the state $\mathbf{k}$ corresponds to a superfluid velocity $\mathbf{V}_s$ given by

$$m\mathbf{V}_s = \hbar k_s \tag{105}$$

where $m$ is the mass of a bare helium atom. This is satisfactory for non-dissipative flows. For then $\Omega(\mathbf{k}_s)$ exists and one can show that

$$\mathbf{j} = m\partial\Omega/\partial\hbar\mathbf{k}_s = \alpha\hbar\mathbf{k}_s.$$

Hence,

$$\Omega(\mathbf{k}_s) = \Omega(0) + \alpha(\hbar k_s)^2/m.$$

If $\mathbf{V}_s$ is defined experimentally from the ratio of $\Omega(k_s) - \Omega(0)$ to $j$, one has (105) as well as

$$\mathbf{j} = \rho_s\mathbf{V}_s, \qquad \Omega(\mathbf{k}_s) = \Omega(0) + \rho_s V_s^2/2m,$$

provided $\rho_s$ is defined by

$$\rho_s = \alpha m.$$

On the usual definition of the normal density $\rho_n$, we have

$$\rho_n = \rho - \rho_s \tag{106}$$

where $\rho$ is the total mass density.

If the excitations drift slowly relative to the superfluid so that $\hbar\mathbf{k}_s \neq 0 (\mathbf{V}_s \neq 0)$ but $\mathbf{j} \neq \rho_s\hbar\mathbf{k}_s$ we can always write

$$\mathbf{j} = \rho_s\hbar k_s/m + \rho_n\mathbf{V}_n \tag{107}$$

where this equation now defines $\mathbf{V}_n$. Superficially this is one of the basic equations of the two fluid model. However, it has not been shown generally that $\mathbf{V}_n$ so defined agrees with observation, nor that the corresponding kinetic energy is

$$\bar{E} = \rho_s(\hbar k_s)^2/2m^2 + \tfrac{1}{2}\rho_n\mathbf{V}_n^2$$

as is assumed in the two fluid model.

It is useful to have a single simple way of characterizing the different states of superfluid flow especially as the two-fluid hydrodynamics needs to be applied to situations where the superfluid velocity varies; for example if superfluid helium flows along a narrowing capillary $\mathbf{V}_s$ must increase in the direction of flow. How do we characterize such a situation in the microscopic theory.

For spacially varying situations it is better to look at the field operator

$$\psi(\mathbf{r}) = \frac{1}{L^{\frac{3}{2}}} \sum_{\mathbf{k}} a_{\mathbf{k}} \exp(i\mathbf{k} \,.\, \mathbf{r}) \tag{108}$$

rather than its Fourier components. Now the equilibrium state has

$$a_0 = \sqrt{N_0}$$

or better

$$\begin{aligned} \langle a_0 \rangle &= \sqrt{N_0} \\ \langle a_{\mathbf{k}} \rangle &= 0, \qquad \mathbf{k} \neq 0. \end{aligned} \tag{109}$$

Hence, in this state

$$\langle \psi(\mathbf{r}) \rangle = \sqrt{\frac{N_0}{L^3}}. \tag{110}$$

Similarly, in the state of flow with superfluid momentum $\mathbf{k}_s$

$$\langle \psi(\mathbf{r}) \rangle = \sqrt{\frac{N_{0k_s}}{L^3}} \exp(i\mathbf{k}_s \,.\, \mathbf{r}).$$

Thus, as is common in quantum mechanics, the phase of $\langle \psi(\mathbf{r}) \rangle$ is related to the momentum. This idea can easily be extended to situations where $\langle \psi(\mathbf{r}) \rangle$ varies slowly in space. We can always write

$$\langle \psi(\mathbf{r}) \rangle = R(\mathbf{r}) \exp[iS(\mathbf{r})]$$

($R$ and $S$ real) and interpret $\hbar \nabla S(\mathbf{r})$ as the local superfluid momentum. As before $\hbar \nabla S(\mathbf{r})/m$ is taken to be the local superfluid velocity. Different superfluid states are characterized by different phase functions $S(\mathbf{r})$; in each case $R(\mathbf{r})$ is determined so that $\Omega$ is a minimum. The expectation value $\langle \psi(\mathbf{r}) \rangle$ which characterizes the different ordered states is called the order parameter of the system.

Formal calculations of $\Omega$ and other properties of the system can be performed by replacing $\psi(\mathbf{r})$ by

$$\psi(\mathbf{r}) = \langle \psi(\mathbf{r}) \rangle + \xi(\mathbf{r})$$

in the hamiltonian and treating $\xi(\mathbf{r})$ as a perturbation. These are difficult calculations to perform when $\langle\psi(\mathbf{r})\rangle$ depends on position, but they are possible in principle.

The description and use of the order parameter in liquid He$^4$ are very similar to their counterparts in the BCS theory[7] of superconductivity. There the particles are electrons which are fermions described by field operators $\psi_\sigma(\mathbf{r})$, where $\sigma$ refers to spin. A superconductive state is described by an order parameter

$$F(\mathbf{r}) = \langle \psi_\downarrow(\mathbf{r})\, \psi_\uparrow(\mathbf{r})\rangle.$$

In the ground state, $F(\mathbf{r})$ is a constant. In a state of flow

$$F(\mathbf{r}) \propto \exp(2i\mathbf{k}\,.\,\mathbf{r})$$

where we have put in the factor 2 because the electrons condense in pairs corresponding to the two field operators appearing in $F(\mathbf{r})$. Different superconducting states correspond to different phases for $F(\mathbf{r})$.

An alternative definition of the order parameter has been given by Penrose and Onsager[23] and discussed by Yang.[24] The definition given above is based on states in which the number of bosons is not fixed. For $\psi(\mathbf{r})$ destroys a boson and $\langle\psi(\mathbf{r})\rangle$ is, therefore, zero if we include in the sum only states with a definite $N$. Since the original hamiltonian conserves $N$, it might be thought unnecessary to include states with varying $N$. The other point of view avoids this and begins instead from the density matrix defined by

$$\rho(\mathbf{r}, \mathbf{r}') = \langle\psi(\mathbf{r})\psi^\dagger(\mathbf{r}')\rangle.$$

In a normal state of the boson system $\rho(\mathbf{r}, \mathbf{r}')$ tends to zero as the separation $|\mathbf{r} - \mathbf{r}'| \to \infty$. There is no long-range order. In the superfluid system, however,

$$\lim_{|\mathbf{r}-\mathbf{r}'|\to\infty} \rho(\mathbf{r}, \mathbf{r}') = F(\mathbf{r})F^*(\mathbf{r}'), \qquad F(\mathbf{r}) \neq 0$$

and we have long range order (usually called off-diagonal long-range order, ODLRO). The different states of the superfluid are described by different functions $F(\mathbf{r})$ which is the order parameter. It can be shown that the two approaches are equivalent and that

$$F(\mathbf{r}) = \langle\psi(\mathbf{r})\rangle.$$

The former approach is usually the more practical in calculations.

One particularly important state of flow is that of superfluid in a ring. Suppose liquid helium flows in an annulus between two cylinders. Clearly the atoms condense into a state of definite angular momentum. If $\theta$ is the polar angle then

$$\langle\psi(\mathbf{r})\rangle \sim \exp(il\theta)$$

where, for the wave function to be single-valued, $l$ is an integer. Then the angular momentum of the condensed atoms is $\hbar l$. If the annulus is thin compared with the inner radius $R$, this corresponds to a superfluid momentum $\hbar l/R$ and so to a superfluid velocity

$$\mathbf{v}_s = \hbar l/mR.$$

The circulation around the annulus is

$$\oint \mathbf{v}_s \,.\, \mathbf{d}s = \frac{2\pi R\hbar l}{mR} = \frac{hl}{m}.$$

Hence, circulation is quantized in units of $h/m$. This effect is analogous to the quantization of magnetic flux in a superconducting ring. In that case, the effect has been quite clearly observed and observation is in agreement with theory. In the case of liquid helium the observation is very difficult but it does now seem that the quantization of circulation is established.[(25)]

One can imagine that the radius of the inner cylinder is decreased until it becomes a line singularity. Then one has a vortex in the liquid. There is no doubt that vortices play an important role in the rotation of buckets of helium (as discussed in other lectures) and it has been suggested[(26)] that they are important in lowering the critical velocity of the superfluid.

## X. Richards–Anderson Experiment

In the case of superconductivity, Josephson[(27)] predicted a number of tunnelling effects which depend on the macroscopic role of the phase of the order parameter and these effects have all now been observed. One expects that analogous effects should be observable in liquid $He^4$ although the practical difficulties make observation much harder. For example, one does not have the benefit of charged particles to help the observer. Also it seems that the definite phase relations necessary for the effects are easily upset by the smallness of the coherence length (the natural length for variations of the wave function) and the presence of stray vortices which are much smaller and much more easily created in liquid $He^4$. An attempt, however, has been made to observe the A.C. Josephson effect, first by Richards and Anderson[(28)] and later by others.[(29)]

In an ideal experiment, two baths of liquid $He^4$ are connected together by a weak link of some kind which allows some transfer of liquid between the baths. The link has to be such that there is a single macroscopic wave function (order parameter) for the two baths. Since liquid can be transferred between the baths, the number of atoms in one bath is not a good quantum number.

Provided the link is sufficiently weak, one can calculate its effect by perturbation theory. (We follow an argument here, first given by Ferrell and Prange[30] for the case of superconductors.) If states with $N_1$ atoms in one bath are denoted by $|N_1\rangle$ and those in the other by $|N_2\rangle$, then, if $N$ is the total number of atoms present, a general state for the combined system is

$$|\rangle = \sum_{N_1} B_{N_1} |N_1\rangle |N - N_1\rangle, \tag{111}$$

where the $B_{N_1}$ are constants. If the link is weak the interaction in the hamiltonian will transfer only one atom at a time from one bath to another and the matrix element will be, for small transfers, independent of $N_1$. If the baths are in similar states, the energies of the baths are essentially independent of $N_1$. Hence, the increase in total energy in the state (111) due to the weak link is

$$E = W \Sigma (B^*_{N_1} B_{N_1 - 1} + B^*_{N_1 - 1} B_{N_1}) \tag{112}$$

where $W$ is the matrix element for the transfer of an atom and where it is assumed that the wave function is normalized, i.e.

$$\sum_{N_1} |B_{N_1}|^2 = 1. \tag{113}$$

$E$ has now to be minimized subject to the restriction (113). The problem is now entirely analogous to that of finding the energies of an electron band in a one-dimensional solid in the tight-binding approximation. The result is that

$$B_{N_1} \propto \exp(iN_1 \phi),$$

and

$$E = 2W \cos \phi, \tag{114}$$

provided that the relative phases of the states $|N_1\rangle |N - N_1\rangle$ are chosen so that the matrix elements are independent of this phase.

The phase in $|\rangle$ is related to the phase of the order parameter discussed in Section IX. To see this consider

$$\langle | \phi^\dagger(\mathbf{r}) \psi(\mathbf{r}') | \rangle \tag{115}$$

where $\phi^\dagger(\mathbf{r})$ creates an atom at position $\mathbf{r}$ in bath "1" and $\psi(\mathbf{r}')$ destroys an atom at position $\mathbf{r}'$ in bath "2". If $\mathbf{r}$ and $\mathbf{r}'$ are separated by a large distance this becomes

$$F_1^\dagger(\mathbf{r}) F_2(\mathbf{r}') \propto \exp[i(\phi_2 - \phi_1)]. \tag{116}$$

But if we evaluate (115) directly from (111) we have

$$\langle | \phi^\dagger(\mathbf{r}) \psi(\mathbf{r}') | \rangle \propto \exp(-i\phi).$$

Hence

$$\phi = \phi_1 - \phi_2$$

the difference of phases of the order parameters in the two baths.

Now in a single bath of $He^4$ a variation of phase corresponds to a superflow (cf. Section IX). Hence, one might expect that a phase difference between two baths corresponds to a superflow between them. To see that this is so we note first that $N_1$ and $\phi$ are canonically conjugate variables (like photon number and phase for a light field). This follows from the fact that

$$\begin{aligned} N_{1\text{op}}|\rangle &= \Sigma N_1 |N_1\rangle |N - N_1\rangle \exp(iN_1\phi) \\ &= -i\frac{\partial}{\partial\phi}\Big|\Big\rangle . \end{aligned}$$

Hence,

$$[\hbar\phi, N_{1\text{op}}] = i\hbar \tag{117}$$

and $N_{1\text{op}}$ and $\hbar\phi$ form a canonically conjugate pair of operators like $p$ and $x$. In a situation where $N_1$ is changing with time

$$\begin{aligned} \langle \dot{N}_1 \rangle &= \langle i[H, N_{1\text{op}}]\rangle/\hbar \\ &= \frac{i}{\hbar}\left\langle\left[E, -i\frac{\partial}{\partial\phi}\right]\right\rangle \\ &= (-2W/\hbar)\sin\phi. \end{aligned} \tag{118}$$

Thus if $\phi \neq 0, \pi$ there is a net rate of change in the number of atoms in the first bath (and a corresponding change in the other bath) and a net superflow. There is no dissipation because we have considered only the flow of atoms in the condensed phase. If there were no weak link the states of the two baths would be indefinitely degenerate (one degenerate state for each phase difference). The weak link lifts the degeneracy according to (114) and most of the now non-degenerate states carry current. This means that in thermodynamic equilibrium no current will flow. If a small current is forced through the link the phase difference will adjust until the current given by (118) is just equal to that being forced through. This adjustment takes place without dissipation. The phase can be adjusted until there is a maximum superflow of $2|W|/\hbar$ through the link. This current maximum is called the Josephson current. If a greater current is forced through the link this can be accommodated only by the transfer of excitations as well as supercurrent and dissipation takes place. In superconductors it is possible to modulate $\phi$, and so the supercurrent, with a magnetic field and a number of interesting quantum effects can be observed. In the case of $He^4$ it has not been possible to perform analogous experiments. There are, however, related time-dependent effects which can be observed.

If we somehow produce a difference of chemical potential $\mu$ between the baths, the total energy also depends on the number of atoms $N_1$ in the first bath. Then (114) has to be replaced by

$$E = \mu N_1 + 2W \cos \phi. \tag{119}$$

Since $N_1$ and $\phi$ do not commute $E$ acts as a hamiltonian in a problem involving supercurrents. It follows from this equation that

$$\dot{\phi} = i[E, \phi]/\hbar = \mu/\hbar.$$

Hence,

$$\phi = \phi_0 + \int_0^t \mu(t')\,\mathrm{d}t'/\hbar \tag{120}$$

and the corresponding current is

$$\langle \dot{N}_1 \rangle = (-2W/\hbar)\sin\left[\phi_0 + \int_0^t \mu(t')\,\mathrm{d}t'/\hbar\right]. \tag{121}$$

If the difference in chemical potential is constant in time the current oscillates with frequency $\mu/\hbar$. This oscillating supercurrent is known as the A.C. Josephson effect.

It is usually best observed by imposing a chemical potential with a small oscillating part itself so that

$$\mu = \mu_0 + \mu_1 \sin 2\pi\nu t$$

and adjusting $\mu_0$. Then the supercurrent of equation (121) has a D.C. component whenever $\mu_0$ is $nh\nu$.

In the experiments performed to date the change of phase is believed to be produced by the generation and motion of vortices at an orifice of about 15 μm diameter between the baths. The chemical potential $\mu_0$ is provided by having a level difference $Z$ between the two baths so that $\mu_0$ is $mgZ$, where $g$ is the acceleration due to gravity. For $Z \sim 1$ mm, the Josephson frequency $mgZ/h$ is close to 100 kHz. Ultrasonic transducers operating near this frequency are used to generate the oscillating part of the chemical potential. A schematic diagram of the apparatus used in the Richards–Anderson experiment is shown in Fig. 20. The actual role of the transducer in the experiment and the way in which it affects the chemical potential is not completely clear. In any case, the smallness of vortices in liquid $He^4$ and the ease with which they are produced are bound to affect considerations of phase and so make the interpretation difficult. In particular, they can lead to resonance at sub-harmonic frequencies.

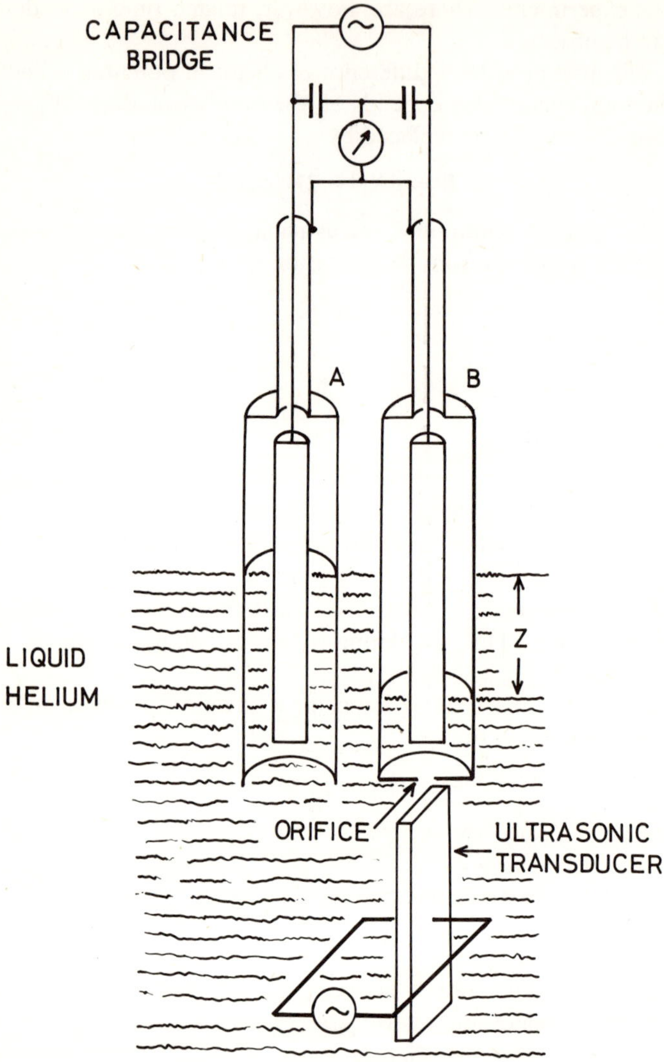


FIG. 20. A schematic diagram of the apparatus used by Richards and Anderson. The two coaxial capacitors A and B form the arms of a capacitance bridge which measures the helium head differences (after ref. 28).

To observe the effects Richards and Anderson observed the decay of a pressure head with time. The decay shows plateaux where the liquid seems to stick. Plots of their values of $dz/dt$ for three transducer frequencies are shown in Fig. 21 where it is seen that the sticking points do occur when the

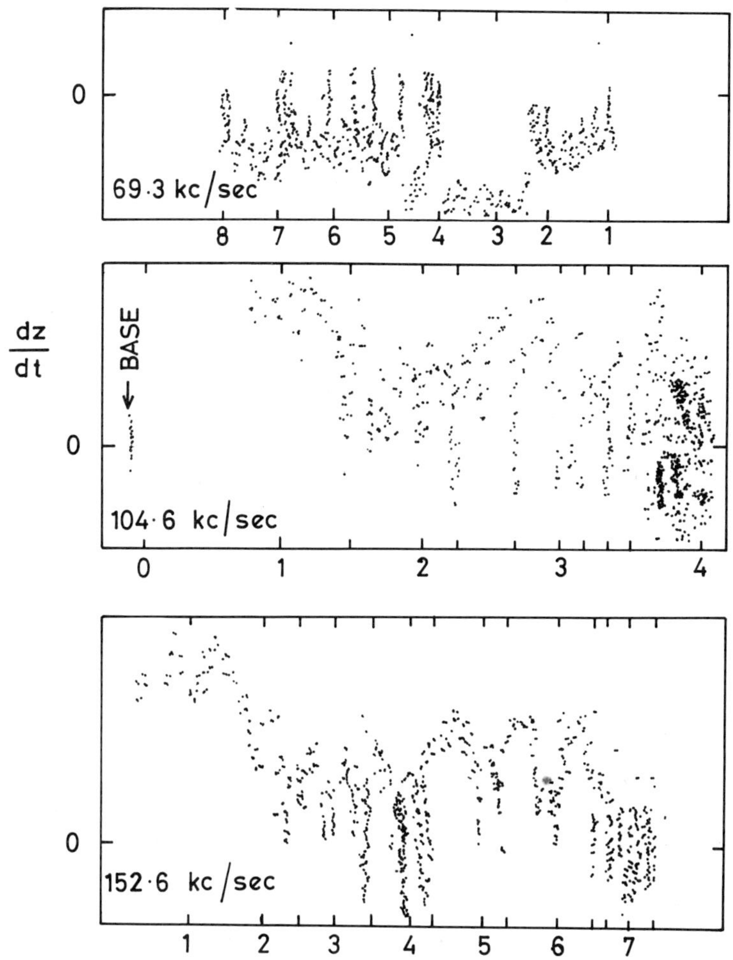


FIG. 21. The rate of change of head with time plotted versus head for three ultrasonic drive frequencies. The intervals of head-difference expected to satisfy the Josephson frequency relation are indicated. Subfrequencies are also shown (after ref. 28).

Josephson frequency relation is satisfied (provided one allows subharmonics). This experiment has been repeated[(29)] to obtain more convincing results. However, all these experiments have been criticized by Musinski and Douglass[(31)] and by Leiderer and Pobell[(32)] who point out that the evidence is consistent with the existence of standing waves of ultra-sound in the reservoir. In fact, in the original experiment of Richards and Anderson the helium level in the reservoir fell during the course of the experiment and it is unlikely that the observation was due to this particular resonance. All the same, a completely convincing observation of the A.C. Josephson effect has yet to be performed.

## ACKNOWLEDGEMENT

The author gratefully acknowledges the benefit he has received from discussions with Dr. W. A. B. Evans.

## APPENDIX

### SECOND QUANTIZATION

In co-ordinate space the wavefunctions of a system of bosons are symmetric functions $\Phi(\mathbf{r}_1, \mathbf{r}_2, \ldots \mathbf{r}_N)$ of the co-ordinates of the particles. If one uses as a basic set of states the states of independent bosons, the basic many-body states $\Phi$ are symmetric functions of the single-particle states $\phi_\alpha(\mathbf{r}_i)$ where $\alpha$ denotes a quantum number. A typical basis wave function is

$$\Phi(\mathbf{r}_1, \mathbf{r}_2 \ldots \mathbf{r}_N) = A \sum_P \phi_{\alpha_1}(\mathbf{r}_1)\phi_{\alpha_2}(\mathbf{r}_2) \ldots \phi_{\alpha_N}(\mathbf{r}_N) \tag{A1}$$

where the sum is over all permutations of the co-ordinates. This means that if we have fixed on a particular complete set of single particle states $\phi_\alpha(r)$, we can construct $\Phi$ completely when we know how many bosons there are in each single-particle state $\alpha$. It is convenient therefore to use a notation which provides us with the limited information we need (the occupation of the single-particle states) and which does not build the state out of superfluous information (the particle at $\mathbf{r}_1$ is in state $\alpha_1$ etc.). This is the purpose of the technique of second quantization.

The only observables in quantum mechanics are matrix elements. We therefore construct a new set of wave functions and operators in one–one correspondence with the old such that all matrix elements are the same in the new and old system. We suppose to start with that we can identify a state in which $n_{\alpha_1}$ bosons are in the single-particle state $\alpha_1$, $n_{\alpha_2}$ bosons are in the state $\alpha_2$ with a normalized ket denoted by $|n_{\alpha_1} n_{\alpha_2} \ldots\rangle$. The states of the form (A1) have a one–one correspondence with these kets but they are not identical with them because the $\Phi$'s are functions of co-ordinates whereas the kets are functions of $n_\alpha$. Because of the freedom from the co-ordinates, it is possible when specifying the kets to consider states containing any number of bosons including a vacuum state $|0\rangle$ with no bosons (to which there is no analogue in (A1)). Just as it is a considerable advantage to use the grand canonical ensemble rather than the canonical ensemble in statistical mechanics, the flexibility arising from states with varying numbers of bosons brings advantages.

We now define a completely new set of operators $a_\alpha$ through their complete sets of matrix elements,

$$\langle n'_{\alpha_1} n'_{\alpha_2} \cdots n'_{\alpha_i} \cdots | a_{\alpha_i} | n_{\alpha_1} n_{\alpha_2} \cdots n_{\alpha_i} \cdots \rangle = \sqrt{n_{\alpha_i}} \tag{A2}$$

if

$$n'_{\alpha_1} = n_{\alpha_1}, \qquad n'_{\alpha_2} = n_{\alpha_2}, \qquad n'_{\alpha_i} = n_{\alpha_i} - 1, \ldots .$$

and is zero otherwise. In other words, the effect of $a_{\alpha_i}$ is to decrease $n_{\alpha_i}$ by one and to leave the occupation number of all other states unaffected. Thus $a_{\alpha_i}$ is called a destruction operator. If $n_{\alpha_i} = 0$, the effect of $a_{\alpha_i}$ on it is to yield zero always. Hence

$$a_{\alpha_i} | n_{\alpha_i} \cdots 0_{\alpha_i} \cdots \rangle = 0.$$

In particular, for the vacuum state

$$a_{\alpha_i} | 0 \rangle = 0 \quad \text{for all } i. \tag{A3}$$

From the complex conjugate of (A2) we see that the hermitian conjugate of $a_\alpha$, $a_\alpha^\dagger$, satisfies

$$\langle n_{\alpha_1} n_{\alpha_2} \cdots n_{\alpha_i} \cdots | a_{\alpha_i}^\dagger | n'_{\alpha_1} n'_{\alpha_2} \cdots n'_{\alpha_1} \cdots \rangle = \sqrt{n_{\alpha_i}} = \sqrt{(n'_{\alpha_i} + 1)} \tag{A4}$$

if

$$n_{\alpha_1} = n'_{\alpha_1}, \qquad n_{\alpha_2} = n'_{\alpha_2}, \qquad n_{\alpha_i} = n'_{\alpha_i} + 1, \ldots$$

and is zero otherwise. Hence, $a_{\alpha_i}^\dagger$ is a creation operator for particles in the state $\alpha_i$.

Since $a_{\alpha_i}$ destroys one boson in state $\alpha_i$ and $a_{\alpha_i}^\dagger$ creates one boson in this state and both do not affect the other single particle states $a_{\alpha_i} a_{\alpha_i}^\dagger$ and $a_{\alpha_i}^\dagger a_{\alpha_i}$ have only diagonal matrix elements. These diagonal elements can be evaluated as follows. (Since only one single-particle state is affected by the operators we indicate only this state.) We then have

$$\begin{aligned} \langle n_\alpha | a_\alpha^\dagger a_\alpha | n_\alpha \rangle &= \sum_{n'_\alpha} \langle n_\alpha | a_\alpha^\dagger | n'_\alpha \rangle \langle n'_\alpha | a_\alpha | n_\alpha \rangle \\ &= \langle n_\alpha | a_\alpha^\dagger | n_\alpha - 1 \rangle \langle n_\alpha - 1 | a_\alpha | n_\alpha \rangle = n_\alpha \end{aligned} \tag{A5}$$

and, similarly,

$$\langle n_\alpha | a_\alpha a_\alpha^\dagger | n_\alpha \rangle = n_\alpha + 1. \tag{A6}$$

Therefore,

$$\langle n'_\alpha | a_\alpha^\dagger a_\alpha | n_\alpha \rangle = n_\alpha \delta_{n'_\alpha, n_\alpha}$$

and

$$a_\alpha^\dagger a_\alpha |n_\alpha\rangle = n_\alpha |n_\alpha\rangle. \tag{A7}$$

Hence, $a_\alpha^\dagger a_\alpha$ is the number operator for the number of bosons in the state $\alpha$. It also follows that

$$\langle n'_\alpha|(a_\alpha a_\alpha^\dagger - a_\alpha^\dagger a_\alpha)|n_\alpha\rangle = \delta_{n'_\alpha, n_\alpha}$$

Hence, the $a$'s satisfy the commutation relations

$$[a_\alpha, a_\alpha^\dagger] = 1.$$

Since $a_\alpha$ and $a_\beta$ affect different states $a_\alpha a_\beta^\dagger$ and $a_\beta^\dagger a_\alpha$ ($\beta \neq \alpha$) have the same effect on any state. Hence

$$[a_\alpha, a_\beta^\dagger] = 0, \qquad \beta \neq \alpha$$

Therefore,

$$[a_\alpha, a_\beta^\dagger] = \delta_{\alpha\beta} \tag{A8}$$

Similarly,

$$[a_\alpha, a_\beta] = 0 \tag{A9}$$

Equations (A8) and (A9) are the basic properties of the creation and annihilation operators.

Now the operator $a_\alpha^\dagger$ creates a boson in state $\alpha$. Hence

$$|n_\alpha + 1\rangle = Ca_\alpha^\dagger|n_\alpha\rangle$$

where $c$ is a constant which can be found from normalization and equation (A6). In fact

$$1 = \langle n_\alpha + 1|n_\alpha + 1\rangle = |c|^2\langle n_\alpha|a_\alpha a_\alpha^\dagger|n_\alpha\rangle = |c|^2(n_\alpha + 1)$$

If we take $c$ to be real,

$$|n_\alpha + 1\rangle = (n_\alpha + 1)^{-\frac{1}{2}}a_\alpha^\dagger|n_\alpha\rangle$$

If this process is continued,

$$|n_\alpha\rangle = (n_\alpha!)^{-\frac{1}{2}}(a_\alpha^\dagger)^{n_\alpha}|0\rangle.$$

Similarly,

$$|n_{\alpha_1}n_{\alpha_2}\ldots\rangle = [n_{\alpha_1}!n_{\alpha_2}!\ldots]^{-\frac{1}{2}}(a_{n_{\alpha 1}}^\dagger)^{n_{\alpha 1}}(a_{n_{\alpha 2}}^\dagger)^{n_{\alpha 2}}\ldots \tag{A10}$$

This equation gives all the states in terms of the vacuum state and the creation operators.

Usually we are interested in the matrix elements of one and two-particle operators. The kinetic energy is an example of the former, the potential energy is an example of the latter. Let us consider these in turn. Suppose $K$ is a single-particle operator of the form

$$K = \sum_i f_i$$

where $f_i$ affects only the co-ordinates of particle $i$. The diagonal matrix element of $K$ in a state of the form of (A1) is

$$\sum_\alpha \int d^3r\, \phi_\alpha^*(\mathbf{r}) f \phi_\alpha(\mathbf{r}) = \sum_\alpha f_{\alpha\alpha} \tag{A11}$$

where the sum is taken over each single-particle state $\alpha$ which is contained in (A1). If $\Phi_1$ and $\Phi_2$ are identical except that the single-particle state $\alpha_2$ is occupied by one more boson in $\Phi_2$ than in $\Phi_1$ and $\alpha_1$ is occupied by one more boson in $\Phi_1$ than in $\Phi_2$, (a boson is scattered from $\alpha_2$ to $\alpha_1$ between them) then

$$(\Phi_1, K\Phi_2) = \int d^3r\, \phi_{\alpha_1}^*(\mathbf{r})\, f \phi_{\alpha_2}(\mathbf{r}) = f_{\alpha_1\alpha_2}. \tag{A12}$$

If $\Phi_1$ and $\Phi_2$ differ in the occupation of more boson states

$$(\Phi_1, K\Phi_2) = 0.$$

Hence the only effect of $K$ on the states (A1) is to leave them unchanged or to scatter a boson from one single-particle state to another.

Can we construct an operator in terms of creation and annihilation operators which has the same matrix elements between the kets as $K$ has between the $\Phi$'s We need an operator which scatters at most one boson from one state to another, that is destroys a boson in one state and creates it in another. The appropriate operator is

$$K_0 = \sum_{\alpha,\beta}' f_{\alpha\beta} a_\alpha^\dagger a_\beta. \tag{A13}$$

This has the same matrix elements as the original $K$ between corresponding states. Notice that the new representation emphasizes the effect of the operator in scattering a boson from one state to another.

If the basic states are plane-wave states $\phi(\mathbf{r}) = L^{-3} \sum_k \exp(i\mathbf{k}.\mathbf{r})$ (where $L^3$ is the normalization volume), we can write $K$ explicitly for a number of cases. If $K$ is the kinetic energy,

$$\begin{aligned} K_0 &= -\sum_{\mathbf{k},\mathbf{k}'} L^{-3} \int d^3r \exp(-i\mathbf{k}'.\mathbf{r}) \frac{\hbar^2\nabla^2}{2m} \exp(i\mathbf{k}.\mathbf{r}) a_\mathbf{k}^\dagger a_{\mathbf{k}'} \\ &= \sum_\mathbf{k} \frac{\hbar^2 k^2}{2m} a_\mathbf{k}^\dagger a_\mathbf{k} \end{aligned} \tag{A14}$$

and is diagonal. If $K$ is a one-particle potential,

$$K = \sum_i V(\mathbf{r}_i)$$

then

$$K_0 = \sum_{\mathbf{k}, k'} V(\mathbf{k} - \mathbf{k}')a_{\mathbf{k}}^{\dagger}a_{\mathbf{k}'}, \qquad V(\mathbf{k} - \mathbf{k}') = L^{-3} \int d^3r\, V(\mathbf{r}) \exp[i(\mathbf{k}' - \mathbf{k}).\mathbf{r}].$$

If $K$ is the particle density at $\mathbf{r}$,

$$K = \rho(\mathbf{r}) = \sum_i \delta(\mathbf{r} - \mathbf{r}_i)$$

then

$$K_0 = \rho_0(\mathbf{r}) = L^{-3} \sum_{\mathbf{k}, k'} a_{\mathbf{k}}^{\dagger}a_{\mathbf{k}'} \exp[i(\mathbf{k}' - \mathbf{k}).\mathbf{r}],$$

with Fourier transform

$$\rho_0(\mathbf{q}) = \sum_{\mathbf{k}} a_{\mathbf{k}}^{\dagger}a_{\mathbf{k}-\mathbf{q}}. \tag{A15}$$

the operator for the total number of particles present is given by

$$N_{\text{op}} = \int d^3r\, \rho_0(\mathbf{r}) = \sum_{\mathbf{k}} a_{\mathbf{k}}^{\dagger}a_{\mathbf{k}} = \sum_{\mathbf{k}} n_{\mathbf{k}} \tag{A16}$$

as it should be. If $K$ is the current density at $\mathbf{r}$

$$K = \mathbf{j}(\mathbf{r}) = \tfrac{1}{2} \sum_i \delta(\mathbf{r} - \mathbf{r}_i)\mathbf{V}_i + \mathbf{V}_i^{\dagger}\delta(\mathbf{r} - \mathbf{r}_i)$$

$$= \frac{-i\hbar}{2m} \sum_i [\delta(\mathbf{r} - \mathbf{r}_i)\nabla_i - \nabla_i\delta(\mathbf{r} - \mathbf{r}_i)],$$

then

$$K_0 = \mathbf{j}_0(\mathbf{r}) = \frac{\hbar}{2m} \sum_{\mathbf{k}, \mathbf{k}'} (\mathbf{k} + \mathbf{k}')a_{\mathbf{k}}^{\dagger}a_{\mathbf{k}'} \exp[i(\mathbf{k}' - \mathbf{k}).\mathbf{r}]. \tag{A17}$$

The diagonal part of this is

$$\mathbf{j}_0 = \sum_{\mathbf{k}} \frac{\hbar\mathbf{k}}{m} a_{\mathbf{k}}^{\dagger}a_{\mathbf{k}}.$$

and shows that each single particle state contributes $\hbar\mathbf{k}/m$ to the average flow.

The main two-body operator for which we require the matrix elements is the potential,

$$V = \tfrac{1}{2} \sum_{i \neq j} V(\mathbf{r}_i - \mathbf{r}_j). \tag{A18}$$

Now this will have non-zero matrix elements between states (A1) only if its effect is to scatter two particles leaving the others unchanged or to scatter one particle or none. If we consider the plane-wave representation, the matrix element of (A18) for scattering bosons in states $\mathbf{k}, \mathbf{q} - \mathbf{k}$ to states $\mathbf{k}', \mathbf{q} - \mathbf{k}'$ (where we have automatically allowed for the conservation of momentum) is

$$V(\mathbf{q}) = L^{-3} \int \mathrm{d}^3 r \, V(\mathbf{r}) \exp(i\mathbf{q} \cdot \mathbf{r}).$$

Hence, we expect the corresponding operator to be a sum of terms each of which is of the form of a product of $V(\mathbf{q})$, two creation operators and two destruction operators. Explicit calculation shows the correct form to be

$$V_0 = \tfrac{1}{2} \sum_{\mathbf{k}, \mathbf{k}', \mathbf{q}} V(\mathbf{q}) a^{\dagger}_{\mathbf{k}'} a^{\dagger}_{\mathbf{q}-\mathbf{k}'} a_{\mathbf{q}-\mathbf{k}} a_{\mathbf{k}}. \tag{A19}$$

This provides us with the form of the hamiltonian and, in principle, allows a complete calculation of the properties of the system.

For use in the text we note a number of useful results. We need to evaluate

$$\mathrm{Tr} \exp(-\alpha a^{\dagger}_{\mathbf{k}} a_{\mathbf{k}}) = \mathrm{Tr} \exp(-\alpha n_{\mathbf{k}})$$

where $\alpha$ is a constant. This is best evaluated using the basic states $|n_{\mathbf{k}}\rangle$. Then

$$\begin{aligned} \mathrm{Tr} \exp(-\alpha n_{\mathbf{k}}) &= \sum_{n_k} \langle n_{\mathbf{k}} | \exp(-\alpha n_{\mathbf{k}}) | n_{\mathbf{k}} \rangle \\ &= \sum_{n_k} \exp(-\alpha n_{\mathbf{k}}) \\ &= [1 - \exp(-\alpha)]^{-1}, \qquad \alpha > 0. \end{aligned} \tag{A20}$$

We also require

$$a(\tau) = \exp(H_0 \tau) a \exp(-H_0 \tau), \qquad H_0 = \varepsilon a^{\dagger} a.$$

Then, by differentiation,

$$\begin{aligned} \frac{\mathrm{d}a(\tau)}{\mathrm{d}\tau} &= \exp(H_0 \tau)[H_0, a] \exp(-H_0 \tau) \\ &= -\varepsilon \exp(H_0 \tau) a \exp(-H_0 \tau) \\ &= -\varepsilon a(\tau) \end{aligned}$$

and

$$a(0) = a$$

Hence

$$a(\tau) = \mathrm{e}^{-\varepsilon \tau} a \tag{A21}$$

Finally it is of interest to look at the actual connection between the kets and the states (A1). First construct a field operator

$$\psi(\mathbf{r}) = L^{-3/2} \sum_{\mathbf{k}} a_{\mathbf{k}} \exp(i\mathbf{k}.\mathbf{r})$$

and its hermitian conjugate

$$\psi^{\dagger}(\mathbf{r}) = L^{-3/2} \sum_{\mathbf{k}} a_{\mathbf{k}}^{\dagger} \exp(-i\mathbf{k}.\mathbf{r}).$$

Then the particle density and current density operators can be written

$$\rho_0(\mathbf{r}) = \psi^{\dagger}(\mathbf{r})\psi(\mathbf{r}) \tag{A22}$$

and

$$j_0(\mathbf{r}) = \frac{-i\hbar}{2m}[\psi^{\dagger}(\mathbf{r})\nabla\psi(\mathbf{r}) - \nabla\psi^{\dagger}(\mathbf{r}).\psi(\mathbf{r})]. \tag{A23}$$

The field operator is thus analogous to the one-particle wavefunction and one can indeed obtain the rules given here by quantizing Schrödinger's equation. Hence the name, second quantization. The commutation relations of the $\psi$'s are obtained from those of the $a$'s and are

$$[\psi(\mathbf{r}), \psi^{\dagger}(\mathbf{r}')] = \delta(\mathbf{r} - \mathbf{r}'), \qquad [\psi(\mathbf{r}), \psi(\mathbf{r}')] = 0 \tag{A24}$$

Now each term in $\psi^{\dagger}(\mathbf{r})$ creates a single boson. Hence $\psi^{\dagger}(\mathbf{r})$ creates a single boson. Indeed

$$N_{\text{op}}\, \psi^{\dagger}(\mathbf{r})|0\rangle = [N_{\text{op}}, \psi^{\dagger}(\mathbf{r})]|0\rangle = \psi^{\dagger}(\mathbf{r})|0\rangle.$$

Also, from the commutation relations of the $\psi$'s,

$$\rho_0(\mathbf{r})\psi^{\dagger}(\mathbf{r}')|0\rangle = \delta(\mathbf{r} - \mathbf{r}')\psi^{\dagger}(\mathbf{r}')|0\rangle. \tag{A25}$$

Hence $\psi^{\dagger}(\mathbf{r}')$ creates a boson at the position $\mathbf{r}'$. Therefore the state

$$\psi^{\dagger}(\mathbf{r}_1)\psi^{\dagger}(\mathbf{r}_2)\ldots\psi^{\dagger}(\mathbf{r}_N)|0\rangle$$

has one boson at $\mathbf{r}_1$, one boson at $\mathbf{r}_2$ and so on. Hence,

$$\langle 0|\psi(\mathbf{r}_N)\psi(\mathbf{r}_{N-1})\ldots\psi(\mathbf{r}_1)|\rangle \tag{A26}$$

is the probability amplitude for finding one boson at $\mathbf{r}_1$, another at $\mathbf{r}_2$ etc., when the system is in the state $|\rangle$. But this is the meaning of Schrödinger's many-particle wave function. Hence $\Phi$ of equation (A1) is the representative (A26). This establishes the exact link between the Schrödinger picture and the abstract one. For any state $|s\rangle$ of $N$ bosons

$$\Phi_s(\mathbf{r}_1.\,\mathbf{r}_2\ldots\mathbf{r}_N) = \langle 0|\psi(\mathbf{r}_N)\ldots\psi(\mathbf{r}_1)|s\rangle.$$

## REFERENCES


1. A. D. B. Woods and R. A. Cowley. *Rep. Prog. Phys.* **36**, 1135 (1973).
2. F. London. "Superfluids." John Wiley and Sons, New York (1954). Vol. II.
3. R. P. Feynman. *Phys. Rev.* **94**, 262 (1954).
4. R. P. Feynman and M. Cohen. *Phys. Rev.* **102** 1189 (1956).
5. N. N. Bogoliubov. *J. Phys. (USSR)* **11**, 23 (1947).
6. T. D. Lee, K. Huang and C. N. Yang. *Phys. Rev.* **106**, 1135 (1957).
7. J. Bardeen, L. N. Cooper and J. R. Schrieffer. *Phys. Rev.* **108**, 1175 (1957).
8. T. D. Lee and C. N. Yang. *Phys. Rev.* **112**, 1419 (1958).
9. L. D. Landau. *J. Phys. (USSR)* **5**, 71 (1941).
10. K. Màki. "Superconductivity," R. D. Parks (Ed.), Marcel Dekker, New York, (1969). Vol. II.
11. D. R. Tilley and J. Tilley. "Superfluidity and Superconductivity," Van Nostrand Reinhold Co., London (1974), p. 64f.
12. H. A. Mook. *Phys. Rev. Lett.* **32**, 1167 (1974).
13. W. E. Parry. "The Many-body Problem," Clarendon Press, Oxford (1973).
14. A. A. Abrikosov, L. P. Gor'kov and I. Ye. Dzyaloshinskii. "Quantum Field Theoretical Methods in Statistical Physics," Pergamon Press, Oxford (1965).
15. R. Kubo. "Statistical Mechanics," North-Holland Publishing Co., Amsterdam (1971).
16. G. C. Wick, *Phys. Rev.* **80**, 268 (1950).
17. G. Rickayzen. "Theory of Superconductivity," John Wiley and Sons, New York (1965).
18. E. C. Titchmarsh. "Theory of Functions," Oxford University Press (1939).
19. J. de Boer and A. Michels. *Physica* **5**, 945 (1938).
20. S. T. Beliaev. *Soviet Phys. JETP* **7**, 289 and 299 (1958).
21. J. Gavoret and P. Nozieres. *Ann. Phys.* (N.Y.), **28**, 349 (1964).
22. D. Pines and P. Nozieres. "The Theory of Quantum Liquids", W. A. Benjamin, Inc., New York (1966), p. 90.
23. O. Penrose and L. Onsager. *Phys. Rev.* **104**, 576 (1956).
24. C. N. Yang. *Rev. Mod. Phys.* **34**, 694 (1962).
25. W. F. Vinen. *Proc. Roy. Soc.* **A260**, 218 (1961); G. W. Rayfield and F. Reif. *Phys. Rev.* **136A**, 1194 (1964); S. C. Whitmore and W. Zimmermann, Jr. *Phys. Rev. Lett.* **15**, 389 (1965); W. A. Steyert, R. D. Taylor and T. A. Kitchens. *Phys. Rev. Lett.* **15**, 546 (1965).
26. R. P. Feynman. *In* "Progress in Low Temperature Physics" (C. H. Gorter Ed.), North-Holland Publishing Co., Amsterdam (1955), Vol. I, Ch. 2.
27. B. D. Josephson. *Phys. Lett.* **1**, 251 (1962).
28. P. L. Richards and P. W. Anderson. *Phys. Rev. Lett.* **14**, 540 (1965).
29. B. M. Khorana and B. S. Chandrasekhar. *Phys. Rev. Lett.* **18**, 230 (1967); J. P. Hulin, C. Laroche, A. Libchaber and B. Perrin. *Phys. Rev.* **A5**, 1830 (1972).
30. R. A. Ferrell and R. E. Prange. *Phys. Rev. Lett.* **10**, 479 (1963).
31. D. L. Musinski and D. H. Douglass. *Phys. Rev. Lett.* **29**, 1541 (1972).
32. P. Leiderer and F. Pobell. *Phys. Rev.* **A7**, 1130 (1973).

# 3. QUANTUM FLUID HYDRODYNAMICS

SETH PUTTERMAN

*Physics Department, University of California*
*Los Angeles 90024, California, USA*

## I. INTRODUCTION

The physical description of nature can proceed on various levels. On the coarsest or macroscopic level the characteristic length which enters the description is the size $L$ of the vessel containing the system. If the system is say water or alcohol then the physical laws are simply those of thermodynamics and hydrodynamics. If one tries to apply these laws to situations where the macroscopic variables such as temperature vary quickly, say over a length $\lambda_M$, then it is found that at some point the macroscopic description yields incorrect results. This happens when

$$\lambda_M \approx \lambda_{\text{MFP}},$$

where $\lambda_{\text{MFP}}$ is the mean free path of the constituent atoms (or excitations). The reason for the breakdown is that the macroscopic variables are averaged and have no meaning over such short distances. For such quick variations one must describe the system with Newton's laws of motion for the microscopic degrees of freedom. The description according to Newton's laws also has a limit on its domain of applicability and breaks down when the distance, $\lambda_N$, over which the Newtonian degrees of freedom change becomes comparable with the de Broglie wavelength

$$\lambda_N \approx \lambda_{\text{deB}}.$$

At this point the quantum theory must be invoked for the microscopic degrees of freedom. In general one has

$$L \gg \lambda_{\text{MFP}} \gg \lambda_{\text{deB}}, \tag{1}$$

so that there is no trace of quantum effects in the macroscopic description.

There exist in nature systems in which a *macroscopic* fraction of the particles condense into a single quantum state below a critical temperature. For these superfluids (e.g. HeII, superconductors) orders of magnitude are completely reversed in that

$$L \approx \lambda_{\mathrm{deB}} \gg \lambda_{\mathrm{MFP}}. \tag{2}$$

No matter how slowly any macroscopic variable is changing, it varies quickly compared with the de Broglie wavelength of the condensed particles. Therefore quantum effects must dominate the macroscopic description even in the adiabatic approximation. For these systems the quantum theory is applied on a scale quite different from that upon which it was motivated.

The problem posed by superfluidity is that of understanding the description of a system for which (2) holds for a finite fraction of the fluid. In these lectures we will discuss the macroscopic or phenomenological description of superfluids especially as contained in two-fluid hydrodynamics. The point of view developed is that the two-fluid hydrodynamics is for superfluids what the Navier–Stokes hydrodynamics is for classical fluids. Emphasis will be placed on a concise development of the foundations of superfluid hydrodynamic theory. The reader is referred to reference (1) for detailed proofs of assertions presented.

## II. THE LANDAU TWO-FLUID HYDRODYNAMICS

As the temperature of liquid helium is lowered below $T_\lambda = 2{\cdot}172\,\mathrm{K}$ a number of changes take place in its dynamical properties. In this HeII phase of helium a number of different experimental arrangements led to widely varying results when attempts were made to measure the viscosity. In forced flow the viscosity appeared to be zero whereas the torque transmitted by oscillation was finite and continuous with that of HeI.[(2)]

The earliest experiment to indicate that on the reversible level (i.e. viscous effects aside) HeII required different hydrodynamics from classical fluids was the fountain effect.[(3)] When two vessels of HeII are connected by a narrow constriction the steady state is characterized by equality of chemical potentials across the constriction. Thus a heating up of one vessel not only increases its temperature but causes fluid to enter and its level to increase. Kapitza found that this fluid flow transported no entropy, in conflict with classical hydrodynamics where entropy always flows with the centre-of-mass velocity. In contrast with these experiments it was found that in some situations (e.g. the propagation of sound) HeII behaved quite classically.

In order to explain that in some cases HeII behaved classically and in others differently, Landau[(4)] postulated that for HeII the velocity of entropy flow ($\equiv \mathbf{v}_n$) is independent of the velocity of mass flow $\mathbf{v}$. Thus the laws of mass and entropy conservation take the form:

$$\frac{\partial \rho}{\partial t} + \nabla \cdot \rho \mathbf{v} = 0 \tag{3}$$

$$\frac{\partial \rho s}{\partial t} + \nabla \cdot \rho s \mathbf{v}_n = 0. \tag{4}$$

The mass density to move with the so-called normal fluid velocity is $\rho_n$. As this must be equal to $\rho$ (in order to have superfluidity) the difference is called the superfluid density

$$\rho_s = \rho - \rho_n,$$

which flows without transport of entropy. The total mass flux density is the sum of those for the separate fluids,

$$\mathbf{J} = \rho \mathbf{v} = \rho_n \mathbf{v}_n + \rho_s \mathbf{v}_s,$$

which defines $\mathbf{v}_s$. The law of conservation of momentum is a direct generalization of that of an Euler fluid and takes the form

$$\frac{\partial J_i}{\partial t} + \frac{\partial}{\partial r_\alpha}\{p\delta_{i\alpha} + \rho_n v_{ni} v_{n\alpha} + \rho_s v_{si} v_{s\alpha}\} = 0, \tag{5}$$

where $p$ is the pressure. In order to describe the fountain effect Landau takes the chemical potential $\mu$ as the driving force of the superfluid:

$$\frac{D_s \mathbf{v}_s}{Dt} \equiv \frac{\partial \mathbf{v}_s}{\partial t} + (\mathbf{v}_s \cdot \nabla)\mathbf{v}_s = -\nabla \mu, \tag{6}$$

so that in a steady state $\mu$ is constant. Equations (3), (4), (5), (6) form eight independent equations for the eight variables $\rho$, $s$, $\mathbf{v}_n$, $\mathbf{v}_s$ which characterize the fluid. These must be supplemented by equations of state for $\rho_n$, $p$, $\mu$. In general the equations of state now depend on three internal variables since $(v_n - v_s)^2$ is also a scalar Galilean invariant. For instance

$$\mathrm{d}\mu = -s\mathrm{d}T + \frac{1}{\rho}\mathrm{d}p - \frac{\rho_n}{2\rho}\mathrm{d}(v_n - v_s)^2, \tag{7}$$

where $T$ is the temperature.

These equations have been the basis of many successful predictions, such as a thermal propagating wave (second sound) and the torque on a Rayleigh disc when the fluid centre of mass is at rest. We return to these shortly.

The question always arises as to what are the minimum number of phenomenological assumptions from which a theory follows. In the event of conflict between experiment and theory one always then knows where to put the blame. Furthermore the goal of any microscopic first principles description is then set.

For the Landau two-fluid hydrodynamics the basic assumptions are:

(A) Eight variables are a complete set; they can be chosen to be $p$, $T$, $\mathbf{v}_n$, $\mathbf{v}_s$ (note that a vector counts for three variables).

(B) The superfluid has a conserved circulation, i.e.

$$\frac{D_s}{Dt}\oint \mathbf{v}_s \cdot \mathrm{d}\mathbf{l} = 0.$$

(C) The stresses obey Pascal's law.

When these assumptions are combined with the macroscopic first principles of

(1) energy, momentum, and mass conservation,

(2) Galilean covariance,

(3) the first and second laws of thermodynamics for reversible processes, one is led uniquely to the equations (3), (4), (5), (6).

The generalization of these equations to the dissipative level requires no further physical assumptions but use is made of

(4) the second law of thermodynamics for irreversible processes. We return to the dissipative theory in Section IV.

As the eight two-fluid equations are a complete set they imply (as mentioned above) a law of energy conservation

$$\frac{\partial U}{\partial t} + \boldsymbol{\nabla}\cdot\mathbf{Q} = 0, \tag{8}$$

where

$$Q_i = \left(\mu + \frac{v_s^2}{2}\right)J_i + \rho s T v_{ni} + \rho_n v_{ni}[\mathbf{v}_n\cdot(\mathbf{v}_n - \mathbf{v}_s)] \tag{9}$$

and the energy density $U$ obeys the first and second laws of thermodynamics

$$\mathrm{d}U = \left(\mu + \frac{v_s^2}{2} - v_n\cdot v_s\right)\mathrm{d}\rho + T\mathrm{d}\rho s + \mathbf{v}_n\cdot\mathrm{d}\mathbf{J} + \rho_s(\mathbf{v}_s - \mathbf{v}_n)\cdot\mathrm{d}\mathbf{v}_s. \tag{10}$$

Due to the existence of two dynamical equations there are for HeII *two* Kelvin, Lagrange and Bernoulli theorems. The Lagrange theorems are for $\mathbf{v}_s$ and $\mathbf{A}$ where

$$\mathbf{A} \equiv \frac{\rho_n}{s\rho}(\mathbf{v}_n - \mathbf{v}_s)$$

and state that if $\boldsymbol{\nabla}\times\mathbf{v}_s$ (or both $\boldsymbol{\nabla}\times\mathbf{v}_s$ and $\boldsymbol{\nabla}\times\mathbf{A}$) are zero everywhere at time zero then $\nabla \times \mathbf{v}_s$ (or both $\nabla \times \mathbf{v}_s$ and $\nabla \times \mathbf{A}$) remain zero for all time. This suggests looking at solutions to the two-fluid equations for which

$$\nabla \times \mathbf{v}_s = \mathbf{0} \tag{11}$$

or

$$\nabla \times \mathbf{v}_s = \mathbf{0} \quad \text{and} \quad \nabla \times \mathbf{A} = \mathbf{0}. \tag{12}$$

In the case (11) one obtains the Bernoulli law

$$\frac{\partial \phi}{\partial t} + \mu + \frac{v_s^2}{2} = \text{constant} \tag{13}$$

where $\mathbf{v}_s = \nabla \phi$. In the case (12) one obtains in addition to (13) the Bernoulli law

$$\frac{\partial \chi}{\partial t} + T + \mathbf{v}_n \cdot \mathbf{A} = \text{constant} \tag{14}$$

where $\mathbf{A} = \nabla \chi$. One finds also from the two-fluid equations that the circulation of $\mathbf{v}_s$ is unchanged as one follows the motion of superfluid whereas the circulation of $\mathbf{A}$ is unchanged as one follows the motion of normal fluid,

$$\frac{D_n}{Dt} \oint \mathbf{A} \cdot \mathbf{dl} = 0.$$

Vortex lines are of particular interest since they have a circulation and still are irrotational (see equations (11), (12)). So according to the circulation theorems they are "indestructible". In an infinite medium super and $\mathbf{A}$ vortices take the form

$$\mathbf{v}_s = \frac{\gamma}{r} \boldsymbol{\theta}$$

$$\mathbf{A} = \frac{\gamma_A}{r} \boldsymbol{\theta},$$

where $\boldsymbol{\theta}$ is a unit circulation vector about the vortex line located at $r = 0$ and $\gamma, \gamma_A$ equal the circulation divided by $2\pi$. When $\gamma_A = 0$ a superfluid vortex is a pressure vortex, as is a classical vortex

$$p = p_\infty - \rho_s \frac{v_s^2}{2}$$

where $p_\infty$ is the pressure at infinity. When $\gamma = 0$, a thermal or $\mathbf{A}$ vortex involves a temperature drop which also follows from the Bernoulli laws,

$$T = T_\infty - \frac{\rho s}{\rho_n} \frac{\gamma_A^2}{r^2}.$$

The thermal vortices and the thermal Bernoulli law have not been investigated from an experimental point of view.

We remark that, according to the circulation theorems, superfluid vortices move with the superfluid velocity and thermal vortices move with the normal fluid velocity. Landau always supplemented his theory with (11).

Due to the presence of an extra dynamical equation the dispersion law implied by the linearized two-fluid equations possesses two propagating-mode solutions. One corresponds to a pressure wave and propagates with the velocity of ordinary sound

$$u_1 = [(\partial p/\partial \rho)_s]^{\frac{1}{2}}.$$

The other is a thermal wave, second sound, which has a velocity

$$u_2 = [\rho_s s^2 T/\rho_n C]^{\frac{1}{2}},$$

where $C$ is the specific heat. Observation[5] of this new mode yields the basic parameter $\rho_n/\rho$ of the two-fluid theory.

In second sound the centre of mass stays at rest while the two fluids flow relative to each other. This motion can be seen by means of a device (the Rayleigh disc) which responds to the square of the velocity field in a classical fluid. In HeII it responds then to the sum of the squares of the super and normal velocity fields. By use of this device in HeII[6] one obtains a check of the two-fluid theory in the quadratic approximation.

Turning to the boundary conditions implied by the two-fluid hydrodynamics we first note that at a stationary solid wall the mass flow must vanish

$$v_\perp = 0 \tag{15}$$

as in a classical fluid. This implies that the weighted sum of super and normal velocities is zero

$$\rho_n v_{n\perp} + \rho_s v_{s\perp} = 0 \tag{16}$$

but the separate velocities can be different from zero if there is an entropy flow. In this case the entropy conservation law (4) implies that

$$\frac{\dot{W}}{T_B} = \rho s v_{n\perp}, \tag{17}$$

where $\dot{W}/T_B$ is the entropy per second to flow from boundary to fluid, $\dot{W}$ is the heat per second and $T_B$ is the temperature of the boundary. From the energy conservation law (8) we also find that the heat flow from boundary to liquid must satisfy

$$\dot{W} = \rho s(T + \mathbf{v}_n \cdot \mathbf{A}) v_{n\perp}. \tag{18}$$

Consistency of (17) and (18) implies a thermal discontinuity at a boundary where heat enters HeII:

$$T_B - T = \mathbf{v}_n \cdot \mathbf{A}. \tag{19}$$

Observation of this small effect will be seriously complicated by the Kapitza resistance.

From the momentum conservation law we see that the force exerted in the *i*th direction on a boundary surface perpendicular to the *j*th direction is per unit area

$$p\delta_{ij} + \rho_n v_{ni} v_{nj} + \rho_s v_{si} v_{sj}. \tag{20}$$

As a result of heat flow from boundary to fluid $v_{n\perp}$ and $v_{s\perp}$ are non-zero, with the result that the heat transport can lead to reaction forces and reaction torques which have in fact been observed.[7] This constitutes another check of the two-fluid theory at the quadratic level.

## III. Quantization of Vorticity

As motivated in the introduction the unusual hydrodynamic description of superfluid helium is due to the quantum theory. Yet the two-fluid hydrodynamics does not contain Planck's constant, so that from the strictly macroscopic point of view one can perhaps question whether HeII is a quantum fluid. This situation changed when Onsager proposed that the superfluid circulation, which is an adiabatic invariant of the motion, be quantized;[8] thus

$$\oint \mathbf{v}_s \cdot \mathrm{d}\mathbf{l} = nh/m, \tag{21}$$

where $h$ is Planck's constant, $m$ the mass of a helium atom and $n$ an integer. Onsager remarked that this step for the macroscopic system was analogous to the Old Quantum Theory where also the adiabatic invariants of the motion were quantized as

$$\oint p \, \mathrm{d}q = nh, \tag{22}$$

where $p$ and $q$ are canonical variables. But according to (21) one now uses the quantum theory to restrict the number of allowed macroscopic states.

An important consequence of (21) is that the strength of vortex lines in the superfluid is now restricted to a discrete set of values. For instance in the line vortex

$$\mathbf{v}_s = \frac{\gamma}{r} \boldsymbol{\theta} \tag{23}$$

discussed previously the circulation $2\pi\gamma$ is now restricted to

$$2\pi\gamma = nh/m. \tag{24}$$

The quantization of vorticity has been most accurately verified in experiments involving the motion of superfluid vortex rings at low temperature.[9] Experiments concerning the Magnus lift exerted on a moving wire with circulation[10] also suggest the validity of (21).

Imposing (21) as a restriction on the two-fluid hydrodynamics (with which it is consistent) leaves us with a theory with striking similarities to the Old Quantum Theory. Namely the quantization principle is used to select a discrete set of allowed states from the deterministic dynamics. As the Old Quantum Theory had to be replaced by a theory including wave-particle duality one can with superfluids also ask if there is required a more complete theory including a macroscopic wave-particle duality.

## IV. ROTATING HeII

Much of the early work in HeII was motivated by the fact that in different experimental arrangements HeII appeared to demonstrate different viscosities. For instance, comparison with the Navier–Stokes hydrodynamics led to a vanishing viscosity in the forced-flow experiments and to a non-zero viscosity as regards the torque on an oscillating disc.[2] This viscosity paradox is resolved in terms of the two-fluid theory by stipulating that on the dissipative level, the normal fluid has a shear viscosity just like a classical fluid, but that the superfluid has no shear corrections to its flow. Then in the flow through a capillary the normal fluid is held back by its viscous interaction with the wall, while the superfluid flows right through without friction. An oscillating disc is similarly not damped by the superfluid but the normal fluid which sticks to it is forced to oscillate and in this way damps the motion.

Basing his argument on this resolution of the viscosity paradox Landau was led to reason that in a cylindrical vessel brought into rotation the superfluid should stay at rest while the normal fluid (like a classical viscous fluid) executed solid-body rotation. Experiments did not bear out this picture and instead indicated that somehow the superfluid partook in the rotation. Also the rotating liquid was still superfluid since it still displayed the fountain effect.[11] So we are confronted with the rotation paradox. How can the frictionless flow of the superfluid be understood as consistent with its twisting into rotation?

Since rotation is an equilibrium state one can first ask what thermodynamics says about it, before investigating the more difficult question of the dynamical approach to equilibrium.

According to thermodynamics the equilibrium state will maximize the entropy for the closed system of given angular momentum. For the two-fluid thermodynamics this leads to the extremum conditions

$$T = \text{constant} \tag{25}$$

$$\mathbf{v}_n = \boldsymbol{\omega} \times \mathbf{r} \tag{26}$$

$$\mu + \frac{v_s^2}{2} - \mathbf{v}_n \cdot \mathbf{v}_s = \text{constant}, \tag{27}$$

so that temperature is constant, the normal fluid is in solid-body rotation and the chemical potential provides for the centripetal acceleration.

The consistency of these restrictions with the Landau two-fluid equations (which must be exact in equilibrium) leads to

$$\boldsymbol{\nabla} \times \mathbf{v}_s = \mathbf{0} \tag{28}$$

$$\boldsymbol{\nabla} \cdot \rho_s(\mathbf{v}_s - \mathbf{v}_n) = 0 \tag{29}$$

and the fact that the equilibrium state is stationary in the rotating reference frame.

If one permits the presence of vortex lines then the superfluid can rotate in equilibrium (but not as a solid body) and still be irrotational (i.e. satisfy (28)) provided that the vortex cores contain no superfluid. This is a partial resolution of the rotation paradox. By considering the stability conditions which follow from the fact that the calculated extremum is a maximum one can say more. One obtains the results

$$\left(\frac{\partial s}{\partial T}\right)_{\rho, k} > 0, \qquad \left(\frac{\partial p}{\partial \rho}\right)_{T, k} > 0 \qquad \rho > 0 \tag{30}$$

$$1 > \frac{\rho_n}{\rho} > 0, \qquad 1 > \frac{\rho_n}{\rho} + w^2 \left[\frac{\partial \rho_n/\rho}{\partial w^2/2}\right]_{p, T} > 0 \tag{31}$$

$$\frac{\rho_s}{\rho} + 2\rho_s w^2 \left[\frac{\partial \rho_n/\rho}{\partial p}\right]_{w^2, T} - w^2 \left[\frac{\partial \rho_n/\rho}{\partial w^2/2}\right]_{p, T} - \left(\frac{\rho_s}{\rho} w\right)^2 \left(\frac{\partial \rho}{\partial p}\right)_{w^2, T} > 0 \tag{32}$$

$$\int \left(\tfrac{1}{2}\rho_s v_s^2 - \rho_s \mathbf{v}_n \cdot \mathbf{v}_s\right) \mathrm{d}^3 r = \text{Minimum}, \tag{33}$$

where

$$\mathbf{w} \equiv \mathbf{v}_n - \mathbf{v}_s$$

$$\mathbf{k} \equiv \frac{\rho_n}{\rho}(\mathbf{v}_n - \mathbf{v}_s).$$

There are many possible superfluid equilibrium velocity fields consistent with (28) and if one tries to satisfy (33) without an additional restriction then one will find that the superfluid rotates around a very dense distribution of very weak vortex lines which approximate solid-body rotation with arbitrary accuracy. So at this point one again imposes the Onsager restriction on the vorticity quantization (21) and in this way one is led to a self-consistent thermodynamical description of the equilibrium state of a rotating superfluid. For each angular velocity there is now present a given number and distribution of quantized vortices determined by (33). In particular, for very low angular velocities

$$\omega < \omega_{cr} \equiv \frac{\hbar}{mR^2} \log \frac{R}{a} \tag{34}$$

there should be no vortices present ($v_s = 0$). In (34) $R$ is the radius of the bucket and $a$ the radius of the vortex core where $\rho_s = 0$. For large angular velocities the superfluid motion approximates solid-body rotation due to a dense uniform array of vortices.

In the state where $v_s$ is zero the superfluid is at rest not with respect to the laboratory but with respect to an inertial reference frame. This quantum Mach paradox has not yet been observed. The quantum jumps in rotation (i.e. number of vortices present) which result as $\omega$ is increased have been observed by Packard and Sanders.[12]

So the resolution of the rotation paradox is that although the superfluid acquires angular momentum and rotates it is not in solid-body rotation as would be implied if it had viscosity. The problem still remains to understand how the vortex lines get into the fluid, since if the fluid has no viscosity the circulation is still an invariant of the motion. This dynamical problem has not been resolved but it should be remarked that the Kelvin circulation theorem can be consistent with a change in number of vortex lines if they enter from the wall.

The detailed generalization of the reversible equations to the dissipative level follows from the added requirement that when the fluid is nearly in equilibrium the dissipative effects act to drive it towards equilibrium. As $\nabla \times \mathbf{v}_s = \mathbf{0}$ in equilibrium the dissipative $v_s$ equation must still conserve circulation. The most general equations consistent with these requirements on the level of linear response involve 13 independent transport coefficients and were first presented by Clark.[13] If one makes the added simplification (not justified as regards rotation) that one linearizes also in $v_n - v_s$ then one is led to the Khalatnikov equations[14]

$$\frac{\partial \rho}{\partial t} + \nabla \cdot (\rho_n \mathbf{v}_n + \rho_s \mathbf{v}_s) = 0$$

$$\frac{\partial \rho s}{\partial t} + \nabla \cdot \left[\rho s \mathbf{v}_n + \frac{\mathbf{q}}{T}\right] = \Sigma/T \tag{34}$$

$$\frac{\partial J_i}{\partial t} + \frac{\partial}{\partial r_\alpha}\{p\delta_{i\alpha} + \rho_n v_{ni} v_{n\alpha} + \rho_s v_{si} v_{s\alpha} + \tau_{i\alpha}\} = 0 \tag{36}$$

$$\frac{D_s \mathbf{v}_s}{Dt} = -\nabla(\mu + H) \tag{37}$$

$$\Sigma = -\frac{\mathbf{q}\cdot\nabla T}{T} - \tau_{\alpha\beta}\frac{\partial v_{n\alpha}}{\partial r_\beta} - H\nabla\cdot\rho_s(\mathbf{v}_s - \mathbf{v}_n) \tag{38}$$

$$\tau_{ij} = -\eta\left[\frac{\partial v_{ni}}{\partial r_j} + \frac{\partial v_{nj}}{\partial r_i} - \tfrac{2}{3}\delta_{ij}\nabla\cdot\mathbf{v}_n\right] - \delta_{ij}[\zeta_1\nabla\cdot\rho_s(\mathbf{v}_s - \mathbf{v}_n) + \zeta_2\nabla\cdot\mathbf{v}_n] \tag{39}$$

$$\mathbf{q} = -\kappa\nabla T \tag{40}$$

$$H = -\zeta_3\nabla\cdot\rho_s(\mathbf{v}_s - \mathbf{v}_n) - \zeta_4\nabla\cdot\mathbf{v}_n \tag{41}$$

and by the Onsager symmetry requirement one has

$$\zeta_1 = \zeta_4.$$

The second law for irreversible processes requires

$$\zeta_2, \zeta_3, \kappa, \eta > 0$$

$$(\zeta_1 + \zeta_4)^2 \leqslant 4\zeta_2\zeta_3,$$

so that there are in this approximation five independent transport coefficients.

We remark that in the presence of an external potential $\Omega$ per unit mass the appropriate equations are obtained by replacing $\mu$ with $\mu + \Omega$ in (37), (10), and adding $-\rho\nabla\Omega$ to the right hand side of (36). The equilibrium conditions are obtained with $\mu + \Omega$ replacing $\mu$ in (27). The fundamental relation (7) is unmodified by the external field.

These equations form the basic hydrodynamic description of HeII just as do the Navier–Stokes equations for classical fluids such as water. When supplemented by the Onsager quantization they comprise our best phenomenological description of HeII.

According to them one gets the hydrodynamic description of sound attenuation (but for sound attenuation with respect to a heat current one must use the Clark theory). They also describe the superthermal conductivity[(15)] of HeII and the absence of a drop in the stand-pipe level due to accelerated flow.[(16)] Further they contain a viscous interaction between superfluid vortices and the local normal fluid motion. Finally they predict that it should be possible to generate first and second sound by transverse oscillations of a plate situated in a counterflow. This last effect is yet to be investigated.

On the dissipative level the new boundary condition to be imposed is that the tangential component of the normal fluid velocity vanishes at a stationary boundary. The stress tensor

$$p\delta_{ij} + \rho_n v_{ni} v_{nj} + \rho_s v_{si} v_{sj} + \tau_{ij}$$

still gives the force per unit area on the boundary; equation (16) still is valid and (17) and (18) are modified to

$$\frac{\dot{W}}{T_{\mathrm{B}}} = \rho s v_{n\perp} + \frac{q_\perp}{T}$$

$$\dot{W} = \rho s(T + \mathbf{v}_n \cdot \mathbf{A}) v_{n\perp} + q_\perp - \rho v_{n\perp} H + \tau_{\perp\perp} v_{n\perp}.$$

From these conditions one finds that the thickness of the thermal boundary shock wave (19) is

$$\delta = \sqrt{(\Lambda\kappa/\rho^2 s^2 T)},$$

where

$$\Lambda \equiv \tfrac{4}{3}\eta + \zeta_2 - \rho(\zeta_1 + \zeta_4) + \rho^2\zeta_3.$$

## V. REMAINING PUZZLES

A question regarding the two-fluid hydrodynamics which has received no attention (in sharp contrast with the Navier–Stokes fluid) is that of the stability of stationary solutions. The problem for HeII will be more interesting because of the new dynamic role of temperature. A situation where this is brought out is the Couette flow which takes place in a liquid confined to the region between two cylinders rotating at different angular velocities. For this case the normal fluid velocity takes the form

$$\mathbf{v}_n = \left(ar + \frac{b}{r}\right)\boldsymbol{\theta}$$

where $r$ is the distance to the rotation axis and $a$, $b$ are determined by the boundary condition that the normal fluid sticks to the walls. The pressure gradient turns out to be

$$\frac{\partial p}{\partial r} = \rho_n \frac{v_n^2}{r},$$

which for $v_n$ is just the behaviour exhibited in a classical fluid. However, in contrast with classical behaviour one finds that there exists a temperature gradient

$$\frac{\partial T}{\partial r} = \frac{\rho_n}{s\rho}\left[\frac{2ab}{r} - \frac{2b^2}{r^3}\right] \tag{42}$$

where we work with the simplest case $v_s = 0$. We know that HeII is very sensitive to temperature gradients and thus this solution might become (or be) unstable with respect to internal convection. Because of the two dynamical laws and the new role of temperature there might for HeII be additional stability criteria. To see the importance of such a question one need only realize that in the two-fluid heat flow experiments a temperature gradient analogous to (42) exists across the width of the capillary and may in fact be leading to the critical velocities. By extending to HeII stability calculations which have received enormous attention in classical fluids one might be able to understand many of the critical velocity observations.

A situation in which I don't see in principle how to solve the two-fluid hydrodynamics is one which involves shock-waves or surfaces of discontinuity. Classically the problem is set up as follows. There are five variables which constitute a complete set. In order to have a complete description of the motion one needs five relations across the surface of discontinuity. These follow by integrating the five conservation laws

$$\frac{\partial \rho}{\partial t} + \nabla\cdot\mathbf{J} = 0$$

$$\frac{\partial J_i}{\partial t} + \frac{\partial}{\partial r_\alpha}P_{i\alpha} = 0$$

$$\frac{\partial U}{\partial t} + \nabla\cdot\mathbf{Q} = 0$$

across the surface, to obtain in the reference frame where the shock is at rest

$$[\mathbf{J}(1) - \mathbf{J}(2)]\cdot \mathrm{d}\mathbf{S} = 0$$

$$[P_{i\alpha}(1) - P_{i\alpha}(2)]\cdot \mathrm{dS}_\alpha = 0$$

$$[\mathbf{Q}(1) - \mathbf{Q}(2)]\cdot \mathrm{d}\mathbf{S} = 0,$$

where $\mathrm{d}\mathbf{S}$ is an element of the shocks area, the points "1", "2" refer to the two sides of the shock and $P_{ij}$ is the momentum flux tensor and $\mathbf{Q}$ is the energy flux density. For HeII eight variables are a complete set yet there are still only five conservation laws. Thus it seems as though the theory is incomplete. One might try to argue that with the Landau condition

$$\nabla \times \mathbf{v}_s = 0 \tag{43}$$

the superfluid equation (37) takes the form (13) and that (43) and (13) constitute the three missing conditions. This is in fact a satisfactory approach as regards longitudinal shocks. But classically a shock incident *obliquely* on potential flow is a source of vorticity. For HeII the same must be the case and in this situation (13) and (43) can no longer be maintained. There are *macroscopic* degrees of freedom in the fluid which are not yet described by the hydrodynamics and that is the problem.

The electron fluid in many metals displays a behaviour quite similar to superfluidity in HeII. In these superconductors the coupling to the electromagnetic field makes practical many additional experiments and in these systems it might soon be possible to observe effects of relativistic hydrodynamics. As the third law of thermodynamics is dominant in superfluids one might think that the observation of relativistic effects is out of the question. However, the low impedance can make possible measurements of great accuracy especially by use of the Josephson effects.

Relativistic effects enter hydrodynamic theories according to the size of the dimensionless parameters $v_s^2/c^2$ and $\mu/c^2$. The first depends on the speed of the convection and is wholly negligible, being of the order of $10^{-17}$. The other term involves the internal stresses and for superconductors this is of the order of $10^{-5}$ so that with five-place accuracy one might in principle see relativistic hydrodynamic effects wherein the macroscopic stresses affect the inertia.

The London equations of the superelectron fluid

$$\mu + \frac{v_s^2}{2} + \frac{q\Phi}{m} = -\frac{\partial \chi}{\partial t}$$

$$\mathbf{v}_s + \frac{q}{mc}\mathbf{A} = \nabla \chi$$

are a generalization of the Landau equations (11), (13) of superfluid $He^4$ to include the charge. They must be supplemented with the Maxwell equations and the relations

$$\mathbf{B} = \nabla \times \mathbf{A}$$

$$\mathbf{E} = -\nabla \Phi - \frac{1}{c}\frac{\partial \mathbf{A}}{\partial t}$$

where **E**, **B** are the electric and magnetic fields.

The two London equations are contained in the single relativistic equation:

$$\left(1 + \frac{\mu}{c^2}\right) v_{s,\nu} + \frac{q}{mc} A_\nu = \frac{\partial \chi}{\partial r_\nu}$$

where the four-vectors are

$$A_\nu = (\mathbf{A}, i\Phi)$$

$$\frac{\partial}{\partial r_\nu} = \left(\nabla, -\frac{i}{c}\frac{\partial}{\partial t}\right)$$

$$v_{s,\nu} = \left\{\frac{\mathbf{v}_s}{\sqrt{[1-(v_s^2/c^2)]}}, \frac{ic}{\sqrt{[1-(v_s^2/c^2)]}}\right\}.$$

Now London already emphasized that if a superconductor is rotated the superelectrons try to lag behind the lattice, thus creating a current and a magnetic field. The degree to which they lag behind depends on their inertia, which relativistically will be modified by the $\mu/c^2$ term. In fact from these equations the relativistic London moment inside a superconductor rotating at angular velocity $\boldsymbol{\omega}$ is

$$\mathbf{B} = -\frac{2mc}{q}\omega\left(1 + \frac{\mu}{c^2}\right)$$

(for $\omega \sim 10^3\ \text{sec}^{-1}$, $B \sim 10^{-4}$ gauss). A measurement of this field to 5- or 6-place accuracy will test relativistic hydrodynamics. There should also be other tests and perhaps interesting questions might arise when one tries to understand relativistic quantum hydrodynamics from a microscopic relativistic quantum theory.

Finally it should be emphasized that the two-fluid hydrodynamics plus Onsager quantization is on the level of the Old Quantum Theory and the question naturally arises as to whether a further generalization to include a macroscopic wave-particle duality will be required. Also will there be macroscopic quantum effects corresponding to the irreversible interaction between observer and quantum system that takes place during a measurement?

## REFERENCES


1. S. J. Putterman. "Superfluid Hydrodynamics," North-Holland, Amsterdam (1974).
2. J. F. Allen, A. D. Misener. *Nature* **141**, 75 (1938); P. Kapitza. *Nature* **141**, 74 (1938); A. D. B. Woods and A. C. Hollis-Hallett. *Can. J. Phys.* **41**, 596 (1963); W. H. Keesom and G. E. MacWood. *Physica* **5**, 737 (1938).
3. J. F. Allen and J. Jones. *Nature* **141**, 243 (1938); P. Kapitza. *J. Phys. USSR* **5**, 59 (1941); *Phys. Rev.* **60**, 354 (1941).
4. L. D. Landau. *J. Phys.* (*USSR*) **5**, 71 (1941); L. D. Landau and E. M. Lifshitz, "Fluid Mechanics," Pergamon, London (1959), Ch. 16.
5. V. P. Peshkov. *J. Phys.* (*USSR*) **8**, 381 (1944).
6. J. R. Pellam and W. B. Hanson. *Phys. Rev.* **85**, 216 (1952).
7. H. E. Hall. *Proc. Phys. Soc.* (*London*) **A67**, 485 (1954); T. K. Hunt. *Phys. Rev.* **170**, 245 (1968).

8. L. Onsager. *Nuovo Cimento* (Suppl. no. 2 to Vol. **6**), 249 (1949).
9. G. W. Rayfield and F. Reif. *Phys. Rev. Lett.* **11**, 305 (1963); *Phys. Rev.* **136**, A1194 (1964).
10. W. F. Vinen. *Proc. Roy. Soc.* **A260**, 218 (1961); S. C. Whitmore and W. Zimmermann Jr. *Phys. Rev.* **166**, 181 (1968).
11. E. L. Andronikashvili and Yu. G. Mamaladze. *Rev. Mod. Phys.* **38**, 567 (1966).
12. R. E. Packard and T. M. Sanders Jr. *Phys. Rev. Lett.* **22**, 823 (1969).
13. A. Clark Jr. PhD Thesis MIT (1963).
14. I. M. Khalatnikov. "An Introduction to the Theory of Superfluidity," Benjamin, New York (1965), Ch. 9.
15. F. London. "Superfluids," Wiley, New York (1954), Vol. 2, p. 155; D. F. Brewer and D. O. Edwards. *Phil. Mag.* **6**, 775 (1961).
16. W. M. van Alphen, R. de Bruyn Ouboter, K. W. Taconis and E. van Spronsen. *Physica* **39**, 109 (1968).

# 4. SUPERFLUID FILM FLOW*

L. J. CAMPBELL

*Los Alamos Scientific Laboratory, University of California, Los Alamos, New Mexico 87544, USA*

## INTRODUCTION

Historically, the phenomenological two-fluid hydrodynamic description has prove more useful than microscopic theory as a means for understanding superfluid flow, especially when supplemented by the concept of quantized vorticity. The major shortcoming of the earlier version of the theory was its inability to predict the behaviour of flow at high velocities where resistance to further acceleration becomes relatively very strong. This was the realm of the "critical velocity", which in the laboratory ranked among the fundamental measurable flow properties as befitted its fundamental theoretical definition by Landau. Within approximately the last decade theoretical considerations and some surprising observations have provided the two-fluid theory with another supplement which, for the first time, extends its range of application past the "critical velocity" barrier which, in the process, has been reduced to an unmeasurable flow property(!). (Of course, habits of speech and notation persist.)

This new freedom to easily calculate flow behaviour is exploited in part of Section I and all of Sections II and III. Section I.3 and Section IV are devoted to the more sobering subject of what is poorly understood. The only systems considered here are those for which the normal fluid component is stationary.

## I. DISSIPATION—THE LIMITS ON SUPERFLUID FLOW

### 1. INTRINSIC DISSIPATION

Just as for classical fluids the quest for microscopic understanding of dissi-

* Work performed under the auspices of the US Atomic Energy Commission.

pation in superfluids leads quickly to unknown and largely inhospitable territory. To some extent this is because of the necessity to wade in the statistical marsh between reversible dynamics and irreversible phenomena. The difficulty is compounded for superfluids by the lack of a useful formalism to describe local quantum changes-in-state within a much larger system and by lack of certainty of even the size of the locale relevant to dissipation. Contrast this with the classical system where, as a last resort, one can put a few hundred particles with their mutual interaction on the computer and (often) get meaningful answers.†

However, for distance scales much larger than the inter-atomic spacing, evidence and consensus for a particular phenomenological description of microscopic dissipation has accumulated to the point that it appears pre-emptive of all others. According to this description the superfluid component supports vortices whose cores interact with the normal component which in turn interacts with the boundaries in the manner of a classical liquid. There is presumably no other way for the superfluid component to lose its energy. Moreover, as discussed by Anderson,[1] steady-state dissipative flow must somehow be accompanied by vortex motion across all the streamlines in a specified sense at a specified rate. This result is very basic, depending only on a sine qua non for superfluidity, the off-diagonal long-range order of the phase of the single-particle density matrix. Actually, as also shown by Anderson[1] and elaborated by Huggins,[2] there is a similar theorem for ideal classical fluids which has not found important application in classical hydrodynamics since the vorticity is not quantized. Although the vortex core structure and its interaction with the excitations that constitute the normal fluid require a quantum mechanical treatment (see review by Fetter[3]), the response, on a larger scale, of the vortex to flow conditions and forces at its core is assumed to be classical. Even so, the resulting motion of the vortex line has no classical analog: instead of moving parallel to the local superfluid streamlines as would a classical vortex line, it moves at a fixed (but temperature dependent) angle with respect to them, in the frame where the normal component is stationary.[4] The frictional interaction of vortices with the normal fluid thus provides the mechanism by which vortices cross the superfluid streamlines and superfluid energy (and momentum) is transferred to the normal fluid.

The well-known problem is to account for the production of vortex lines.[5] A purely quantum mechanical solution analogous to the appearance of nodes in the wave function of an ideal rotating Bose gas[6] would be aesthe-

† Ground-state properties of quantum liquids can be obtained by computer simulation. While the success of these techniques is encouraging their complexity has prevented application to the more complicated problem of dissipation. See, for instance, M. H. Kalos, D. Levesque and L. Verlet, *Phys. Rev.* **A9**, 2178 (1974).

tically pleasing and some progress[7] has been made for $T = 0$. However, at $T = 0$ the *entire* dissipation process requires a quantum mechanical description since vortices move parallel to the stream lines in the absence of normal fluid.

At finite temperatures there is, in principle, a finite probability for a vortex line to be produced by thermal fluctuations.† For lines long compared to the atomic spacing a two-step process is imagined. The first step is presumably quantum mechanical and results in a very small vortex ring which in the second step is gradually buffeted into a larger ring through interaction with other excitations, the same mechanism responsible for the decay of a ring (!). Vinen[8] was the first to estimate the probability of this process: Iordanskii[9] and then Langer and Fisher[10] considered the problem in more detail and more generality, respectively. Later Chester and Ziff,[11] Donnelly and Roberts,[12] and E. R. Huggins (unpublished) addressed the subject.‡

One consensus in the above work is the prediction of dissipation for all superfluid velocities, i.e. there is no "critical" velocity, uniquely dependent on the channel geometry, but only a limiting velocity dependent upon the particular geometry, driving force, and temperature. Experimental evidence for this was found in the spontaneous decay of circulating currents in 500 Å filter material,[13] in pressure driven flow through pores[14] (200–2000 Å diameter), in thermally driven film flow,[15] in pressure and temperature driven flow through packed powder,[16] and in gravitational film flow.[17, 18] All these experiments confirm dissipation proportional to $\exp(-v_B(T)/v_s)$, some agree well with the theoretical temperature dependence of $v_B(T)$, and some[17, 18] report a strong temperature dependence in the proportionality constant. Contrary to the above, an experiment using wide channels ($10^{-2}$ cm) found a different $v_s$ dependence.[19] A useful expression for the intrinsic dissipation was introduced by Langer and Reppy[20]

$$\frac{\partial \mathbf{v}_s}{\partial t} = -\frac{\mathbf{v}_s}{|v_s|}\mathscr{A}\kappa\nu\exp(-v_B/|v_s|) \tag{I.1}$$

where $\mathscr{A}$ is the flow area and $\kappa$ is the unit of circulation (0·001 cm sec$^{-1}$). The phenomenological parameters $\nu$ and $v_B$ are functions of the temperature. Some of the theories of intrinsic dissipation suggest specific forms for $\nu$ and $v_B$ but the general expression is sufficient here.

It is not difficult to understand why earlier experiments seemed to show the existence of a critical velocity. The decay constant

† Considerations of energy balance in steady-state, dissipative flow imply that vortex production must arise from fluctuations since the available stream energy is totally exhausted by the necessary vortex motion. See L. J. Campbell, *J. Low Temp. Phys.* **8**, 105 (1972).

‡ See also S. J. Putterman, "Superfluid Hydrodynamics", (North Holland, 1974). This interesting treatment relates the viscosity coefficient $\zeta_3$ to vortex motion and nucleation.

$$\alpha \equiv \frac{1}{v_s}\frac{dv_s}{dt}$$

due to the above process, which will hereafter be called intrinsic dissipation, is a very steep function of $v_s$. Figure 1 shows $\alpha/\mathscr{A}$ versus $v_s$ (where $\mathscr{A}$ is the

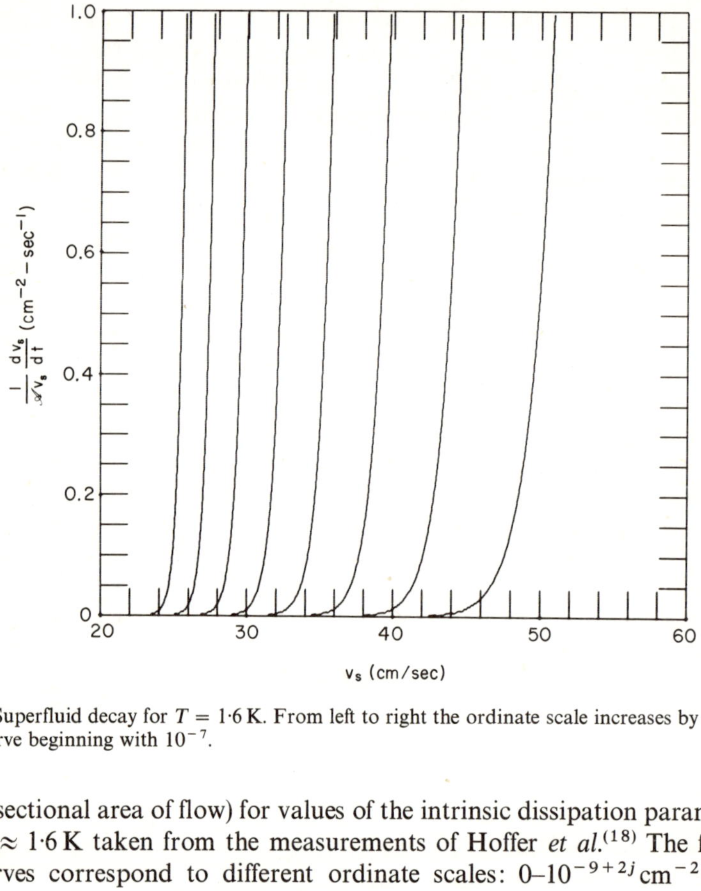


FIG. 1. Superfluid decay for $T = 1{\cdot}6$ K. From left to right the ordinate scale increases by $10^2$ for each curve beginning with $10^{-7}$.

cross sectional area of flow) for values of the intrinsic dissipation parameters for $T \approx 1{\cdot}6$ K taken from the measurements of Hoffer *et al.*[18] The family of curves correspond to different ordinate scales: $0\text{–}10^{-9+2j}\,\text{cm}^{-2}\,\text{sec}^{-1}$ where $j = 1, 2, \ldots, 8$ from left to right. A good case for the existence of a "critical velocity" could always be made as the decay rate fell below the sensitivity of the apparatus. The effect of temperature is shown in Fig. 2 where $1 \leqslant \alpha/\mathscr{A} \leqslant 10^{10}$ and $T = 2{\cdot}1, 1{\cdot}6, 1{\cdot}1$ K from left to right.

If taken seriously, (1) predicts that for sufficiently large $v_s$ the rate of velocity decay should saturate or, in steady state, the velocity should increase dramatically with a small increase in driving force. The velocity needed is something in excess of $v_B(T)$ which varies[18] from over 1000 cm sec$^{-1}$ near 1 K to

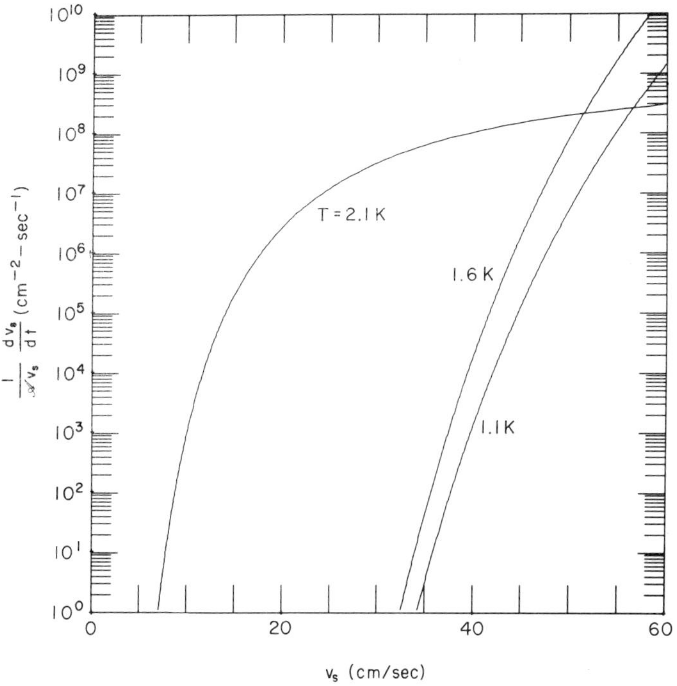


FIG. 2. Superfluid decay for $T = 2{\cdot}1$, $1{\cdot}6$, $1{\cdot}1$ K from left to right.

less than 100 cm sec at 2·12 K. This is such a striking and unique prediction of the intrinsic dissipation model that one might expect many experiments had put it to the test. Surprisingly, there has only been one, by Liebenberg,[21] but it supports the existence of saturation.

The barrier velocity $v_B$ can be interpreted as the velocity of the vortex ring population which constitutes the thermal bath at a given $T$. The higher the temperature the larger are the rings (and the smaller is $v_B$) in this steady state population because their growth is aided by excitations (phonons and rotons). This suggests that for channels smaller than the radius of a "bath" vortex ring a different and probably faster mechanism will be responsible for dissipation. Therefore, it might be expected that this critical channel size is closely related to the channel size for which Telschow, Rudnick and Wang[22] found a maximum in the limiting velocity. The radius of a hollow-core ring is, implicitly, $R_c = (\kappa/4\pi v_B)[\ln(8R_c/a) - \frac{1}{2}]$. The core radius $a$ is 1·28 Å and smoothed values of $v_B$ from the Los Alamos data[18] are 2050, 1490, 1380, 1040 and 488 cm sec$^{-1}$ for $T = 0$, 1·5, 1·6, 1·8, 2·0 K. The results are shown in Fig. 3 together with the data of Telshow *et al.* The $T = 0$ value of $v_B$ required, of course, an extrapolation.

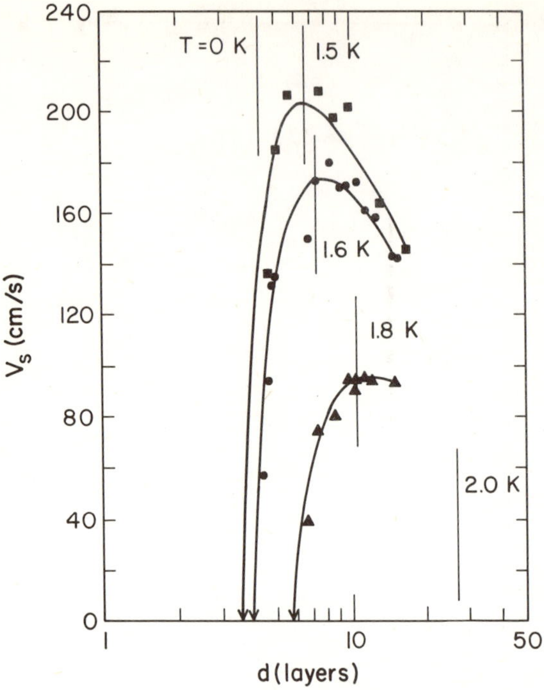


FIG. 3. Predicted thickness of maximum $v_{sc}$ for $T = 0$, 1·5, 1·6, 1·8, 2·0 K. The length and vertical positions of the lines have no significance. Data and curves from ref. 22.

## 2. THERMAL DISSIPATION

Saturated film flow between reservoirs of bulk liquid introduces relative pressure and temperature differences, resulting in a loss of free energy. This will be called thermal dissipation. It is important, not because of any fundamental questions it raises, but because it is experimentally inescapable.

The starting point of film flow analysis is the superfluid equation of motion

$$\dot{\mathbf{v}}_s + \nabla\left[\mu_0(P, T) + \frac{1}{2}\frac{\rho_s}{\rho}v_s^2 - \frac{\alpha}{\delta^n} + gh\right] = -\frac{\mathbf{v}_s}{|v_s|}\mathscr{A}\kappa\nu\exp(-v_B/|v_s|) \tag{I.2}$$

which includes the van der Waals energy for a film thickness $\delta$, the gravitational energy for height $h$ and the intrinsic dissipation. As usual, $\mu_0(P, T)$ is taken to be the chemical potential of the bulk liquid. The presence of the intrinsic dissipation term in (I.2) is permissible only if $v_s$ is regarded as the spatially coarse-grained superfluid velocity.

Integrating (I.2) along a streamline between the surfaces of the reservoirs gives

$$\frac{\partial}{\partial t}\int_1^2 \mathbf{v}_s \cdot \mathrm{d}\mathbf{y} + \mu_0\big|_1^2 + gh\big|_1^2 = -\int_1^2 \frac{\mathbf{v}_s}{|v_s|}\mathscr{A}\kappa\nu \exp(-v_B/|v_s|)\cdot \mathrm{d}\mathbf{y}. \qquad (\mathrm{I.3})$$

The difference in chemical potentials between the reservoirs is

$$\Delta\mu_0 = -s\Delta T + \frac{1}{\rho}\Delta P \quad \text{where} \quad \Delta T = T_2 - T_1, \text{etc.}$$

Assuming an insignificant net amount of fluid is exchanged between the film and vapour implies $v_s(y) = v_s\mathscr{A}/\mathscr{A}(y)$ where $v_s$ and $\mathscr{A}$ are the velocity and area at a particular point. The same assumption for the reservoirs implies $\rho_s\mathscr{A}v_s = \rho A\dot{z}$ where $A$ is the reduced surface area of the reservoirs, $1/A = 1/A_1 + 1/A_2$, and $z = h_2 - h_1$. The effect of evaporation–condensation on the vapour pressure will be discussed later. Incorporating the above, (I.3) becomes

$$\frac{\partial}{\partial t}\frac{\rho A\ell}{\rho_s\mathscr{A}}\dot{z} - s\Delta T + \frac{1}{\rho}\Delta P + gz = -\operatorname{sign}(\dot{z})\kappa\nu\int_1^2 \mathscr{A}(y) \times \exp\left(-\frac{v_B\rho_s\mathscr{A}(y)}{|\dot{z}|\rho A}\right)\mathrm{d}y. \qquad (\mathrm{I.4})$$

where $\ell$ is the effective kinetic length defined by

$$\ell = \int_1^2 \mathrm{d}y\, \mathscr{A}/\mathscr{A}(y). \qquad (\mathrm{I.5})$$

All that remains is to find equations for $\Delta T$ and $\Delta P$. This is mostly bookkeeping. Only the simplest case of heat conduction, gas flow, and evaporation will be treated.

There are three contributions to the temperature change.

(i) *Superfluid flow*

The superfluid carries no entropy so $s[T_1(t)]v_1(t) = \text{const.}$ where $s$ is the entropy ($\mathrm{erg\,g^{-1}\,K^{-1}}$) and $v_1$ the liquid volume of reservoir 1. The time derivative of this conservation equation gives

$$\dot{T}_1 = -\frac{s\dot{v}_1}{Cv_1} \qquad (\mathrm{I.6})$$

where $C \equiv \mathrm{d}s/\mathrm{d}T$ is the liquid heat capacity ($\mathrm{erg\,g^{-1}\,K^{-2}}$) at saturated vapour pressure. A similar equation holds for reservoir 2. Note, $\dot{v}_1 = -\dot{v}_2 = -A\dot{z}$.

(ii) *Heat conduction*

The heat flow from 1 to 2, $\dot{Q}$ (erg sec$^{-1}$), is related to the temperature difference by

$$\dot{Q} = -K\Delta T \tag{I.7}$$

where $K$ is the effective conductivity (erg sec$^{-1}$ K$^{-1}$). The resulting temperature change is

$$\dot{T}_1 = -\frac{\dot{Q}}{v_1 \rho T_1 C} = \frac{K(T_2 - T_1)}{v_1 \rho T_1 C} \tag{I.8}$$

and similarly for reservoir 2 with a change of sign and subscripts.

(iii) *Evaporation*

It is assumed that the temperatures of the vapour and liquid are always equal. The rate of evaporation is then proportional to $P_0(T) - P(T)$ where $P_0(T)$ is the saturated vapour pressure. The change in gas density $\dot{n}_e$ due to evaporation will be written

$$\dot{n}_{e1} = A_1 \gamma (P_{01} - P_1) \tag{I.9}$$

where $\gamma$ is the effective evaporation rate (number sec$^{-1}$ dyne$^{-1}$ cm$^{-3}$) and $A_1$ the effective evaporation area. The change in temperature (neglecting the heat capacity of the vapour) is

$$\dot{T}_1 = -\frac{\mathcal{v}_1 L}{v_1 \rho T_1 C} \dot{n}_{e1} \tag{I.10}$$

where $\mathcal{v}_1$ is the vapour volume of reservoir 1 and $L$ is the latent heat per atom (erg).

Lastly, the pressure change is derived from the ideal gas equation, $\dot{P} = \dot{n}kT + nk\dot{T}$. There are two contributions to $\dot{n}$: gas flow and evaporation,

$$\begin{aligned} \dot{n}_1 &= \frac{\beta}{\mathcal{v}_1} \Delta P + \dot{n}_{e1} \\ \dot{n}_2 &= -\frac{\beta}{\mathcal{v}_2} \Delta P + \dot{n}_{e2} \end{aligned} \tag{I.11}$$

where $\beta$ is the effective gas flow conductivity (number.cm$^2$ sec$^{-1}$ dyne$^{-1}$)

For convenience, suppose $\mathcal{v}_1 = \mathcal{v}_2 = \mathcal{v}$, $v_1 = v_2 = V$, and $A_1 = A_2 = 2A$. The total temperature change, $\dot{T}_2 - \dot{T}_1$, from (I.6), (I.8) and (I.10) for both reservoirs becomes, to first order in $\Delta T$ and $\Delta P$,

$$\Delta\dot{T} = -\frac{2As}{VC}\dot{z} - \frac{2K}{T\rho VC}\Delta T - \frac{\upsilon AL\gamma}{T\rho VC}\left[\frac{\mathrm{d}P_0}{\mathrm{dT}}\Delta T - \Delta P\right]. \tag{I.12}$$

The relative pressure change is found from $P = nkT$ and (I.11)

$$\Delta\dot{P} = -\frac{2\beta kT}{\upsilon}\Delta P + A\gamma kT\left[\frac{\mathrm{d}P_0}{\mathrm{d}T}\Delta T - \Delta P\right] + \frac{P}{T}\Delta\dot{T}. \tag{I.13}$$

The behaviour of the system is now completely described by (I.4), (I.12), (I.13) plus the initial values of $z$, $\dot{z}$, $\Delta T$, $\Delta P$.

By inspection of the equations for $\Delta\dot{T}$ and $\Delta\dot{P}$ four different† relaxation times can be identified for the $\Delta T$, $\Delta P$ system alone; they are proportional to $\beta$, $K$, $\gamma$, and $\gamma L$. Had we not assumed equal temperatures for gas and liquid at each reservoir there would be a fifth. Equation (I.12) is coupled to (I.13) through the $L\gamma$ term which is equal to 0·0089 $(\upsilon/V)$ for $T = 1{\cdot}6$ K indicating effective pressure-temperature decoupling if the relative vapour volume is not large. The common assumptions that either $\Delta P = 0$ or $\Delta P/\Delta T = \mathrm{d}P_0/\mathrm{d}T$ correspond to the $\beta$ or $\gamma$ term, respectively, being dominant in (I.13).

It is impossible to assign exclusive credit for the theory of thermal damping; its concepts have been public domain since the discovery by J. F. Allen and J. Jones of the fountain pressure effect. Workers have found no difficulty in rederiving whatever aspect of the theory they needed. In 1951 Robinson[23] published a useful, systematic analysis of the solutions of (I.4) and (I.12) for the case of low velocity and $\Delta p/\Delta T = \mathrm{d}P_0/\mathrm{d}T$. The equivalent 3rd order equation admits periodic and aperiodic behaviour under either adiabatic or isothermal conditions which he delineated. After Manchester and Brown,[24] further tests of the theory received little attention until 1972.[18, 25] Since then Hallock and Flint[26] have reported observations of various features of the theory including the predicted temperature oscillations. The agreement with experiment is especially gratifying because there is very little uncertainty in the equations for thermal dissipation.

## 3. PROBLEMS AND OPPORTUNITIES

Intrinsic and thermal dissipation explain much, but not yet all, of small channel and saturated film flow phenomena. The question of film thinning may have an answer within the above theory and one type of observed anomalous dissipation definitely has (Section III). Other observations seem more challenging. Although they are probably related, four types can be recognized, of which only the last will be discussed.

† Different but not independent, due to the Onsager symmetry relation for kinetic coefficients.

(i) *Instability*

Switching of the flow among apparently preferred rates has been reported for film flow by Turkington and Harris-Lowe[27] and for flow through small holes by Banton.[28] In the former case the transitions were often between rates that differed as little as 1·5% while in the latter the transitions were relatively large, often between rates having an approximate integer ratio. It would be interesting to repeat the experiment for film surfaces having a very small perimeter to see if the transitions increase in size.

(ii) *Pressure-sensitive flow rates*

The intrinsic dissipation theory predicts a limiting velocity due to a pressure head $z$,

$$v_c = v_B \Big/ \ln \frac{l\mathscr{A}\kappa v}{gz}. \tag{I.14}$$

Although this is clearly $z$-dependent it is much weaker than that observed by Hulin *et al.*[29] and Banton[28] for flow through micron-size holes, viz.

$$v_c \propto \sqrt{z}. \tag{I.15}$$

(iii) *Boundary-sensitive velocities*

The intrinsic dissipation theory does not, in its simplest form, distinguish between rough and smooth boundaries. Hess[30] has found the limiting velocity can be greatly reduced by rough boundaries and has suggested an explanation based on localized, high velocity fields present at such boundaries. The intrinsic dissipation theory should also predict more vortex production from such regions.

(iv) *High velocities in filtered flow*

Hess[31] discovered that inserting a superleak near a micron-size hole resulted in flow velocities of several hundred cm sec$^{-1}$ under modest pressures (35 μm of He). Such velocities are without precedent except for those implicit in Liebenberg's[21] saturation experiment where the driving force was greater by $10^5$. Perrin *et al.*[32] showed that these high velocities also occurred in the absence of any dissipative driving force. The tendency of fast flow to follow the slow flow mode of item (ii) above was observed by Hulin *et al.*[29] They also found the slow flow mode to be preferred at lower temperature.

It appears that the superleak filter severely inhibits the growth of vortex rings beyond the size characteristic of $v_B$. Moving the filter farther from the orifice or making its pore size greater should continuously reduce this effect, causing the measured velocity to continuously return to its unfiltered value. If true, then there is no clear distinction between "intrinsic" and "extrinsic" velocities, the latter designation often used for the slower velocities.[31] The fact that the magnitude of the fast velocity is not unique is clear from the data of Hulin *et al.*[29] which shows the velocity from the filtered side of the orifice is 25% larger than that from the unfiltered side. In no case should the fast velocity exceed $v_B$ and, indeed, the values reported so far are approximately half the $v_B(T)$. (Hereafter, all empirical values of $v$ and $v_B$ will refer to ref. 18.) The temperature dependence of the fast velocities is intermediate between that for $v_B(T)$ and the slow velocities, as expected.

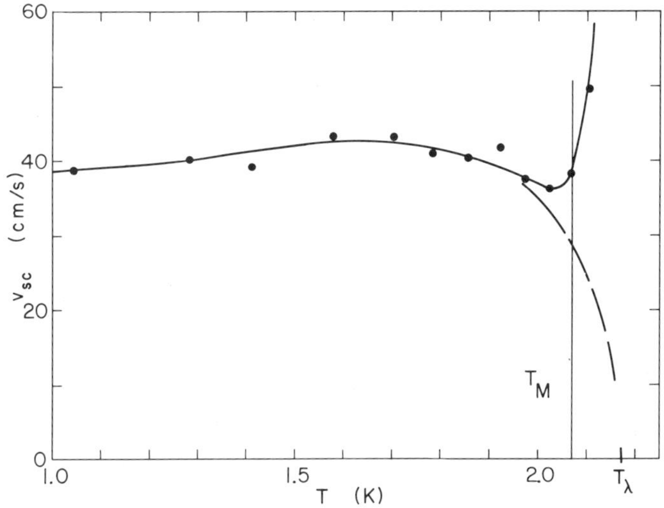


FIG. 4. Predicted limiting velocity for channel 10 μm diam., 10 μm long, with 35 μm He pressure head. (See ref. 31a.) The curves are qualitative. For $T > T_M$ the prediction is invalidated by size effects.

The velocity predicted from (I.14) for Hess's experiment[31a] is shown in Fig. 4 as a function of temperature. For $T < 2$ K it agrees reasonably with the unfiltered results implying the vortex growth conditions (i.e. $v(T)$) in the film and the unfiltered orifice are about the same. The predicted $v_c$ diverges above 2·1 K because in that temperature range the critical vortex radius corresponding to $v_B$ exceeds the film thickness (in ref. 18), resulting in effective $v$ and $v_B$ that are inappropriate for the case of a much larger flow area and smaller dissipative volume. This phenomenon arose earlier in the discussion of Fig. 3.

In Fig. 5 is shown the prediction of (I.14), with $z = l$, for $v_c d$ versus $d$ compared to the $v_c = d^{-1/4}$ rule.[33] These two figures illustrate that the intrinsic

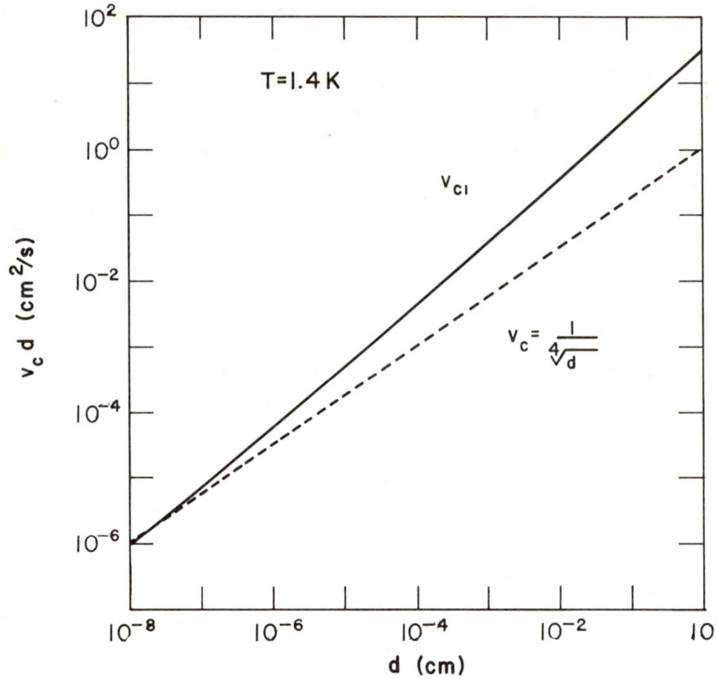


FIG. 5. Comparison of $v_s d$ from intrinsic dissipation formula and $v_s = d^{-1/4}$ rule. The former is not valid for $d < 10^{-7}$ cm where size effects become important.

dissipation theory can predict extrinsic type flow behaviour with respect to both temperature and channel size dependence. Conversely, dynamic response of extrinsic flow can reveal intrinsic processes. In spite of the common belief that the fast velocity represents an intrinsic process, it has not, in this way, been tested against (I.1) although suggestive data exists.[31b]

In the following sections the phenomenological form of the intrinsic dissipation given by (I.1) will often be used to represent the limiting-velocity feature of film flow. This should be a reliable description of film behaviour within the dynamical range of forces that characterized the original film flow experiments from which $\nu$ and $v_B$ were determined. In such application (I.1) is simply a convenient formula to represent a measurable phenomenon quite independent of any theoretical underpinning.

## II. Beaker Film Flow

Albert Einstein once said that physics should be made as simple as possible but not one bit simpler. It is in this spirit that the properties of beaker film flow will be discussed, beaker film flow being steady state flow over an arbitrary film path between reservoirs of different chemical potentials.

### 1. THEORY

The starting point is the local, but averaged, superfluid equation of motion, (I.2).

It will be assumed that beaker flow is a steady-state process, i.e. determined far more by the instantaneous level difference and rim height than by the history of the run-in. This is justified except in the case of very small level differences *and* a very small reservoir area, the typical arrangement for oscillation experiments for which (I.4) is appropriate. The simplicity gained by neglecting time dependence in (I.2) applied to beaker flow is lost by the new necessity to solve that equation at each point on the flow path between the bulk reservoirs whose large level difference precludes foreknowledge of the film thickness profile. The temperature and pressure will be assumed constant over the flow path. Of course, the potential energy of the higher liquid level is transformed into heat at the region of intrinsic dissipation along the flow path. However, due to the high latent heat the rate of liquid loss from evaporation (even assuming no heat flow to the substrate) is quite small compared to the mass flux in the film. That experiments can be designed in which the factors neglected above become important is possible but irrelevant to the essence of beaker flow.

With the steady-state assumption $\dot{v}_s = 0$, the integral of (I.2) along a path connecting points at a reservoir surface $y_1$ and in the film $y$ gives

$$\frac{1}{2}\frac{\rho_s}{\rho} v_s^2(y) - \frac{\alpha}{\delta^n(y)} + gz(y) + \int_{y_1}^{y} c\mathscr{A}(l)\kappa\nu \exp\left(-\frac{v_B}{|v_s(l)|}\right) \mathrm{d}l = 0 \qquad \text{(II.1)}$$

where $c$ is the cosine between $\mathbf{v}_s(l)$ and $\mathrm{d}\mathbf{l}$. It is conceptually simpler to take the film point $y$ at the surface rather than in the interior. This is not a new approximation. The total chemical potential and $v_s$ are constant across the film so for an interior point a distance $x$ from the substrate,

$$\mu_0(P(x), T) - \frac{\alpha}{\delta(y,x)^n} = \mu_0(P, T) = \frac{\alpha}{\delta(y)^n} \tag{II.2}$$

where $P(x)$ is the local pressure at $x$ and $\delta(y, x)$ some function representing the van der Waals potential energy at $x$ which, in general, is not just equal to $x$ except when $x$ is at the surface. To be very proper, perhaps the film point $y$ should be a coherence length below the surface to avoid quantum effects which presumably do not invalidate the assumption that the total chemical potential is constant across the film.

It is useful to write (II.1) in terms of the flow rate $M \equiv v_s(y)\delta(y)p(y)$ where $p(y)$ is the local perimeter of the beaker. This gives

$$\frac{1}{2}\frac{\rho_s}{\rho}\left(\frac{M}{\delta(y)p(y)}\right)^2 - \frac{\alpha}{\delta^n(y)} + gz(y) + \int_{y_1}^{y} cp(l)\delta(l)\kappa v \exp\left(-\frac{v_B\delta(l)p(l)}{M}\right) \mathrm{d}l = 0 \tag{II.3}$$

which is to be solved for $\delta(y)$ given $M$ and the known functions specific to the experiment, $p(y)$ and $z(y)$. Since the surface tension has not been included the film thickness $\delta(y)$ will not be accurately predicted by (II.3) for heights $z(y)$ like the capillary constant ($\approx 0{\cdot}07$ cm) above the bulk liquid. In special cases the effect of the surface tension at other points in the flow path may be detectable[34] and could be included in (II.3).

## 2. CONSEQUENCES, IMMEDIATE

For convenience in discussion the terms in (II.3) will be labelled, from left to right, $K$, $V$, $G$, $D$ for kinetic, van der Waals, gravitational, and dissipative components. For $D = 0$, which will occur for small $M$ or $y$ near $y_1$, the usual relation between $\delta$, $v_s$, and $z$ is obtained.[35] For $M$ sufficiently large to cause intrinsic dissipation along the path, $D(y)$ will be a monotonically increasing positive or decreasing negative function depending on whether $\mathbf{v}_s \cdot \mathbf{dl} > 0$ or $< 0$. In the former case it is clear that the height $z(y_2)$ of the second reservoir must be below $z(y_1) \equiv 0$ by the amount

$$z(y_2) = -D(y_2)/g.$$

At this height, $z(y_2)$, $K + V = 0$, implying $\delta \to \infty$ which is the definition of reservoir. Likewise for $\mathbf{v}_s \cdot \mathbf{dl} < 0$, it is easy to see that $z(y_2)$ must be above $z(y_1)$.

Another feature of (II.3) is obvious by inspection: if $D > 0$, then over the region between $y_1$ and the beaker rim, the film is thinner than it would be if $D = 0$ at the same $v_s$. In this case, $D$ adds to the effective height of the film. If $D < 0$ it subtracts from the effective height of the film. Therefore, *dissipation thins a rising film and thickens a falling film.*

The case of dissipation in the falling film is interesting because it explains the common observation[36] that inflow rates are insensitive to $L_O - L_I$, if $L_O$ is not too large. (See Fig. 6 for definitions of $L_O$, $L_I$.) If the change in $G$

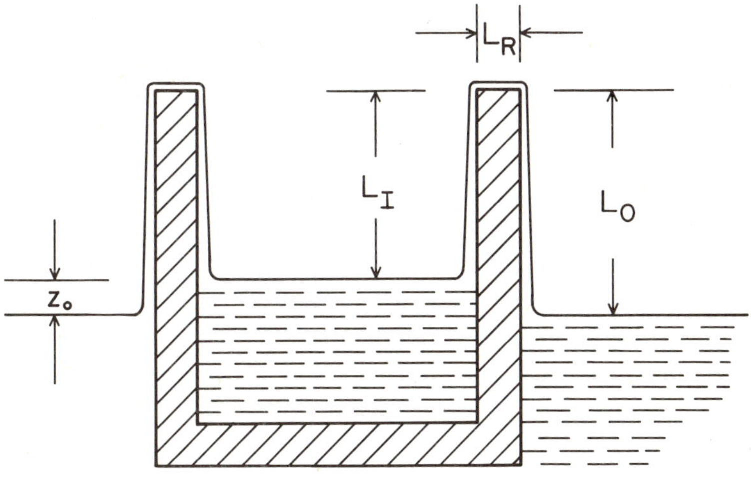


FIG. 6. Beaker film flow.

over a path element d$y$ is exactly cancelled by the change in $D$ then (II.3) will be satisfied by $\delta(y) =$ const. until $p(y)$ changes. That is, when the local energy lost due to intrinsic dissipation exactly cancels the energy gained by falling in the gravitational field, then energy balance (II.3) can be satisfied without a change in film thickness. This predicts that the inflow rate will be totally independent of $L_I$ once conditions are such that the integrand of $D$ is equal to $-g \cos \theta$ where $\theta$ is the angle of the path from vertical. Of course, the same effect can occur for falling films on the outside of the beaker, i.e. outflow, but it is not as easy to observe because (i) there is much less dissipation on the outside wall making the requisite condition difficult to achieve and (ii) it is awkward to observe rates for a fixed high inner level and a variable low outer level.

Note that (II.3) also predicts the formation of bulk liquid due to a contraction of perimeter which is the case of a beaker or parts of it suspended above the bath. Consider outflow with $y_1$ at the inside surface so $D$ is positive.

Upon pursuing the solution of (II.3) to the outside the end of the beaker is indicated by $p(y) \to 0$ which rapidly increases $K$ and precipitates $\delta \to \infty$ for sufficiently large $M$.

Although the rate and one level do not always uniquely determine the second level, it appears that two levels always determine a unique rate. It can be prove that if $\delta(y, y_1)$ is the solution starting from $y_1$ and ending at $y_2$ then $\delta(y_2 - y, y_1)$ is the solution starting from $y_2$ and ending at $y_1$.

### 3. CONSEQUENCES, NUMERICAL

Since self-consistent predictions of various film parameters along a flow path have not been shown before, it is appropriate to do so here. Besides illustrating some of the previous remarks it will offer insight into more subtle features of (II.3).

The calculations are for a cylindrical beaker 2 cm o.d. with 0·2 cm thick walls ($L_R$). The dissipation parameters, $\nu = 5{\cdot}14\ 10^{22}\ \text{cm}^{-3}\ \text{sec}^{-1}$ and $v_B = 1260\ \text{cm sec}^{-1}$ correspond to a temperature of roughly 1·7 K. The van der Waals exponent $n$ is taken to be 4 and the thickness constant $\alpha$ corresponds to a static film thickness of 300 Å at 1 cm height.

Figures 7 and 8 illustrate outflow for large and small level differences. Figure 9 shows inflow for large level difference only since the case of small difference is similar to Fig. 8, only reversed. The parts of the figure groups show, as functions of path length (cm),

(a) $K$, $V$, $G$, $D$ terms of (II.3) ($\text{cm}^2\ \text{sec}^{-2}$),
(b) static ($a/z^{0{\cdot}25}$) and kinetic film thicknesses (Å), $a \equiv 3 \times 10^{-6}$,
(c) local dissipation ($\text{cm sec}^{-2}$), and
(d) film velocity ($\text{cm sec}^{-1}$).

In each case the path is measured from the *lowest* level whether it be inside or outside; the distance from the rim, $L_I$ or $L_O$, of this lowest level is always 5 cm.

In all three cases the kinetic energy $K$ is of minor importance compared to $V$, $G$, and $D$. The $b$ figures support the maxim that dissipation thickens (thins) falling (rising) films. In spite of the thinning effect of $v_s$, the flowing film is predicted to be *thicker* than the static film in Fig. 7(b) (near the rim) and in Fig. 9(b) (over a distance of 2·5 cm below the inside rim). Even for relatively small dissipation the falling film is still observably thickened relative to the rising film as shown in Fig. 8(b). In all cases the dissipation is strong at the inside rim and extends toward the lowest level a little for outflow, Fig. 7(c), and a lot for inflow, Fig. 9(c).

The dissipation profile can be measured by potential probes[37] which are sensitive to the local chemical potential, $K + V + G$. The qualitative features of the dissipation profile inferred from such measurements agree with the above predictions.

There is also evidence for the predicted film thickness change in the dissipative region which will be discussed in Section IV.

The theory of film flow given here is the absolute minimum required to consistently relate level differences and geometry with flow rate and film profile. Even so, it successfully explains a great deal of film flow phenomena and so would seem to be a logical starting point for more detailed theories.

The application of this theory to flow rate dependence on temperature for $n = 3$ has been previously made.[(38)]

## III. Flow in Multiply Connected Geometries

Multiply connected geometries have long been employed for studying superfluid flow under quite diverse conditions except, conspicuously, the saturated film. Indeed, at this writing, observation of persistent current lying *entirely* within the saturated film has yet to be reported.† By detailing here the quantitative behaviour of parallel-path flow *I* hope to increase its usefulness both as a specific experimental tool and as a means of recognizing the cause of otherwise mysterious phenomena that can occur when persistent circulation is unknowingly generated in an experiment.

The specific subject to be considered is gravitational flow between bulk fluid reservoirs joined by parallel, superleak flow paths that are assumed to be sufficiently macroscopic that quantum effects are negligible. It is further assumed that the dissipation region of the flow path is well-defined and the level difference between the reservoirs is small so that, in the case of film flow paths, it is not necessary to perform a numerical integration between reservoirs, as in Section II, to find the instantaneous flow rate. A last assumption, applicable only to films, is that the kinetic energy at any point in the flow path is small compared to gravitational potential energy with respect to the bulk reservoirs. This avoids film thinning effects which will be discussed in Section IV.

### 1. Basic Equations

Consider two bulk liquid reservoirs at points 1 and 2 joined by a single path from 1 to 3 and by $N$ parallel paths from 3 to 2. Let the area and effective

† Added in proof. Such currents have now been seen: R. K. Galkiewicz and R. B. Hallock. *Phys. Rev. Lett.* **33**, 1073 (1974); H. J. Verbeek, E. van Spronsen, H. van Beelen, R. de Bruyn Ouboter and K. W. Taconis. *Physica* (to be published); M. Banton (to be published).

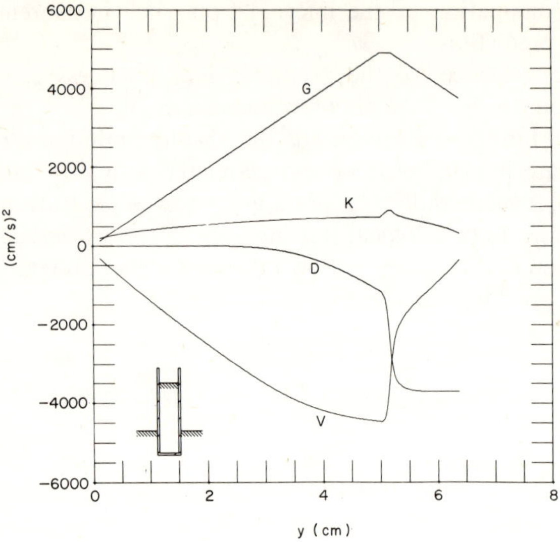
6000
4000
2000
0
−2000
−4000
−6000
(cm/s)$^2$
G
K
D
V
0
2
4
6
8
y (cm)

(a)

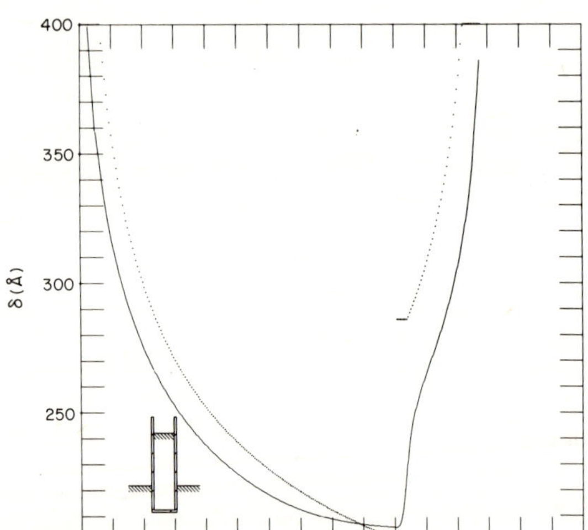
400
350
300
250
200
δ(Å)
0
2
4
6
8
y (cm)

(b)

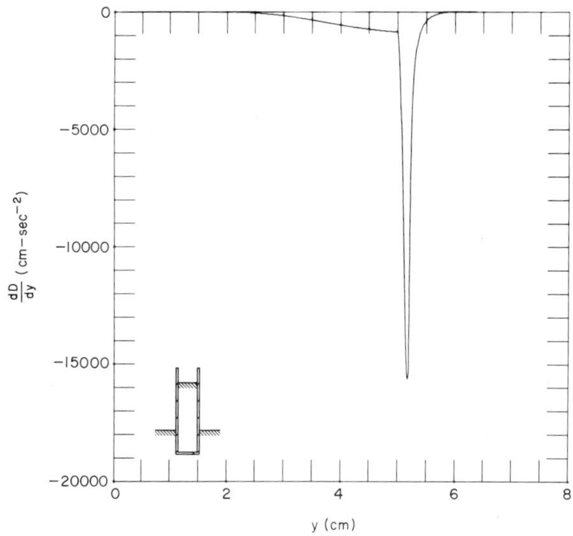


(c)

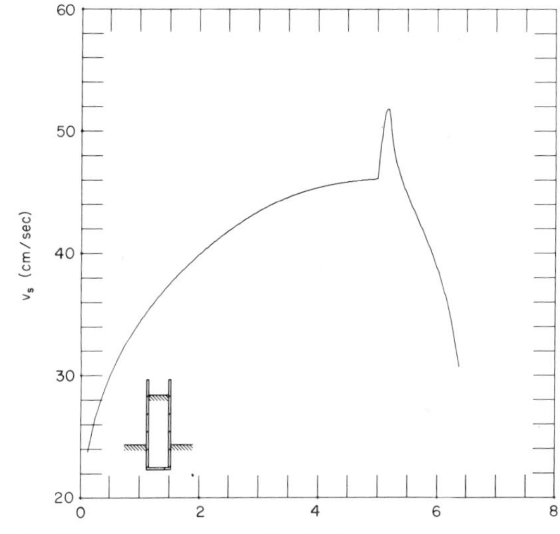


(d)

FIG. 7. Film properties for outflow. $L_I = 1{\cdot}22$ cm, $L_0 = 5$ cm, the relative rate, normalized to the smallest perimeter is $R = \rho_s M(\rho p)^{-1} = 8 \times 10^{-5}$ cm$^2$/sec. The origin of the abscissa is the lowest liquid level.

(a) Kinetic, Van der Waals, gravitational, and dissipative terms of Eqn (II.3), respectively *K*, *V*, *G*, *D* (erg/gm).

(b) Film thickness profile. Static: dotted line. Kinetic: solid line (Å).

(c) Dissipation profile (erg/gm-cm).

(d) Velocity profile (cm/sec).

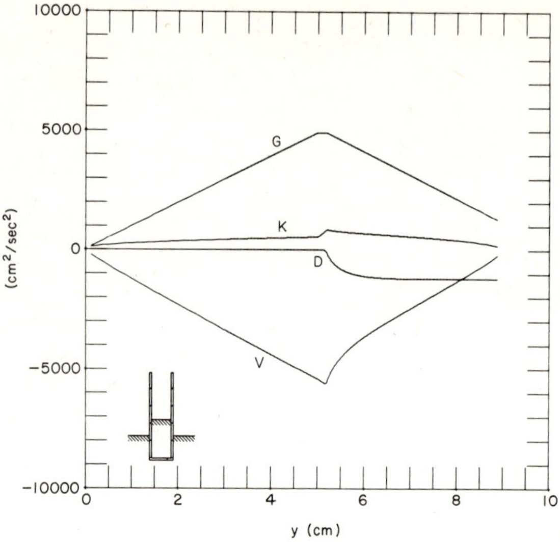
10000
5000
0
−5000
−10000
(cm²/sec²)
G
K
D
V
0
2
4
6
8
10
y (cm)

(a)

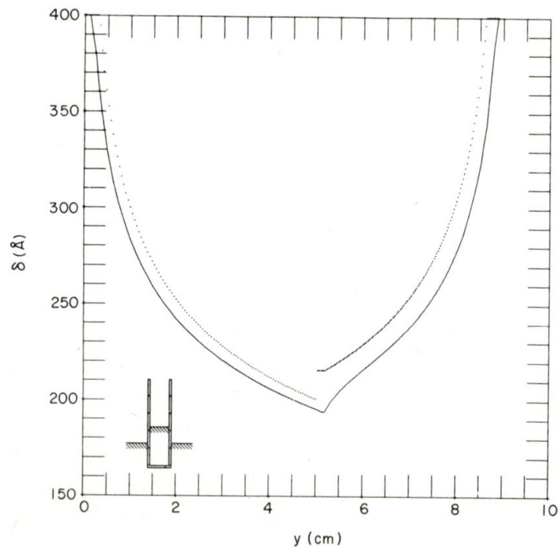
400
350
300
250
200
150
δ (Å)
0
2
4
6
8
10
y (cm)

(b)

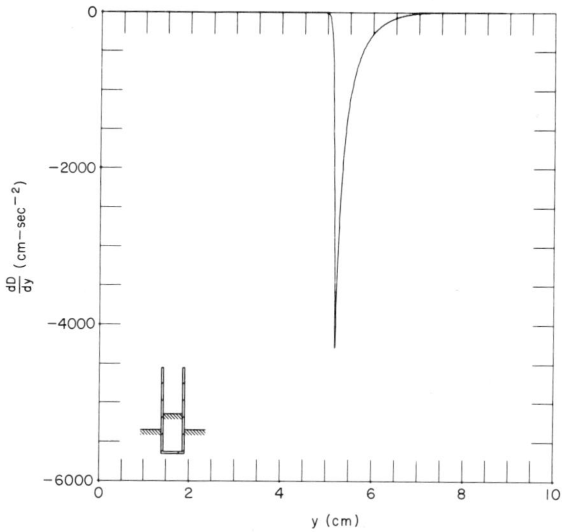


(c)

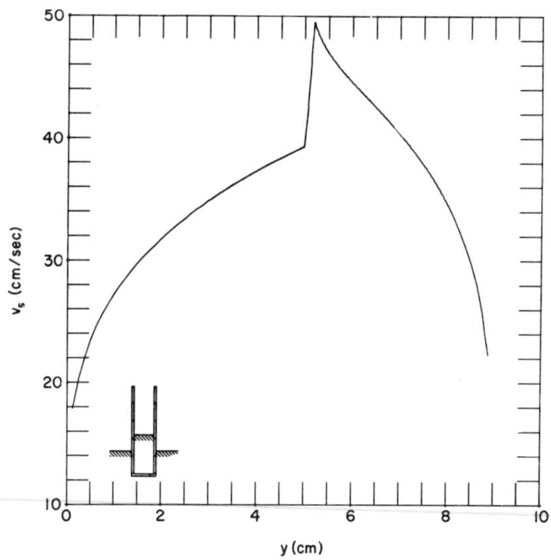


(d)

FIG. 8. Film properties for outflow. $L_I = 3{\cdot}75$ cm, $L_0 = 5$ cm, $R = 6{\cdot}5 \times 10^{-5}$ cm$^2$/sec. See Fig. 7.

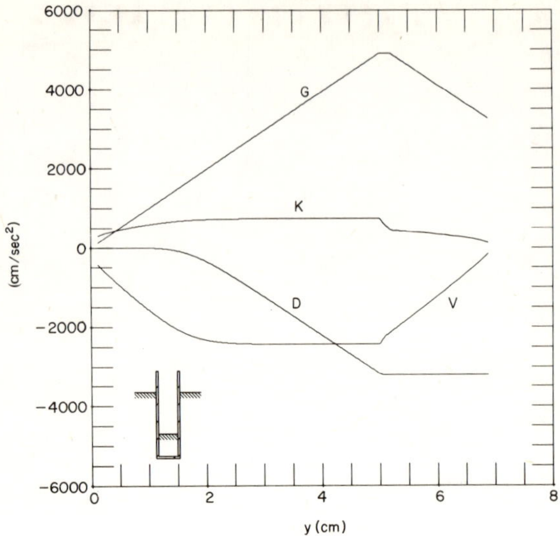

6000
4000
2000
0
−2000
−4000
−6000
(cm/sec$^2$)
G
K
D
V
0
2
4
6
8
y (cm)


(a)

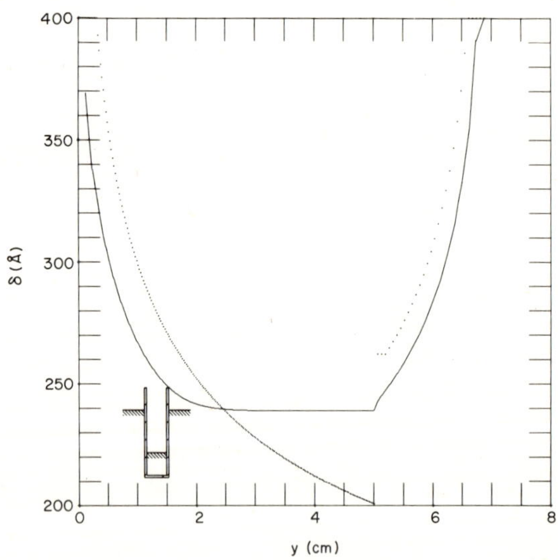

400
350
300
250
200
δ (Å)
0
2
4
6
8
y (cm)


(b)

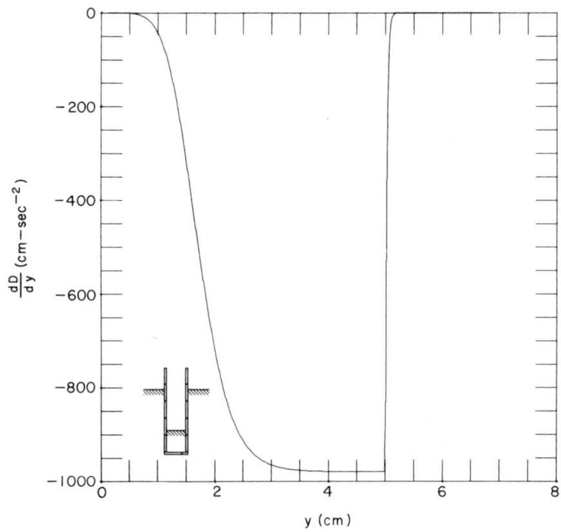


(c)

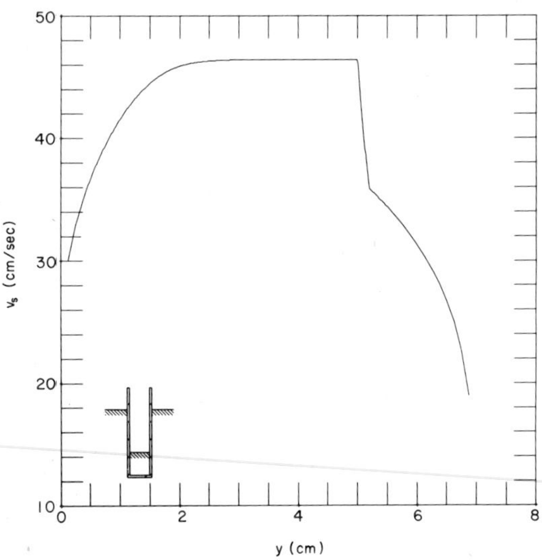


(d)

FIG. 9. Film properties for inflow. $L_0 = 1{\cdot}72$ cm, $L_I = 5$ cm, $R = 7{\cdot}5 \times 10^{-5}$ cm$^2$/sec. See Fig. 7.

length of the single path be $\mathscr{A}_0$, $\ell_0$ and those of the parallel paths be $\mathscr{A}_j$, $\ell_j$ ($j = 1 - N$), respectively. The previous assumptions allow the equation of motion for the superfluid to be integrated from 1 to 2 for each of the $N$ different paths,

$$\int_{1(j)}^{2} \dot{\mathbf{v}} \cdot d\mathbf{l} + \mu\Big|_1^2 = -\int_{1(j)}^{2} s(v)\mathscr{A}\kappa\nu \exp(-v_B/|v|)\, dl, \qquad s(v) \equiv \frac{v}{|v|}. \tag{III.1}$$

Defining the effective length $\ell_j$ and velocity $v_j$ of a path to be

$$\ell_j = \int_{(j)} [\mathscr{A}_j/\mathscr{A}(l)]\, dl$$

$$v_j\mathscr{A}_j = v(l)\mathscr{A}(l),\ l \text{ any point on path } j,$$

and evaluating the chemical potential at the reservoirs, (III.1) becomes

$$\dot{v}_0\ell_0 + \dot{v}_j\ell_j + gz - s\Delta T = -D_0\ell_0 - D_j\ell_j \tag{III.2}$$

where $z$ is the level difference and $\Delta T$ the temperature difference between reservoirs (the fluid kinetic energy in the reservoirs is ignored). Positive flow velocity is chosen to correspond to increasing $z$. The definititions of the dissipation functions $D_j$ above are obvious by comparing (III.1) and (III.2). As discussed in Section 1, $\Delta T$ and $z$ satisfy an additional equation originating from the thermal coupling between the reservoirs; for present purposes it is sufficient to take $-s\Delta T = q\dot{z}$. The conservation of fluid completes the equations,

$$\rho A\dot{z} = \rho_s \sum_1^N \mathscr{A}_j v_j = \rho_s \mathscr{A}_0 v_0 \tag{III.3}$$

where $A$ is related to the reservoir areas $A_1$, $A_2$ by

$$A = A_1A_2/(A_1 + A_2).$$

Summing (III.2) over all the paths and using (III.3) gives the equation for $z$,

$$\ddot{z} + 2\xi\dot{z} + \omega^2 z = -D \tag{III.4}$$

where

$$\xi = \frac{\rho_s q}{2\rho AF} \sum_1^N \frac{\mathscr{A}_j}{\ell_j},$$

$$\omega^2 = \frac{\rho_s g}{\rho AF} \sum_1^N \frac{\mathscr{A}_j}{\ell_j},$$

$$D = \frac{\rho_s}{\rho AF}\left(D_0\ell_0 \sum_1^N \frac{\mathscr{A}_j}{\ell_j} + \sum_1^N D_j\mathscr{A}_j\right),$$

$$F = 1 + \frac{\ell_0}{\mathscr{A}_0} \sum_1^N \frac{\mathscr{A}_j}{\ell_j}.$$

If $\dot{v}_i = 0$ for some $i = 1, \ldots, N$ then the form of (III.4) remains the same but the $i$th term is omitted from every summation over paths.

If $\mathscr{A}_0$ is sufficiently large, say greater than $\Sigma_1^N \mathscr{A}_i$, the limiting velocity will not be reached in the common path and $D_0$ will be effectively zero. For very large $\mathscr{A}_0$ or small $\ell_0$ the intertial effect of the common path becomes negligible, i.e. $F \approx 1$.

The equation for the circulation $c_{ij}$ around the $i$ and $j$ paths is simply found from (III.2),

$$\dot{c}_{ij} \equiv \dot{v}_i \ell_i - \dot{v}_j \ell_j = -D_i \ell_i + D_j \ell_j. \tag{III.5}$$

This shows that the necessary condition for circulation change is dissipation in the paths (which is obvious from the original superfluid equation of motion).

The total mechanical energy of the system is given by

$$E = \tfrac{1}{2}\rho_s \sum_0^N v_j^2 \mathscr{A}_j \ell_j + \tfrac{1}{2} g\rho A z^2.$$

From the previous equations the time derivative of $E$ can be written as

$$\dot{E} = -\rho_s \sum_0^N v_j D_j \mathscr{A}_j \ell_j - \rho_s q \mathscr{A}_0 v_0 \dot{z}. \tag{III.6}$$

Note that every term in the above expression is negative definite.

## 2. PHENOMENA

In spite of the fact that the preceding equations are highly nonlinear it is possible to gain some insight into their solutions by taking advantage of the all-or-nothing property of the intrinsic dissipation: to good approximation, intrinsic dissipation is either negligible or dominant in determining the flow behaviour. This suggests the approximation $D_j = 0$ for $|v_j| < v_{jc}$ and $\dot{v}_j = 0$ for $|v_j| = v_{jc}$ where $v_{jc}$ is a limiting velocity appropriate to the parameters of the $j$th path and the driving force. The equations can then be solved in closed form in a piecewise manner over time periods separating instants when some $|v_j|$ attains or retreats from $v_{jc}$. Continuity in $z$, $\dot{z}$, and the other $v_i$ joins the solutions. For convenience, thermal damping will be assumed absent in much of the following discussion.

Let the number of paths be restricted to two with $\ell_1 \leqslant \ell_2$ and let $\ell_0$ (and $D_0$) be zero. In the approximate calculations, the limiting velocities will at first be assumed equal, $v_{1c}/v_{2c} \equiv \zeta = 1$, but the results will later be generalized to $\zeta \neq 1$ which, in fact, is of some importance.

Consider a typical experiment. Starting with a given flow in the paths a displacement of the reservoir levels of $z_0 > 0$ causes the fluid to accelerate (negatively) in each path until the limiting velocities are reached. This run-in period persists until $z = 0$ when the driving force on the flow reverses as $z$ is driven negative due to the flow momentum. The level $z$ will continue to oscillate about the equilibrium value with diminishing amplitude until sufficient energy is lost to preclude intrinsic dissipation after which further decay will depend on whatever thermal damping is present.

Using the approximation described above the equation for $z$ is

$$\ddot{z} + \omega_a^2 z = 0 \tag{III.7}$$

where

$$\frac{\rho A}{\rho_s g}\omega_a^2 = \frac{\mathscr{A}_1}{\ell_1} + \frac{\mathscr{A}_2}{\ell_2}, \frac{\mathscr{A}_1}{\ell_1}, \frac{\mathscr{A}_2}{\ell_2}, \text{ or } 0$$

corresponding respectively to neither path, path 2, path 1, and both paths having reached their limiting velocity. Since flow in the shorter path reaches its limiting value first it is useful to calculate the velocity $v_2(j)$ at the end of the $j$th half-cycle (when $z = 0$) given the initial velocities $v_2(j-1)$ and $v_1(j-1) = \pm v_c$. (According to our convention the sign of $v_1(j)$ is opposite that of $z$ during the $j$th half-cycle.) The result, which requires several minutes of algebra ,is

$$(-1)^j \eta_j = -a + \sqrt{[4a\ell + (a - (-1)^{j-1}\eta_{j-1})^2]} \tag{III.8}$$

where

$$\eta_j = -v_2(j)/v_c, \qquad a = \mathscr{A}_1/\mathscr{A}_2, \qquad \ell = \ell_1/\ell_2,$$

and it is assumed that odd numbered half-cycles correspond to negative $z$. At the end of a sufficiently long run-in starting from $z_0 > 0$ the velocity in the long leg will be $-v_c$ so $\eta_0 = 1$. Iteration of (III.8) then gives $v_2$ at the end of any subsequent half-cycle until the $m$th half-cycle defined by $(-1)^m \eta_m < \ell$ after which $\eta_j = (-1)^{j-m}\eta_m$, $j \geqslant m$. The above condition is derived by requiring that $v_1$ not reach its limiting value during the following half-cycle. For $j < m$, $(-1)^j \eta_j$ is always possitive.

The circulation is defined by $c = v_1\ell_1 - v_2\ell_2$ so at the end of the $j$th half-cycle it is given by

$$\begin{aligned} c_j &= v_c \ell_2[\eta_j - (-1)^j \ell], \qquad && j \leqslant m \\ c_j &= c_m, && j \geqslant m \end{aligned} \tag{III.9}$$

The circulation attains its final value $c_m$ at the $m$th half-cycle (defined above) which corresponds to a certain persistent current $v_{2p}$ through the long path, $v_{2p} = -c_m a/\ell_2(a + \ell)$. Expressing this as $J_r \equiv -v_{2p}/v_c$ and using (III.9) the result is,

$$J_r = \frac{a}{a + \ell}[\eta_m - (-1)^m \ell]. \tag{III.10}$$

The quantity $J_r$ was introduced by Van Alphen, de Bruyn Ouboter, Taconis, and van Spronsen[39] to describe the ability of a doubly-connected system to trap circulation. They calculated $J_r$ for the special case that the final circulation is just that generated during the run-in, viz. $c_0 = v_c(\ell_2 - \ell_1)$, which corresponds to $m = 0$ in (III.10): $J_r = a(1 - \ell)/(a + \ell)$. (For $m = 0$ only, the requirement that $|v_1| < v_c$ is not implicitly contained in (III.10). Imposing this gives $v_c \leqslant v_1 = c_0/\ell_2(a + \ell)$ or $J_r = a$ for $a < 1 - 2\ell$.)

The behaviour of $J_r$ given by (III.10) is shown in Fig. 10 as a function of

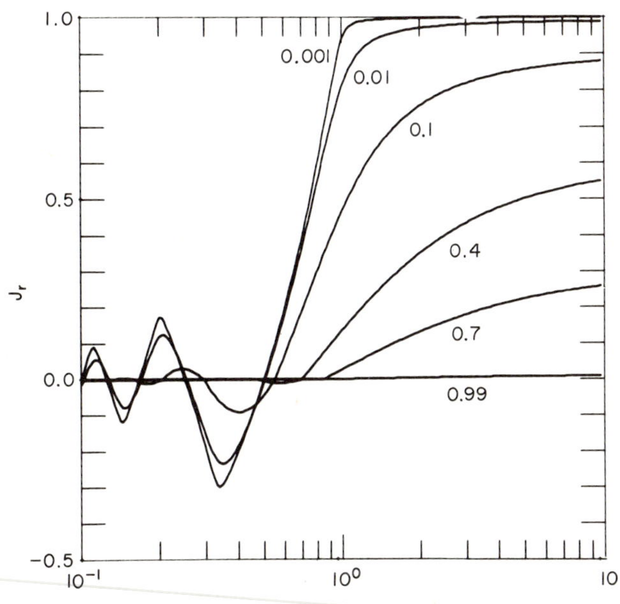


FIG. 10. $J_r$ versus $\zeta a$ for (in descending order of the curves) $\zeta\ell$ = 0·001, 0·01, 0·1, 0·4, 0·7, 0·99, where $a \equiv \mathscr{A}_1/\mathscr{A}_2$, $\ell \equiv \ell_1/\ell_2$, $\zeta \equiv v_{1c}/v_{2c}$.

$a$ for various values of $\ell$. In the limit $\ell \to 0$, $J_r$ becomes a saw-tooth function with zeros at $a = 1/2n$ and extrema at $a = 1/(2n - 1)$ which have the respective values $J_{r,\mathrm{ext}} = (-1)^{n+1}/(2n - 1)$, $n = 1, 2, 3, \ldots$ This limiting behaviour is already evident in the curve for $l = 0{\cdot}001$ in Fig. 10.

Before the above approximate results can be tested against exact numerical solutions of the non-linear differential equations it is necessary to take explicit account of the different limiting velocities. By definition, $v_{jc}$ is the velocity for which $\dot{v}_j = 0$, so from (III.2),

$$\frac{1}{v_{2c}} - \frac{1}{v_{1c}} \cong -\frac{1}{v_B} \ln(a\ell), \tag{III.11}$$

assuming the dissipation is uniform over lengths $\ell_1$ and $\ell_2$. For typical cases, $v_{1c}$ can easily exceed $v_{2c}$ by 15%. The effect this has on the preceding is simple to state: if $\zeta = v_{1c}/v_{2c}$ and $v_c \equiv v_{2c}$ then (III.8), (III.9), (III.10) and everything between and following remain valid if $\zeta a$ and $\zeta\ell$ are substituted for $a$ and $\ell$.

Figure 11 compares $J_r$ for $\ell = 0{\cdot}01$ obtained by exact solution with

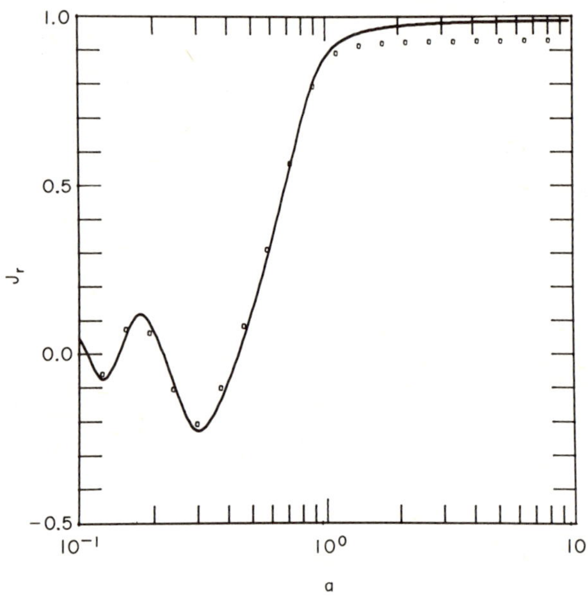


FIG. 11. $J_r$ versus $a$ for $\ell = 0{\cdot}01$. The points are numerical results.

(III.10) including the $\zeta$ factor, which itself is a function of $a$ and $\ell$. (Comparing with the corresponding curve in Fig. 10 shows the effect of $\zeta$.) The discrepancy between the exact and approximate solutions when $a > 1$ illustrates the limitation of replacing the smoothly increasing dissipative force by an abrupt barrier at some fixed velocity $v_c$. In the exact calculation there is some arbitrariness in the choice of the $v_c$ in the definition, $J_r \equiv -v_{2p}/v_c$.

In Fig. 11 the magnitude of $v_2$ at the beginning of the first half-cycle was used for $v_c$.

The effects of thermal damping $\xi$ should be mentioned. If the damping is weak it is of little importance when $m$ (the number of dissipative half-cycles) is small. However, even small thermal damping imposes on $m$ an upper bound defined by $v_2(m)/v_c = \exp(-\xi\pi m/\omega_2) \leqslant \ell$. If the level oscillations are critically damped, i.e. $\xi \gtrsim \omega$ in (III.4), then most of the circulation generated during the run-in will be preserved. This condition was realized in the jewellers rouge experiments of Van Alphen *et al.*[(39)] who also observed and explained the dependence of $\xi$ and $\omega$ on the total number of parallel flow paths (which has been slightly generalized here to be the number of paths for which $\dot{v}_j \neq 0$). The results of exact calculations of $J_r$ for the length ratio $\ell = 0{\cdot}15$ and critical thermal damping $\xi = \omega$ are shown in Fig. 12 together with the experimental result of Van Alphen *et al.* (for the same $\ell$)

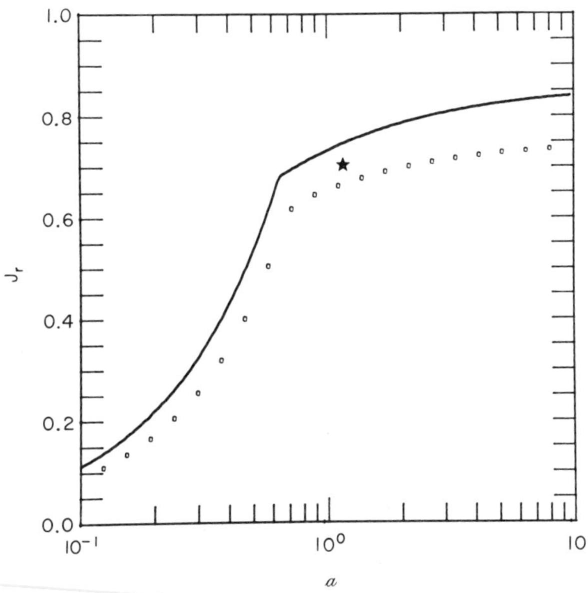


FIG. 12. $J_r$ versus $a$ for $\ell = 0{\cdot}15$ and critical thermal damping. The points represent the numerical lower bound. The star is the experimental value of Van Alphen *et al.* [(39)]

and the prediction of (III.10) for $m = 0$ (including $\zeta$). For this case the magnitude $v_c$ used to normalize the exact $v_{2p}$ was the *maximum* value of $v_2$ during the run-in which yields a lower bound on $J_r$. In the numerical calculations the different behaviour of $J_r$ for $\zeta a < 1 - 2\zeta\ell$ is associated with a half-cycle oscillation in $z(t)$ in spite of nominal critical damping.

So far little has been said about the effect of circulation on the appearance of the level oscillations. This can be striking as shown in Fig. 13 where $z(t)$ is plotted for various area ratios $a$ between 0·05 and 2·5 and $\ell = 1/80$. In all but two of the curves the final circulation is attained during the first

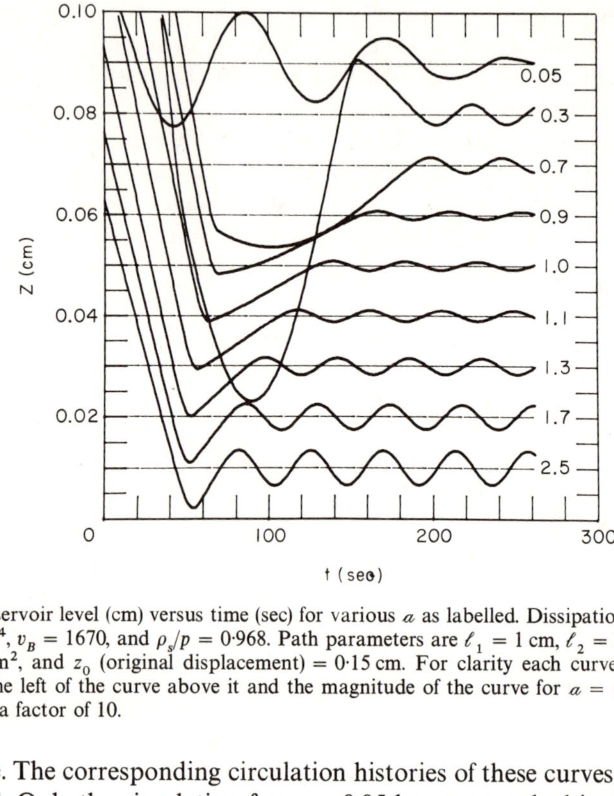


FIG. 13. Reservoir level (cm) versus time (sec) for various $a$ as labelled. Dissipation parameters are $\nu = 10^{24}$, $v_B = 1670$, and $\rho_s/p = 0{\cdot}968$. Path parameters are $\ell_1 = 1$ cm, $\ell_2 = 80$ cm, $\mathscr{A}_1 = 1{\cdot}31\ 10^{-6}$ cm$^2$, and $z_0$ (original displacement) = 0·15 cm. For clarity each curve is displaced 20 sec. to the left of the curve above it and the magnitude of the curve for $a$ = 0·05 has been reduced by a factor of 10.

half-cycle. The corresponding circulation histories of these curves are shown in Fig. 14. Only the circulation for $a$ = 0·05 has not reached its final value. (These plots and all succeeding ones represent exact solutions of the differential equations.) In these two figures the initial displacements are sufficiently large to drive the velocity in each path to its limiting value; larger displacements would not change the shape of the oscillations. However, if the displacement $z_0$ is sufficiently small the maximum circulation will not be generated during the run-in and the oscillations appear quite "normal" unless there is circulation already present. This is illustrated in Fig. 15 which shows seven relatively small successive displacements starting with zero initial circulation for the top curve and continuing to the bottom with the initial circulation for each curve equal to the final circulation of the preceding curve. The circulation is "pumped up" to its maximum value in this manner

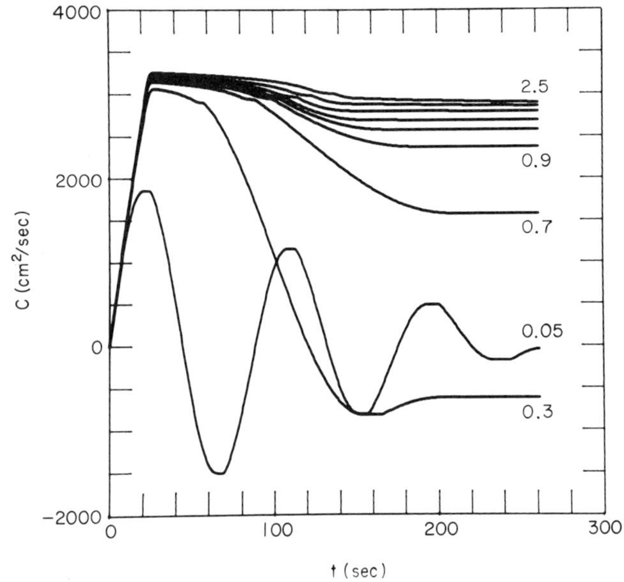


FIG. 14. Circulation ($cm^2$/sec) versus time (sec) accompanying preceding figure. For $t < 100$ sec the curves are in descending order of $a$ ,beginning at the top. ($a$ values as for Fig. 13.)

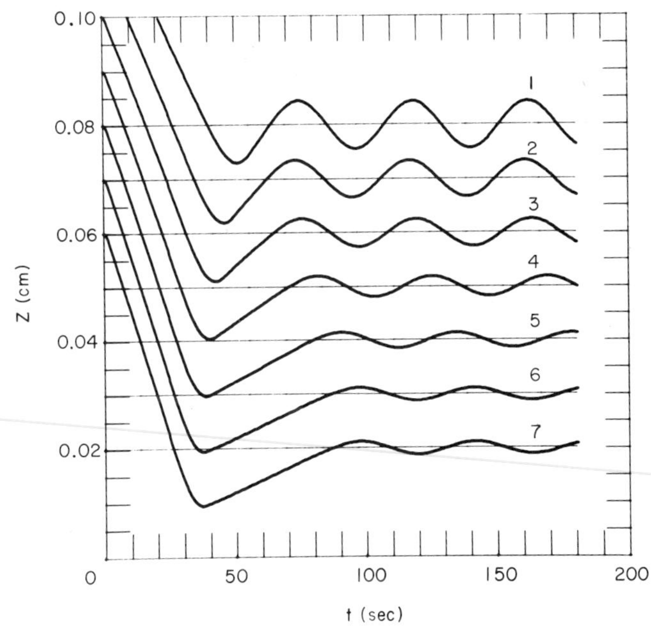


FIG. 14. Circulation ($cm^2$/sec) versus time (sec) accompanying preceding figure. For $t < 100$ sec same direction, beginning with the top curve. Parameters are $\ell = 1/80$, $a = 1{\cdot}1$.

as shown by the corresponding circulation histories in Fig. 16. For suitable $a$ and $\ell$ the distortion in the first half-cycle is a convenient indicator of the circulation present in the circuit whatever its origin. (Of course, circulation can be cancelled by displacement in the opposite direction.)

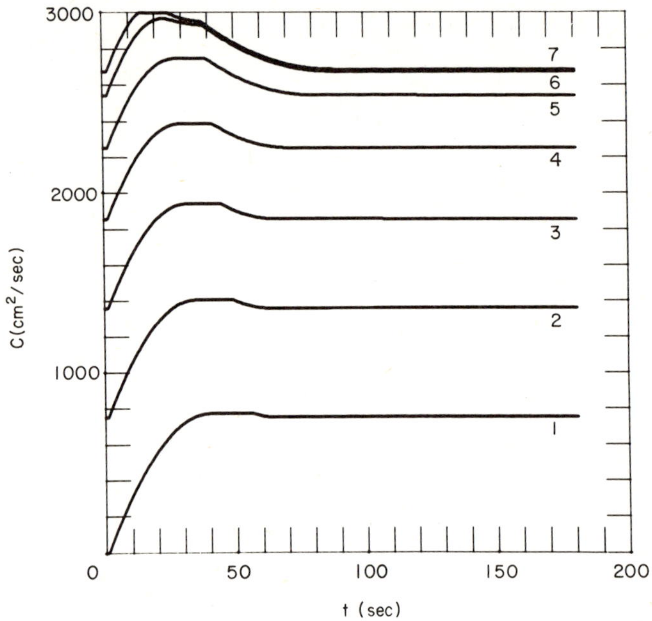


FIG. 16. Circulation ($cm^2$/sec) versus time (sec) accompanying preceding figure. The curves are in order of the successive level displacements beginning at the bottom.

No observations of the above effects which occur in the absence of strong thermal damping have every been reported, presumably because no one ever looked for them. However, for the case of $a \ll 1$, $\ell \ll 1$ and the number $m$ of dissipative half-cycles consequently large, hitherto unexplained observation have been reported by Hammel, Keller and Sherman.[40] They analyzed the decaying level oscillations of two reservoirs connected by the super fluid film and found a large anomalous peak in the decay constant $\alpha$ (defined by $z(t) = \exp(-\alpha t) \sin 2\pi \nu t$) after about 10 cycles. Accompanying this peak was an abrupt increase in the frequency to a new stationary value. This is exactly the behaviour that would be expected if a small leak were in parallel with the principal flow path. To understand the peak in $\alpha$ it is necessary to solve (III.8) which can be done if the difference equation is converted to a differential equation (i.e. $\eta_{j+1} - \eta_j \to \mathrm{d}\eta/\mathrm{d}j$). The solution is (setting $\zeta = 1$),

$$1 - \eta_j + \ell \ln\left(\frac{1-\ell}{\eta_j - \ell}\right) = 2aj. \tag{III.12}$$

Using the equation for the total energy given before, the decrease in $z$ over a half-cycle $\Delta z_j$ can be related to $\eta_j$ by

$$\Delta z_j = \sqrt{\left(\frac{\rho_s \mathscr{A}_2 \ell_2}{\rho g A}\right) v_c 2a\left(\frac{\ell}{\eta_j} - 1\right)} \tag{III.13}$$

which implies an effective decay constant,

$$\alpha_j = -\frac{2}{\tau}\ln\left(1 + \frac{2a}{\eta_j}\left[\frac{\ell}{\eta_j} - 1\right]\right) \tag{III.14}$$

where $\tau$ is the period appropriate to the $j$th half-cycle. (III.14) predicts a maximum in $\alpha\tau$ at $\eta_j = 2\ell$ which has the value

$$(\alpha\tau)_{\max} = -2\ln\left(1 - \frac{a}{2\ell}\right). \tag{III.15}$$

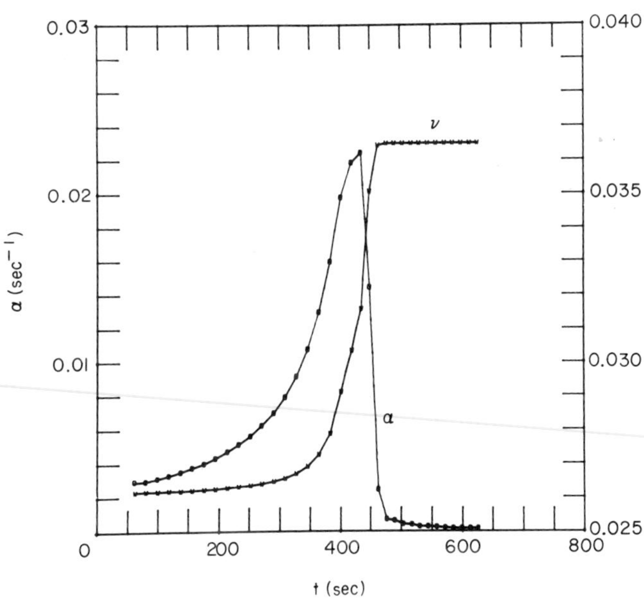


FIG. 17. Decay constant $\alpha$ (sec$^{-1}$) and frequency $\nu$ (sec$^{-1}$) versus time (sec) for $a = 0{\cdot}02$, $\ell = 0{\cdot}02$.

This is a differential maximum and is the upper bound of what can be observed.

An early characteristic signature of this type of damping is the approximately *linear* decrease in the oscillation amplitude in the first few half-cycles (contrasted with *exponential* decay for thermal damping). This can be seen from (III.13) where $\eta_j \approx 1$ for small $j$ and, by assumption, $\ell \ll 1$.

The abrupt change in frequency is easily understood: during the dissipative cycles the velocity $v_1$ (short path) is almost always at its limiting value which implies $\dot{v}_1 = 0$ which effectively cancels the contribution of this path to $\omega^2$ in (III.4).

Taking the apparatus parameters of ref. 40 and estimating the presumed leak area and length ($\mathscr{A}_1 = 5\,10^{-8}\,\text{cm}^2$, $\ell_1 = 0{\cdot}02$ cm), the details of the behaviour reported there are reproduced well by numerical solutions of the double path flow equations. The results for the decay constant $\alpha$ and the frequency $\nu$ averaged over each half-cycle are plotted in Fig. 17 which should be compared with Fig. 2 of ref. 40. There the averaging was over 5 cycles, resulting in a smaller maximum for $\alpha$ and less detail in the curves.

## 3. CONCLUSIONS AND APPLICATIONS

Persistent currents can easily be generated and studied by merely displacing and observing the relative levels of reservoirs joined by parallel flow paths. A wide range of behaviour is possible depending on the relative areas and lengths of the paths. High thermal damping is not required to preserve circulation generated during transfer between reservoirs. While numerical calculations are required for precise prediction, approximate formulae can be derived which adequately describe the systematics of the process. (Additional results are given in ref. 41.)

Aside from being a sophisticated tool for leak detection the theory of parallel path superfluid flow could find various applications. The following questions are offered for film flow in particular:

(i) *Testing the hypotheses.* Is something missing from (III.1)? Can a persistent current exist in the film? At what vapour pressures?

(ii) *Measurement and calibration.* What is the thickness of the film? What are the intrinsic dissipation parameters?

(iii) *Simply-connected geometries.* Can variation in film thickness and path length ever favour effective parallel path flow?

## IV. The Great Film-Thinning Puzzle

### 1. THERMODYNAMIC DESCRIPTIONS

Imagine two large reservoirs of helium connected by a film as shown, in part, in Fig. 18. Let the temperature, the reservoir levels, and the (saturated)

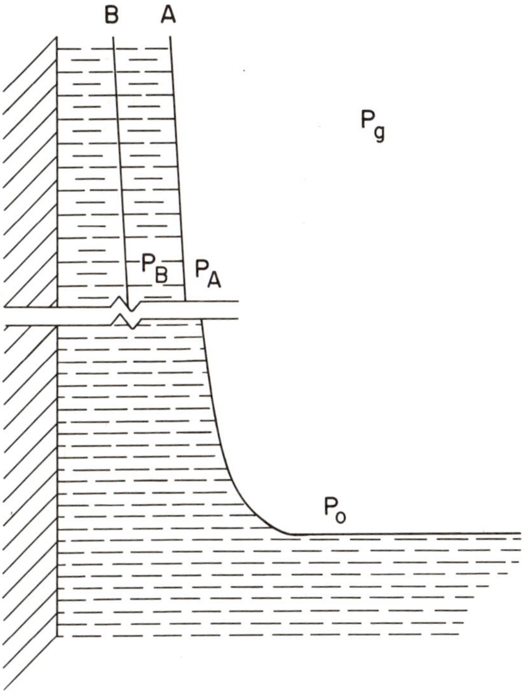


FIG. 18. Thickness profiles of static and flowing film.

vapour pressure $P_0$ at those levels be equal. Subscripts $F$, $R$, $L$, $V$, $\mathscr{A}$, $\mathscr{B}$ will refer to film surface, reservoir surface, liquid, vapour, $v_s = 0$, and $v_s \neq 0$, respectively.

$$\text{Case } \mathscr{A}: v_s = 0$$

The system is in true equilibrium so the chemical potential must be constant between and throughout the two phases. Consider the chemical potential at four places: the liquid surfaces of the reservoir and the film (at height $z$) and the neighbouring vapour regions.

$$\mu_{RL} = \mu_L(P_0), \qquad \text{reservoir liquid}$$

$$\mu_{FL} = \mu_L(P_{\mathscr{A}}) + gz - \frac{\alpha}{\delta^n}, \qquad \text{film liquid}$$

$$\mu_{RV} = \mu_V(P_0), \qquad \text{reservoir vapour}$$

$$\mu_{FV} = \mu_V(P_{\mathscr{A}}) + gz - \frac{\alpha}{\delta^n_{\mathscr{A}}}, \qquad \text{film vapour}$$

Note that the influence of van der Waals forces on the vapour in the vicinity of the film is explicitly included in $\mu_{FV}$. Since $\mu_{FL} = \mu_{FV}$, their difference implies $\mu_L(P_{\mathscr{A}}) = \mu_V(P_{\mathscr{A}})$ which, by definition of the saturated vapour pressure, implies the vapour pressure at the film surface $\mathscr{A}$ is the same as at the reservoir surface, $P_{\mathscr{A}} = P_0$ for all $z$. Of course, in the interior of the vapour, outside the influence of the van der Waals force, the vapour pressure $P_g$ follows the barometric law and is smaller. Substituting $P = P_0$ in $\mu_{FL}$ then gives the static film thickness,

$$\delta_{\mathscr{A}} = \left(\frac{\alpha}{gz}\right)^{1/n}. \tag{IV.1}$$

Case $\mathscr{B}$: $v_s \neq 0$

Let the velocity be low enough to make the intrinsic dissipation negligible. Can exchange of atoms between the film and vapour destroy metastable equilibrium? Three arguments suggest no.

(i) Evaporation–condensation does not affect $v_s$.[42]

(ii) The chemical potentials of the reservoirs and the vapour (far from the film) are still equal. Therefore neither phase can grow at the expanse of the other which would occur if net evaporation or condensation took place at the film surface, assuming neither the film itself nor the vapour in its vicinity totally absorbed the excess or deficit.

(iii) The potential probe measurements of Hammel and Keller[37] show the chemical potential of the flowing film (without intrinsic dissipation) is equal to that of the reservoir so the film volume should remain constant.

Apart from these arguments, what can be said about the values of $P_{\mathscr{B}}$, $\delta_{\mathscr{B}}$ and the $\mu$? As mentioned above, the reservoirs and vapour far from the film are unchanged so $\mu_L(P_0) = \mu_V(P_0)$ as before. Integrating the steady-state superfluid equation of motion between surface points on the film and reservoir gives (see Section II),

$$\mu_L(P_0) = \mu_L(P_{\mathscr{B}}) + \tfrac{1}{2}\frac{\rho_s}{\rho}v_s^2 + gz - \frac{\alpha}{\delta^n_{\mathscr{B}}}. \tag{IV.2}$$

In addition, the chemical potential of the vapour near the film is

$$\mu_{FV} = \mu_V(P_{\mathscr{B}}) + gz - \frac{\alpha}{\delta^n_{\mathscr{B}}}. \tag{IV.3}$$

where it is assumed the vapour and film pressures at surface $\mathscr{B}$ are equal. Three different interpretations of the system based on different sets of assumptions, as listed in the headings, will be discussed.

### *T1.* $P_{\mathscr{B}} = P_0$

The purpose of this assumption, discussed in the review article by de Bruyn Ouboter,[43] is to prevent condensation–evaporation at the film which was expected to occur if $P_{\mathscr{B}} > P_0$ or $P_{\mathscr{B}} < P_0$ respectively. The consequences for $\delta_{\mathscr{B}}$ follow at once from (IV.2) where the $\mu_L$'s cancel giving a thinner film,

$$\delta_{\mathscr{B}} = \left[\frac{\alpha}{\frac{1}{2}(\rho_s/\rho)v_s^2 + gz}\right]^{1/n}. \tag{IV.4}$$

With this thinner $\delta$ it follows from (IV.3) that $\mu_F^V$ will now be less than $\mu_V(P_0)$ so that vapour will flow to the vicinity of the film where the chemical potential is smaller, thus raising $P_{\mathscr{B}}$, causing condensation, thickening the film and finally stopping the flow in an unspecified way. To experimentally realize the thinning predicted by (IV.4) it is necessary,[43] according to this view, to effectively isolate the vapour near the film from the vapour over the reservoirs by somehow blocking the vapour flow path. However it is difficult to understand how the vapour could be sustained at $P_0$ since the chemical potential of the vapour $\mu_{FV}$ is also less than that of the *film* so, in spite of the presence of saturated vapour pressure, there should be evaporation from the film.

### *T2.* $P_{\mathscr{B}F} < P_{\mathscr{B}V} = P_0$

The vapour pressure at the film is assumed to be $P_0$ to prevent condensation, just as before. The additional assumption that the pressure is discontinuous across the film surface allows (IV.2) to be satisfied *beneath* the surface with $\delta_{\mathscr{B}} = \delta_{\mathscr{A}}$ and $P_{\mathscr{B}} = P_{\mathscr{B}L} < P_0$. Note the last relation implies superheated liquid in the film. Likewise (IV.3), with $P_{\mathscr{B}} = P_0$ then predicts equilibrium with the reservoir, $\mu_{FV} = \mu_V(P_0)$. Alternatively one could assume $\delta_{\mathscr{B}} = \delta_{\mathscr{A}}$, $P_{\mathscr{B}V} = P_0$ and infer the pressure discontinuity across the film.

This view is taken by Goodstein and Saffman[44] who argue that the pressure discontinuity could be supported by a surface tension $\gamma$ which depends on $v_s$ and $\delta$ as $\gamma = \gamma_0 - \frac{1}{2}\rho_s\delta v_s^2$.

From this additional energy term the surface experiences an outward pressure

$$P = -\frac{d\gamma}{d\delta} = \tfrac{1}{2}\rho_s v_s^2. \tag{IV.5}$$

which combined with the surface pressure of the liquid, $P_0 - \frac{1}{2}\rho_s v_s^2$, just equals the vapour pressure $P_0$. A serious difficulty with this theory is the lack of any known mechanism to transmit the pressure in (IV.5) to the wall; it obviously cannot be transmitted by the liquid for that would raise the liquid pressure. Something other than the liquid itself is needed to keep the rigid walls of a siphon a fixed distance apart.

Actually, Goodstein and Saffman regard $P_{\mathscr{B}F} < P_{\mathscr{B}V}$ (or $\delta_{\mathscr{A}} = \delta_{\mathscr{B}}$) not as an assumption but a consequence of requiring local thermodynamic equilibrium with respect to variations in $\delta$ and $v_s$. This derivation has been criticized by Bergman[45] on the basis that $v_s$ is a boundary condition and not a legitimate variational parameter.

*T3.* $\mu_{FV} = \mu_V(P_0)$

From this assumption of meta-stable equilibrium, self-consistent values of $P_{\mathscr{B}}$ and $\delta_{\mathscr{B}}$ can be derived as in case $\mathscr{A}$. Subtracting (IV.3) from (IV.2) gives

$$\left(\frac{1}{\rho} - \frac{1}{\rho_v}\right)(P_0 - P_{\mathscr{B}}) \approx \frac{1}{2}\frac{\rho_s}{\rho} v_s^2 \tag{IV.6}$$

where $1/\rho \equiv \partial\mu/\partial P$. Therefore, the pressure at the film exceeds $P_0$. However, this does not imply that vapour will freely condense on the film as commonly supposed, for the simple reason that $\mu_{FV} = \mu_{FL}$. It does, however, imply a region of supercooled vapour outside the liquid but it is estimated that this supercooled vapour is relatively stable against droplet formation.[46]

The predicted film thickness is

$$\delta_{\mathscr{B}} = \left[\frac{\alpha}{(F/2)(\rho_s/\rho)v_s^2 + gz}\right]^{1/n}. \tag{IV.7}$$

where $F = 1/(1 - \rho_v/\rho) \approx 1$. This is essentially the result of Kontorovich[35] embellished only slightly by requiring equilibrium with the vapour.

The third interpretation above is the standard one and the least controversial.[47] However it doesn't agree with all the experiments. Whether or not the first two interpretations can explain all the experiments is somewhat doubtful but still an open question. Two other theories of film thinning are no longer pursued.[47, 48]

The suggestion[49] that there is no thinning of the film because $\rho_s = 0$ at the surface [i.e. $\rho_s = 0$ in (IV.7)] is inconsistent with equilibrium within the film itself which requires a constant chemical potential throughout. The fact that $\rho_s$ (and even $\rho$!) vanishes at the surface is irrelevant to (IV.2) if the surface thickness (region of variation of $\rho_s$) is small compared to the film thickness.

## 2. EXPERIMENTS THAT OBSERVE NO THINNING

### (i) *Keller*

Measuring the film thickness near the inside rim of a metal beaker with a sensitive capacitor, Keller[50] detected no change in thickness due to flow "within an uncertainty of $< 5$ Å ... in some 33 experiments about equally divided between inflows and outflows ... covering the ranges $1{\cdot}12\,\mathrm{K} < T < 1{\cdot}93\,\mathrm{K}$ and $1{\cdot}4\,\mathrm{cm} < z < 8\,\mathrm{cm}$." Although the measurement represented an average over a region where the dissipation was maximum the prediction of film thickening discussed in Section II cannot, alone, account for the results since:

(a) dissipation predicts *additional thinning* on outflow which was not observed and

(b) dissipation effectively ceased at the end of each run when $v_s$ was still appreciable yet no thickness change was observed. (The response time of the instruments was more than sufficient to see this type of transient change.)

The possibility of a leak in the bucket which allowed the liquid to bypass the capacitor cannot explain the results since, as discussed in Section III, $v_s$ in *all* parallel flow paths reaches its limiting value under a sustained driving force.

The possibility of a very rough flow surface whose microstructure held the average film thickness constant due to capillarity must be discounted because the thickness responded over a range of 200 Å to variation of the inside bulk level according to a van der Waals exponent $n \approx 3$. Theoretically, the response of $\delta$ to $\frac{1}{2}(\rho_s/\rho)v_s^2$ should be the same as to $gz$.

This intransigent independence of $\delta$ on $v_s$ is ascribed by $T1$ to the high vapour pressure which condenses on the film, restoring its thickness. It is built into $T2$ and, of course, disagrees strongly with $T3$.

### (ii) *Telschow, Rudnick and Wang*

By measuring the velocity of third sound $c_3$ in an unsaturated film the thickness was inferred[51] from the relation

$$c_3^2 = \frac{\langle \rho_s \rangle}{\rho} f \delta$$

where $f$ is the van der Waals force per unit mass, $f = n\alpha/\delta^{n+1}$ and $\langle \rho_s \rangle$ is the reduced superfluid density appropriate to a thin film. The velocities parallel and anti-parallel to $v_s$ were averaged to cancel the effects of Doppler shift. No change in thickness with flow velocity was detected in eleven acceptable runs at temperatures between 1·5 and 1·8 K and film thicknesses between 10 and 15 atomic layers. At the highest $v_s$ the discrepancy between the *T*3 (Kontorovich) prediction and the measurement was typically 10 times the rms deviation of the latter. The flow was thermally driven up a stepped glass post in the manner of Liebenberg's experiments[15, 21] on saturated films. Recondensation and other effects are considered and discounted by the authors who state, "... careful examination of our experimental procedure and reduction of data has uncovered no way in which we can blame our null result on an artifact."

(iii) *van Spronsen, Verbeek, de Bruyn Ouboter, Taconis and van Beelen*

This experiment[52] consisted of a horizontal rotating torus, the interior of which was covered with film in contact with a bulk reservoir below and on axis with the torus. The small surface area of the reservoir magnified its level response $\Delta z$ so that the volume of liquid in the film could be monitored.

The change in $\Delta z$ as the torus was rotated up to 30 rad sec$^{-1}$ was consistent with prediction assuming the superfluid fully participated in the rotation. However, when the rotation was slowly stopped the reservoir level returned to its pre-rotation value suggesting one of two possibilities:

1. The flow states before and after rotation were identical (presumably, $v_s = 0$).
2. No thinning accompanied the final flow state ($v_s \neq 0$).

Both are consistent with *T*1 since the temperature was 1·34 K, i.e. high enough for significant vapour condensation which should thicken the film and eventually stop its flow.

Support for the first possibility, the absence of persistent current in the film, can be found in two other experiments. Wang and Rudnick, [53] despite repeated attempts, were never able to detect circulating current in an unsaturated film around the exterior of a rotating smooth cylinder. Comparing the transit times of third sound pulses with and against the rotational velocity showed that the film was invariably stationary with respect to the cylinder. Wagner[54] also found no evidence of circulating currents in either saturated or unsaturated films covering the interior of a rotating cell. By measuring the limiting values of thermally driven superflow parallel to the rotation axis *z* it was possible to detect the presence of flow in the azimuthal direction also, since $v_s = \sqrt{(v_{sz}^2 + v_{s\phi}^2)}$. In all cases the film showed no relative motion (in the $\phi$ direction) to the substrate. Both these experiments were sufficiently sensitive to be convincing. However, it is not clear that the same results would be obtained in geometries that were *manifestly doubly-connected*.

Verbeek, van Spronsen, Mars, van Beelen, de Bruyn Ouboter and Taconis[55] have observed persistent current in the saturated film at $T = 1{\cdot}4$ K for periods exceeding 18 hours. The path consisted of the inner surface of a 130 m long glass spiral, 0·035 cm i.d., in series with a superleak, at both ends of which were connections to bulk reservoirs. This will be discussed further (in 3(ii) of this section) in conjunction with an earlier experiment which used a similar flow path.

### 3. EXPERIMENTS THAT OBSERVE OR SUGGEST THINNING

#### (i) *Gribbon and Jackson*

This early test[56] of the Kontorovich prediction measured film thickness by examining the change in polarization of light reflected through the film on the outside of a metal beaker. There was significant scatter and about the same amount of thinning was seen at 1·68 K and 1·83 K. At the higher temperature this corresponded to approximately 63% of the theoretical value but was much less than the theoretical value at the lower temperature.

#### (ii) *van Spronsen, Verbeek, de Bruyn Ouboter, Taconis and van Beelen*

Film thinning of the magnitude expected by the Kontorovich theory was clearly seen[57] when flow between reservoirs was established through a 120 m long glass capillary (0·033 cm i.d.) The film volume in the capillary was sufficiently large to permit its thickness changes to be read from the reservoir levels. The observation of thinning here was attributed to the geometry which presented a high resistance for vapour transport between the reservoirs and the film thereby, according to *T*1, preventing condensation of the vapour at the film surface. However, as pointed out under *T*1, the instability against condensation (which is the relatively weak instability of a supercooled vapour) has been avoided in this view at the expense of instability of a more fundamental sort, viz, the chemical potential of the vapour is less than that of the film. This suggests that the *T*1 argument is not the explanation for the observation of thinning in this experiment and that the same thinning would be observed if the vapour over the film communicated freely with the reservoirs. Likewise in the above experiment of Verbeek *et al.*[55] on the observation of a persistent current in a long capillary the *T*1 argument is probably not the necessary explanation† and the obser-

† The authors do not, in fact, advance the *T*1 argument in this later publication.

vation of persistent currents in saturated films under quite different vapour conditions can be expected. Of course, this questioning of the explanation of these experiments in no way detracts from their considerable success and originality in demonstrating instances of phenomena whose existence had been in doubt.

(iii) *Williams and Packard*

This experiment[(58)] used a capacitor to measure the film thickness on a vertical wall during the run-in and level oscillations of gravitational flow between reservoirs. Quantitatively, twenty-five filling sequences showed thinning within 30% of that expected from (IV.7) using $n = 3$. Qualitatively, the thinning accurately followed the phase of $v_s^2$ which was known from simultaneous measurement of a reservoir level. Because the temperature was always below 0·9 K (while in Keller's[(50)] it exceeded 1·1 K) a lack of sufficient vapour to feed the alleged instability discussed in *T*1 was cited as the cause of the thinning. Since they did not perform measurements at higher temperature and in view of the previous discussion here of the *T*1 argument their conclusion that "the surface contour calculated by Kontorovich is a nonequilibrium state which is detectable only in the limit of vanishing vapour" is questionable.

(iv) *Duthler and Pollack*

Preliminary experiments by van Spronsen *et al.* (cited in ref. 43) lacked sufficient sensitivity to observe thinning because of a large distance between the film and the reservoir. Equation (IV.7) shows thinning is magnified if $z$ is small. Therefore any beaker flow experiment of the standard type discussed in Section II should test thinning if $z$ and the level difference $\Delta$ are kept small. Fortunately, a set of such experiments was reported by Duthler and Pollack[(59)] in 1971. These data for transfer rate vs the height-to-rim distance can be compared with the predictions of the theory in section II for two cases: (i) no thinning, i.e. the first term of (II.1) omitted and (ii) thinning, i.e. (II.1) unmodified. The results are shown in Fig. 19 where the region of the data scatter is bounded by dashed lines (see Fig. 4 of ref. 59) and the solid curves depict the theoretical prediction (using the specific experimental parameters) for a van der Waals exponent $n = 3$ (top three curves) and $n = 4$ (bottom curve) and level differences $\Delta$ as labelled. The no-thinning curve (n.t.) diverges strongly from the data (which represent $\Delta \to 0$ extrapolations) at small $z$ and it does not show the curvature associated with the data and thinning predictions at small $z$. (The no-thinning prediction for $n = 4$ is similar but is omitted for clarity.) Clearly, the transfer rate prediction of (II.1), which also

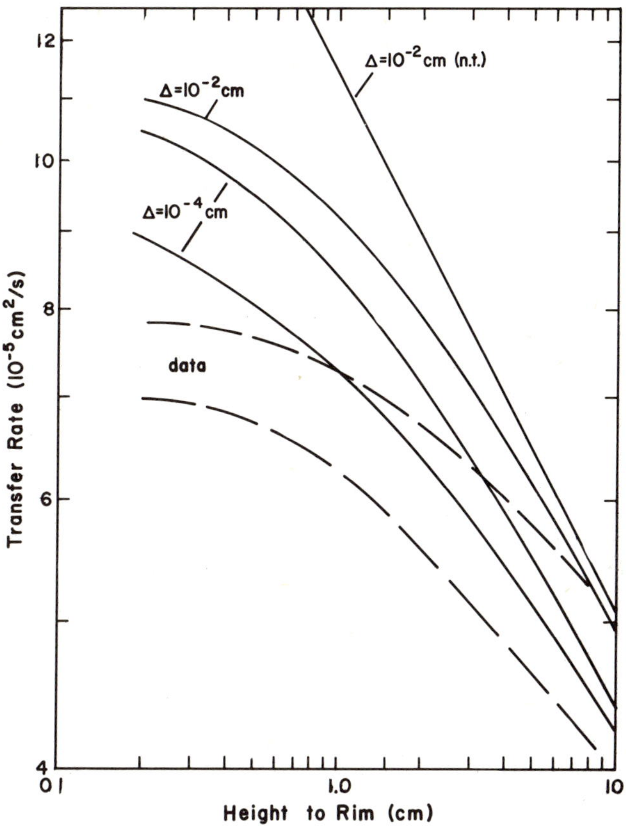


FIG. 19. Beaker film flow rate. Solid lines: theory; curve labelled n.t. assumes no thinning. Dashed lines: range of data, ref. 59. Reservoir level difference is Δ.

predicts thinning, gives much better agreement with the data, quantitatively, and qualitatively, in the region where thinning should become important. The discrepancy between the magnitude of the measured and predicted transfer rates for thinning at small $z$ could have several origins:

(i) Small temperature differences may have been present in the experiments.[60]

(ii) The data extrapolations for larger $z$, which are the least certain, may not correspond to $\Delta = 0$.

(iii) The value of $\alpha$ used $[g(3 \times 10^{-6})^n]$ may be inaccurate.

The temperature of these experiments was 1·65 K and the vapour-reservoir path was unrestricted.

(v) *Unpublished, preliminary and suggestive results*

Banton *et al.*[61a] have reported effects at $T \leqslant 1{\cdot}55$ K analogous to those seen[58] at $T \leqslant 0{\cdot}9$ K. Graham and Vittoratos[46] measured the thickness (by capacitors) at two radii on the horizontal rim of a "beaker" for $T < 1{\cdot}8$ K and report flow-dependent changes indicative of thinning.

Analysis of the film oscillation data of Hoffer *et al.*[18] (unpublished), van Spronsen *et al.* (unpublished) and of Hallock and Flint[26b, 61b] (their appendix C) shows a tendency for the oscillation frequency in a given run to slowly increase. To see how this could be a result of thinning, consider (I.4). If the film thickness obeys (IV.7) (with $F = 1$) then the effective kinetic length $l$ is also time dependent. Using the conservation of flow, $\mathscr{A}(y, t)$ and hence $l(t)$ can be expressed in terms of $z(t)$. For example if $n = 4$,

$$\mathscr{A}(y,t)^2 = \frac{1}{2}\left[-\frac{1}{2}\frac{\rho}{\rho_s}\frac{A^2\dot{z}^2}{gh(y)} + \sqrt{\left(\frac{1}{2}\frac{\rho}{\rho_s}\frac{A^2\dot{z}^2}{gh(y)}\right)^2 + \frac{4\alpha p^4}{gh(y)}}\right] \tag{IV.8}$$

where $h(y)$ is the height above the bulk level and $p(y)$ is the perimeter.

For any $n$ the time derivative of $\mathscr{A}$ to lowest order is

$$\dot{\mathscr{A}} = -\frac{1}{n}\frac{\rho}{\rho_s}\frac{A^2}{\mathscr{A}gh(y)}\dot{z}\ddot{z} \tag{IV.9}$$

When this is used, the equation for $z$ becomes, upon neglecting dissipation and using $q\dot{z} \approx -s\Delta T + (1/\rho)\Delta P$,

$$\ddot{z} + 2\xi z + w^2 z = 0 \tag{IV.10}$$

where

$$\xi = \frac{\rho_s q \mathscr{A}}{2\rho A \mathscr{F} \ell}, \tag{IV.11}$$

$$w^2 = \frac{\rho_s g \mathscr{A}}{\rho A \mathscr{F} \ell}, \tag{IV.12}$$

$$\mathscr{F} = 1 + \dot{z}^2 \frac{\rho}{\rho_s}\frac{\mathscr{A}A^2}{n\ell g}\int_1^2 \frac{\mathrm{d}y}{\mathscr{A}^3(y)h(y)}. \tag{IV.13}$$

(Recall that the kinetic length is proportional to the arbitrarily chosen $\mathscr{A}$, thus $\mathscr{A}/\ell$ is well-defined.) Even in this low-order approximation, (IV.10) is non-linear. Qualitatively, for $\xi$ small compared to $w$, the observed frequency should increase as

$$1/\sqrt{(1 + \overline{\dot{z}^2} b)} \approx 1 - \tfrac{1}{2}b\overline{\dot{z}^2} \qquad \text{as } \dot{z} \to 0,$$

where the constant $b$ is defined in (IV.13). An analysis of accurate oscillation data according to (IV.10)–(IV.13) could test the presence of thinning more conclusively.

## 4. THICKNESS CHANGES DUE TO DISSIPATION

As shown in Section II, films that flow between reservoirs with a large level difference should be thicker on the downward side and thinner on the upward side. A general analysis of stability with respect to vapour condensation–evaporation is greatly complicated by the lack of equilibrium in the vapour. However, the exact profile of $\mu_v$ between the high and low reservoirs may not be very important.[62] In any case, the chemical potential of the thick film, like that of the vapour, is always intermediate between that of the reservoirs so there may be sufficient stability for observation.

In a different beaker apparatus from that which showed no thinning, Keller[63] measured capacitance changes near the inside rim during inflows and outflows starting from level differences of $\approx 2{\cdot}5$ cm. He found for inflows that the film thickness (near the inside rim) was always greater than the static film (corresponding to the same inside level) and for outflows the film thickness was always less than the static value. Since the outside level is above (below) the inside level for inflow (outflow) he interpreted this to mean the outside level always controlled the film thickness near the inside rim. This interpretation together with his expectation that dissipation would affect the film in the same way in inflow and outflow led to the conclusion that dissipation did not affect the film thickness. However, the observed asymmetry in film thickness between inflow and outflow is exactly what is predicted in Section II: in inflow, dissipation thickened the falling film and in outflow it thinned the rising film. These thickness changes are quite independent of any Kontorovich type thinning which may be superimposed on the dissipation thickness variation but cannot be so easily extracted from the data. Figure 2 of ref. 63 vividly illustrates the quick thickness increase at the start of inflow and the severe thinning during outflow. Moreover, the flow rate was constant during most of the inflow, as also predicted in Section II.

## 5. SUMMARY

Three very different experiments[50–52] observe no Bernoulli-type thinning of the flowing film predicted by Kontorovich. One of these[52] can alternatively be interpreted as showing the absence of a remnant persistent current in a doubly connected geometry after cessation of rotation. Two additional experiments[53, 54] also show the absence of persistent film currents but their geometries may be unfavourable. Another experiment[55] observes it.

Two experiments[57, 58] show thinning and a third[59] strongly suggests it. The conditions of the former two differ from the no-thinning experiments in being colder or highly resistant to vapour flow. However the third enjoys neither of these conditions. In addition, preliminary reports of other experiments lacking these conditions suggest the presence of thinning.[46, 61]

The only experiment[63] which specifically looked for the independent thickness changes due to intrinsic dissipation unmistakably observed the predicted qualitative features.

The two proposed explanations[43, 44] for the absence of Bernoulli thinning emphasize the importance of vapour condensation on the surface of a thinned film. According to one, thinning can occur when the chemical potential of the vapour is less than that of the film; according to the other a pressure discontinuity exists across the film surface. At this point neither seems capable of explaining all the experiments, a deficiency shared by the standard theory[35] which predicts thinning at the expense of supercooled vapour in the region vacated by the film. However, the thickness of this region is generally much less than the mean free path of atoms in the vapour. In the standard theory there is no instability for vapour condensation on the film because the chemical potentials of vapour and film are equal.

## Acknowledgments

I wish to thank the following for their interest and willingness to share their accumulated experience in low temperature physics: M. E. Banton, L. Goldstein, E. F. Hammel, J. K. Hoffer, W. E. Keller, D. H. Liebenberg, and W. J. Trela.

## References

1. P. W. Anderson. *Rev. Mod. Phys.* **38**, 298 (1966).
2. E. R. Huggins. *Phys. Rev.* **A1**, 332 (1970).
3. A. L. Fetter. *In* "The Physics of Liquid and Solid Helium", (K. H. Bennemenn and J. B. Ketterson, Eds.), J. Wiley and Sons, New York.
4. L. J. Campbell. *J. Low Temp. Phys.* **3**, 175 (1970).
5. W. F. Vinen. *In* "Progress in Low Temperature Physics". Vol. III. (C. J. Gorter, Ed.), North-Holland, Amsterdam (1961), Chap. I, Sect. 6.
6. S. J. Putterman, M. Kac and G. E. Uhlenbeck. *Phys. Rev. Lett.* **29**, 564 (1972).
7. G. E. Volovik, *JETP Lett.* **15**, 81 (1972).
8. W. F. Vinen. *In* "Liquid Helium, Proceedings International School of Physics XXI", (G. Careri, Ed.), Academic Press, London and New York (1963), pp. 347–350.

9. S. V. Iordanskii. *Soviet Phys. JETP* **21**, 467 (1965).
10. J. S. Langer and M. E. Fisher. *Phys. Rev. Lett.* **19**, 560 (1967).
11. M. Chester and R. Ziff. *J. Low Temp. Phys.* **5**, 285 (1971).
12. R. J. Donnelly and P. H. Roberts. *Phil. Trans. Roy. Soc. Lond.* **A271**, 41 (1971).
13. G. Kukich, R. P. Henkel and J. D. Reppy. *Phys. Rev. Lett.* **21**, 197 (1968).
14. H. A. Notarys. *Phys. Rev. Lett.* **22**, 1240 (1969).
15. D. H. Liebenberg. *Phys. Rev. Lett.* **26**, 744 (1971).
16. D. R. Williams and M. Chester. *Phys. Rev.* **A4**, 707 (1971).
17. W. C. Cannon, M. Chester and B. K. Jones. *J. Low Temp. Phys.* **9**, 307 (1972).
18. J. K. Hoffer, J. C. Fraser, E. F. Hammel, L. J. Campbell, W. E. Keller, and R. H. Sherman. "Low Temperature Physics LT-13", Vol. 1, Plenum Press (1974), p. 253.
19. G. van der Heijden, J. J. Giezen and H. C. Kramers. *Physica* **61**, 566 (1972).
20. J. S. Langer and J. D. Reppy. *In* "Progress in Low Temperature Physics", Vol. VI, (C. J. Gorter, Ed.), North-Holland, Amsterdam (1970), Chap. 1.
21. L. J. Campbell and D. H. Liebenberg. *Phys. Rev. Lett.* **29**, 1065 (1972).
22. K. Telschow, I. Rudnick and T. G. Wang. *Phys. Rev. Lett.* **32**, 1292 (1974).
23. J. E. Robinson. *Phys. Rev.* **82**, 440 (1951).
24. F. D. Manchester. *Can. J. Phys.* **33**, 146 (1955).
F. D. Manchester and J. B. Brown. *Can. J. Phys.* **35**, 483 (1957).
25. J. F. Allen, J. G. M. Armitage and B. L. Saunders. "Low Temperature Physics LT-13", Vol. 1, Plenum Press (1974), p. 258.
26. (a) R. B. Hallock and E. B. Flint. *Phys. Rev. Lett.* **31**, 1383 (1973).
(b) R. B. Hallock and E. B. Flint. *Phys. Rev.* **A10**, 1285 (1974).
27. R. R. Turkington and R. F. Harris-Lowe. *J. Low Temp. Phys.* **4**, 525 (1971); **10**, 369 (1972).
28. M. E. Banton. *J. Low Temp. Phys.* **16**, 211 (1974).
29. J. P. Hulin, D. D'Humières, B. Perrin and A. Libchaber. *Phys. Rev.* **A9**, 885 (1974).
30. G. B. Hess. *Phys. Rev. Lett.* **29**, 96 (1972); *Phys. Rev.* **A8**, 899 (1973).
31. (a) G. B. Hess. *Phys. Rev. Lett.* **27**, 977 (1971);
(b) G. B. Hess. "Low Temperature Physics LT-13", Vol. 1 Plenum Press (1974), p. 302.
32. B. Perrin, D. D'Humières, C. Laroche, J. P. Hulin and A. Libchaber. *Phys. Rev. Lett.* **28**, 1551 (1972).
33. W. M. van Alphen, G. J. van Haasteren, R. de Bruyn Ouboter and K. W. Taconis. *Phys. Lett.* **20**, 474 (1966).
34. P. V. E. McClintock. *Phys. Lett.* **46A**, 145 (1973).
35. V. M. Kontorovich. *Soviet Physics, JETP* **3**, 770 (1956).
36. K. R. Atkins. *Proc. Roy. Soc.* (*London*) **A203**, 240 (1950), Sect. 3.
37. W. E. Keller and E. F. Hammel. *Phys. Rev. Lett.* **17**, 998 (1966); L. J. Campbell, E. F. Hammel, D. M. Jones and W. E. Keller. "Proceedings 11th International Conference on Low Temperature Physics", St. Andrews, 1968, St. Andrews University (1969). Vol. 1, p. 182.
38. L. J. Campbell. "Low Temperature Physics LT-13", Vol. 1, Plenum Press (1974), p. 268.
39. W. M. van Alphen, R. de Bruyn Ouboter, K. W. Taconis and E. van Spronsen. *Physica* **39**, 109 (1968).
40. E. F. Hammel, W. E. Keller and R. H. Sherman. *Phys. Rev. Lett.* **24**, 712 (1970).
41. L. J. Campbell. *Physica* (to be published).
42. R. P. Henkel, E. N. Smith and J. D. Reppy. *Phys. Rev. Lett.* **23**, 1276 (1969).
43. R. de Bruyn Ouboter. *J. Low Temp. Phys.* **12**, 3 (1973).

44. D. L. Goodstein and P. G. Saffman. *Proc. Roy. Soc.* (*London*) **A325**, 447 (1971); "Low Temperature Physics LT-13", Vol. 1, Plenum Press (1974), p. 243.
45. D. J. Bergman. *Proc. Roy. Soc.* (*London*) **A333**, 261 (1973).
46. G. M. Graham and E. Vittoratos. *Phys. Rev. Lett.* **33**, 1136 (1974).
47. J. Tilley. *Proc. Phys. Soc.* (*London*) **84**, 77 (1964).
48. E. R. Huggins. *Phys. Rev. Lett.* **24**, 699 (1970), erratum.
49. S. J. Putterman and I. Rudnick. *Physics Today* **24**, No. 8, 39 (1971).
50. W. E. Keller. *Phys. Rev. Lett.* **24**, 569 (1970).
51. K. Telschow, I. Rudnick and T. G. Wang. *J. Low Temp. Phys.* **18**, 43 (1975).
52. E. van Spronsen, H. J. Verbeek, R. de Bruyn Ouboter, K. W. Taconis and H. van Beelen. *Physica* **61**, 129 (1972).
53. T. Wang and I. Rudnick. "Low Temperature Physics LT-13", Vol. 1, Plenum Press (1974), p. 239.
54. F. Wagner. *J. Low Temp. Phys.* **13**, 185 (1973).
55. H. J. Verbeek, E. van Spronsen, H. Mars, H. van Beelen, R. de Bruyn Ouboter, and K. W. Taconis. *Physica* **73**, 621 (1974).
56. P. W. F. Gribbon and L. C. Jackson. *Can. J. Phys.* **41**, 1047 (1963).
57. E. van Spronsen, H. J. Verbeek, R. de Bruyn Ouboter, K. W. Taconis and H. van Beelen. *Phys. Lett.* **45A**, 49 (1973).
58. G. A. Williams and R. E. Packard. *Phys. Rev. Lett.* **32**, 587 (1974).
59. C. J. Duthler and G. L. Pollack. *Phys. Rev.* **A3**, 191 (1971).
60. K. R. Atkins, B. Rosenbaum and H. Seki. *Phys. Rev.* **113**, 751 (1959), Sect. 4.
61. (a) M. E. Banton, J. K. Hoffer and W. E. Keller. *Bull. Am. Phys. Soc.* **19**, 436 (1974), abstract AK-11.
(b) E. B. Flint and R. B. Hallock. *ibid.* abstract AK-13.
62. O. T. Anderson, D. H. Liebenberg and J. R. Dillinger. *Phys. Rev.* **117**, 39 (1960).
63. W. E. Keller. "Low Temperature Physics LT-12", Academic Press, Japan (1971), p. 125.

# 5. SUPERFLUIDITY IN RESTRICTED GEOMETRIES

ANN EGGINGTON

*Department of Theoretical Physics, University of Oxford, Parks Road, Oxford, England*

## I. INTRODUCTION

### 1. ONE- AND TWO-DIMENSIONAL WORLDS

One might imagine that life in a one-dimensional or two-dimensional world would be very different than in our own. Besides any biological or psychological changes, we might expect that the physics of the matter in these worlds would behave in a peculiar manner. In one dimension there are indications that magnets would not exist,[(1)] that percolation properties[(2)] are very different from three dimensions, (the percolation probability on an infinite lattice is *one* if all bonds are open and *zero* if any at all are closed), that localization always takes place when disorder is present; in two dimensions that crystals would not be stable.[(3)]

In these two lectures I shall try to give a sketch of the way $He^4$ would behave in these one- and two-dimensional worlds and indicate some of the ideas that both theoreticians and experimentalists have had about them.

### 2. CHARACTERISTIC LENGTHS

I shall be looking both at strictly one-dimensional and two-dimensional systems (of dimensions $L$ and $L \times L$ respectively where, in general, $L \to \infty$) and at partially finite quasi one- and two-dimensional systems ($L \times d \times d$ and $L \times L \times d$ with $L \to \infty$ and $d$ small). In the latter case we might distinguish two meanings of $d$ small: (a) $d \ll L$ so that the systems have effectively a dimensionality less than three and (b) $d \sim \lambda_T$ where $\lambda_T$ is the thermal de Broglie wavelength. Since we believe superfluidity to have a quantum mechanical origin in bulk $He^4$, case (b) might lead us to hope that quantum size effects might show up.

To help us to get a feel for the sizes of the systems involved let us look at some of the characteristic lengths and energies in the problem. The thermal de Broglie wavelength $\lambda_T$, is:

$$\lambda_T = \left(\frac{2\pi\hbar^2}{mk_BT}\right)^{\frac{1}{2}}, \tag{I.1}$$

which is about 9 Å at 1 K. The thickness of a helium layer is about 3·6 Å. If we look at the coherence length in the bulk liquid it rises from a few Å at $T = 0$ to infinity at $T_\lambda$ which it approaches as $(T - T_\lambda)^{-\frac{2}{3}}$; this quantity is important in consideration of the $\lambda$-phase transition.

In most experiments the helium is laid down on a substrate. The substrate potential varies typically as

$$V = \frac{\alpha}{r^3} \tag{I.2}$$

where $r$ is the distance from the wall in atomic layers and $\alpha$ is a material dependent constant usually in the range 10–70 K; its effects extends only a few Å from the wall.

What are typical excitation energies? For free particle excitations in a box with $L_1 \times L_2 \times L_3$ where $L_1 \geqslant L_2 \geqslant L_3$ and periodic boundary conditions, the wavevectors have the allowed values:

$$\mathbf{k} = 2\pi\left(\frac{l_1}{L_1}, \frac{l_2}{L_2}, \frac{l_3}{L_3}\right) \qquad l_i = 0, \pm 1, \pm 2 \ldots, \quad i = 1, 2, 3$$

and the energy of the lowest excited state is

$$\varepsilon_1 = \frac{\lambda_T^2 \pi k_B T}{L_1^2}. \tag{I.3}$$

If we look at a collective excitation of the system and consider typical phonon energies for a $L \times L \times d$ system we know

$$\varepsilon = c|k|$$

where $c$ is the sound velocity. The allowed values of the component of $\mathbf{k}$ in the direction with the smallest dimension ($x$, say) are

$$k_x = \frac{m\pi}{d}, \qquad m = 0, 1, 2 \ldots. \tag{I.4}$$

We shall consider the system two-dimensional if $\hbar c\pi d^{-1} \gg k_BT$ (in which case only the $k_x = 0$ modes will be excited). For $d = 100$ Å, this criterion is equivalent to $T \sim 0{\cdot}5$ K.

To summarize the most important lengths

$$\lambda_T \sim 9 \text{ Å (at 1 K)},$$

$$\text{atomic layer} \sim 3{\cdot}6 \text{ Å},$$

$$\text{“change” in phonon spectrum} \sim 100 \text{ Å (at 0·5 K)}.$$

Finally let me mention that typical system sizes so far examined experimentally are: (i) for films (two-dimensional geometry) $d$ in the range 1–100 atomic layers, (ii) for pores (one-dimensional geometry) radii down to about 30 Å.

## 3. GENERAL QUESTIONS

Before starting out to sketch the theoretical and experimental advances to date, let me suggest some questions that we might ask about our restricted geometry systems.

Does superfluidity still exist in restricted geometries? If so should we expect any changes from the bulk case? If we start from a Bose–Einstein condensation viewpoint, will the macroscopic occupation of the ground state, $n_0$, change? Starting from a Landau elementary excitation viewpoint,(4) will changes in the phonon spectrum affect the behaviour? Can we still talk about a superfluid density, $\rho_s$? Will it be altered from the bulk value?

Does superfluidity (a *macroscopic* quantum effect) persist in systems one layer thick; if not how thick must the sample be before superflow sets in? Is there a typical thickness at which there is a cross-over from two- to three-dimensional behaviour? At a solid boundary we expect the wave-function to vanish (and probably $\rho_s$ as well); is there a characteristic distance from the wall in a film in which $\rho_s$ reaches its bulk value—or must we use a totally new description?

We may also examine the normal—superfluid transition—if it exists; is the transition sharp or is it smeared?

We might expect that in studying one- and two-dimensional systems we shall find some new problems and perhaps some new effects. We may hope that in so doing we shall sharpen our understanding of superfluidity in general.

We can distinguish several features which might lead to changes of behaviour between bulk and finite systems:

(i) dimensionality,

(ii) the presence of smooth hard walls,
(iii) the presence of free surfaces,
(iv) the presence of inhomogeneities or a strong substrate potential (breaking translational invariance).

In these lectures I shall discuss primarily the changes brought about by dimensionality, and its effect upon the excitation spectrum and Bose condensation. I shall comment briefly (Section VI.1) on the effect of hard walls and in passing on the implications of a substrate potential (cf. ref. 5). I shall say little about the presence of free surfaces; clearly the study of surface excitations is an interesting problem in its own right[6] but in some ways perhaps a less fundamental one than the others specified above.

## II. Rigorous Results for Bose Condensation

### 1. Ideal Bose Gas

We turn first to the ideal Bose gas. As in the bulk case we do not expect that this idealized model of a non-interacting gas will provide an adequate description of real liquid $He^4$. However (again as in the bulk case[7]) the one- and two-dimensional cases are exactly soluble[8–10] and provide insight into features of the behaviour of interacting systems. We shall find that in the restricted geometry cases that the dimensionality of phase space does play a crucial role in the determination of the behaviour of the system.

In the grand canonical ensemble the total expected number of particles is given by:

$$N = \sum_{\sigma} N_{\sigma} = \sum_{\sigma} [\exp[\beta(\varepsilon_{\sigma} - \mu)] - 1]^{-1} \tag{II.1}$$

where $\sigma$ denotes states of the system, and $\beta = (k_B T)^{-1}$. In a three-dimensional box $(L_1 \times L_2 \times L_3)$ we can solve for $\alpha = \beta\mu$, where $\alpha$ is a function of $(N, T, L_1, L_2, L_3)$. Denoting the lowest energy state by $\sigma = 0$ and choosing our zero of energy appropriately, we have $\varepsilon_{\sigma} = 0$ and $(e^{-\alpha} - 1)^{-1} = N_0 \gg N_{\sigma}$ for $\sigma \neq 0$. When $N_0$ is of "order $N$" or equivalently $\alpha(N, T, L_1, L_2, L_3)$ is of "order $N^{-1}$", it is customary to say that the system is "condensed". In the thermodynamic limit ($N \to \infty$; $L_1$, $L_2$, $L_3 \to \infty$; $N/(L_1 L_2 L_3) = \rho$(finite)) this is made more precise by defining Bose–Einstein condensation as the existence of a non-zero value of $N_0/N$ in this limit.

Having defined Bose–Einstein condensation, let me briefly sketch the arguments which show us that it exists in the thermodynamic limit for a three-dimensional system but not in one or two dimensions. My analysis will be very crude but remember these results can be established completely rigorously.

We argue that $\mu(T) \ll 0$ otherwise we shall have the nonsensical result of negative occupancy. If we apply box boundary conditions the allowed energies are

$$\varepsilon_\sigma = \frac{\hbar^2 k_\sigma^2}{2m} \tag{II.2}$$

where $\mathbf{k}_\sigma$ is the wavevector. In three dimensions it is a three-component vector with components

$$k_{\sigma i} = \frac{\pi n_\sigma^i}{L_i},\ n_\sigma^i = 1, 2, 3 \ldots.;\ i = 1, 2, 3.$$

In one(two) dimension(s) $\mathbf{k}_\sigma$ is a one(two) component vector with its component(s) defined similarly. When $L_i$ are large we can convert the summation to an integral. In three dimensions we obtain

$$N \approx \frac{V4\pi}{(2\pi)^3} \int_0^\infty k^2\, dk \left\{ \exp\left[ \beta\left( \frac{\hbar^2 k^2}{2m} \right) \right] - 1 \right\}^{-1}. \tag{II.3}$$

This is valid in general, but if we have $\mu = 0$, then we have to separate out the ground state occupation

$$N \approx N_0 + \frac{V4\pi}{(2\pi)^3} \int_0^\infty k^2 dk \left\{ \exp\left[ \beta\left( \frac{\hbar^2 k^2}{2m} \right) \right] - 1 \right\}^{-1}. \tag{II.4}$$

In a full analysis London[7] found that $\mu = 0$ for $T \leqslant T_\beta$ and below this temperature there was a macroscopic occupancy of the ground state, $N_0 \sim O(N)$.

What happens in two dimensions? We might expect Bose condensation if (in the thermodynamic limit) we could satisfy (II.1) with $\mu = 0$ for some $T$.

However in two dimensions the relevant integral is

$$I = \frac{L^2 2\pi}{(2\pi)^2} \int_0^\infty k\, dk \left\{ \exp\left[ \beta\left( \frac{\hbar^2 k^2}{2m} \right) \right] - 1 \right\}^{-1}. \tag{II.5}$$

This is divergent.

In one dimension the situation is even worse

$$I = \frac{L}{2\pi} \int_0^\infty dk \left\{ \exp\left[ \beta\left( \frac{\hbar^2 k^2}{2m} \right) \right] - 1 \right\}^{-1}. \tag{II.6}$$

This is even more strongly divergent. It appears that (2.1) can never be satisfied in one and two dimensions with $\mu = 0$, which was precisely the condition which we had expected for the presence of Bose condensation. We conclude that in the thermodynamic limit there is no Bose condensation. This conjecture is borne out by more detailed and rigorous analyses of a number of authors.[8–10]

## 2. INTERACTING SYSTEMS

These free Bose gas results give us an indication of what we may find in other (interacting) Bose systems. Some rigorous and elegant results have been obtained[11–15] for one- and two-dimensional systems and quasi one- and two-dimensional systems of infinite extent which prove that $N_0/N$ is vanishingly small for $T > 0$.[16] The proof relies on a well-known operator inequality due to Bogoliubov (which can be established from statistical mechanics). I shall quote it here

$$\tfrac{1}{2}\langle\{A, A^\dagger\}\rangle \geqslant k_B T\,|\langle(A,C)\rangle|^2/\langle[C^\dagger,(\mathscr{H}, C)]\rangle \tag{II.7}$$

where $\mathscr{H}$ is the hamiltonian and $A$ and $C$ are arbitrary operators (provided that the relevant traces exist) and $\langle\,\rangle$ indicates an ensemble average. We work in an ensemble with broken symmetry i.e. we employ the usual trick of introducing a symmetry-breaking field into the hamiltonian

$$\mathscr{H} = \sum_i \frac{p_i^2}{2m} + \sum_{\substack{\text{all pairs}\\ \{ij\}}} V(\mathbf{r}_i - \mathbf{r}_j) + \mathscr{H}_{\text{symm}}, \tag{II.8}$$

where

$$\mathscr{H}_{\text{symm}} = -\int_\Omega \{\xi\psi^\dagger(\mathbf{r}) + \eta^*\psi(\mathbf{r})\}\,\mathrm{d}^3r \qquad (\Omega \text{ denotes the system}) \tag{II.9}$$

where $\psi^\dagger$ and $\psi$ are particle creation and annihilation operators. This stabilizes the anomalous average

$$\psi_0 = \frac{\langle a_0\rangle}{V_\Omega^{\frac{1}{2}}} = V_\Omega^{-1}\int_\Omega \mathrm{d}^3r\,\langle\psi(\mathbf{r})\rangle \tag{II.10}$$

$a_0$ is the ground state particle annhiliation operator.[17] Now substituting for the operators

$$A = a_{\mathbf{k}} = V_\Omega^{-\frac{1}{2}}\int_\Omega \mathrm{d}^3r\,\exp(i\mathbf{k}\ \ \mathbf{r})\psi(\mathbf{r}) \tag{II.11}$$

and

$$C = \int_\Omega \mathrm{d}^3r\,\exp(-i\mathbf{k}\cdot\mathbf{r})\psi^\dagger(\mathbf{r})\psi(\mathbf{r}), \tag{II.12}$$

it can be shown that

$$n_{\mathbf{k}} \equiv \langle a_{\mathbf{k}}^{\dagger} a_{\mathbf{k}} \rangle \geqslant \frac{mn_0 k_B T}{\rho k^2 + |\eta| mn_0^{\frac{1}{2}}} - \tfrac{1}{2}, \mathbf{k} \neq 0 \tag{I..13}$$

(where the inequality $|\psi_0| \equiv \sqrt{n_0} \geqslant \mathscr{R}(\psi_0 \, e^{i\alpha})$ has been used).

So we now have an inequality linking the number density of particles in a state labelled by wave vector $\mathbf{k}$, $n_{\mathbf{k}}$, with the ground state occupation density $n_0$. (If condensation is to occur then $n_0$ is of the order of $\rho$ the total number density.) But we know that the total density of particles is

$$\rho \geqslant \lim_{|\eta| \to 0} \lim_{V_\Omega \to \infty} \frac{1}{V_\Omega} \sum_{\mathbf{k} \neq 0} n_{\mathbf{k}}$$

$$\geqslant \lim_{|\eta| \to 0} \lim_{V_\Omega \to \infty} \frac{1}{V_\Omega} \sum_{\mathbf{k} \neq 0} \left( \frac{mn_0 k_B T}{\rho k^2 + |\eta| mn_0^{\frac{1}{2}}} - \frac{1}{2} \right). \tag{II.14}$$

Converting this to an integral

$$\rho \geqslant \lim_{|\eta| \to 0} \lim_{V_\Omega \to \infty} \frac{1}{V_\Omega} \int_0^\infty \left\{ -\frac{1}{2} + \left( \frac{mn_0 k_B T}{\rho k^2 + |\eta| mn_0^{\frac{1}{2}}} \right) \right\} \mathrm{d}^s k, \tag{II.15}$$

where $s$ is the dimensionality.

In three dimensions this integral is well-behaved and the inequality can be satisfied. However, in one and two dimensions it diverges at the lower end (when $|\eta| \to 0$). Hence the inequality cannot be satisfied for $T > 0$ unless $n_0 = 0$. Hence no Bose condensation can take place, even for interacting systems.

Now I have only sketched the proof rather scrappily here but it seems to me that if one looks at the proofs in detail the theorem is rock hard. A number of theoreticians[12, 14, 15, 18, 19] have sharpened and extended the conditions and I think one can now say that it has been established

(i) not only for strictly but also for quasi one- and two-dimensional geometries.

(ii) both in the presence and absence of a symmetry breaking field.

(iii) in the presence of an external potential (for example a substrate potential), provided that the density is bounded.

Let me state the theorem again as it is a very powerful one: any system of interacting bosons in one or two dimensions with a real hamiltonian cannot have a condensate in the thermodynamic limit for $T > 0$ if the density of particles is everywhere bounded.

Note that if (as we certainly do in bulk $He^4$) we associate the existence of the lambda phase transition with the presence of a Bose condensate, then what we have just proved is equivalent to the nonexistence of superfluidity in restricted geometries.[20]

## III. The Experimental Situation

### 1. Experimental Systems

The ideal restricted geometry experiment would measure the properties of liquid $He^4$ as a function of size and temperature. Using a series of tubes or boxes with a range of dimensions from 10–1000 Å with smooth walls, a series of experiments would be carried out as a function of container size, without concern about effects of a strong substrate potential, substrate inhomogeneities, or complications arising from the presence of free liquid surfaces. (We should like to be able to distinguish clearly between size effects and interfacial effects.)

Unfortunately such trouble-free containers are not available, and such systems as are available are in many ways complex and 'dirty'. It is somewhat surprising that we can obtain any meaningful information from them at all. The experimental systems can be divided into (a) films and (b) filled pores. In case (a) the film may be laid down on a plate substrate or on the surface of powder particles. An advantage of films is the ease with which the sample thickness may be varied;[21] a strong complicating feature is the presence of a free surface and the possibility of surface excitations. The pore experiments, case (b), are usually carried out in a porous glass such as Vycor which has an average pore diameter of about 50 Å. Its internal structure is complex; a maze of interconnected channels approximately 100 Å wide but with a fairly wide scatter about this mean, and of two main types, some of irregular circular shapes and some that appear to be about three times as long as they are wide. Samples can be obtained with slightly varying average pore size but few experiments have been carried out on more than a single pore size. The alternative system is a tube packed with powder; variation of grain size and packing density is used to alter the average channel size (which, however, cannot easily be measured). The structure and geometry of the packed powder systems seem very complicated.

These comments may indicate the rather unsatisfactory nature of the systems in which restricted geometry investigations have to be carried out. Further the substrate potential is often not well-known and the surfaces are often rough, at least on an atomic scale.

Turning to the experimental results, it is encouraging to find sufficient agreement between results from different substrates and types of system to hope that, as far as surface heterogeneity is concerned, the effects may not reach further than the first monolayer. (However, a word of caution. Recent results on Grafoil (to be discussed later) may lead us to question this opinion.)

A number of experiments (including adsorption isotherm,[21] specific heat[21] and accommodation coefficient[22] measurements) indicate that for the first monolayer the large helium-substrate attraction outweighs the zero-point energy and that a fairly strongly bound solid-like structure is formed; beyond this first layer the helium seems to remain liquid.† The properties of the first layer have been the focus of much theoretical investigation;[23–25] but it (definitely) seems not to be superfluid and I shall not discuss it here. For most purposes it seems possible to treat it as an extension to the wall whose contribution to such properties as specific heat can be subtracted out from experimental data.‡

What have the experiments shown us? Most surprising in the light of our previous theoretical results, is the experimental finding that superfluidity does indeed exist both in films and in filled pores. From a theoretical viewpoint this places us in something of a dilemma; I shall discuss later how we theoreticians try to wriggle out of this apparent contradiction between theory and experiment. For the moment, however, let us look in more detail at the experimental results.

The properties measured so far fall into the broad categories (a) specific heat, (b) density, (c) superflow onset and (d) $\rho_s/\rho$. Arising from these we may distinguish three characteristic temperatures; (i) of the specific heat maximum, (ii) of the density maximum and (iii) of the onset of superflow.

## 2. SPECIFIC HEAT

In bulk $He^4$ the specific heat manifests a logarithmic divergence at $T_\lambda$ = 2·17 K. For films the situation is somewhat different—the specific heat peak rounds and shifts to a lower temperature $T_C$. With decreasing film thickness the height of the peak falls and shifts to even lower temperatures, eventually (at a thickness somewhere between one and two layers) the peak disappears altogether. Agreement on these points between experiments on different substrates (jeweller's rouge,[26] rutile,[27] Vycor[19]) is reasonably good. At low temperatures the specific heat per atom of the film is greater than that of the bulk.

The shift and rounding of the peak are also present in full pores[19, 28] but they seem less pronounced than in films. With filled pores it is hard to say

† In Grafoil the first two layers appear to be solid.

‡ Of course it is somewhat speculative to assume a structure of independent layers, and that the first layer(s) remains solid even in the presence of higher layers.

whether the effects are less marked because the pores are relatively wide or because no free liquid surfaces are present.

### 3. DENSITY

The density[19] exhibits a maximum which takes place at a temperature well above that of the onset of superflow and the specific heat temperature maximum, and very close to the bulk $T_\lambda$.

### 4. SUPERFLOW ONSET

In bulk liquid the $\lambda$-transition is marked by the concurrent appearance of the specific heat divergence and the onset of superflow at $T_\lambda$. In restricted geometries we have seen that the temperature of the specific heat maximum, $T_C(d)$ (where $d$ is the film thickness), is depressed below the bulk value. The superflow onset occurs at yet another temperature, $T_0(d)$, which is lower than $T_C(d)$ and is again a function of film thickness.

Two fairly direct methods have been used to study superflow onset. Long and Meyer[29] used a direct observation method for films: a series of superleaks were sealed to a chamber that contained an adsorbent; the exit of any given superleak was connected to a measuring system so that flow through the leak could be investigated under a range of conditions of the saturation ($P/P_0$) in the adsorption system, the temperature and the pressure gradient across the superleak. The system did not quite permit direct measurement of the film flow itself since it depended on flow through the leak and subsequent measurement of the gas pressure in the measuring system at room temperature. In method I, the measurements system and the exit of the superleak were initially pumped to high vacuum with the leak and adsorption system at a given temperature and saturation ($P/P_0$); the saturation determines the film coverage. Once these conditions had been attained the pumping was stopped and the pressure rise in a known volume was measured over a period of time. Superflow was detected. For any given $P/P_0$ (and therefore for any given film thickness over the range investigated) there existed a sharply defined temperature below which superflow occurred, the flow increasing suddenly from the very small gas flow rate to a rate which could only be due to the superfluid behaviour of the unsaturated film. The onset temperature dropped from 2·04 K at a thickness of 17 layers to around 1·39 K for 2 layers; it appeared to be independent of the channel geometry.

[As an aside, let me briefly describe Long and Meyer's[29] method II; they repeated the above experiment with a small pressure gradient across the superleak (in place of the large gradient from $P/P_0$ to a vacuum). Superflow was again observed, but (in contrast to method I) it always started at the bulk value of $T_\lambda$. Flow was extremely sensitive to slight fluctuations of the bath temperature. No satisfactory explanation of this discrepancy has yet been given.]

The second fairly direct observation of superflow[28] utilized the onset of the anomalous heat transport. The simplest experimental arrangement was that of two chambers each provided with a sensitive thermometer and connected by a thin-walled tube. One chamber was in good thermal contact with a heat sink and the other was provided with a heater and insulated by a vacuum. If helium gas were then admitted to the system adsorption occurred, and the film coverage could be controlled by regulating the saturation. The thermal conductivities of the tube and the contained helium gas were so small that any heat transport due to superflow in the adsorbed film easily became the dominant effect. Anomalously high thermal conductivity was observed which set in at temperatures $T_0(d)$ in reasonable agreement with the superflow measurements of Long and Meyer.

Somewhat more sophisticated experiments which pinpointed the onset of superflow by the appearance of anomalously high thermal conductivity have shown that for a fixed temperature very thin films cannot support superflow. This critical thickness varied from a couple of atomic layers at about 1·37 K to twenty or more layers at temperatures about 2·13 K (near $T_\lambda$).

The superflow onset has also been observed directly by Brewer *et al.*[28] and by Atkins *et al.*;[30] their results suggest that the film thickness at which, for a given temperature, superflow set in was considerably less than the corresponding (filled) pore diameter. Perhaps this was caused by attraction to the walls. Another feature of Brewer *et al.*'s filled pore results was that $T_0(d)$ and $T_C(d)$ seemed to be approximately the same. It is not clear whether this observation has some significance, highlighting some intrinsic difference between films and pore systems (e.g. the presence of surface excitations in the former) or whether the pore size (which is relatively large at 50 Å) was such that the system was beginning to approach the bulk limit. To resolve the question we would need to carry out our experiments in narrower pores.

## 5. SUPERFLUID DENSITY

We have so far briefly described a series of experiments which establish the existence of superfluidity in restricted geometries. Let us now turn to a number of experiments which claim to measure the superfluid density $\rho_s$ in films and filled pores. From a theoretical viewpoint I have some reservations about these experiments. They rely upon a two-fluid hydrodynamical interpretation; essentially bulk two-fluid hydrodynamics is applied under the condition that the normal fluid is clamped. It seems to me that the application of bulk hydrodynamics to such small systems (a few atomic layers thick) is somewhat questionable. However in defence of these methods it must be said that the theoretically predicted modes are observed experimentally.

## (a) *Third sound*

Third sound is essentially second sound in layers with a free surface, with few enough layers so that the normal fluid is held rigid by its viscosity, and superfluid ripples are seen on the surface. Two-fluid hydrodynamics gives the velocity of propagation of third sound as

$$C_3^2 = (\overline{\rho_{sf}}/\rho)3\mu$$

where $\overline{\rho_{sf}}/\rho$ is the superfluid fraction of the film, $\mu$ is the chemical potential given by $\mu = \alpha/d^3$ and $d$ is the film thickness in atomic layers.

The bar is placed over $\overline{\rho_{sf}}$ to indicate that the superfluid density may well vary over the film and it is the average of $\rho_s$ which appears in the formula. These experiments establish the variation of $\overline{\rho_{sf}}$ with $d$ and $T$. (I will discuss this variation and its interpretation briefly later.)

One interesting feature of the data is the region in which the third sound signal disappears. Rudnick and Fraser[(31)] found that at certain characteristic ($T$-dependent) thicknesses the third sound signal which still had a large $c_3$ value (and hence finite $\rho_s$) became strongly attenuated. This might be interpreted as the disappearance of superfluidity (at least in the sense of convective heat flow) at finite $\rho_s$; $\overline{\rho_{sf}}/\rho$ is about 0·38 at this point. Ratnam and Mochel[(32)] followed the third sound signal into this region of strong attenuation and found an abrupt (but continuous) drop to zero of $c_3$, which suggests that at a critical $T$-dependent thickness $\overline{\rho_{sf}}$ drops abruptly and continuously to zero. However the meaning of third sound results in this strongly attenuated region is by no means clear.

## (b) *Fourth sound*

Fourth sound is an essentially unattenuated wave which propagates in HeII in which only the superfluid component moves, a condition which will be attained in a superleak when the viscous wavelength of the normal component exceeds the pore diameter. The velocity is given to 1% accuracy by

$$\frac{c_4}{c_1} = \left(\frac{\rho_s}{\rho}\right)^{\frac{1}{2}},$$

where $\overline{\rho_s}/\rho$ is the superfluid fraction averaged over the pore, and $c_1$ is the velocity of first sound.

The value of $\overline{\rho_s}/\rho$ is difficult to calculate because of the uncertainty in the effective paths, which depend on the geometry and structure of the samples. In Vycor measurements[33] this problem was circumvented by a somewhat arbitrary normalization procedure. $\overline{\rho_s}/\rho$ was found to follow the bulk liquid roughly up to about 1·9 K and then to drop sharply below it. The measurements were not sufficiently sensitive to define a transition temperature, but they were consistent with superfluid onset at $T_0 = 2{\cdot}0$ K, the value found by other methods.

Powder measurements[34, 35] over a range of grain sizes determined a well-defined $T_0$ but did not support the sharp drop-off in $\overline{\rho_s}/\rho$ seen in the Vycor measurements. In some of these measurements an arbitrary normalization was adopted; in the later measurements an attempt was made to determine the pore sizes but this is difficult in such a complicated system. As far as one can see the powder measurements indicate a depression of $\overline{\rho_s}/\rho$ below the bulk value.

### (c) *Microbalance*

Chester *et al.*[36, 37] have measured $\overline{\rho_{sf}}/\rho$ for films as a function of thickness, *d*, and temperature by using a shear-mode oscillating quartz crystal microbalance which was employed to measure adsorption and which responds only to the *non*-superfluid content of the film. One feature of the results was a linear dependence of the total film superfluid density as a function of *d* (at fixed *T*) for large thickness and a precipitate drop to zero at a temperature dependent critical thickness.

### (d) *Persistent current*

A persistent current is formed in an unsaturated film of thickness *d* (or full pores). The angular momentum $\mathbf{L}_p$ of the persistent current is a function of *T*, the superfluid velocity and thickness, *d*. In this experiment *T* was held constant and *d* was increased by carefully condensing additional helium gas on the unsaturated film.[38, 39] The result was a crystal-like growth of the superfluid state with the additional liquid assuming the same superfluid velocity field as the liquid already present. An increase in *d* was found to produce an increase in $\mathbf{L}_p$ proportional to the change *d* and the superfluid density of the additional mass. Measurements were made of $\overline{\rho_s}$ for temperatures from 1 K to the $\lambda$ point. The filled pore measurements were made of a wide range of radii from 20 Å to 2000 Å; in this case it was again necessary to use an arbitrary normalization procedure.

(e) *Low temperature measurements*

Comparison[40] of a number of filled pore and powder experiments in the low temperature region below 1 K (persistent current, fourth sound and U-tube oscillations) showed that (i) as pore size was decreased so was $\rho_s$ and its temperature dependence was enhanced and (ii) with small pore sizes the temperature dependence of $\rho_s$ became nearly linear at low $T$. (In the analysis of these results there appears to be an arbitrary normalization of $\rho_s/\rho$ to unity as $T$ tends to zero.)

## 6. GRAFOIL SUBSTRATE

The film measurements so far described at least have the reassuring feature that superflow onset temperatures and specific heat data show consistency between experiments carried out on a variety of substrates.† We might well assert that the results are substrate-independent. However recent measurements by Herb and Dash[41] and by Bretz[42] using a graphite substrate (Grafoil) seem to yield somewhat different results. Grafoil is claimed to be a substrate of much higher chemical and crystallographic purity than that of materials previously used; it consists of laminas of exfoliated natural graphite oriented and interleaved as in a loose stack of cards. The differences found in these measurements may indicate that the first few helium layers do not smooth any surface heterogeneity (as we had hoped) so that the upper layers remain undisturbed but that substrate properties continue to affect distant layers. If this is so, our task of theoretical description will be much complicated.

Here is a summary of the results.

(i) A specific heat anomaly was found, a steep rise at approximately the bulk $T_\lambda$, but instead of rounding (compared to bulk behaviour) the curves are truncated. This occurred for films thicker than (approximately) three layers. Plotting the incremental specific heat ($\Delta C/\Delta N$) it can be seen that for 9·77 to 8·64 layers the specific heat appears bulk like, as the coverage is further reduced rounding of the peak occurs and a shoulder remains. The specific heat curves tend to bulk values at both the highest and lowest temperatures and exhibit an abrupt fall at bulk $T_\lambda$.

(ii) Mass transport showed a marked change in flow rates of two different types ("onset $A$" and "onset $B$"). Onset $B$ was a function of thickness $d$ similar to that found on other substrates. At a higher temperature, approaching from the high temperature region, the mass flow underwent a relatively

† At least qualitatively.

abrupt increase (onset $A$) at $T_A \approx T_\lambda$. Although $T_A$ is effectively constant, the initial rate of increase and variation at lower temperatures change markedly and non-monotonically with $d$. This aspect is almost completely novel (but compare Long and Meyer's results[(29)] for mass transport under a small pressure head when they found superflow onset at bulk $T_\lambda$).

(iii) The thermal conductivity showed an abrupt rise at a temperature, $T_K$, less than $T_\lambda$ where $T_K$ was a function of $d$. The $T_K$ values correspond closely to values of the superflow onset temperature $T_B$ but lie about 0·05 K below them. This may have been due to a lack of sensitivity in the measuring technique.

There are (at least) two positions that may be adopted concerning the discrepancy between the results from Grafoil and other substrates. One possibility is that Grafoil is a much smoother substrate than those used previously and that results of previous substrates are blurred because of inhomogeneities. The other is the strength of the van der Waal's potential—for previous substrates $\alpha$ lay in the range 15–30 K, for Grafoil it is much stronger, about 60 K. This difference might well lead, for example, to a change in the superflow onset temperature as a function of film thickness but it is hard to explain the distinct qualitative differences (two onset temperatures in the Grafoil case) and the specific heat results (truncation as opposed to rounding of the peak).

## 7. SUMMARY OF EXPERIMENTAL FEATURES

As I mentioned earlier our experimental systems are far from clear as far as the theoretical interpretation is concerned, but let me summarize those features which have emerged and require explanation.

(i) The very existence of superflow.

(ii) The depression of the temperature of superflow onset with decreasing film thickness. The vanishing of superflow at a critical thickness $d_C(T)$.

(iii) The variation of $\rho_s/\rho$ for films and pores as a function of $d$ and $T$; in particular the depression below the bulk value.†

(iv) The low temperature filled pore behaviour of $\rho_s/\rho$; its linear dependence on $T$. $\rho_s = a - bT$, ($a$, $b$ constants as $T \to 0$).

† Another problem is the possible vanishing of superflow at finite $\rho_s$—either in the sense of Rudnick's finding of the disappearance of convective flow at finite $\overline{\rho_{sf}}$ or in the sense that between $T_0$ and $T_c$ (specific heat maximum) there is a region in which short range order may exist but long range order probably does not.

(v) The rounding of the specific heat peak and the depression of its position below bulk $T_\lambda$.

(vi) The Grafoil results, their divergence from those on other substrates, the truncation of the specific heat peak, the nature of the superflow region between $T_A$ and $T_B$.

## IV. THE DILEMMA

### 1. STATEMENT OF THE DILEMMA

We stand now in a dilemma. From a theoretical viewpoint we have shown rigorously that for one- and two- and quasi one- and two-dimensional geometries of infinite extent there is no macroscopic occupation of the groundstate; in analogy with the usual treatment of the bulk situation we have argued that the absence of a finite $n_0/\rho$ implies that the system is not superfluid. Yet experimentally we have seen manifestations of superflow in a number of systems. How can we wriggle out of this apparent contradiction between theory and experiment?

### 2. ESCAPE ROUTES

#### (a) *Finite systems*

One possible approach is to take the attitude that the Bogoliubov inequality results, while certainly rigorous, only have formal interest. They have been established for quasi one- and two-dimensional systems of infinite extent ($\infty \times d \times d$, $\infty \times \infty \times d$ respectively); our experimental systems are of course finite not in one (or two) but all three dimensions, although two (one) dimensions may be large.

In a two-dimensional free Bose gas of $N$ particles of finite extent ($L \times L$) we find[(43)] a macroscopically occupied state with $N_0 \approx N$ when the temperature is less than $T_1^{[2]}$, where $T_1^{[2]} = T_\beta/\log N$ where $T_\beta$ is the three-dimensional bulk condensation temperature.† (Of course the thermodynamic limit involves letting $N \to \infty$, so that the transition temperature $T_1^{[2]}$ tends to zero, consistent with our previous Bogoliubov inequality results.)

Jasnow and Fisher[(14, 15)] have used a rather sophisticated analysis to extend the Bogoliubov inequality method to the case of finite systems (including interactions). They have established upper bounds for $n_0$ in

† For a finite system there is, of course, not a sharply defined transition to the macroscopically occupied state. We would expect such sharp transitions only in the thermodynamic limit.

quasi one- and two-dimensional (finite) systems. (The property they bound is actually rather subtle but in simple cases it reduces to $n_0$.) For a one-dimensional system $(L \times d \times d)$

$$n_0 \leqslant \rho^{\frac{3}{2}} \frac{4}{\sqrt{\pi}} \lambda_T \frac{d}{L^{\frac{1}{2}}}$$

i.e. $n_0 \leqslant A(\rho, T, d)L^{-\frac{1}{2}}$ where $A$ is a constant at fixed $\rho$, $T$, $d$ and for a two-dimensional system $(L \times L \times d)$

$$n_0 \leqslant B(\rho, T, d)[\ln(L \times L \times d)]^{-1}$$

where $B$ is a constant at fixed $\rho$, $T$, $d$.

Hence in finite one- and two-dimensional interacting systems we cannot say whether or not there is a macroscopic occupancy density $n_0$, but we do know if we make the system very large then $n_0$ can be (at most) rather small.

Let us consider further our argument that the helium system is finite and hence below a certain temperature there will exist a finite (and indeed large) $n_0$, which then accounts for the (observed) presence of superflow. There seems to be some difficulty here. If the interacting system is anything like the free Bose gas the macroscopic occupancy will set in at temperatures $T_1^{[2]} \sim T_\beta / 1nN$ which will vary with assembly size. This would seem to imply superflow onset temperatures dependent on the sizes of the plates used in the experiments. The experimental results up to now are fairly reproducible but, admittedly, $T_1^{[2]}$ is only logarithmically dependent on $N$ which will probably not be a sufficiently strong variation to visibly affect the experimental results.

### (b) *Metastability*

Another possible explanation may be that in fact a macroscopic $n_0$ is not a necessary condition for the existence of superfluidity.†

This is a question of some fundamental interest in our general understanding of superfluidity, and now is perhaps the moment at which to emphasize that "superfluidity" is a rather amorphous term even when applied to bulk $He^4$. It is an umbrella term which covers the appearance of a wide range of "anomalous" properties. We shall pick out two here for detailed consideration.

(i) *Hess-Fairbank effect.* For a bucket of HeII rotated sufficiently slowly, the equilibrium state is one in which at least part of the HeII stays at rest in the

† It can be shown(44, 45) to be a sufficient condition for superfluidity (at least in the "Hess–Fairbank" effect sense) for bulk systems.

frame of the laboratory. Hess and Fairbank[(46)] cooled the bucket of $He^4$ from the normal phase (where all the $He^4$ rotated with the bucket) and showed that as the temperature was lowered through $T_\lambda$ the $He^4$ gradually stopped rotating until as $T \to 0$ it was stationary in the laboratory. (Note that Hess and Fairbank did not just start the container from rest and verify that the HeII did not rotate. This would have just shown that the He had low or zero viscosity.) The Hess–Fairbank effect is an equilibrium effect; it is the analogue of the Meissner effect (flux expulsion) in superconductors.

It can be shown[(44, 45)] that a necessary and sufficient condition for the Hess–Fairbank effect to occur is that the free energy as observed from the frame of the rotating walls is a non-trivially periodic (i.e. non-constant) function of the rotational velocity, $\omega$. Further, at least in three dimensions, it can be shown[(44)] that for $He^4$ in an annulus of mean radius $R$, height $h$ and thickness $d\,(R \gg d \gg a_0$, the interatomic spacing), Bose–Einstein condensation is a necessary and sufficient condition for the Hess–Fairbank effect. The case of one- and two-dimensional systems (e.g. when $d \sim a_0$) is not very clear.

(ii) *Persistent currents.* Another property (and in a sense the most obvious) associated with "superfluidity" is the possibility of the system sustaining persistent (superfluid) currents while the walls are at rest. This is a metastable phenomenon and hence from the theoretical viewpoint somewhat more complicated than the Hess–Fairbank effect.

As in the superconducting case the persistent currents may be explained in terms of high free energy barriers separating circulating current states of different circulation; in this case the formal result of applying equilibrium statistical mechanics may not be a sensible description of the physical system over times of the order of (say) a few minutes. It has been suggested[(47)] that there might well exist systems in which, while application of true equilibrium thermodynamics would yield a zero result for $n_0$, the possibility of occupation of a long-lived metastable state might allow a free energy periodic in $\omega$ in certain circumstances. The fluctuations which destroy off-diagonal long-range order ($n_0$) are small fluctuations which do not require the system to pass over high free energy barriers; hence the time scale for the destruction of $n_0$ should be microscopic. Under these conditions it is not clear[(47)] whether the criterion $d \gg a_0$ alone is sufficient to guarantee the result[(44)] that Bose condensation is a necessary condition for the free energy to be periodic in $\omega$ (and hence for the Hess–Fairbank effect). Setting aside the uncertainties in deciding whether or not the Hess–Fairbank effect can exist in such systems, a detailed understanding of the nature of the persistent current states and transitions between them is itself difficult. However it can be made plausible[(48, 49)] that a sufficient condition for the existence of persistent currents is that we have quantized circulation† and that the mechanisms for passing from one value of the circulation to another involve crossing sufficiently high free energy barriers to be negligibly important in practice. While the question of whether or not such transitions from one circulation value to another are negligibly improbable is in principle an extremely complicated and subtle one, in a quasi one-dimensional system we should expect the answer to depend only on the width and local geometry of the sample and to be quite insensitive to the length which, however, plays a crucial role in the arguments leading to the destruction of $n_0$.

### (c) *An experimental test*

This suggests an experiment[47] which might give us some information about the connection between a finite $n_0$ and superfluidity. Suppose we carried out a series of measurements of superflow through a number of long narrow tubes of different lengths. (This type of superflow would presumably be closely connected to the persistent current situation.) From Jasnow and Fisher's work[14, 15] we have an upper bound for the possible value of $n_0$ in the system

$$n_0 \leqslant A(\rho, T, d)L^{-\frac{1}{2}}.$$

As we make the tube longer and longer we can be sure that $n_0$ must be very small. My suggestion is that it is at least plausible that if $n_0$ is necessary for superflow then we would expect the measured superflow to fall as the system length is increased; if it is not then we would expect no strong variation of superflow with length. So here is a chance for an experimenter to tell us something interesting about superfluidity. The snag is that he will require rather long and narrow tubes; for a pore diameter of 50 Å one would really like a length of several centimetres, but even if he could obtain a long narrow pore 50 Å × 50 Å × 5000 Å there are indications that he might see something interesting.

† We shall say that the circulation is quantized if there exists a velocity **v** such that in any given pure quantum state it satisfies the condition

$$\oint \mathbf{v}\cdot d\mathbf{l} = n\kappa$$

(where $\kappa$ is a constant independent of the size of the system, $n$ is an integer and the integral is taken around a closed path) and if, furthermore, a finite mass of the liquid is associated with this velocity. In the usual theories of superfluidity **v** is, of course, the "superfluid velocity" $\mathbf{v}_s$ and the constant $\kappa$ is $h/m$.

### 3. A COMMENT

In this section we have suggested that if we look at the $He^4$ system allowing for metastable states then the fact that $n_0$ is zero is perhaps irrelevant for the Hess–Fairbank effect and almost certainly irrelevant for persistent currents. The experiments to date (third sound, fourth sound and so on) seem to match more closely the persistent current (metastable) type of experiment. It might be interesting to try experimentally to observe the Hess–Fairbank effect in films or pores. If it exists, then we have the intriguing situation of free energy periodicity in a system where $n_0$ is probably zero.† On the other hand if no effect is found then we have systems in which superfluidity in the sense of criterion (ii) but not in the sense of criterion (i) is present. This would not be totally surprising since for the free Bose gas in bulk we know that superfluidity in the sense of (i) but not (ii) is expected. However, in bulk interacting $He^4$ it is usually assumed that (i) and (ii) are intimately related properties.

In summary there seem to be (at least) two possible escape routes for theoreticians from the experimental–theoretical conflict regarding the existence of superflow in restricted geometries: one approach (a) emphasises the finite nature of the experimental systems and suggests that $n_0$ is finite in such systems, the other (b) suggests that the problem should be seen as a metastable rather than a true equilibrium one so that the existence or non-existence of $n_0$ is an irrelevancy. Perhaps some experiments in quasi one-dimensional systems might be of some help in clarifying this question. If we may draw an analogy again with the free Bose gas the "condensation temperature" for the one-dimensional case is $T_1^{[1]} \sim T_\beta/N$ (where $T_\beta$ is the bulk condensation temperature). Here the variation of onset temperature with sample size should be observable if explanation (a) is correct (as previously discussed, in two dimensions the variation of onset temperature might well be masked as it falls only as $(\ln N)^{-1}$).

Let us now turn to a more detailed description of the specific heat and variation of $\rho_s/\rho$ and $T_0$ with system size.

## V. CRITICAL PHENOMENA

### 1. BULK $\lambda$-TRANSITION

Let us first discuss the $\lambda$-transition. In bulk(50, 51) this is characterized by the "anomalous" behaviour of a number of properties. These include

† I say probably zero because of the difficulties of having a finite (rather than infinite) system.

(i) a singularity and discontinuity in the specific heat $C_p$,

$$C_p(T) \sim \text{constant} + \text{constant} \times \ln|T_\lambda - T|. \qquad \text{(V.1)}$$

(ii) the vanishing of $\rho_s$,

$$\rho_s(T) \sim (T_\lambda - T)^{\frac{2}{3}} \qquad \text{(V.2)}$$

(iii) the vanishing of the critical velocity,

$$v_C(T) \sim (T_\lambda - T)^{\frac{2}{3}}. \qquad \text{(V.3)}$$

The transition is of second order.

The field of critical phenomena is a wide and quite complex study in its own right, which would require a series of lectures (if not a whole summer school) to do it justice. I shall confine myself here to a few remarks to indicate (rather superficially) the differences between transitions in the bulk and in a finite geometry and the extent to which they are understood.

Deeply inbuilt into the investigation of critical phenomena is a belief in the crucial nature of a characteristic length, the correlation length, $\xi$, which determines the range of correlation (and the typical size of regions of the system which fluctuate into the opposite phase as we approach the transition). A possible definition of the coherence length is through the correlation function

$$g(\mathbf{r}, \mathbf{r}') = [\langle\psi^\dagger(\mathbf{r})\psi(\mathbf{r}')\rangle - \langle\psi^\dagger(\mathbf{r})\rangle\langle\psi(\mathbf{r}')\rangle] \qquad \text{(V.4)}$$

which is expected to vary in the limit of infinite separation as

$$g(\mathbf{r}, \mathbf{r}') \sim G(\mathbf{r}, \mathbf{r}') \exp(-|\mathbf{r} - \mathbf{r}'|/\xi), \qquad \text{(V.5)}$$

where $G(\mathbf{r}, \mathbf{r}')$ varies more slowly than exponentially.

As $T_\lambda$ is approached (from above or below) in the bulk liquid there is evidence[50] (which I will not discuss here) that $\xi$ diverges to infinity as:

$$\xi \propto |T - T_\lambda|^{-\frac{2}{3}}$$

It appears that critical phenomena in many different systems[50] (such as liquid–gas, magnetic–nonmagnetic transitions) fit into a very general scheme of analysis in which properties such as the symmetry of the hamiltonian are of importance but the microscopic nature of the interactions are not. Bulk $He^4$ behaviour at the $\lambda$-transition appears to fit quite well into this general scheme[50] where the order parameter (analogous, say, to the magnetization) is taken to be $\langle\psi(\mathbf{r})\rangle$.

The film and pore transition seem to differ from the bulk by two main features.

(a) In the bulk all anomalies take place at $T_\lambda$ (cf (V.1), (V.2) and (V.3) in finite systems (at least those with free surfaces) there seems to be a splitting between the onset temperature, $T_0$, and the specific heat maximum, $T_C$. I have found no convincing explanation of this. However perhaps the following rather vague comments might be made.

(i) At present the measurements in filled pores have not been made in sufficiently small systems to establish whether or not this splitting takes place. If it were found not to occur then we might attribute the splitting to some surface excitation effect.

(ii) Possibly at $T_C$ short-range order sets in, with an accompanying specific heat increase, but the "necessary condition" for superflow (whatever that may be) is achieved only at a lower temperature. The problem remains unsolved (but cf. Section VI.3).

(b) The specific heat peak is shifted to $T_C$ (which is below $T_\lambda$) and instead of a logarithmic divergence as in the bulk, the peak is rounded and its height decreases with decreasing film width. These features are somewhat better understood. It is known[52, 53] from the study of finite Ising models and indeed also on very general grounds that in finite systems no singularities will be found in the free energy or its derivatives; singularities will only appear in the thermodynamic limit. In bulk systems in this limit there is a logarithmic singularity in $C_p$, in the finite systems the singularity is rounded. Model calculations[10, 54] on the free Bose gas in quasi two-dimensional geometry $[\infty \times \infty \times d]$ indicate a similar rounding.

A crude (scaling-type) argument estimates the temperature, $T^*$, at which significant deviations from bulk behaviour begin to appear, as that temperature at which the (bulk) coherence length is equal to the film thickness:

$$\xi_0(T_\lambda - T^*)^{-\frac{2}{3}} \approx d$$

where $\xi_0$ is the zero-temperature coherence length (a few Å). This estimate of $T^*$ as a function of $d$ suggests that rounding sets in earlier, the thinner the film (as is observed experimentally).

Model free Bose gas calculations for restricted geometries also indicate (at least when free surfaces are present) that, $T_C$, the temperature of the specific heat maximum, is reduced below the bulk value. The argument is slightly easier to see in the magnetic case[53] where a naive mean-field-type theory indicates that $T_C$ is depressed below the bulk $T_{CB}$:

$$\frac{k_B T_C}{k_B T_{CB}} \approx \frac{E(n_1, n_2, n_3)}{n_1 n_2 n_3 E_0}$$

where $E(n_1, n_2, n_3)$ is the ground state energy of the system with boundaries while $E_0$ is the limiting groundstate energy/spin in an infinite system. $E(n_1, n_2, n_3)$ falls below the bulk value energy $(n_1 \times n_2 \times n_3 \times E_0)$ by an amount proportional to the surface energy.

Moore[(55)] has shown that the available experimental data on $He^4$ films on Vycor does indeed fit the scaling predictions developed by more sophisticated arguments along these lines.

We note that it does not seem possible to fit the truncation of the specific heat peak found in Grafoil measurements into the usual framework of critical phenomena analysis.

Doniach[(56)] has suggested that the truncated specific heat peak might be expected in films and is characteristic of a combination of a partially suppressed bulk specific heat singularity (associated with $T_\lambda$) and some sort of singularity arising from a transition in a two-dimensional (but finite) system (associated with a lower [2] transition temperature). Are measurements in other film systems (with less homogeneous substrates than Grafoil) merely examples of such a transition smeared because of substrate inhomogeneities...?

## VI. Theoretical Calculations

In this section we consider theoretical attempts to understand the reduction of $\rho_s$ below bulk value and the depression of the temperature of superflow onset.

From a theoretical viewpoint we might consider three effects which would lead to a depression of $\rho_s$:

(i) the presence of a smooth hard wall boundary (or a free surface) at which the macroscopic wavefunction is zero and only rises to a substantial value over some characteristic length (the healing length). (cf. Section VI.1 and Chester's discussion in his lectures.)

(ii) the presence of a spatially varying substrate potential or other inhomogeneities in the system (cf. Chester's lectures in these Proceedings).

(iii) an alteration in the energy spectrum of the elementary excitations due to the dimensionality of the system. (cf. Section VI.2.)

Note that both effects (i) and (ii) could give rise to a reduction of $\rho_s$ to a value less than 1 even for $T = 0$. In particular effect (ii) might be substantial and is well known for the analogous properties in superconductivity. Effect (i) at $T = 0$ would probably be small since the healing length is then only a few Å.

## 1. HEALING LENGTH

Let us now turn to a consideration of these experiments which claim to measure $\rho_s$; in films these use techniques of (i) third sound, (ii) the microbalance and (iii) the gyroscope (described in Section II). The experiments have (generally) been analyzed under the assumption that the superfluid density in the film takes the bulk value everywhere except in a region near the wall (or surface) whose size $\mathscr{L}(T)$ is temperature dependent:

$$\overline{\rho_{sf}} = \rho_s(\text{bulk})\left[1 - \frac{\mathscr{L}(T)}{d}\right]. \quad \text{(VI.1)}$$

A plot of the experimentally determined $\overline{\rho_{sf}}$ values against film thickness have been found to yield a straight line for large $d$ whose intercept is taken to be $\mathscr{L}(T)$.

Note that, although the experimental data does fit to a straight line plot it is not possible (at least for methods (i) and (iii) to check that the slope is indeed given by bulk value of $\rho_s$; an arbitrary normalization procedure has to be used. Microbalance results[36] give the constant a value 15% below bulk $\rho_s$ but the authors have attributed this to inaccurate features of their apparatus.

The length $\mathscr{L}(T)$ is also analyzed into the form:

$$\mathscr{L}(T) = d_0 + \ell(T), \quad \text{(VI.2)}$$

$d_0$ is considered to be the region near the wall in which no superflow can take place—it is usually taken to be one atomic layer, the solid layer next to the wall.† $\ell(T)$ is that length in which the superfluid wavefunction rises to its bulk value from its value at the wall or first solid layer (usually assumed to be zero). At this point the analysis is very theory-dependent. Usually a Ginzburg–Pitaeveskii[57] (GP) type of approach is employed which yields an $\ell(T)$ proportional to the bulk coherence length $\xi(T)$ i.e. as $T \to T_\lambda$, $\xi(T) \to \infty$ and the superfluid density requires a longer and longer healing length before it reaches its bulk value. If the system thickness is of the order of $\xi$ and the boundary condition is that the order parameter, $\psi(\mathbf{r})$, vanish at both boundaries, then $\psi$ is expected to be zero everywhere. Hence qualitatively we have explained the vanishing of superflow in thin films.

From a theoretical viewpoint an analysis of film superfluidity in terms of a healing length seems (at least qualitatively) quite plausible, if we associate it with a macroscopic wavefunction which must vanish at the boundary and which we would expect to build up to a non-zero value away from the wall. It is not clear to me that there is a general argument that this value

† c.f. ref. 58 for a more detailed discussion of $d_0$.

should necessarily yield the bulk superfluid density even in films ten or twenty layers thick, although presumably it might be expected in really thick films which we would associate with the bulk limit.

I do not intend to discuss the GP theory here. From a theoretical viewpoint it seems to me to be of rather doubtful validity except perhaps near the normal-superfluid transition.† In our restricted geometry problem it seems troublesome to identify a reasonable spatially varying order-parameter and particularly awkward to make the association that

$$\rho_s = |\psi|^2 \tag{VI.3}$$

which even in bulk $He^4$ is known to be too naïve.[59] Usually in critical phenomena problems we define a spatially varying order-parameter, $\psi$, by some coarse grained averaging procedure, e.g. an average over small volumes. For films of only a few layers thickness this procedure seems rather difficult to establish.

However, it must be remarked that the experiments have been fitted fairly self-consistently to this type of theory. Further GP theory does predict[32] some deviation from the form for $\overline{\rho_{sf}}$ for rather thin films (but before superflow cessation) which would lead to some change from linear behaviour in a $\overline{\rho_{sf}}$ versus $d^{-1}$ plot. Some deviation from linearity has been observed in microbalance studies[36] and third sound measurements[32] but quantitatively the agreement is not good.

## 2. LANDAU APPROACH

We have as yet not progressed very far with any quantitative understanding of the superfluid density in restricted geometries. Some progress[60, 61] has been made in understanding the importance of restricted geometries in determining the low temperature behaviour, by following Landau's arguments[4] for the contribution of the elementary excitation spectrum to the normal density, $\rho_n$. For $T < 1$ K, the elementary excitations will be phonons (no rotons will be present) with energy $\varepsilon_0(k) = \hbar ck$ where $c$ is the sound velocity. Suppose the background fluid has a velocity

$$\mathbf{v}_s = v_s\hat{\mathbf{z}} \quad (\hat{\mathbf{z}} \text{ is always taken in the direction of an infinite dimension})$$

then the momentum density is

$$\mathbf{j} = \rho\mathbf{v}_s + V^{-1}\sum_{\mathbf{k}} \hbar\mathbf{k}n_{\mathbf{k}}, \tag{VI.4}$$

† c.f. Chester's lectures in these Proceedings for a discussion of the establishment of similar equations at low temperatures and the difficulties involved.

where the second term gives the momentum of the excitations, $n_{\mathbf{k}}$ denotes the number of states with wave number $\mathbf{k}$ and $V$ is the volume.

We can show that $n_{\mathbf{k}} = n(\varepsilon_{\mathbf{k}})$ where

$$n(\varepsilon_{\mathbf{k}}) = \left[\exp\left(\frac{\varepsilon_{\mathbf{k}}}{k_B T}\right) - 1\right]^{-1} \tag{VI.5}$$

and $\varepsilon_{\mathbf{k}}$ is the spectrum in the frame of the walls:

$$\varepsilon_{\mathbf{k}} = \varepsilon_0(\mathbf{k}) + \hbar\mathbf{k}\cdot\mathbf{v}_s. \tag{VI.6}$$

Provided $v_s \ll c$ we can linearize (VI.4) to give

$$\mathbf{j} = \rho\mathbf{v}_s + \hbar^2 V^{-1}\sum_{\mathbf{k}}(\mathbf{v}_s\cdot\mathbf{k})\mathbf{k}\left.\frac{\mathrm{d}n(\varepsilon)}{\mathrm{d}\varepsilon}\right|_{\varepsilon=\varepsilon_0} \tag{VI.7}$$

$$\equiv (\rho - \rho_n)\mathbf{v}_s.$$

This defines the normal density:

$$\rho_n = -\hbar^2 V^{-1}\sum_{\mathbf{k}} k_z^2 \left.\frac{\mathrm{d}n(\varepsilon)}{\mathrm{d}\varepsilon}\right|_{\varepsilon=\varepsilon_0}. \tag{VI.8}$$

At this point we are able to see how the system dimensionality will affect the normal density. In bulk liquid $\Sigma_{\mathbf{k}}$ goes over to $V(2\pi)^{-3}\int \mathrm{d}^3k$; for a quasi-two-dimensional system ($\infty \times \infty \times d$) it goes to $A(2\pi)^{-2}\Sigma_{k_x}\int \mathrm{d}k_y\,\mathrm{d}k_z$; and in the quasi one-dimensional case it becomes

$$\frac{L}{2\pi}\sum_{k_x}\sum_{k_y}\int \mathrm{d}k_z.$$

In the filled pore case (hard wall boundary conditions) the $k_x(k_y)$ summation is over values

$$k_x = \frac{m\pi}{d}, \qquad m = 0, 1, 2 \ldots.$$

For simplicity we restrict ourselves to the limit in which $d_i$ is very small; our criterion for $d_i$ being small is

$$\hbar c\pi d_i^{-1} \geqslant k_B T. \tag{VI.9}$$

(for $d_i = 100$ Å this energy corresponds to about 0·5 K). If $d_i$ satisfies this inequality for one (two) dimensions we shall refer to the system as "two" ("one") dimensional. If the inequality is satisfied then the summation over $k_x$ (say) reduces to one term, the $k_x = 0$ mode.

Let us look at the one-dimensional situation ($\infty \times d \times d$):

$$\rho_n = -\frac{L}{2\pi}\frac{\hbar^2}{V}\sum_{k_x}\sum_{k_y}\int \mathrm{d}k_z k_z^2 \left.\frac{\mathrm{d}n(\varepsilon)}{\mathrm{d}\varepsilon}\right|_{\varepsilon=\varepsilon_0}$$

$$= -\frac{L}{2\pi}\frac{\hbar^2}{V}\int \mathrm{d}k_z k_z^2 \frac{\exp[\hbar c k_z/(k_B T)]}{\{\exp[\hbar c k_z/(k_B T)]-1\}^2}\frac{1}{k_B T}$$

$$= \frac{\hbar}{2\pi c d^2 \lambda_P^2}\int \mathrm{d}x \frac{x^2\,\mathrm{e}^x}{(\mathrm{e}^x-1)^2} \qquad \text{(VI.10)}$$

where $x = \hbar c k_z/k_B T$ and $\lambda_p = \hbar c/k_B T$.

Hence on integrating, in one dimension

$$\rho_n^{[1]} = (\tfrac{1}{3}\pi)\hbar c^{-1}\lambda_p^{-2}d^{-2}. \qquad \text{(VI.11)}$$

Similarly it may be shown: in three dimensions (bulk)

$$\rho_n^{[3]} = \left(\frac{2\pi^2}{45}\right)\hbar c^{-1}\lambda_p^{-4}, \qquad \text{(VI.12)}$$

and in two dimensions

$$\rho_n^{[2]} = \left[\frac{3}{2}\frac{\zeta(3)}{\pi}\right]\hbar c^{-1}\lambda_p^{-3}d^{-1}. \qquad \text{(VI.13)}$$

The most interesting features of these results are (a) the temperature dependence

$$\rho_n^{[3]} \propto T^4$$
$$\rho_n^{[2]} \propto T^3$$
$$\rho_n^{[1]} \propto T^2 \qquad \text{(VI.14)}$$

and (b) the enhancement of $\rho_n$ in restricted geometries over its bulk value.

The specific heat for these situations may be calculated from the expression of the free energy and gives temperature dependence as follows:

$$C_V^{[3]} \sim T^3,$$
$$C_V^{[2]} \sim T^2,$$
$$C_V^{[1]} \sim T. \qquad \text{(VI.15)}$$

No specific heat measurements have been made at low enough temperatures to enable us to check these predictions.

This theory is quite a nice extension of our ideas about the bulk superfluid, the only theory (except for *GP*) developed far enough to give us a quantitative estimate of $\rho_n$. The results explain (at least qualitatively) the (experimentally observed) enhancement of $\rho_n$ over the bulk value and its increase with narrowing pore size. Unfortunately the powder experiments give a linear temperature dependence[40] of $\rho_n$—even in one-dimension the theory predicts a $T^2$ variation. Padmore[60] has suggested that the powder situation is quasi zero-dimensional (no dimension is large compared to $\lambda_p$ and the variation of the substrate potential along the tube plays a vital role) and has put forward arguments leading to a $\rho_n \propto T$. He pointed out that a linear $T$ dependence for $\rho_n$ could be obtained from an assumption of few momentum carrying states with energies $\sim k_B T$ and a large density of states for lower energies. He suggested that the above requirement would probably be satisfied by the assumption that the powder geometry comprised open volumes fairly uniform in size and connected by substantially smaller cracks or channels. However the arguments seem to me to be not totally clearcut and rather complex to formulate in detail, and I would say that anyway a full analysis of the powder situation remains open. It would be useful to see the one- and two-dimensional predictions of this theory tested experimentally by measurements of $\rho_s$, and the specific heat in pores and between plates.

## 3. VORTICES

Another approach[62–64] to the restricted geometry situation, this time in the higher temperature region, emphasizes the importance of vortex excitations and their crucial role in destroying superflow. This approach allows an estimate of the superflow onset temperature $T_0$ to be made.

The argument suggests that although a finite fraction of the particles cannot occupy a single quantum state for the infinite system, it is reasonable to surmise the presence of a fair degree of short-range order—local Bose–Einstein condensation. Superflow may be sustained (no detailed description of it is given) provided the system is in a metastable state from which large energy fluctuations are required for escape, the destruction mechanism being the formation of free vortices.

The destruction of superflow in the bulk by vortex formation has been discussed by Langer;[49, 65] if a singly quantized vortex line crosses a moving superfluid the average velocity changes by $\pm(\hbar/mL)$, that is the system jumps from one metastable state to another.

In two dimensions in the low temperature region it is energetically unfavourable to form free vortices, although it is possible to form bound pairs of equal and opposite vorticity. For free vortices the energy $E = \alpha \ln (A/A_0)$ and the entropy $S = k_B \ln (A/a^2)$ (where $A$ is the area of the system, $A_0$ is an area of order $a^2$ where $a$ is the lattice spacing) and the free energy is, of course, given by

$$F = E - TS. \tag{VI.16}$$

At a certain temperature ($T_\alpha$ say) it becomes energetically favourable to form free vortices (i.e. the entropy term dominates the free energy) and fluctuations between metastable states become probable. (We could imagine in a plot of the free energy against some parameter characterizing the states of the system, that at temperatures near $T_\alpha$ the free energy barriers between metastable states start to fall.) We associate $T_\alpha$ with the experimentally observed superflow onset.

Using these ideas we can carry out a (suggestive but non-rigorous) dimensional analysis. There are two characteristic energies in the problem, the thermal energy $k_B T$ and a typical flow energy. In a three-dimensional situation we suppose that the flow pattern is restricted to a three-dimensional volume $R^3$. The energy must have the form:

$$E_a \sim \rho_s v^2 R^3 \tag{VI.17}$$

where $\mathbf{v}$ is some characteristic velocity—but the only velocity in the theory is $\mathbf{v}_s$ and the only other quantities with dimensions are $\hbar$ and $m$. So the only characteristic length (other than the correlation length $\xi \propto \rho_s^{-1}$) that can be formed is $\hbar/mv_s$.

Substituting this for $R$ we obtain

$$E_a^{[3]} \propto \frac{\rho_s}{v_s}\left(\frac{\hbar}{m}\right)^3. \tag{VI.18}$$

By a similar argument for fluctuations constrained in a film of thickness $d$ we find

$$E_a^{[2]} \propto \rho_s v^2 \, \mathrm{d}R^2 \propto \rho_s \, \mathrm{d}(\hbar/m)^2. \tag{VI.19}$$

It is assumed that when $E_a$ reaches some critical value a transition will occur. This critical value varies from theory to theory but is of $O(1)$, thus in three dimensions we expect

$$\frac{\rho_s(T_\alpha)}{V_s(T_\alpha)}\left(\frac{\hbar}{m}\right)^3 \sim \text{constant}, \tag{VI.20}$$

and in two dimensions

$$\rho_s(T_\alpha)\, d\left(\frac{\hbar}{m}\right)^2 \sim \text{constant}. \tag{VI.21}$$

Note the interesting features that (i) the critical temperature, $T_\alpha$, for two dimensions is thickness dependent; (ii) while for three dimensions from (VI.20) we might anticipate a similar temperature dependence $\rho_s$ for the critical velocity, no such association is implied for the two dimensional case i.e. superflow may disappear at finite $\rho_s$. A fuller treatment of the two-dimensional situation has been given by Kosterlitz and Thouless[62, 63] and by Penrose.[64] I follow the latter's treatment here. In a two-dimensional ideal Bose gas, the number of particles with momentum $\hbar\mathbf{k}$ is

$$n_{\mathbf{k}} = \left\{\exp\left[\frac{\hbar^2}{2mk_BT}\left(k^2 + \frac{1}{l^2}\right)\right] - 1\right\}^{-1}$$

$$\sim \frac{4\pi}{\lambda_T^2(k^2 + 1/l^2)} \qquad \text{for } k \ll \lambda_T^{-1} \tag{VI.22}$$

where $\lambda_T$ is the thermal wavelength, $m$ is the atomic mass and the length, $l$, which is related to the chemical potential, is the reciprocal of the range of $\mathbf{k}$-values for which $n_{\mathbf{k}}$ is not far from its maximum.

We can calculate $l$ from the formula for the number of particles per unit area

$$\frac{N}{A} = \lim_{A\to\infty} \frac{1}{A} \sum_{\mathbf{k}} n_{\mathbf{k}}$$

$$= (2\pi)^{-2} \int_0^\infty n_{\mathbf{k}} 2\pi k \, dk$$

$$= \lambda_T^{-2} \ln\left\{\left[1 - \exp\left(-\frac{\hbar^2}{2mk_BTl^2}\right)\right]^{-1}\right\} \tag{VI.23}$$

which gives

$$l \sim (4\pi)^{-\frac{1}{2}} \lambda_T \exp\left(\tfrac{1}{2}\rho'\lambda_T^2\right),$$

where $\rho' = N/A$ is the number of particles/unit area. It is a condition of validity of the formula that $l \gg \lambda_T$ so that $\frac{1}{2}\rho'\lambda_T^2$ is significantly greater than 1. $l$ may be interpreted as the range of coherence of the single particle wave-functions. (In three dimensions this length would be infinite—off-diagonal long-range order.) For two dimensions this may be illustrated by taking the Fourier transform of (VI.23) to obtain an approximation to the one-particle density matrix

$$\langle \mathbf{q} | \sigma_1 | \mathbf{q} + r \rangle \sim \lambda_T^{-2} \left(\frac{2\pi l}{r}\right)^{\frac{1}{2}} e^{-r/l} \quad \textit{if } l \ll r. \tag{VI.24}$$

The above calculation is appropriate for a strictly two-dimensional gas, but it also works for a three-dimensional gas contained between infinite plane walls of separation $d$, provided that $d \ll \lambda_T$. If $d \gg \lambda_T$ the calculation is more complicated but very similar—$\rho' = N/A$ is replaced by

$$\rho' = \left(1 - \frac{2\cdot 612}{\rho\lambda_T^3}\right)\rho d \tag{VI.25}$$

where $\rho$ is the number density and the new term $(2\cdot 612)/\rho\lambda_T^3$ is the contribution from particles with non-zero momentum perpendicular to the slab. It is now assumed that (VI.23) holds in the presence of interactions and that the above formula for $\rho'$ can be replaced by

$$\rho' = \rho_s d \, (d \gg \lambda_T) \tag{VI.26}$$

where $\rho_s$ is the bulk superfluid number density. This seems to me to be a rather strong assumption.

Now consider the formation of a vortex—because of the finite coherence length the largest possible quantized vortex in the liquid may be expected to have a radius about $l$. Taking the core radius to be $\lambda_T$ and the velocity at a distance $r$ from the centre to be

$$v = \frac{\hbar}{mr}$$

we find that such a vortex has an energy

$$\begin{aligned} E &= \tfrac{1}{2}m\rho' \int_{\lambda_T}^{l} \left(\frac{\hbar}{mr}\right)^2 2\pi r \, \mathrm{d}r \\ &= \pi\rho' \left(\frac{\hbar^2}{m}\right) \ln\left(\frac{l}{\lambda_T}\right) \\ &\sim \tfrac{1}{2}k_B T \rho' \lambda_T^2 (\tfrac{1}{2}\rho' \lambda_T^2 - \tfrac{1}{2}\ln 4\pi). \end{aligned} \tag{VI.27}$$

A rough estimate of the number of large vortices/unit area is given by $\lambda_T^{-2}\exp(-E/k_B T)$ where it is assumed that the number of linearly independent vortex states/unit area is $\lambda_T^{-2}$. The condition for just one large vortex in the entire system is:

$$A\lambda_T^{-2}\exp(-E/k_B T) \sim 1 \tag{VI.28}$$

which gives

$$\rho_s \frac{d\hbar^2 2\pi}{mk_B T} \sim 7 \quad \text{for} \quad A \sim 10^{-10}\,\text{cm}^2,$$

i.e. a transition will be expected at a temperature $T_0$ where

$$k_B T_0 \sim \frac{\rho_s d\hbar^2 2\pi}{7m}. \tag{VI.29}$$

Above $T_0$ free vortices are present and the system can easily fluctuate between states of different circulation—superflow is destroyed.

Note (as mentioned above) that this approach suggests disappearance of superflow at finite $\rho_s$. Penrose has checked his $T_0$ values for different film thicknesses on Grafoil and finds reasonable agreement.

This treatment is rather nice, but it certainly does not fully solve the problem—a number of difficulties remain e.g. the assumption that $\rho_s$ in (VI.26) is in fact bulk $\rho_s$ and a reconciliation between Penrose's value for $\rho'\lambda_T^2 \sim 7$ and Kosterlitz and Thouless' value of $\rho'\lambda_T^2 \sim 4$. If indeed this theory is correct, what is the nature of the region just above $T_0$? Is there a temperature region in which local condensation is present, but superflow is absent and if so does this correspond to the region between $T_A$ and $T_B$ in the Grafoil measurements?[41, 42]

## VII. Conclusion

Our study of superfluidity in restricted geometries has thrown up some problems which suggest that our understanding of superfluidity even in bulk is far from perfect. The major dilemma has been that theoretically there is a rigorous theorem that Bose condensation is absent from one and two-dimensional systems of infinite extent while experimentally superflow is certainly observed. We have suggested two possible explanations: (i) that the finiteness of all real systems means that Bose condensation cannot be totally excluded and (ii) that superflow in films and filled pores is essentially a metastable problem and the presence or absence of a finite $n_0$ is irrelevant. I would favour the second suggestion. An experimental test perhaps in a one-dimensional system might throw some light on this question.

We have shown that some of the features of the experimental results can be explained theoretically at least qualitatively. Into this category fall the specific heat measurements. We have discussed how the excitation spectrum (both for phonons and vortices) might be affected by a restricted geometry and have indicated the effect this might have on the superfluid properties. While these suggestions for calculating, say, $\rho_s$ and $T_0$ are plausible, the theories are far from being conclusive. In finite geometries we seem even further away from a microscopic calculation of $\rho_s$ than in the bulk.

The experimental scene does not appear very satisfactory either. The experimental systems seem rather messy and the recent Grafoil results shake our hope that the results are substrate independent. More experimental work, particularly if possible in filled smooth pores (to eliminate free surface effects), is called for.†

I hope I have managed to convey to you some of the pecularities that superfluid $He^4$ would exhibit in a one- or two-dimensional world. In conclusion let me remark that perhaps the study of films of superfluid $He^3$ (for which even in the bulk we expect an anisotropic superfluid density[(66)]) would reveal even knottier problems and stronger boundary effects than those I have discussed here for $He^4$.

## References

1. L. D. Landau and E. M. Lifshitz. "Statistical Physics", Pergamon Press, London (1958), p. 482.
2. V. K. S. Shante and S. Kirkpatrick. *Adv. Phys.* **20**, 325 (1971).
3. R. E. Peierls. *Ann. Inst. Henri Poincare* **5**, 177 (1935).
4. L. Landau. *J. Phys.* (*U.S.S.R.*) **5**, 71 (1941).
5. See also Chester's lectures in these Proceedings.
6. M. W. Cole and W. F. Saam. *Phys. Rev. Lett.* **32**, 985 (1974) and preprint.
7. F. London. "Superfluids", Wiley, New York (1954), Vol. II, Chap. C.
8. M. F. M. Osborne. *Phys. Rev.* **76**, 396 (1949).
9. D. A. Krueger. *Phys. Rev.* **172**, 211 (1968).
10. M. N. Barber and M. E. Fisher. *Phys. Rev.* **A8**, 1124 (1972).
11. P. C. Hohenberg. *Phys. Rev.* **158**, 383 (1967).
12. G. V. Chester, M. E. Fisher and N. D. Mermin. *Phys. Rev.* **185**, 760 (1969).
13. D. A. Krueger. *Phys. Rev. Lett.* **19**, 563 (1967).
14. D. Jasnow and M. E. Fisher. *Phys. Rev. Lett.* **23**, 286 (1969).
15. D. Jasnow and M. E. Fisher. *Phys. Rev.* **B3**, 895 (1971).
16. Strictly speaking these proofs deny the possibility of a non-zero value for the anomalous average

$$\psi_0 = V_\Omega^{-1} \int_\Omega \mathrm{d}^3 r \, \langle \psi(\mathbf{r}) \rangle,$$

but for our purposes this is equivalent to the absence of Bose condensation.
17. For those unfamiliar with the introduction of an anomalous field it may be helpful to consider the analogy of a magnetic system. In a three-dimensional Heisenberg ferromagnet, for example, a spontaneous magnetisation is expected below the Curie temperature, but the direction in which it will point is undetermined, all directions are equally likely. To define a direction (break the symmetry) a magnetic field (in the $z$-direction, say), is introduced, whose magnitude may be arbitrarily small but which will cause the magnetisation to align in the $z$-direction.

† A possible experimental system might be an irradiated mica sheet in which fairly smooth straight pores of diameter 100 Å are not hard to obtain for sheet thicknesses up to 5 μm. In two dimensions it might be interesting to carry out measurements (for example fourth sound measurements of $\rho_s$) for helium trapped between the leaves of Grafoil. It would also be useful to carry out film experiments on a number of new substrates to explore the extent of the substrate dependence of film properties.

18. G. V. Chester. *In* "Lectures in Theoretical Physics", (K. T. Mahanthapa, Ed.), Gordon and Breach, New York (1968), Vol. XI, p. 253.
19. J. F. Fernandez. *Phys. Rev.* **A3**, 1104 (1971).
20. We note the wide ranging applicability of the Bogoliubov approach to phase transitions. Very similar approaches to that employed here establish the impossibility of spontaneous magnetisation (for $T > 0$) in one- and two-dimensional systems with a broken continuous symmetry (e.g. in Heisenberg ferromagnets, but not of course in Ising systems).
21. D. F. Brewer. *J. Low Temp. Phys.* **3**, 205 (1970).
22. T. G. Wang. *Phys. Rev. Lett.* **31**, 344 (1973).
23. J. G. Dash. *J. Low Temp. Phys.* **1**, 173 (1969).
24. C. E. Campbell and M. Schick. *Phys. Rev.* **A**3, 691 (1971).
25. M. Bretz and J. G. Dash. *Phys. Rev. Lett.* **27**, 647 (1971).
26. H. P. R. Frederikse. *Physica* **15**, 860 (1949).
27. S. V. R. Mastrangelo and J. G. Aston. *J. Chem. Phys.* **19**, 1370 (1951).
28. D. F. Brewer, D. C. Champeney and K. Mendelssohn. *Cryogenics* **1**, 108 (1960).
29. E. A. Long and L. Meyer. *Phys. Rev.* **85**, 1030 (1952).
30. K. R. Atkins, H. Seki and E. U. Condon. *Phys. Rev.* **102**, 582 (1956).
31. I. Rudnick and J. C. Fraser. *J. Low Temp. Phys.* **3**, 225 (1970).
32. B. Ratnam and J. Mochel. *J. Low Temp. Phys.* **3**, 239 (1970).
33. D. F. Brewer, G. W. Leppelmeier, C. C. Lim, D. O. Edwards and J. Landau. *Phys. Rev. Lett.* **19**, 491 (1967).
34. I. Rudnick, E. Guyon, K. A. Shapiro and S. A. Scott. *Phys. Rev. Lett.* **19**, 488 (1967).
35. M. Kriss and I. Rudnick. *J. Low Temp. Phys.* **3**, 339 (1970).
36. M. Chester and L. C. Yang. *Phys. Rev. Lett.* **31**, 1377 (1973).
37. M. Chester and L. C. Yang. *Phys. Rev.* **A9**, 1475 (1973).
38. R. P. Henkel, G. Kukich and J. D. Reppy. *In* "Proceedings of the Eleventh International Conference on Low Temperature Physics, St. Andrews, 1968", (J. F. Allen, D. M. Finlayson and D. M. McCall, Eds), University of St. Andrews Press, St. Andrews, Scotland (1968), p. 178.
39. R. P. Henkel, E. N. Smith and J. D. Reppy. *Phys. Rev. Lett.* **23**, 1276 (1969).
40. F. D. M. Pobell, H. W. Chan, L. R. Corruccini, R. P. Henkel, S. W. Schwenterly and J. D. Reppy. *Phys. Rev. Lett.* **28**, 542 (1972).
41. J. A. Herb and J. G. Dash. *Phys. Rev. Lett.* **29**, 846 (1972).
42. M. Bretz. *Phys. Rev. Lett.* **31**, 1447 (1973).
43. Y. Imry. *Ann. Phys. (New York)* **51**, 1 (1969).
44. W. Kohn and D. Sherrington. *Rev. Mod. Phys.* **42**, 1 (1970).
45. A. J. Leggett. *Phys. Fenn.* **8**, 125 (1973).
46. G. B. Hess and W. M. Fairbank. *Phys. Rev. Lett.* **19**, 216 (1967).
47. M. A. Eggington and A. J. Leggett. "Collective Phenomena" (to be published).
48. W. A. Little. *Phys. Rev.* **156**, 396 (1967).
49. J. S. Langer. *Phys. Rev.* **167**, 183 (1968).
50. L. P. Kadanoff, W. Götze, D. Hamblen, R. Hecht, E. A. S. Lewis, V. V. Palciauskas, M. Rayl and J. Swift. *Rev. Mod. Phys.* **39**, 395 (1967).
51. J. Wilks. "The Properties of Liquid and Solid Helium", Clarendon Press, Oxford, (1967).
52. M. E. Fisher and A. E. Ferdinand. *Phys. Rev. Lett.* **19**, 169 (1967).
53. A. E. Ferdinand and M. E. Fisher. *Phys. Rev.* **185**, 832 (1969).

54. M. N. Barber. *Aust. J. Phys.* **26**, 483 (1973).
55. M. A. Moore. *Phys. Lett.* **A37**, 345 (1971).
56. S. Doniach. *Phys. Rev. Lett.* **31**, 1450 (1973).
57. V. L. Ginzburg and L. P. Pitaevskii. *Zh. Eksp. Teor. Fiz.* **34**, 1240 (1958); *Sov. Phys. JETP* **7**, 858 (1958).
58. E. A. Sabisky and C. H. Anderson. *Phys. Rev. Lett.* **30**, 1122 (1973).
59. B. D. Josephson. *Phys. Lett.* **21**, 608 (1966).
60. T. C. Padmore. *Phys. Rev. Lett.* **28**, 1512 (1972).
61. H. Haug. *J. Low Temp. Phys.* **12**, 479 (1973).
62. J. M. Kosterlitz and D. J. Thouless. *J. Phys. C* **5**, L124 (1972).
63. J. M. Kosterlitz and D. J. Thouless. *J. Phys. C* **6**, 1181 (1973).
64. O. Penrose. *In* "Quantum Statistical Mechanics in the Natural Sciences", (S. Mintz *et al.*, Eds), Plenum, New York (to be published).
65. J. S. Langer and J. D. Reppy. "Progress in Low Temperature Physics", (C. J. Gorter, Ed.), North Holland, Amsterdam (1970), Vol. VI, p. 1.
66. W. M. Saslow. *Phys. Rev. Lett.* **31**, 870 (1973).

# 6. THEORY OF FERMI LIQUIDS

A. RON

*Physics Department, Technion-Israel Institute of Technology, Haifa, Israel*

## INTRODUCTION

A system of interacting particles obeying Fermi–Dirac statistics is called a Fermi liquid. The theory of Fermi liquids, which is mainly due to Landau, is concerned with the low temperature behaviour of the Fermi liquid, i.e. with temperatures much less than the kinetic energy per particle. In these lectures we consider only *normal* Fermi liquids, when the liquid retains the essential properties of the non-interacting Fermi gas and the interactions do not drastically modify these properties. Examples of normal Fermi liquids are neutral $He^3$ above a few millidegrees and the charged conduction electrons in non-superconducting metals.

The properties of normal Fermi liquids are dominated by Fermi–Dirac statistics, while the interactions between particles may cause only gradual changes of the properties and thus perturbation theory is in principle applicable. If the interactions cause *drastic* changes of the system, in which condensation occurs and the ground state is fundamentally different from the non-interacting state, the perturbation approach is out of the question. The Landau theory assumes that the transition from the non-interacting gas to the liquid is smooth, and we do not need to ask the details of how the quantum states of the liquid were reached from those of the gas.

In the first part of these lectures we review the properties of the Fermi gas, indicate the pre-Landau attempts to understand interacting Fermi systems (e.g. Hartree approximation), and introduce the basic concepts of the theory. Then we study the phenomenological theory of Fermi liquids, starting with the neutral system and covering both equilibrium properties and transport phenomena. We close with a microscopic derivation of the Landau theory.

# I. The Fermi Gas

## 1. THE FERMI DISTRIBUTION

Consider a system of non-interacting particles which obey Fermi–Dirac statistics. Each particle is characterized by its mass $m$, momentum $\mathbf{p}$, and spin $\sigma$. The system is assumed to be in equilibrium with a thermal bath at temperature $T$, and a particle bath at chemical potential $\mu$. The probability that a particle has energy $\varepsilon$ is given by the Fermi distribution $n^0(\varepsilon)$, i.e.

$$n^0(\varepsilon) = \frac{1}{1 + \exp\{(\varepsilon - \mu)/kT\}} \tag{I.1}$$

where $k$ is Boltzmann's constant. In the absence of a magnetic field, the energy of a particle depends only on the momentum $\mathbf{p}$, i.e. $\varepsilon = \varepsilon_{\mathbf{p}} = p^2/2m$, and the momentum spin distribution, $n^0_{\mathbf{p}\sigma}$, is

$$n^0_{\mathbf{p}\sigma} = \frac{1}{1 + \exp\{(\varepsilon_{\mathbf{p}} - \mu)/kT\}}. \tag{I.2}$$

At low temperatures, if the density $n$ is given, the chemical potential, $\mu(n, T)$, can be solved from the equation

$$n = \sum_{\mathbf{p}\sigma} n^0_{\mathbf{p}\sigma}(T, \mu), \tag{I.3}$$

and is given by

$$\mu(n, T) = \mu_0(n)\left\{1 - \frac{\pi^2}{12}\left(\frac{kT}{\mu_0}\right)^2 + O(T^4)\right\}, \tag{I.4}$$

where $\mu_0(n) = \mu_0$ is the zero-temperature value of the chemical potential.

The ground state of the system is described in terms of a Slater determinant (to take account of the Pauli principle) made up of single-particle plane waves of momentum $p \leqslant p_f$, where $p_f$ is the Fermi momentum. Thus, the ground-state distribution is

$$n^0_{\mathbf{p}\sigma} = \begin{cases} 1 & p \leqslant p_f \\ 0 & p > p_f \end{cases} \tag{I.5}$$

and $\mu = \varepsilon_f = p_f^2/2m$, where $\varepsilon_f$ is the Fermi energy. The excited states of the gas are also described by Slater determinants, but $n_{\mathbf{p}\sigma}$ is then different from (I.5). The energy of the system is

$$E = \sum_{\mathbf{p}\sigma} n_{\mathbf{p}\sigma}\, \varepsilon_{\mathbf{p}} \tag{I.6}$$

and the momentum

$$\mathbf{P} = \sum_{\mathbf{p}\sigma} \mathbf{p}\, n_{\mathbf{p}\sigma}. \tag{I.7}$$

## 2. ELEMENTARY EXCITATIONS

It is convenient to describe the properties of the gas in terms of the one-particle states:

(a) The ground state has all the single-particle states up to $p_f$ occupied, and those above empty.

(b) An excited state is described in terms of the ground state plus the number of excitations, i.e. single-particle states occupied above $p_f$ and unoccupied below it. The amount of excitations is specified by the departure of the excited-states distribution function from that of the ground state

$$\delta n_{\mathbf{p}\sigma} = n_{\mathbf{p}\sigma} - n^0_{\mathbf{p}\sigma}, \tag{I.8}$$

where $\delta n_{\mathbf{p}\sigma}$ positive is a particle excitation, and negative is a hole excitation.

(c) The energy and momentum of the gas are given in terms of the excitations. If we denote by $E_0$ the ground state energy

$$E_0 = \sum_{\mathbf{p}\sigma} \varepsilon_{\mathbf{p}} n_{\mathbf{p}\sigma} = 2 \sum_{\mathbf{p} \leqslant \mathbf{p}_f} \varepsilon_{\mathbf{p}} = n\varepsilon_0, \tag{I.9}$$

where $\varepsilon_0 = \frac{3}{5}\mu$ is the average energy per particle, we can characterize the excited states by

$$E = E_0 + \sum_{\mathbf{p}\sigma} \varepsilon_{\mathbf{p}} \delta n_{\mathbf{p}\sigma} \tag{I.10}$$

and

$$\mathbf{P} = \sum_{\mathbf{p}\sigma} \mathbf{p} \delta n_{\mathbf{p}\sigma}. \tag{I.11}$$

(d) Since we deal with a many-body system the number of excitations (even for low-lying excited states and low $T$) is enormous, and it is convenient to introduce the concept of the *density of states of the excitation.* In p-space the one-particle state is a plane wave $V^{-\frac{1}{2}} \exp(i\mathbf{p}\cdot\mathbf{r}/\hbar)$ ($V$ is the volume of the system); and thus the number of states of excitations per unit volume is $1/(2\pi\hbar)^3$ per spin. The density of energy states per unit volume per spin is

$$\mathcal{N}(\varepsilon)\,\mathrm{d}\varepsilon = \frac{1}{(2\pi\hbar)^3} 4\pi p^2 \frac{\mathrm{d}p}{\mathrm{d}\varepsilon}\,\mathrm{d}\varepsilon, \tag{I.12}$$

i.e. with $p = \sqrt{(2m\varepsilon)}$

$$\mathcal{N}(\varepsilon) = \frac{mp}{2\pi^2\hbar^3}. \tag{I.13}$$

(e) The energy of an elementary excitation is conveniently measured with respect to the Fermi energy

$$\delta\varepsilon_{\mathbf{p}} = |\varepsilon_{\mathbf{p}} - \varepsilon_f|, \tag{I.14}$$

and the energy of an excited state of the gas is given by

$$E - E_0 = \sum_{\mathbf{p}\sigma} \varepsilon_f \delta n_{\mathbf{p}\sigma} + \sum_{\mathbf{p}\sigma} \delta\varepsilon_{\mathbf{p}} \delta n_{\mathbf{p}\sigma}. \tag{I.15}$$

## 3. EQUILIBRIUM PROPERTIES

We can now summarize the equilibrium properties of the Fermi gas at low temperatures and relatively high density

$$kT \ll \frac{\hbar^2}{m} n^{2/3}. \tag{I.16}$$

(a) The chemical potential is given by (I.4).

(b) The energy (I.10) of the gas can be written as

$$E - E_0 = 2\int \mathrm{d}\varepsilon\ \mathcal{N}(\varepsilon)\ \varepsilon\ \delta n(\varepsilon, T). \tag{I.17}$$

The integrand is different from zero only in an interval $kT$ around the Fermi energy $\varepsilon_f$, and thus we expect

$$E - E_0 \approx \mathcal{N}(\varepsilon_f)(kT)^2 \tag{I.18}$$

and the specific heat $C_V = \partial E/\partial T$ linear in $T$. Doing the integration one obtains $C_V = \gamma T$, where

$$\gamma = \frac{2\pi^2 k^2}{3}\ \mathcal{N}(\varepsilon_f) = \frac{k^2 m p_f}{3\hbar^3}. \tag{I.19}$$

(c) The compressibility $\kappa$ of the gas is defined by

$$\frac{1}{\kappa} = -V\left(\frac{\partial P}{\partial V}\right)_s$$

where $P$ is the pressure. At low temperatures $\mathrm{d}P = n\mathrm{d}\mu$, and with $n = N/V$ (where $N$ is the total number of particles) we obtain

$$\frac{1}{\kappa} = n^2\left(\frac{\partial \mu}{\partial n}\right)_V. \tag{I.20}$$

To increase the density of the non-interacting gas by $\delta n$ we must add $2\mathcal{N}(\mu)\delta\mu$ particles per unit volume, and thus

$$\frac{1}{\kappa} = \frac{n^2}{2\mathcal{N}(\mu)}. \tag{I.21}$$

(d) The paramagnetic susceptibility of the gas is $\partial M/\partial\mathcal{H}$, where $M$ is the (spin) magnetic moment of the system in an applied magnetic field $\mathcal{H}$. The extra energy due to the field is $-\beta\mathcal{H}$ for spin parallel to the field (up) and $+\beta\mathcal{H}$ for spin anti-parallel (down), where $\beta$ is the spin magnetic moment of a particle. To maintain the same chemical potential for the spin-up and spin-down particles the number of spin-up particles is increased by $\beta\mathcal{H}\mathcal{N}(\mu)$ and the number of spin-down particles is decreased by the same amount. The net magnetic moment is $M = 2\beta\mathcal{H}\mathcal{N}(\mu)\cdot\beta$ and the paramagnetic susceptibility

$$\chi_0 = 2\beta^2\mathcal{N}(\mu). \tag{I.22}$$

Notice that the equilibrium properties depend on the density of states on the Fermi surface $\mathcal{N}(\mu)$. This is a characteristic feature of Fermi systems at low $T$, and was considered as the main *a posteriori* justification of the independent models of electrons in metals. The success of the independent model of Bloch was a puzzle for many years, and before we turn to Landau's theory which explains it, we review some previous attempts to understand the success of the model.

## II. Interacting Fermi Gas

As an introduction to the Fermi liquid theory, where interactions are important, we review some aspects of the Fermi gas with interaction. The purpose of this short excursion to a model calculation is to motivate the ideas underlining the phenomenological theory. Thus we pick up only two results of self-consistent approximations, namely the Hartree–Fock corrections to the energy and the time-dependent self-consistent-field approach to collective excitations.

### 1. Self-consistent energy

We start with a system of interacting fermions with the hamiltonian

$$H = \sum_i \frac{\mathbf{p}_i^2}{2m} + \frac{1}{2}\sum_{i\neq j}\phi(\mathbf{r}_i - \mathbf{r}_j), \tag{II.1}$$

where $\mathbf{p}_i$ is the momentum of the $i$th fermion, $m$ the mass and $v(\mathbf{r}_i - \mathbf{r}_j)$ the interaction potential. Using the second quantization representation for fermions, with $a_{\mathbf{p}\sigma}$ and $a^\dagger_{\mathbf{p}\sigma}$ the annihilation and creation operators of momentum $\mathbf{p}$ and spin $\sigma$ we have

$$H = \sum_{\mathbf{p}\sigma} \varepsilon_{\mathbf{p}} a^\dagger_{\mathbf{p}\sigma} a_{\mathbf{p}\sigma} + \tfrac{1}{2} \sum_{\substack{\mathbf{pp}' \\ \sigma\sigma'}} \phi_{\mathbf{p}-\mathbf{p}'} \sum_{\mathbf{p}''} a^\dagger_{\mathbf{p}',\sigma} a^\dagger_{\mathbf{p}''+\mathbf{p}-\mathbf{p}',\sigma'} a_{\mathbf{p}'',\sigma'} a_{\ ,\sigma}, \tag{II.2}$$

and $a$, $a^\dagger$ obey the anticommutation rules. Since the number operator is $n_{\mathbf{p}\sigma} = a^\dagger_{\mathbf{p}\sigma} a_{\mathbf{p}\sigma}$, the first part of the hamiltonian can be written as

$$H^0 = \sum_{\mathbf{p}\sigma} \varepsilon_{\mathbf{p}} n_{\mathbf{p}\sigma}, \tag{II.3}$$

which corresponds to a free gas of fermions. The question of taking interactions into account, and still having a kind of free-fermion hamiltonian, is an old one, stemming from the fact that experiments on interacting Fermi systems (like metals) show manifestations of free-fermion behaviour. Formally we ask if it is possible to approximate $H$ of (II.1) as

$$H = \sum_{\mathbf{p}\sigma} E_{\mathbf{p}\sigma} n_{\mathbf{p}\sigma} \tag{II.4}$$

with

$$E_{\mathbf{p}\sigma} = \varepsilon_{\mathbf{p}} + \delta\varepsilon_{\mathbf{p}\sigma}, \tag{II.5}$$

where $\delta\varepsilon_{\mathbf{p}\sigma}$ carries the information about the interaction $v$.

The well-known self-consistent Hartree–Fock approximation does essentially what we want. The four $a$-operators in the interaction term of (II.2) are lumped into pairs and their momenta are restricted in such a way that they form $a^\dagger_{\mathbf{p}\sigma} a_{\mathbf{p}\sigma} = n_{\mathbf{p}\sigma}$ pairs. The result of this procedure is that (II.4) is obtained and $\delta\varepsilon$ of (II.5) takes the form

$$\delta\varepsilon_{\mathbf{p}\sigma} = \sum_{\mathbf{p}\sigma} f^{HF}(\mathbf{p}, \sigma; \mathbf{p}', \sigma') n_{\mathbf{p}'\sigma'} \tag{II.6}$$

with

$$f^{HF}(\mathbf{p}, \sigma; \mathbf{p}'\sigma') = (\delta_{\mathbf{p},\mathbf{p}'} - \delta_{\sigma,\sigma'})\phi_{\mathbf{p}-\mathbf{p}'}. \tag{II.7}$$

We do not study this approximation any further, but keep its form in mind when we formulate Landau's theory.

## 2. COLLECTIVE EXCITATIONS

We now turn to another aspect of the self-consistent approximation and ask what kind of collective modes are expected in the interacting Fermi gas. The momentum occupation operator $n_{\mathbf{p}\sigma} = a^\dagger_{\mathbf{p}\sigma} a_{\mathbf{p}\sigma}$ is now replaced in our study by the density matrix operator

$$\rho_{\sigma\sigma'}(\mathbf{p}, \mathbf{k}) = a^{\dagger}_{\mathbf{p}-\hbar\mathbf{k}/2,\sigma} a_{p+\hbar\mathbf{k}/2,\sigma'}. \tag{II.8}$$

The time dependence of this operator is obtained from the equation of motion of $a_{\mathbf{p}\sigma}$, i.e.

$$i\hbar \frac{\partial}{\partial t} a_{\mathbf{p}\sigma} = \{a_{\mathbf{p}\sigma}, H\} = \varepsilon_{\mathbf{p}} a_{\mathbf{p}\sigma} + \sum_{\mathbf{q}} \phi_{\mathbf{q}} \sum_{\mathbf{p}',\sigma'} a^{\dagger}_{\mathbf{p}'+\mathbf{q},\sigma'} a_{\mathbf{p}'\sigma'} a_{\mathbf{p}-\mathbf{q},\sigma}, \tag{II.9}$$

and it takes the form (with $\bar{\mathbf{k}} = \hbar\mathbf{k}$),

$$\left\{i\hbar \frac{\partial}{\partial t} + (\varepsilon_{\mathbf{p}-\bar{\mathbf{k}}/2} - \varepsilon_{\mathbf{p}+\bar{\mathbf{k}}/2})\right\} \rho_{\sigma\sigma'}(\mathbf{p}, \mathbf{k})$$

$$= \sum_{\mathbf{q}} \phi_{\mathbf{q}} \sum_{\mathbf{p}'\sigma''} \{a^{\dagger}_{\mathbf{p}-\bar{\mathbf{k}}/2,\sigma} a^{\dagger}_{\mathbf{p}'+\mathbf{q},\sigma''} a_{\mathbf{p}'\sigma''} a_{\mathbf{p}+\mathbf{q}+\bar{\mathbf{k}}/2,\sigma'}$$

$$- a^{\dagger}_{\mathbf{p}+\mathbf{q}-\bar{\mathbf{k}}/2,\sigma} a^{\dagger}_{\mathbf{p}'+\mathbf{q},\sigma''} a_{\mathbf{p}'\sigma''} a_{\mathbf{p}+\bar{\mathbf{k}}/2,\sigma'}\}. \tag{II.10}$$

Now consider the system under the influence of a spatially and temporally dependent disturbance. We can construct an ensemble of systems with the field present and define a "distribution function"

$$F_{\sigma\sigma'}(\mathbf{p}, \mathbf{k}) = \langle \rho_{\sigma\sigma'}(\mathbf{p}, \mathbf{k}) \rangle, \tag{II.11}$$

where $\langle\ \rangle$ indicates the ensemble average. If, instead of having an external field, we assume a fluctuation to be taking place, the same procedure can be applied. Introducing the thermal equilibrium distribution function

$$f_{\mathbf{p},\sigma} = \langle a^{\dagger}_{\mathbf{p}\sigma} a_{\mathbf{p}\sigma} \rangle_T, \tag{II.12}$$

where $\langle\ \rangle_T$ denotes the thermal equilibrium average, and pairing the four $a$-operators of (II.10), we can approximate the equation of motion for $F$ and obtain

$$\left(i\hbar \frac{\partial}{\partial t} + E_{\mathbf{p}-\bar{\mathbf{k}}/2,\sigma} - E_{\mathbf{p}+\bar{\mathbf{k}}/2,\sigma'}\right) F_{\sigma\sigma'}(\mathbf{p}, \mathbf{k})$$

$$= \delta_{\sigma\sigma'} \phi_{\mathbf{k}} \{f_{\mathbf{p}-\bar{\mathbf{k}}/2,\sigma} - f_{\mathbf{p}+\bar{\mathbf{k}}/2,\sigma}\} \sum_{\mathbf{p}''\sigma''} F_{\sigma''\sigma''}(\mathbf{p}'', \mathbf{k})$$

$$- \{f_{\mathbf{p}-\bar{\mathbf{k}}/2,\sigma} - f_{\mathbf{p}+\bar{\mathbf{k}}/2,\sigma'}\} \sum_{\mathbf{p}'} \phi_{\mathbf{p}-\mathbf{p}'} F_{\sigma\sigma'}(\mathbf{p}', \mathbf{k}). \tag{II.13}$$

In (II.13) $E_{\mathbf{p}\sigma}$ is the "Hartree–Fock" energy (II.5). Equation (II.13) for the averaged density matrix is homogeneous in $F$ and thus has the usual structure of a normal-mode equation; equating to zero the "operator" that is operating on $F$ we get the condition for this normal mode to survive in time.

We consider, for simplicity, the case of a $\delta$-function interaction $\phi_{\mathbf{k}} = \phi_{\mathbf{0}}$, and assume that the exchange term can be neglected. Equation (II.13), with spin indices suppressed, is then

$$\left(i\hbar\frac{\partial}{\partial t} - \hbar\mathbf{k}\cdot\mathbf{v}\right)F(\mathbf{p},\mathbf{k}) + \hbar\mathbf{k}\cdot\frac{\partial f}{\partial \mathbf{p}}\phi_0\sum_{\mathbf{p}'}F(\mathbf{p}',\mathbf{k}) = 0; \tag{II.14}$$

here $\mathbf{k}$ is assumed to be much smaller than $\mathbf{p}$, so that

$$E_{\mathbf{p}-\bar{\mathbf{k}}/2} - E_{\mathbf{p}+\bar{\mathbf{k}}/2} \to -\hbar\mathbf{k}\cdot\frac{\partial E_{\mathbf{p}}}{\partial \mathbf{p}} = -\hbar\mathbf{k}\cdot\mathbf{v} \tag{II.15}$$

and similarly for $f_{\mathbf{p}-\bar{\mathbf{k}}/2} - f_{\mathbf{p}+\bar{\mathbf{k}}/2}$. If we now assume an oscillatory time dependence for $F$, i.e. $F \propto e^{-i\omega t}\bar{F}$, we get

$$(\omega - \mathbf{k}\cdot\mathbf{v})F(\mathbf{p},\mathbf{k}) + \mathbf{k}\cdot\frac{\partial f}{\partial \mathbf{p}}\phi_0\sum_{\mathbf{p}'}F(\mathbf{p}',\mathbf{k}) = 0. \tag{II.16}$$

Equation (II.16) for the collective excitation of the system is to be kept in mind for comparison with the normal modes of quasiparticles in Fermi liquids.

## III. Phenomenological Theory of Fermi Liquids

We now consider a system of $n$ fermions per unit volume interacting with each other by a short-range potential. We assume that the system is *normal*, i.e. that if the interaction is turned on adiabatically the transition from the gas to the normal liquid is smooth and the classification of all states of the gas remains invariant. Thus we exclude formation of bound states and phase transitions to superfluidity. It was observed by Landau that the ground state and the low lying excited states of the normal liquid behave in much the same way as the corresponding states of the Fermi gas. The role of the elementary excitations of the non-interacting particles of the gas is now played by *quasiparticles*—the real low lying excitations of the liquid. The first important observation is to realize that these excitations obey Fermi statistics, and thus they are *almost* independent. The quasiparticles can be classified by their momentum $\mathbf{p}$ and energy $\varepsilon_{\mathbf{p}}$. If we denote by $n_{\mathbf{p}}$ the number of quasiparticles in the liquid with energy $\varepsilon_{\mathbf{p}}$, the energy of an excited state of the liquid is given by

$$E = E_0 + \sum_{\mathbf{p}}\varepsilon_{\mathbf{p}}n_{\mathbf{p}}, \tag{III.1}$$

where $E_0$ is the energy of the real ground state. If $\varepsilon_{\mathbf{p}}$ were independent of the number $n_{\mathbf{p}}$ of quasiparticles, then the analogy with the Fermi gas would be complete; however, it is the dependence of $\varepsilon$ on $n$ that makes the Landau theory a distinguished forward step to the understanding of Fermi liquids. The quasi-particle energy $\varepsilon_{\mathbf{p}}$, which is the change in the energy of the liquid due to the addition of a single quasiparticle, depends on the presence of other quasiparticles, i.e.

$$\varepsilon_{\mathbf{p}} = \varepsilon_{\mathbf{p}}^{0} + \sum_{\mathbf{p}'} f_{\mathbf{pp}'} n_{\mathbf{p}'}. \tag{III.2}$$

Equations (III.1) and (III.2) should be considered as expressions for $E$ and $\varepsilon_{\mathbf{p}}$ as power series in the small parameter $n_{\mathbf{p}}$, and the expansion is explicit only to first order. The success of the theory hinges on the fact that at low temperatures the number of quasiparticles is rather small, and can be used as a small parameter for the theory of Fermi liquids.

## 1. QUASI-PARTICLE INTERACTIONS

Before we turn to a more detailed study of the interaction between quasi-particles, we make (III.1) and (III.2) more precise and take into account the spin of the fermions. The main difficulty is that, though the ground state is well defined, the interaction of the quasiparticles rules out a precise definition of the notion of quasiparticle. An "elementary excitation" given in terms of a number of quasiparticles has a finite lifetime. A consistent treatment of the "gas" of quasiparticles should yield a lifetime decreasing with the distance from the Fermi momentum. The validity of the theory is thus restricted to small numbers of quasiparticles with energies not far from the Fermi energy.

We now consider the gas of interacting elementary excitations. The quasi-particle interactions are not small, but the gas is assumed to be dilute, so the expansion parameter, $\alpha$, is

$$\alpha = \frac{\text{No. of quasiparticles}}{\text{Total No. of particles}}.$$

The gas is described in terms of two quantum-mechanical operators for the quasiparticles:

(a) The density (matrix) operator $\rho_{\sigma\sigma'}(\mathbf{r}, \mathbf{r}')$

(b) The energy operator $E_{\sigma\sigma'}(\mathbf{r}, \mathbf{r}')$,

where $\mathbf{r}$ and $\sigma$ are the position and spin indices. The ensemble average of the spatial diagonal element $\rho_{\sigma\sigma'}(\mathbf{r}, \mathbf{r})$ plays the role of the distribution function of the quasiparticles. If we use spin operator notation [e.g. $\hat{\rho}(\mathbf{p}, \mathbf{r})$ stands for $\rho_{\sigma\sigma'}(\mathbf{p}, \mathbf{r})$] we can study the dynamics of the system using the equation of motion of the density operator

$$i\hbar \frac{\partial}{\partial t} \hat{\rho} = [\hat{E}, \hat{\rho}]. \tag{III.3}$$

It is convenient to introduce the mixed representation of the operators in analogy with the Wigner representation, e.g.

$$\rho_{\sigma\sigma'}(\mathbf{r}, \mathbf{r}') = \sum_{\mathbf{p}} \exp\left[i\mathbf{p}\cdot(\mathbf{r} - \mathbf{r}')/\hbar\right] \rho_{\sigma\sigma'}(\mathbf{p}, \mathbf{R}), \tag{III.4}$$

where $\mathbf{R} = \frac{1}{2}(\mathbf{r} + \mathbf{r}')$ and $\mathbf{p}$ play the role of position and momentum of the quasiparticles. We can generalize (III.2) and write

$$E(\mathbf{p}, \mathbf{r}, \boldsymbol{\sigma}) = \varepsilon(\mathbf{p}, \boldsymbol{\sigma}) + \mathrm{Tr}_{\sigma'} \sum_{\mathbf{p}'} f(\mathbf{p}, \boldsymbol{\sigma}; \mathbf{p}', \boldsymbol{\sigma}')\, \rho(\mathbf{p}', \mathbf{r}, \boldsymbol{\sigma}'), \tag{III.5}$$

where $f$ is assumed to be a "local interaction". $E$ of (III.5) is the "effective" hamiltonian of the quasiparticle.

The "hamiltonian" $E$ plays an important role in the phenomenological theory. $\varepsilon(\mathbf{p}, \boldsymbol{\sigma})$ is interpreted as the energy of a single quasiparticle, i.e. the energy added to the liquid in its ground state by adding a particle with momentum $\mathbf{p}$ and spin $\hat{\boldsymbol{\sigma}}$. In the absence of a magnetic field the energy of the quasiparticle is $\varepsilon = \hat{I}\varepsilon_{\mathbf{p}}$, where $\hat{I}$ is a unit matrix in the spin space, and $\varepsilon_{\mathbf{p}}$ is a continuous function of $\mathbf{p}$ and can be written as

$$\varepsilon_{\mathbf{p}} = \mu + v_f(p - p_f). \tag{III.6}$$

In (III.6) $\mu$ is the chemical potential, $p_f$ is the Fermi momentum and $v_f = p_f/m^*$ is the Fermi velocity, i.e. the velocity of a quasiparticle on the Fermi surface. The effective mass, $m^*$, can be related to the interaction parameter $f$ by Galilean invariance. When a magnetic field $\mathscr{H}$ is present, we should add to the energy a term $-\beta\hat{\boldsymbol{\sigma}}\cdot\mathscr{H}$, where $\hat{\boldsymbol{\sigma}}$ is the Pauli spin vector, and $\beta$ is the fermion magnetic moment.

The interaction energy, $f$, of two quasiparticles is assumed to be continuous when $\mathbf{p}$ and $\mathbf{p}'$ cross the Fermi surface. Only $f$ on the Fermi surface is needed and thus we take $f(\mathbf{p}, \boldsymbol{\sigma}; \mathbf{p}', \boldsymbol{\sigma}')$ to depend on the direction of $\mathbf{p} = p_f\mathbf{i}_p$ and $\mathbf{p}' = p_f\mathbf{i}_{p'}$ (where $\mathbf{i}_p$ is a unit vector in the direction of $\mathbf{p}$) and on the spins. We further assume that spin-orbit effects may be neglected and take

$$f(\mathbf{p}, \boldsymbol{\sigma}; \mathbf{p}'\boldsymbol{\sigma}') = \phi(\mathbf{i}_p \cdot \mathbf{i}_{p'}) + \hat{\boldsymbol{\sigma}}\cdot\hat{\boldsymbol{\sigma}}'\, \zeta(\mathbf{i}_p \cdot \mathbf{i}_{p'}); \tag{III.7}$$

with $\mathbf{i}_p \cdot \mathbf{i}_{p'} = \cos\xi$ we expand $\phi$ and $\zeta$ in Legendre polynomials

$$\begin{aligned} \phi(\mathbf{i}_p \cdot \mathbf{i}_{p'}) &= \sum_{l=0}^{\infty} \phi_l P_l(\cos\xi) \\ \zeta(\mathbf{i}_p \cdot \mathbf{i}_{p'}) &= \sum_{l=0}^{\infty} \zeta_l P_l(\cos\xi). \end{aligned} \tag{III.8}$$

It is convenient to define dimensionless quantities

$$F_l = 2\mathscr{N}_f\phi_l, \qquad Z_l = 2\mathscr{N}_f\zeta_l, \tag{III.9}$$

in terms of the density $\mathcal{N}_f$ of quasiparticle states per spin on the Fermi surface.

Using Galilean invariance we can now find the relationship between $m^*$ and the interaction $f$. Consider an arbitrary distribution $n_{\mathbf{p}}$ of quasiparticles (spin suppressed). The momentum $2\Sigma_{\mathbf{p}}\mathbf{p}n_{\mathbf{p}}$ of the system must be equal to the mass-density current $2\Sigma_{\mathbf{p}}m\mathbf{v}_{\mathbf{p}}n_{\mathbf{p}}$, where $\mathbf{v}_{\mathbf{p}}$ is the velocity of the quasiparticle. The velocity is obtained from (III.5) by assigning the momentum $\mathbf{p}$ to the hamiltonian

$$E(\mathbf{p}) = \varepsilon(\mathbf{p}) + 2\sum_{\mathbf{p}'} f_{\mathbf{pp}'}n_{\mathbf{p}'} \tag{III.10}$$

and taking the derivative with respect to $\mathbf{p}$.

Equation (III.6) gives

$$\frac{\partial}{\partial \mathbf{p}}\varepsilon(\mathbf{p}) = \frac{1}{m^*}\mathbf{p}, \tag{III.11}$$

where $\mathbf{p}$ is essentially $p_f$, and thus

$$2\sum_{\mathbf{p}}\mathbf{p}n_{\mathbf{p}} - 2\sum_{\mathbf{p}} m\frac{1}{m^*}\mathbf{p}n_p = 2\sum_{\mathbf{p}} mn_{\mathbf{p}}\frac{\partial}{\partial \mathbf{p}}2\sum_{\mathbf{p}'} f(\mathbf{p},\mathbf{p}')n_{p'}. \tag{III.12}$$

Integrating the right hand term by parts and exchanging $\mathbf{p}$ and $\mathbf{p}'$ we have

$$-m\,2\sum_{\mathbf{p}} n_{\mathbf{p}}\,2\sum_{\mathbf{p}'} f(\mathbf{p},\mathbf{p}')\frac{\partial}{\partial \mathbf{p}'}n_{\mathbf{p}'}.$$

But $n_{\mathbf{p}'}$ depends on energy only, i.e.

$$\frac{\partial}{\partial \mathbf{p}'}n_{\mathbf{p}'} = \frac{\partial n_{p'}}{\partial \varepsilon_{\mathbf{p}'}}\frac{\partial}{\partial \mathbf{p}'}\varepsilon_{\mathbf{p}'} = \frac{1}{m^*}\mathbf{p}'\frac{\partial n_{\mathbf{p}'}}{\partial \varepsilon_{\mathbf{p}'}};$$

since $n(\varepsilon_{\mathbf{p}})$ is significant only near the Fermi surface, we have

$$\frac{\partial n_{\mathbf{p}}}{\partial \varepsilon_{\mathbf{p}}} \approx -\delta(\varepsilon_{\mathbf{p}} - \mu).$$

The summation over $\mathbf{p}'$ is now carried out:

$$\begin{aligned}
2\sum_{\mathbf{p}'}\mathbf{p}' f(\mathbf{p}',\mathbf{p})\,\delta(\varepsilon_{\mathbf{p}'} - \mu) &\approx \mathbf{p}\,2\sum_{\mathbf{p}'}\mathbf{i}_p\cdot\mathbf{i}_{p'}\phi(\mathbf{i}_p\cdot\mathbf{i}_{p'})\,\delta(\varepsilon_{\mathbf{p}'} - \mu)\\
&= \mathbf{p}\,2\int \mathrm{d}\varepsilon\,\mathcal{N}(\varepsilon)\,\delta(\varepsilon - \mu)\int \frac{\mathrm{d}\Omega}{4\pi}\cos\xi\,\phi(\cos\xi)\\
&= \mathbf{p}\,\tfrac{1}{3}2\,\mathcal{N}_f\phi_1\\
&= \mathbf{p}\,\tfrac{1}{3}F_1.
\end{aligned}$$

We return to (III.12) to find

$$2\sum_{\mathbf{p}} \mathbf{p} n_{\mathbf{p}} = \frac{m}{m^*} 2\sum_{\mathbf{p}} \mathbf{p} n_{\mathbf{p}}(1 + F_1/3).$$

But since $n_{\mathbf{p}}$ is arbitrary, we have

$$m^* = m(1 + F_1/3). \tag{III.13}$$

This is an exact expression for the effective mass, and it is given in terms of only *one* of the interaction parameters of (III.9), namely in terms of $F_1$.

## 2. EQUILIBRIUM PROPERTIES

At thermal equilibrium, when the temperature $T$ and the chemical potential $\mu$ are given, the density operator $N$ and the energy operator $E$ take their ensemble average values $n(\mathbf{p}, \boldsymbol{\sigma})$ and

$$\tilde{\varepsilon}(\mathbf{p}, \boldsymbol{\sigma}) = \varepsilon(\mathbf{p}, \boldsymbol{\sigma}) + \mathrm{Tr}_{\sigma'} \sum_{\mathbf{p}'} f(\mathbf{p}, \boldsymbol{\sigma}; \mathbf{p}', \boldsymbol{\sigma}')\, n(\mathbf{p}', \boldsymbol{\sigma}'). \tag{III.14}$$

The equilibrium distribution of the quasiparticles (in the absence of a magnetic field) is given by

$$n_{\mathbf{p}}^{eq}(T, \mu) = n^0(\tilde{\varepsilon}_{\mathbf{p}}, T, \mu) - n^0(\tilde{\varepsilon}_{\mathbf{p}}, O, \mu), \tag{III.15}$$

where

$$n^0(\varepsilon, T, \mu) = \frac{1}{1 + \exp{(\varepsilon - \mu)/kT}} \tag{III.16}$$

is the Fermi distribution for particles with energy $\varepsilon$. Notice that in (III.15) the full energy $\tilde{\varepsilon}$ of (III.14) appears, and the two equations can be solved in principle for both the temperature-dependent energy $\tilde{\varepsilon}$ and the equilibrium distribution $n_{\mathbf{p}}^{eq}(T, \mu)$. This procedure enables us to find the thermal properties of the system to lowest order in $kT/\mu$, which is equivalent to our expansion parameter $\alpha$. From (III.15) we see that $\Sigma_{\mathbf{p}} n^{eq}$ is linear in $T$, and so is $\tilde{\varepsilon} - \mu$, and therefore the two terms of (III.14) with $\varepsilon$ replaced by $\varepsilon - \mu$ are of the same order in $T$, i.e. linear.

### (a) *Specific heat*

The low temperature specific heat is defined as

$$C_V = \left(\frac{\partial E}{\partial T}\right)_{V,N} = \frac{\partial}{\partial T} 2\sum_{\mathbf{p}} \tilde{\varepsilon}_{\mathbf{p}} n_{\mathbf{p}}^{eq}, \tag{III.17}$$

where the spin dependence is suppressed. The momentum summation is of order $T$, and thus to lowest order in $T$ we can write

$$C_V = 2\sum_{\mathbf{p}} \varepsilon_{\mathbf{p}} \left( \frac{\partial n_{\mathbf{p}}^{eq}}{\partial T} \right)_N . \tag{III.18}$$

This is completely analogous to the Fermi gas with $\varepsilon_{\mathbf{p}}$, the quasiparticle energy, replacing the free-fermion energy $p^2/2m$.

Thus the only difference is that now

$$2\sum_{\mathbf{p}} \rightarrow 2\int \mathrm{d}\varepsilon\, \mathcal{N}^*(\varepsilon), \tag{III.19}$$

where $\mathcal{N}^*(\varepsilon)$ is the quasiparticle density of states, and $C_V = \gamma^* T$ with

$$\gamma^* = \frac{2\pi k^2}{3} \mathcal{N}_f, \tag{III.20}$$

where $\mathcal{N}_f$ is the density of states on the Fermi surface.

Using (III.6) we can also write

$$\mathcal{N}_f = \frac{m^* p_f}{2\pi^2 \hbar^3}, \tag{III.21}$$

i.e. $m^*$ replaces the bare mass of the fermions, or

$$\mathcal{N}_f = \frac{p_f m}{2\pi^2 \hbar^3}\left(1 + \frac{F_1}{3}\right). \tag{III.22}$$

From the measurement of the specific heat of the liquid we can find $\mathcal{N}_f$ and $F_1$.

### (b) *Compressibility*

Compressing the liquid involves a change of the radius $p_f$ of the Fermi surface but no distortion from the form of the surface as a sphere centred on the origin in **p**-space. This is in contrast with the calculation of $m^*$, where the Fermi surface is distorted by the flow, and thus $F_1$ occurs in the expression for $m^*$, (III.13). To calculate the compressibility, (I.20), we have to evaluate $(\partial\mu/\partial n)_V$, i.e. the change of the chemical potential due to the change in density. Consider a system with density $n$ and chemical potential $\mu(n)$. If we add now $\Delta n$ particles per unit volume the Fermi sphere swells, and the energy of a quasiparticle on the new Fermi surface—i.e. $\mu(n + \Delta n)$—must be equal to that of a quasiparticle on $p_f(n + \Delta n)$ due to the addition of $\Delta n = \Sigma_{\mathbf{p}} n_{\mathbf{p}}$ extra quasiparticles to the old system (of density $n$). Using (III.10) and (III.6) we write (spin suppressed)

$$\mu(n + \Delta n) = \mu(n) + v_f\{p_f(n + \Delta n) - p_f(n)\} + 2\sum_{\mathbf{p}'} (p_f, \mathbf{p}')n_{\mathbf{p}'} \tag{III.23}$$

with the condition

$$\Delta n = 2\sum_{\mathbf{p}} n_{\mathbf{p}}; \tag{III.24}$$

here $f(\mathbf{p}, \mathbf{p}')$ is assumed not to be changed by the increase in density. Rewriting (III.23)

$$\Delta n \frac{\Delta\mu}{\Delta n} = v_f \frac{\Delta p_f}{\Delta n} \Delta n + 2\sum_{\mathbf{p}'} (\mathbf{p}, \mathbf{p}') n_{\mathbf{p}'} \tag{III.25}$$

we see that in the second term on the right $n_{\mathbf{p}'}$ does not depend on directions on the Fermi surface, while $f$ depends only on the angle $\mathbf{p}\cdot\mathbf{p}'$ (see (III.8)). We thus have

$$\begin{aligned}\sum_{\mathbf{p}'} f(\mathbf{p}, \mathbf{p}') n_{\mathbf{p}'} &= 2\int \mathrm{d}\varepsilon' \ \mathcal{N}(\varepsilon') n(\varepsilon') \int \frac{\mathrm{d}\Omega}{4\pi} \phi \ (\cos \xi) \\ &= \Delta n \phi_0 \end{aligned} \tag{III.26}$$

and since

$$n = \frac{1}{6\pi^2\hbar^3} p_f^3$$

we find

$$v_f \frac{\partial p_f}{\partial n} = \frac{1}{\mathcal{N}_f};$$

thus

$$\frac{\partial\mu}{\partial n} = \frac{1}{\mathcal{N}_f} + \phi_0 = \frac{1}{\mathcal{N}_f}(1 + F_0), \tag{III.27}$$

and the compressibility $\kappa$ is given by

$$\frac{1}{\kappa} = \frac{n^2}{2\mathcal{N}_f}(1 + F_0). \tag{III.28}$$

Comparing the Fermi liquid compressibility to that of the gas, we see two changes: (1) the density of states of the fermions is replaced by that of the quasiparticles, and (2) the interaction between quasiparticles causes a renormalization by a factor $(1 + F_0)$.

A way of measuring the compressibility is by propagating sound through the liquid. If local equilibrium is established, the velocity of sound is given by

$$c^2 = \frac{1}{mn\kappa} = \frac{n}{m}\frac{\partial\mu}{\partial n} = \frac{n}{m}\frac{1}{2\mathcal{N}_f}(1 + F_0). \tag{III.29}$$

The sound velocity of the Fermi gas is

$$c_0^2 = \frac{n}{m}\frac{1}{2\mathcal{N}(\mu)}$$

and thus

$$\frac{c^2}{c_0^2} = \frac{m}{m^*}(1 + F_0). \tag{III.30}$$

Equation (III.30) with (III.22) would yield $F_0$.

(c) *Spin susceptibility*

In a magnetic field $\mathcal{H}$ we can split the quasiparticle energy

$$\tilde{\varepsilon} = \varepsilon + \sum f n \tag{III.14}$$

and distribution into two parts:

$$\begin{aligned} \hat{\tilde{\varepsilon}}(\mathbf{p}, \boldsymbol{\sigma}) &= \hat{1}\tilde{\varepsilon}_p + \hat{\boldsymbol{\sigma}}\cdot\boldsymbol{\eta}_p \\ \hat{n}(\mathbf{p}, \boldsymbol{\sigma}) &= \hat{1}n_p + \hat{\boldsymbol{\sigma}}\cdot\mathbf{m}_p, \end{aligned} \tag{III.31}$$

where $\boldsymbol{\eta}_p$ and $\mathbf{m}_p$ play an equivalent role to $\tilde{\varepsilon}_\mathbf{p}$ and $n_\mathbf{p}$. Using

$$\hat{\varepsilon}(\mathbf{p}, \boldsymbol{\sigma}) = \hat{1}\varepsilon_\mathbf{p} + \hat{\boldsymbol{\sigma}}\cdot(-\beta\mathcal{H})$$

and (III.7) we find, by equating the coefficients of matrices in the spin space,

$$\tilde{\varepsilon}_\mathbf{p} = \varepsilon_\mathbf{p} + 2\sum_{\mathbf{p}'}\phi(\mathbf{p}, \mathbf{p}')n_{\mathbf{p}'} \tag{III.10A}$$

and, with

$$\mathrm{Tr}_{\sigma'}\boldsymbol{\sigma}\cdot\boldsymbol{\sigma}'\,\boldsymbol{\sigma}'\cdot\mathbf{m}_p = 2\boldsymbol{\sigma}\cdot\mathbf{m}_p,$$

$$\boldsymbol{\eta}_p = -\beta\mathcal{H} + 2\sum_{\mathbf{p}'}\zeta(\mathbf{p}, \mathbf{p}')\mathbf{m}_{\mathbf{p}'}. \tag{III.32}$$

The magnetization per unit volume is

$$\begin{aligned} \mathbf{M} &= \sum_\mathbf{p} \mathrm{Tr}_\sigma \beta\boldsymbol{\sigma}\hat{n}(\mathbf{p}, \boldsymbol{\sigma}) \\ &= 2\beta\sum_\mathbf{p}\mathbf{m}_\mathbf{p}. \end{aligned} \tag{III.33}$$

For a small magnetic field $\mathcal{H}$ the spin density distribution function $\mathbf{m}_\mathbf{p}$ is proportional to the quasiparticle magnetic energy

$$\boldsymbol{\eta}_p = -\beta^*\mathcal{H}, \tag{III.34}$$

and is significant only near the Fermi surface

$$\mathbf{m}_\mathbf{p} = -\delta(\varepsilon_p - \mu)(-\beta^* \mathscr{H}). \tag{III.35}$$

Thus

$$\mathbf{M} = 2\beta\beta^* \mathscr{N}_f \mathscr{H}, \tag{III.36}$$

where (III.32) yields

$$-\beta^* = -\beta + \beta^* Z_0$$

i.e.

$$\beta^* = \frac{\beta}{1 + Z_0} \tag{III.37}$$

and thus

$$\chi = 2\beta^2 \frac{\mathscr{N}_f}{1 + Z_0}. \tag{III.38}$$

Here we see again that the interaction between quasiparticles—and this time it is the "exchange" $\zeta$ that is involved—causes a renormalization of the equilibrium property aside from the replacement of the density of states. The measurement of the susceptibility yields $Z_0$. If $Z_0$ is negative (e.g. in $He^3$) the exchange interaction gives rise to enhancement of the spin susceptibility. The system will be ferromagnetic if the denominator becomes zero.

## IV. Kinetic and Transport Equations

We now turn to the nonequilibrium properties of Fermi liquids, and consider the system when a weak time-dependent inhomogeneous perturbation occurs. This perturbation may be due to an external field applied to the system, or may be a consequence of an internal fluctuation; it is assumed to be weak, so that the system responds linearly. The dynamical variables of the interacting dilute gas of quasiparticles are the density operator $\hat{N}$ and the energy operator $\hat{E}$ of (III.4) and (III.5). The objects of the present study are the ensemble averages of the following quantities:

(a) The distribution function

$$n_{\mathbf{p},\boldsymbol{\sigma}}(\mathbf{r}, t) = \langle N(\mathbf{p}, \mathbf{r}, \boldsymbol{\sigma}, t)\rangle$$

and (b) the energy of a quasiparticle

$$\varepsilon_{\mathbf{p},\boldsymbol{\sigma}}(\mathbf{r}, t) = \langle E(\mathbf{p}, \mathbf{r}, \boldsymbol{\sigma}, t)\rangle,$$

where $\langle \ldots \rangle$ indicates the ensemble average of the quasiparticle gas. In previous lectures we have been concerned with the equilibrium values of $n$

and $\varepsilon$ and thus $\mathbf{r}$ and $t$ were not indicated. Here we are mainly interested in the deviations from thermal equilibrium, and we divide our discussion into two parts: collective behaviour and effects of collisions between quasiparticles.

## 1. DERIVATION OF THE KINETIC EQUATION

We start by writing the equation of motion (III.3) of the density operator explicitly in terms of space and spin indices:

$$i\hbar \frac{\partial}{\partial t} N_{\sigma_1\sigma_2}(\mathbf{r}_1, \mathbf{r}_2, t) = \sum_{\sigma_3} \int d^3r_3 \{E_{\sigma_1\sigma_3}(\mathbf{r}_1, \mathbf{r}_3, t) N_{\sigma_3\sigma_2}(\mathbf{r}_3, \mathbf{r}_2, t) - N_{\sigma_1\sigma_3}(\mathbf{r}_1, \mathbf{r}_3, t) E_{\sigma_3\sigma_2}(\mathbf{r}_3, \mathbf{r}_2, t)\}. \tag{IV.1}$$

If we now use the transformation (III.4) to the mixed representation and further introduce Fourier transforms in space, e.g.

$$N_{\sigma_1\sigma_2}(\mathbf{p}, \mathbf{R}, t) = \sum_{\mathbf{k}} \exp(i\mathbf{k}\cdot\mathbf{R}) N_{\sigma_1\sigma_2}(\mathbf{p}, \mathbf{k}, t), \tag{IV.2}$$

equation (IV.1) becomes (in spin matrix notation, and with $t$ suppressed)

$$i\hbar \frac{\partial}{\partial t} \hat{N}(\mathbf{p}, \mathbf{k}) = \sum_{\mathbf{k}'} \left\{ \hat{E}\left(\mathbf{p} + \frac{\hbar}{2}\mathbf{q}, \mathbf{k}'\right) \hat{N}\left(\mathbf{p} - \frac{\hbar}{2}\mathbf{k}, \mathbf{q}\right) - \hat{N}\left(\mathbf{p} + \frac{\hbar}{2}\mathbf{q}, \mathbf{k}'\right) \hat{E}\left(\mathbf{p} - \frac{\hbar}{2}\mathbf{k}', \mathbf{q}\right) \right\}, \tag{IV.3}$$

where

$$\mathbf{q} = \mathbf{k} - \mathbf{k}'. \tag{IV.4}$$

The ensemble average of (IV.3) should give the proper kinetic equation. In the present lectures we do not dwell upon the detailed study of correlations between quasiparticles. We just notice that if $N$ and $E$ are divided into

$$N = n + \delta n, \qquad E = \varepsilon + \delta E, \qquad n = \langle N \rangle, \qquad \varepsilon = \langle E \rangle$$

where $\delta N$ and $\delta E$ are the fluctuating parts of $N$ and $E$ respectively, equation (IV.3) can be written as

$$\frac{\partial}{\partial t} \hat{n}_{\mathbf{p}}(k) - \frac{1}{i\hbar} \sum_{\mathbf{k}'} \{\hat{\varepsilon}_{\mathbf{p}+\hbar\mathbf{q}/2}(k') \hat{n}_{\mathbf{p}-\hbar\mathbf{k}'/2}(\mathbf{q}) - \hat{n}_{\mathbf{p}+\hbar\mathbf{q}/2}(k') \hat{\varepsilon}_{\mathbf{p}-\hbar\mathbf{k}'/2}(\mathbf{q})\} = \hat{I}, \tag{IV.5}$$

where $\hat{n}$ and $\hat{\varepsilon}$ are matrices in the spin space and the collision integral $\hat{I}$ is given in terms of the $\delta N$ and $\delta E$,

$$\hat{I}_{\mathbf{p}}(k) = \frac{1}{i\hbar}\sum_{\mathbf{k}'}\left\langle \delta\hat{E}\left(\mathbf{p}+\frac{\hbar}{2}\mathbf{q},\mathbf{k}'\right)\delta\hat{N}\left(\mathbf{p}-\frac{\hbar}{2}\mathbf{k}',\mathbf{q}\right) - \delta\hat{N}\left(\mathbf{p}+\frac{\hbar}{2}\mathbf{q},\mathbf{k}'\right)\right.$$
$$\left.\times\ \delta\hat{E}\left(\mathbf{p}-\frac{\hbar}{2}\mathbf{k}',\mathbf{q}\right)\right\rangle. \quad \text{(IV.6)}$$

We first concentrate on the left hand side of (IV.5), and later on we shall discuss the effect of collisions (or rather correlations) between quasiparticles.

We recall that the very notion of a quasiparticle in the Fermi liquid is meaningful only for momenta $\mathbf{p}$ whose absolute value is close to the Fermi momentum. Furthermore, only small deviations from homogeneity are considered, so that the wavelength of spatial perturbation is much longer than the de Broglie wavelength, i.e. $\mathbf{k} \ll \mathbf{p}$ in (IV.5). We expand both $n$ and $e$ for small $\mathbf{k}$ and $\mathbf{q} = \mathbf{k} - \mathbf{k}'$, e.g.

$$\hat{n}_{\mathbf{p}-\frac{1}{2}\hbar\mathbf{k}'}(\mathbf{q}) \approx \hat{n}_{\mathbf{p}}(\mathbf{q}) - \frac{\hbar}{2}\mathbf{k}'\cdot\frac{\partial}{\partial\mathbf{p}}\hat{n}_{\mathbf{p}}(\mathbf{q}), \quad \text{(IV.7)}$$

substitute in (IV.5), and obtain

$$\frac{\partial}{\partial t}\hat{n}_{\mathbf{p}}(\mathbf{k}) - \frac{1}{i\hbar}\sum_{\mathbf{k}'}[\hat{\varepsilon}_{\mathbf{p}}(\mathbf{k}'), \hat{n}_{p}(\mathbf{k}-\mathbf{k}')]$$
$$-\tfrac{1}{2}\sum_{\mathbf{k}'} i\mathbf{k}'\cdot\left\{\hat{\varepsilon}_{\mathbf{p}}(\mathbf{k}'), \frac{\partial}{\partial\mathbf{p}}\hat{n}_{\mathbf{p}}(\mathbf{k}-\mathbf{k}')\right\}$$
$$+\tfrac{1}{2}\sum_{\mathbf{k}'} i\mathbf{k}'\cdot\left\{\frac{\partial}{\partial\mathbf{p}}\hat{\varepsilon}_{\mathbf{p}}(\mathbf{k}-\mathbf{k}'), \hat{n}_{\mathbf{p}}(\mathbf{k}')\right\}$$
$$= \hat{I}_{\mathbf{p}}(\mathbf{k}). \quad \text{(IV.8)}$$

In (IV.8) we use the usual notation of commutators $[A, B] = AB - BA$, and anti-commutators $\{A, B\} = AB + BA$. Notice that the summations over $\mathbf{k}'$ are convolutions in $\mathbf{k}'$-space and thus in real space the kinetic equation for $n_{\mathbf{p}}(\mathbf{r}, t)$ takes the form of Silin's equation

$$\frac{\partial}{\partial t}\hat{n} - \frac{1}{i\hbar}[\hat{\varepsilon}, \hat{n}] + \tfrac{1}{2}\left\{\frac{\partial\hat{\varepsilon}}{\partial\mathbf{p}}, \frac{\partial\hat{n}}{\partial\mathbf{r}}\right\} - \tfrac{1}{2}\left\{\frac{\partial\hat{\varepsilon}}{\partial\mathbf{r}}, \frac{\partial\hat{n}}{\partial\mathbf{p}}\right\} = \hat{I}, \quad \text{(IV.9)}$$

and the energy operator

$$\hat{\varepsilon}_{\mathbf{p}}(\mathbf{r}, t) = \hat{\varepsilon}_{\mathbf{p}} + \sum_{\mathbf{p}'}\mathrm{Tr}_{\sigma'} f(\mathbf{p}, \boldsymbol{\sigma}; \mathbf{p}', \boldsymbol{\sigma}')n_{\mathbf{p}'}(\mathbf{r}, t) \quad \text{(IV.10)}$$

is local in space and time. In passing, we point out that in the absence of magnetic fields (IV.9) reduces to the usual kinetic equation of Landau, i.e.

$$\frac{\partial}{\partial t} n_{\mathbf{p}}(\mathbf{r}, t) + \frac{\partial \varepsilon_{\mathbf{p}}}{\partial \mathbf{p}} \cdot \frac{\partial n_{\mathbf{p}}}{\partial \mathbf{r}} - \frac{\partial \varepsilon_{\mathbf{p}}}{\partial \mathbf{r}} \cdot \frac{\partial n_{\mathbf{p}}}{\partial \mathbf{p}} = I. \tag{IV.11}$$

Our next problem is the linearization of the kinetic equation around thermal equilibrium. We denote here the equilibrium values of the density and the energy operators by $\hat{n}_{\mathbf{p}}$ and $\hat{\varepsilon}_{\mathbf{p}}$ and write

$$\hat{n}_{\mathbf{p}}(\mathbf{r}, t) = \hat{n}_{\mathbf{p}}^{(0)} + \hat{n}_{\mathbf{p}}^{(1)}(\mathbf{r}, t)$$
$$\hat{\varepsilon}_{\mathbf{p}}(\mathbf{r}, t) = \hat{\varepsilon}_{\mathbf{p}}^{(0)} + \hat{\varepsilon}_{\mathbf{p}}^{(1)}(\mathbf{r}, t). \tag{IV.12}$$

The linearized kinetic equation is now

$$\frac{\partial}{\partial t} \hat{n}_{\mathbf{p}}^{(1)} - \frac{1}{i\hbar}[\hat{\varepsilon}_{\mathbf{p}}^{(0)}, \hat{n}_{\mathbf{p}}^{(1)}] - \frac{1}{i\hbar}[\hat{\varepsilon}_{\mathbf{p}}^{(1)}, \hat{n}_{\mathbf{p}}^{(0)}]$$
$$+ \tfrac{1}{2}\left\{\frac{\partial \hat{\varepsilon}_{\mathbf{p}}^{(0)}}{\partial \mathbf{p}}, \frac{\partial \hat{n}_{\mathbf{p}}^{(1)}}{\partial \mathbf{r}}\right\} - \tfrac{1}{2}\left\{\frac{\partial \hat{\varepsilon}_{\mathbf{p}}^{(1)}}{\partial \mathbf{r}}, \frac{\partial n_{\mathbf{p}}^{(0)}}{\partial \mathbf{p}}\right\} = \hat{I}^{(1)}. \tag{IV.13}$$

## 2. COLLECTIVE EXCITATIONS

Using the kinetic equation and neglecting collisions between quasiparticles, we can study the possible normal modes which can propagate in the quasiparticle gas. These are the collective excitations of the interacting quasiparticles, which can be spontaneously excited by fluctuations. We now consider the system in the absence of external fields, and substitute

$$\hat{n}^{(1)} = \hat{1}\tilde{n}_{\mathbf{p}}(\mathbf{r}, t) + \hat{\boldsymbol{\sigma}} \cdot \mathbf{m}_{\mathbf{p}}(\mathbf{r}, t)$$
$$\hat{\varepsilon}^{(1)} = \hat{1}\tilde{\varepsilon}_{\mathbf{p}}(\mathbf{r}, t) + \hat{\boldsymbol{\sigma}} \cdot \boldsymbol{\eta}_{\mathbf{p}}(\mathbf{r}, t) \tag{IV.14}$$

into (IV.13). Using (III.7) for $f$ and decoupling the equations, we obtain

$$\frac{\partial}{\partial t} \tilde{n}_{\mathbf{p}} + \frac{\partial \varepsilon_{\mathbf{p}}}{\partial \mathbf{p}} \cdot \frac{\partial}{\partial \mathbf{r}} \tilde{n}_{\mathbf{p}} - \frac{\partial \tilde{\varepsilon}_{\mathbf{p}}}{\partial \mathbf{r}} \frac{\partial}{\partial \mathbf{p}} n_{\mathbf{p}} = 0 \tag{IV.15}$$

and

$$\frac{\partial}{\partial t} \mathbf{m}_{\mathbf{p}} + \frac{\partial \varepsilon_{\mathbf{p}}}{\partial \mathbf{p}} \cdot \frac{\partial}{\partial \mathbf{r}} \mathbf{m}_{\mathbf{p}} - \frac{\partial n_{\mathbf{p}}}{\partial \mathbf{p}} \cdot \frac{\partial}{\partial \mathbf{r}} \boldsymbol{\eta}_{\mathbf{p}} = 0, \tag{IV.16}$$

where

$$\tilde{\varepsilon}_{\mathbf{p}} = 2 \sum_{\mathbf{p}'} \phi(\mathbf{i}_p \cdot \mathbf{i}_{p'}) \tilde{n}_{\mathbf{p}'} \tag{IV.17}$$

and

$$\boldsymbol{\eta}_{\mathbf{p}} = 2 \sum_{\mathbf{p}'} \zeta(\mathbf{i}_p \cdot \mathbf{i}_{p'}) \mathbf{m}_{\mathbf{p}'}. \tag{IV.18}$$

First we discuss the quasiparticle density modes (IV.15) and (IV.17). We Fourier transform the equations in space and time, with

$$\tilde{n}_{\mathbf{p}} \propto \bar{n}_{\mathbf{p}}(\mathbf{k}, \omega) \exp(i\mathbf{k} \cdot \mathbf{r}) \exp(-i\omega t), \tag{IV.19}$$

and substitute the equilibrium values

$$\mathbf{v} = \frac{\partial \varepsilon_{\mathbf{p}}}{\partial \mathbf{p}}, \qquad \frac{\partial n_{\mathbf{p}}}{\partial \mathbf{p}} = \mathbf{v}\frac{\partial n^{eq}}{\partial \varepsilon} \tag{IV.20}$$

to find

$$(\omega - \mathbf{k} \cdot \mathbf{v})\bar{n}_{\mathbf{p}} + \mathbf{k} \cdot \mathbf{v}\frac{\partial n^{eq}}{\partial \varepsilon}\bar{\varepsilon}_{\mathbf{p}} = 0, \tag{IV.21}$$

or, allowing $\omega$ to be slightly complex,

$$\bar{n}_{\mathbf{p}} = \frac{\partial n^{eq}}{\partial \varepsilon}\frac{\mathbf{k} \cdot \mathbf{v}}{\mathbf{k} \cdot \mathbf{v} - \omega}\bar{\varepsilon}_{\mathbf{p}}. \tag{IV.22}$$

Since $\tilde{n}_{\mathbf{p}} \propto \partial n^{eq}/\partial \varepsilon$, we can integrate (IV.17) over $\varepsilon'$ and find that $\bar{\varepsilon} = \bar{\varepsilon}(\mathbf{i}_p)$ depends only on the direction $\mathbf{p}$ on the Fermi surface

$$\bar{\varepsilon}(\mathbf{i}_p) = -\mathcal{N}_f \int \frac{d\Omega}{4\pi} \phi(\mathbf{i}_p \cdot \mathbf{i}_{p'}) \frac{\mathbf{i}_k \cdot \mathbf{i}_{p'}}{\mathbf{i}_k \cdot \mathbf{i}_{p'} - s}\bar{\varepsilon}(\mathbf{i}_{p'}), \tag{IV.23}$$

where $\mathbf{k} = k\mathbf{i}_k$ and $s = \omega/kv_f$. We have a homogeneous integral equation for $\varepsilon(\mathbf{i}_p)$, whose solutions in terms of the "eigenvalues" $s$ give, in turn, the dispersion relation for the propagating modes. A general solution of (IV.23) is beyond the scope of these lectures, but to demonstrate the nature of these collective modes we limit ourselves to a simple model with $\phi$ constant. In this case (IV.23) reduces to

$$\varepsilon = -F_0 \int \frac{d\Omega}{4\pi} \frac{\mathbf{i}_k \cdot \mathbf{i}_{p'}}{\mathbf{i}_k \cdot \mathbf{i}_{p'} - s}\varepsilon \tag{IV.24}$$

where $\varepsilon$ is independent of $\mathbf{i}_p$, and thus the dispersion relation of the collective mode (for $\varepsilon \neq 0$) is

$$1 = -F_0 \int_{-1}^{1} \frac{d\cos\theta}{2} \frac{\cos\theta}{\cos\theta - s} \tag{IV.25}$$

with $\cos\theta = \mathbf{i}_k \cdot \mathbf{i}_{p'}$. After elementary integration we find the dispersion relation

$$\tfrac{1}{2} s \ln\frac{s+1}{s-1} - 1 = \frac{1}{F_0}. \tag{IV.26}$$

The solutions of (IV.26) for $s$ yield the frequency versus wavelength relation for this mode, which was named by Landau "zero-sound". If $F_0 > 0$ (repulsive interaction between quasiparticles) there is *one* real root, such that $s > 1$. This mode is undamped ($\omega$ is real) and its phase velocity $\omega/k$ is larger than the quasiparticle Fermi velocity $v_f$. If $F_0$ is negative, which corresponds to attraction, we find complex roots for $s$. We now discuss the different solutions:

(a) *Undamped zero sound*

In this case the perturbation in the distribution function is a "kind of undamped wave" on the Fermi surface,

$$\bar{n}_p \propto \frac{\partial n^{eq}}{\partial \varepsilon} \frac{\mathbf{k} \cdot \mathbf{v}}{\mathbf{k} \cdot \mathbf{v} - \omega}. \tag{IV.27}$$

In contrast with the normal sound (first sound), where the density of the quasiparticles changes in space and a local equilibrium occurs in the system, here the zero-sound mode is essentially a "wave" of the distribution $n_{\mathbf{p}}$ oscillating on the Fermi surface. Though the density itself, $\Sigma_{\mathbf{p}} \bar{n}_{\mathbf{p}}$, oscillates (this is the reason for the term "sound"), the main feature of the zero sound is the coherent wave motion of the momentum distribution. The velocity of the zero sound can be obtained in the limiting cases of strong and weak interaction. When $F_0 \ll 1$, $s$ goes to 1 and the mode propagates with velocity slightly larger than $v_f$. When $F_0 \gg 1$ we find $s \approx \sqrt{(F_0/3)}$ and $\hat{n}_{\mathbf{p}} \approx \mathbf{k} \cdot \mathbf{v}$. Thus the velocity of weakly interacting (quasiparticle) zero sound $c_0 = \omega/k \approx v_f$ is by a factor of $\sqrt{3}$ larger than non-interacting (local equilibrium) first sound, where $c_1^2 = p_f^2/3m^2$.

(b) *Landau damped zero sound.*

If

$$-1 < F_0$$

we find that a complex solution of (IV.26) for $s$ has a negative imaginary part, and the mode is damped. This damping is due not to the relaxation time $\tau$ of the collision integral I of the kinetic equation, but rather to a "resonance" motion of some quasiparticles moving with the collective mode. This mechanism is known as Landau damping and it occurs when for a group of quasiparticles the denominator ($\mathbf{k} \cdot \mathbf{v} - \omega$) of (IV.27) is very small, and the quasiparticles "surf-ride" on the zero-sound wave. It is clear that once Landau damping becomes significant the collective mode is not a well defined excitation.

(c) *Unstable modes*

If the interaction is too attractive, i.e. $F_0 < -1$, the mode becomes unstable, and $s$ has a complex value with positive imaginary part. This gives rise to exponentially growing fluctuations of the system, and the ground state becomes unstable.

To conclude this section we indicate that the quasiparticle spin-density modes of (IV.16) and (IV.18) can be studied in an analogous manner. If we use the same simple model, i.e. $\zeta$ constant, we end up with a similar dispersion relation (IV.26), but with $F_0$ replaced by $Z_0$ (see (III.9)). Now, since in $He^3$ $Z_0$ is negative, we do not expect spin waves to propagate (because of Landau damping) in the system. However, if a constant magnetic field is present spin waves can propagate. These waves have been seen in experiments on simple non-ferromagnetic metals.

## 3. TRANSPORT PHENOMENA

In order to find the transport coefficients of Fermi liquids the lifetime effects of the collision integral (IV.6) should be studied. We present here a short review of a method of deriving the collision integral by a correlation approach, which is much in the spirit of Landau's phenomenological theory. For simplicity, we consider only $Z_0$ to be different from zero, so that the effect of spin fluctuations, or paramagnons, is demonstrated.

We consider (IV.6) for $\langle \delta N \delta E \rangle$ to lowest order in the correlation, so that the equation of motion for $\delta N$ is (with $\bar{k} = \hbar k$)

$$i\hbar \frac{\partial}{\partial t} \partial N_{\alpha\beta}(\mathbf{p}, \mathbf{k}) - (\varepsilon_{\mathbf{p}+\bar{\mathbf{k}}/2,\alpha} - \varepsilon_{\mathbf{p}-\bar{\mathbf{k}}/2,\beta})\delta N_{\alpha\beta}(\mathbf{p}, \mathbf{k}) + (n_{\mathbf{p}+\bar{\mathbf{k}}/2,\alpha} - n_{\mathbf{p}-\bar{\mathbf{k}}/2,\beta})\delta E_{\alpha\beta}(\mathbf{p}, \mathbf{k}) = 0, \quad \text{(IV.28)}$$

where $n_{\mathbf{p},\sigma}$, $\varepsilon_{\mathbf{p},\sigma}$ are the diagonal components of the distribution function and of the energy respectively. The energy fluctuation in our simple case is just

$$\delta E_{\alpha\beta}(\mathbf{p}, \mathbf{k}) = \zeta_0 \hat{\boldsymbol{\sigma}}_{\alpha\beta} \cdot \sum_{\mathbf{p}'} \sum_{\alpha'\beta'} \hat{\boldsymbol{\sigma}}_{\alpha'\beta'} \delta N_{\beta'\alpha'} = \delta E_{\alpha\beta}(\mathbf{k}), \quad \text{(IV.29)}$$

i.e. $\delta E$ is independent of $\mathbf{p}$. Assuming an initial fluctuation $\delta N^0$ we can solve (IV.28) and (IV.29) for $\delta N$ and $\delta E$ as functions of time. Substituting these solutions into (IV.6), which now takes the form

$$I_{\alpha\beta}(\mathbf{p}, \mathbf{k}) = \frac{2}{\hbar} \operatorname{Im} \sum_{\mathbf{q}} \sum_{\alpha'\beta'} \langle \delta E_{\alpha'\beta'}(-\mathbf{q}) \delta N_{\beta'\alpha'}(\mathbf{p} + \mathbf{q}/2, \mathbf{k}) \rangle, \quad \text{(IV.30)}$$

and taking the ensemble average over "initial conditions"

$$\langle \delta N_{\alpha\beta}(\mathbf{p}_1, -\mathbf{k}) \delta N_{\beta\alpha}(\mathbf{p}_2, \mathbf{k}) \rangle = \delta_{\mathbf{p}_1 \mathbf{p}_2} n_\beta(\mathbf{p}_1 + \bar{\mathbf{k}}/2)\{1 - n_\alpha(\mathbf{p}_1 - \bar{\mathbf{k}}/2)\}, \quad \text{(IV.31)}$$

we find for example

$$I_{\uparrow\uparrow}(p) = 2\pi \sum_{\mathbf{k}} \sum_{\mathbf{p}'} \left| \frac{2\zeta_0}{\chi_{\downarrow\uparrow}} \right|^2 \{ n_{\mathbf{p}\uparrow}(1 - n_{\mathbf{p}+\bar{\mathbf{k}}\downarrow}) n_{\mathbf{p}'+\bar{\mathbf{k}}\downarrow}(1 - n_{\mathbf{p}'\uparrow})$$

$$- n_{\mathbf{p}+\bar{\mathbf{k}}\downarrow}(1 - n_{\mathbf{p}\uparrow}) n_{\mathbf{p}'\uparrow}(1 - n_{\mathbf{p}'+\bar{\mathbf{k}}\downarrow})\} \, \delta(\varepsilon_{\mathbf{p}'+\bar{\mathbf{k}}\downarrow} - \varepsilon_{\mathbf{p}'\uparrow} - \varepsilon_{\mathbf{p}+\bar{\mathbf{k}}\downarrow} + \varepsilon_{\mathbf{p}\uparrow}). \quad \text{(IV.32)}$$

In (IV.32) we express explicitly only the up–up collision integral, but other components of $I$ can be obtained similarly. The function $\chi_{\downarrow\uparrow} \equiv \chi_{\downarrow\uparrow}(\mathbf{k}, \varepsilon_{\mathbf{p}+\mathbf{k}\downarrow} - \varepsilon_{\mathbf{p}\uparrow})$ is the off-diagonal element of the susceptibility function

$$\chi_{\alpha\beta}(\mathbf{k}, \omega) = 1 - \zeta_0 \sum_{\mathbf{p}} \frac{n_{\mathbf{p}+\bar{\mathbf{k}}/2,\alpha} - n_{\mathbf{p}-\bar{\mathbf{k}}/2,\beta}}{\varepsilon_{\mathbf{p}+\bar{\mathbf{k}}/2,\alpha} - \varepsilon_{\mathbf{p}-\bar{\mathbf{k}}/2,\beta} - \omega\hbar}. \quad \text{(IV.33)}$$

The collision integral (IV.32) has the following features:

(a) The "Fermi-like" distribution functions $n_{\mathbf{p}\sigma}$ of the quasiparticles appear explicitly in the proper combination for two-fermion collisions.

(b) The conservation of energy, as indicated by the $\delta$-function, carries the proper quasiparticle energy $\varepsilon_{\mathbf{p}\sigma}$.

(c) The transition probability $|2\zeta_0/\chi|^2$ is explicitly expressed in terms of the Landau parameters (in our case only $\zeta_0$).

(d) The $\chi$ in the denominator is a manifestation of dynamic screening between quasiparticles, i.e. the "quasiparticle clouds" around the colliding quasiparticles are of many-body origin.

To calculate transport coefficients one must solve the kinetic equation (IV.9) with the proper collision integral. If $\chi$ is neglected in the denominator of (IV.32) and the kinetic equation is linearized around the equilibrium Fermi distribution for $n_{\mathbf{p}\sigma}$, one finds that the phase space available for $\mathbf{p}'$ and $\mathbf{k}$ integration is of order $kT$, and thus the asymptotic value

$$I \sim T^2,$$

due to Landau, occurs. However, if $\chi$ is maintained, a correction of order $T$ (and not $T^2$) is obtained, and thus

$$I \sim \tau^{-1} = T^2(a + bT). \quad \text{(IV.34)}$$

The effect of spin fluctuation is to make the temperature correction stronger than is otherwise expected.

## V. Microscopic Derivation of Landau's Theory

In previous lectures the phenomenological theory of Fermi liquids has been developed. We close these lectures by reviewing the microscopic theory which justifies the basic assertions of Landau. We concentrate on the notion of quasiparticles and on the two main parameters: the quasiparticle energy, $\varepsilon_{\mathbf{p}}$, of (III.6) and the interaction function $f(\mathbf{p}, \boldsymbol{\sigma}; \mathbf{p}', \boldsymbol{\sigma}')$, (and, in particular, its meaning).

## 1. THE QUASIPARTICLE ENERGY

Consider a system of interacting fermions at zero temperature with hamiltonian $H$. The one-particle Green's function ($\hbar = 1$)

$$G_{\sigma\sigma'}(x, x') = -i\langle T\{\psi_\sigma(x), \psi^\dagger_{\sigma'}(x')\}\rangle \tag{V.1}$$

is defined, as usual, in terms of the creation and annihilation operators $\psi^\dagger$ and $\psi$, and the time ordering operators $T$. Here $x$ denotes $\mathbf{r}, t$ and $\sigma$ is the spin index of the fermion fields. The average $\langle \ldots \rangle$ is taken with respect to the exact ground state of the system, $H|\,0\rangle = E_0|\,0\rangle$, and the operators are given in the Heisenberg representation $\psi(t) = \exp(iHt)\psi \exp(-iHt)$. In the absence of external fields the system is homogeneous in space, spin and time, i.e.

$$G_{\sigma\sigma'}(x, x') = \delta_{\sigma\sigma'}G(\mathbf{r} - \mathbf{r}', t - t'); \tag{V.2}$$

this can be Fourier transformed as

$$G(x - x') = \int_{-\infty}^{\infty} \frac{d\omega}{2\pi} \exp[-i\omega(t - t')] \frac{1}{V} \sum_{\mathbf{p}} \exp[i\mathbf{p}\cdot(\mathbf{r} - \mathbf{r}')]\, G(\mathbf{p}, \omega). \tag{V.3}$$

For reference we indicate that for non-interacting fermions we find

$$G^0(\mathbf{p}, \omega) = [\omega - \varepsilon^0_{\mathbf{p}} + i\delta \operatorname{sgn}(p - p_f)]^{-1}, \tag{V.4}$$

where $\varepsilon^0_{\mathbf{p}} = p^2/2m$, $\delta$ is a small positive parameter, $p_f$ is the Fermi momentum ($p_f = (3\pi^2 n)^{1/3}$, where $n$ is the total density) and sgn stands for the sign of the argument.

If we write $G(x - x')$ explicitly in terms of the exact eigenstates of the $N$-particle system, $H|\alpha\rangle = E_\alpha|\alpha\rangle$ we can find the Lehmann representation of $G$, i.e.

$$G(\mathbf{p}, \omega) = \int_0^{\infty} dE \left(\frac{A(\mathbf{p}, E)}{\omega - E - \mu + i\delta} + \frac{B(\mathbf{p}, E)}{\omega + E - \mu - i\delta}\right), \tag{V.5}$$

where $\mu = E_0(N + 1) - E_0(N)$ is the chemical potential, and where

$$A(\mathbf{p}, E)\,dE = (2\pi)^3 \sum_\alpha |\langle 0|\psi|\alpha\rangle|^2 \delta(\mathbf{p} - \mathbf{p}_\alpha), \qquad E < E_\alpha < E + dE, \tag{V.6}$$

and also $B(\mathbf{p}, E)$, defined similarly, are positive definite. The function $G(\mathbf{p}, \omega)$

is not analytic in $\omega$-space but has a singularity at $\omega = \mu$, and its imaginary part changes sign there, i.e.

$$\operatorname{Im} G(\mathbf{p}, \omega) = \begin{cases} -\pi A(\mathbf{p}, \omega - \mu) & \omega > \mu \\ \pi B(\mathbf{p}, \mu - \omega) & \omega < \mu. \end{cases} \tag{V.7}$$

To interpret the Green's function, and to extract the relevant information for the Fermi liquid, we consider

$$G(\mathbf{p}, \tau) = \int_{-\infty}^{\infty} \frac{d\omega}{2\pi} \exp(-i\omega\tau)\, G(\mathbf{p}, \omega) \tag{V.8}$$

for positive $\tau = t - t'$. Starting from the $N$-particle ground state $|0\rangle$ and adding one fermion of momentum $\mathbf{p}$, we have the wave function of the system at time $\tau$ as $\psi(\tau) = \exp(iH\tau)\, \psi_{\mathbf{p}}^{\dagger}|0\rangle$. Thus the probability $W_{\mathbf{p}}(\tau)$ of finding the system at $\tau$ with one particle at $\mathbf{p}$ in the state $\psi_{\mathbf{p}}^{\dagger}|0\rangle$ is

$$W_{\mathbf{p}}(\tau) = \langle 0|\psi_{\mathbf{p}} \exp(iHt)\psi_{\mathbf{p}}^{\dagger}|0\rangle \equiv -iG(\mathbf{p}, \tau), \tag{V.9}$$

or by using (V.5) and (V.9)

$$W_{\mathbf{p}}(\tau) = \exp(-i\mu\tau) \int_{0}^{\infty} dE \exp(-iE\tau)\, A(\mathbf{p}, E). \tag{V.10}$$

For free particles $A$ is a $\delta$-function and $W_{\mathbf{p}}^{0}(\tau) \propto \exp(-i\varepsilon_{\mathbf{p}}^{0}\tau)$; but for interacting particles $A$ should be replaced by a function having a sharp maximum near $E = \varepsilon_{\mathbf{p}} - \mu$, where $\varepsilon_{\mathbf{p}}$ is the energy of the quasiparticle. That this is so can be shown by continuing $A(\mathbf{p}, E)$ analytically to the lower half plane of $E$ for $\operatorname{Re} E > \mu$ and deforming the contour of integration further below the real axis. The first pole, $E = -\mu + \varepsilon_{\mathbf{p}} - i\gamma_{\mathbf{p}}$, we meet will survive the longer time, and (V.10) yields

$$W_{\mathbf{p}}(\tau) = a_{\mathbf{p}} \exp(-i\varepsilon_{\mathbf{p}}\tau - \gamma_{\mathbf{p}}\tau), \tag{V.11}$$

where $a_{\mathbf{p}}$ is the residue at the pole, and $\gamma_{\mathbf{p}}$ is the "lifetime" of the excitation. Thus the energy and "lifetime" of the quasiparticle of momentum $\mathbf{p}$ are determined by the analytical continuation of $G(\mathbf{p}, \omega)$, i.e.

$$G(\mathbf{p}, \omega) \approx \frac{a_{\mathbf{p}}}{\omega - \varepsilon_{\mathbf{p}} + i\gamma_{\mathbf{p}}}. \tag{V.12}$$

It is clear that (V.11) and (V.12) are valid only if the contribution of the integration along the deformed contour is negligible compared to that of the pole. It turns out that this is correct only if $\varepsilon_{\mathbf{p}}$ is close to $\mu$. Thus the quasiparticle concept is valuable only near $p = |\mathbf{p}| = p_f$, where $\varepsilon_f = \mu$; and with $v_f = (\partial\varepsilon/\partial p)_{p_f}$ we have

$$\varepsilon_{\mathbf{p}} = \mu + v_f(p - p_f). \tag{V.13}$$

From phase space arguments we see that the interaction of two quasiparticles causes a lifetime that is proportional to $(p - p_f)^2$; thus $\gamma_{\mathbf{p}} \ll \varepsilon_{\mathbf{p}} - \mu$, as should be. Finally we remark that the distribution of particles

$$n_{\mathbf{p}} = \langle \psi_{\mathbf{p}}^{\dagger} \psi_{\mathbf{p}} \rangle = iG(\mathbf{p}, \tau)\big|_{\tau \to -0} = \lim_{\tau \to -0} \int_{-\infty}^{\infty} \frac{d\omega}{2\pi} \exp(-i\omega\tau)\, G(\mathbf{p}, \omega), \qquad \text{(V.14)}$$

can be obtained using a deformed contour in $\omega$-space. From (V.12) and (V.7) we see that near $p = p_f$

$$G(\mathbf{p}, \omega) = \frac{a}{\varepsilon_{\mathbf{p}} - \omega - i\gamma_{\mathbf{p}}} + R(\mathbf{p}, \omega), \qquad \text{(V.15)}$$

where $R$ is regular, and $\gamma_{\mathbf{p}}$ is positive for $p > p_f$ and negative for $p < p_f$. Therefore the $\omega$-integration yields

$$n_{p_f - 0} - n_{p_f + 0} = a; \qquad \text{(V.16)}$$

that is, the discontinuity of the particle distribution on the Fermi surface is the residue of the pole of the Green's function, which is positive by (V.5).

To conclude we write the quasiparticle propagator as $(\varepsilon = \omega - \mu)$

$$G(\mathbf{p}, \varepsilon) = \frac{a}{\varepsilon - v_f(p - p_f) + i\delta \operatorname{sgn}(p - p_f)}, \qquad \text{(V.17)}$$

and notice that the discontinuity of the momentum distribution of the interacting system takes place at the same $p_f$ as for the non-interacting system.

## 2. THE QUASIPARTICLE INTERACTION

Starting from the quasiparticle propagator we now turn to the interactions between quasiparticles, and in particular to the microscopic origin of Landau's function $f$. What is involved in the interaction of two particles is the motion in time and space of two-particle Green's functions.

To find how $f(\mathbf{p}, \boldsymbol{\sigma}; \mathbf{p}', \boldsymbol{\sigma}')$ is related to the real interaction between particles we use the following device: we assign to each particle an infinitesimal charge $\delta e$ and apply a homogeneous magnetic field $\mathscr{H} = (\partial/\partial \mathbf{r}) \times \mathbf{A}$, where the vector potential $\mathbf{A}(\mathbf{r})$ varies slowly in space (and does not affect the spin). The hamiltonian $H$ of the system is changed to $H + \delta H$, where

$$\delta H = -\frac{\delta e}{mc} \sum_{\sigma} \int d^3 r\, \psi_{\sigma}^{\dagger}(\mathbf{r})(\mathbf{p} \cdot \mathbf{A}(\mathbf{r}))\, \psi_{\sigma}(\mathbf{r}) \qquad \text{(V.18)}$$

is a small perturbation. The one-particle Green's function (V.1) is changed to $G + \delta G$, where

$$\delta G_{\sigma\sigma'}(x, x') = -\sum_{\sigma''} \int d^4 x'' \left\{ -\frac{\delta e}{mc} \mathbf{p}'' \cdot \mathbf{A}(\mathbf{r}'') \right\}$$

$$\times \left(\langle T\{\psi_\sigma(x)\psi^\dagger_{\sigma''}(x'')\psi_{\sigma''}(x'')\psi_{\sigma'}(x')\}\rangle - \langle T\{\psi_\sigma(x)\psi^\dagger_{\sigma'}(x')\}\rangle\langle T\{\psi^\dagger_{\sigma''}(x'')\psi_{\sigma''}(x'')\}\rangle\right). \tag{V.19}$$

The first term in the bracket is related to the two-particle Green's function

$$G^{(2)}_{\sigma_1\sigma_2\sigma_3\sigma_4}(x_1x_2x_3x_4) = -\langle T\{\psi_{\sigma_1}(x_1)\psi_{\sigma_2}(x_2)\psi^\dagger_{\sigma_3}(x_3)\psi^\dagger_{\sigma_4}(x_4)\}\rangle, \tag{V.20}$$

which, in turn, is given in terms of the vertex $\Gamma$, i.e.

$$G(1, 2, 3, 4) = iG(1, 3)G(2, 4) - iG(1, 4)G(2, 3) + i\sum_{5,6,7,8} G(1, 5)G(2, 6)G(3, 7)G(4, 8)\Gamma(5, 6, 7, 8). \tag{V.21}$$

The vertex function plays the role of an effective interaction between the quasiparticles.

Since $\mathbf{A}$ depends on $\mathbf{r}$ only, the Fourier transform of (V.19) can be written as

$$\delta G(p, k) = G(p)\left\{-\frac{\delta e}{mc}\mathbf{p}\cdot\mathbf{A}\right\}G(p + k) - iG(p)G(p + k)\int\frac{d^4p'}{(2\pi)^4}\sum_{\sigma_1\sigma_2}\Gamma_{\sigma_1\sigma_2,\sigma_1\sigma_2} \times (p, p'; k)G(p)\left\{-\frac{\delta e}{mc}\mathbf{p}'\cdot\mathbf{A}\right\}G(p' + k), \tag{V.22}$$

where $p \equiv (\mathbf{p}, \omega)$, $k \equiv (\mathbf{k}, 0)$ are the four-momenta and $\Gamma(p, p'; \mathbf{k})$ is the Fourier transform of the vertex function $\Gamma$. We notice that in the presence of the field $\mathbf{p} \to \mathbf{p} - (\delta e/c)\mathbf{A}(\mathbf{r})$, and thus in the homogeneous-field limit $\mathbf{k} \to 0$ we have

$$\frac{c}{\mathbf{A}}\frac{\delta G}{\delta e} \to -\frac{\partial}{\partial \mathbf{p}}G, \tag{V.23}$$

or, using (V.17),

$$\frac{\partial}{\partial \mathbf{p}}G^{-1}(p) = -\frac{1}{am^*}\mathbf{p}. \tag{V.24}$$

From (V.22) we obtain

$$\frac{c}{\mathbf{A}}\frac{\delta G}{\delta e} \to G^2(p)\left\{\frac{1}{m}\mathbf{p} - i\int\frac{d^4p'}{(2\pi)^4}\frac{1}{m}\mathbf{p}'[G^2(p')]\right\}_k \sum_{\sigma_1\sigma_2}\Gamma^k_{\sigma_1\sigma_2,\sigma_1\sigma_2}(p, p'), \tag{V.25}$$

where

$$\{G^2(p)\}_k = \lim_{k\to 0}\lim_{\omega\to 0} G(p)G(p + k) \tag{V.26}$$

and

$$\Gamma^k(p, p') = \lim_{k\to 0}\lim_{\omega\to 0}(p, p'; k), \tag{V.27}$$

and thus

$$-\frac{1}{am^*}\mathbf{p} = \frac{1}{m}\mathbf{p} + i\frac{1}{2m}\int \frac{d^4p'}{(2\pi)^4}\mathbf{p}'\{G^2(p')\}_k \sum_{\sigma_1\sigma_2} \cdot \Gamma^k_{\sigma_1\sigma_2,\,\sigma_1\sigma_2}(p, p'). \tag{V.28}$$

This equation, when compared to (V.12), should give us the relation between $\Gamma$ and $f$.

We now make a digression to study $\Gamma(p_1, p_2; k)$ for $|\mathbf{k}| \ll p$ and $\omega \ll \mu$. The lowest-order diagrams of $\Gamma$ are

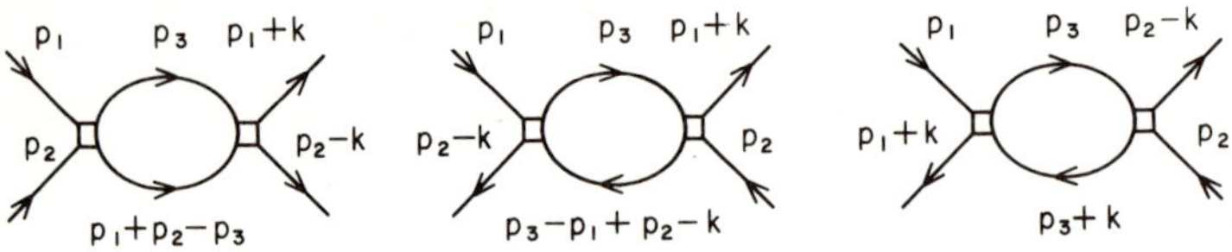


where the lines stand for the one-particle Green's function and the squares represent the real interaction between the particles. When we do perturbation theory these diagrams repeat many times, and an integral equation for $\Gamma$ would look like

$$\Gamma = \Gamma^0 + \int d^4p'\Gamma^0 G(p')G(p'')\Gamma,$$

where $p''$ is one of the four-momenta opposite to $p'$ in the diagrams. In the limit $k \to 0$ the diagram with $G(p')G(p' + k)$ has a double pole and it leads to a singularity in $\Gamma$. The integral equation for $\Gamma$ can be written for small $k$ as

$$\Gamma(p_1, p_2; k) = \Gamma^{(1)}(p_1, p_2) - i\int \frac{d^4p_3}{(2\pi)^4}\Gamma^{(1)}(p_1, p_3)G(p_3)G(p_3 + k)\Gamma(p_3, p_2; k), \tag{V.29}$$

where $\Gamma^{(1)}$ is the solution of the non-singular part (summation of the non-singular diagrams) with $k = 0$. Using (V.17) for $G$ and considering only four-momenta near the Fermi surface (where the notion of a quasiparticle holds) we have

$$G(p)G(p + k) \approx -2\pi i\frac{a^2}{v_f}\frac{\mathbf{k}\cdot\mathbf{v}}{\mathbf{k}\cdot\mathbf{v} - \omega}\delta(\varepsilon)\delta(|\mathbf{p}| - p_f) + \chi(p), \tag{V.30}$$

where $p \equiv (\mathbf{p}, \varepsilon)$, $\mathbf{v} = \mathbf{p}/m^*$, and $\chi(p)$ is regular as $k \to 0$. Now that we have explicit dependence of (V.29) on $\mathbf{k}$ and $\omega$ we can study the two limits:

(1) the $\omega$-limit, where first $\mathbf{k}$ goes to zero and then $\omega$, and
(2) the $k$-limit, where the order is reversed.

In the $\omega$-limit $G(p)G(p + k) = \chi(p)$ is regular and (V.29) gives

$$\Gamma^{\omega}(p_1, p_2) = \Gamma^{(1)}(p_1, p_2) - i \int \frac{d^4 p_3}{(2\pi)^4} \Gamma^{(1)}(p_1, p_3)\chi(p_3)\Gamma^{\omega}(p_3, p_2), \qquad \text{(V.31)}$$

and thus $\Gamma^{\omega}$ can, in principle, be solved. If we rather solve for $\Gamma^{(1)}$ in terms of $\Gamma^{\omega}$ we find

$$\Gamma(p, p'; k) = \Gamma^{\omega}(p, p') - a^2 \mathcal{N}_f \int \frac{d\Omega''}{4\pi} \Gamma^{\omega}(p, p'') \frac{\mathbf{v}'' \cdot \mathbf{k}}{\mathbf{v}'' \cdot \mathbf{k} - \omega} \Gamma(p'', p', k) \qquad \text{(V.32)}$$

where $\mathcal{N}_f$ is the density of quasiparticle states and the integration is over all directions on the Fermi surface. Finally the $k$-limit that occurs in (V.28) is found to be

$$\Gamma^{k}(p, p') = \Gamma^{\omega}(p, p') - a^2 \mathcal{N}_f \int \frac{d\Omega''}{4\pi} \Gamma^{\omega}(p, p'')\Gamma^{k}(p'', p'). \qquad \text{(V.33)}$$

Since $\{G^2(p)\}_{\omega} = \chi(p)$, we can write (V.30) as

$$\{G^2(p)\}_k = \{G^2(p)\}_{\omega} - 2\pi i \frac{a^2}{v} \delta(e)\delta(|\mathbf{p}| - p_f). \qquad \text{(V.34)}$$

If we now substitute these results into (V.28) we obtain

$$\frac{m^*}{m} = 1 + \mathcal{N}_f \int \frac{d\Omega}{4\pi} a^2 \Gamma^{\omega}(\cos\xi) \cos\xi. \qquad \text{(V.35)}$$

Comparison with (III.13) yields

$$f(\mathbf{p}, \boldsymbol{\sigma}; \mathbf{p}', \boldsymbol{\sigma}') = a^2 \Gamma^{\omega}(\mathbf{p}, \boldsymbol{\sigma}; \mathbf{p}', \boldsymbol{\sigma}');$$

that is, $f$ is equal to the forward-scattering matrix of the two quasiparticles. This completes Landau's proof of his phenomenological theory.

## General References


1. The original papers of Landau: L. D. Landau. *Soviet Phys. JETP* **3**, 920 (1957); **5**, 101 (1957); **8**, 70 (1959).
2. An old review, much in the spirit of Landau's papers: A. A. Abrikosov and I. M. Khalatnikov. *Rept. Progr. Phys.* **22**, 329 (1959).
3. An excellent detailed account of the phenomenological theory: D. Pines and P. Nozières. "The Theory of Quantum Liquids", Benjamin, New York (1966).
4. The microscopic derivations of Fermi-liquid theory are presented in the book: A. B. Abrikosov, L. P. Gor'kov and I. Ye. Dzyaloshinskii. "Quantum Field Theoretical Methods in Statistical Physics", Pergamon, Oxford (1965), pp. 49–62, 152–176.
5. Theoretical studies of transport properties, which go beyond the asymptotic results of Landau, are presented by, for example: M. J. Rice. *Phys. Rev.* **159**, 153 (1967); V. J. Emery. *Phys. Rev.* **170**, 205 (1968); C. J. Pethick. *Phys. Rev.* **177**, 391 (1969); A. Ron. *Phys. Rev.* **A1**, 1211 (1970).

# 7. THREE LECTURES ON THE EXPERIMENTAL PROPERTIES OF LIQUID $He^3$ †

JOHN C. WHEATLEY

*Department of Physics, University of California, San Diego, La Jolla, California 92037*

These lectures constitute an introduction to the experimental properties of liquid $He^3$. The selection of material has not been at all comprehensive, neither with respect to the experimental results presented nor with respect to the depth of theoretical interpretation. Rather we present key experimental results and confine ourselves to a level of presentation of theoretical models and concepts which allows an understanding of the motivation and design of the experiments. In Part I we present concepts and experimental results regarding the normal Fermi liquid with emphasis on those items which can be used immediately in understanding the experiments on the newly discovered superfluid phases of $He^3$. Part II is concerned primarily with a discussion of experiments on thermodynamic properties of the superfluid phases, emphasizing the phase diagram and what is known about both the thermal and static magnetic properties. The superfluid density near $T_c$ is also discussed together with concepts regarding the possible anisotropy and orientation of the superfluid. In Part III we discuss experiments on the nuclear spin dynamics as they manifest themselves in a manner peculiar to superfluid $He^3$. We also include a discussion of superfluid flow in the hydrodynamic transfer of heat.

† Work supported by the US Atomic Energy Commission under Contract No. AT(04–3)–34, P.A. 143.

## I. Experimental Properties of Fermi Liquid $He^3$

This lecture is not intended to be a comprehensive survey of the experimental properties of pure Fermi liquid $He^3$. Rather it is intended to serve as both a qualitative and a quantitative base on which we can stand to view the remarkable properties of superfluid $He^3$. It will be helpful for us to realize just how large the viscosity and thermal conductivity of normal liquid $He^3$ actually are when the fluid is about to make its superfluid transition. It will also be useful for us to have an appreciation of the size of the magnetic susceptibility and how sound propagates in both the hydrodynamic and collisionless regimes in the normal liquid. Further, we need to know how the strong interactions between the $He^3$ quasiparticles are accounted for, using Landau's theory of the Fermi liquid and how large the corresponding phenomenological parameters are. Thus before we begin our discussion of the newly discovered superfluid phases of liquid $He^3$ we introduce the properties of the normal liquid.

The specific heat of the normal liquid has been measured at very low temperatures only at low pressure and at a pressure somewhat below the minimum in the melting curve. The experimental measurements[(1)] are shown in Fig. I.1 where $C/nRT^*$ has been plotted versus $T^*$. Owing to the very interesting changes of superfluid properties with pressure it is now highly desirable to make accurate measurements of normal liquid specific heat over the full pressure range from zero to the melting pressure of 34·4 Bar. However, until new and better data are available we will have to be content with an interpolation and extrapolation of results obtained[(2)] at intermediate pressures over a decade ago. The temperatures indicated in Fig. I.1 are magnetic. They are based on measurements of the susceptibility of a powdered sample of cerous magnesium nitrate (CMN) in the shape of a cylinder with diameter equal to height. Thanks to the absolute noise thermometry work of Webb *et al.*[(3)] these temperatures, down to about 5 mK, are thought to be within one or two tenths of a millikelvin of the absolute temperature. The data seem to be represented best by the simple law

$$\frac{C}{nRT} = \gamma - \lambda T, \tag{I.1}$$

where $C/n$ is the molar specific heat, $R$ is the gas constant and $\gamma$ and $\lambda$ are functions of pressure. Although the temperature dependence of $C/nRT$ is more rapid than that expected for a Fermi gas there does not appear to be a divergence of this quantity at low temperatures.

The most accurate measurements of magnetic susceptibility at low temperatures have been made by Ramm *et al.*[(4)] Their measurements are shown in part in Fig. I.2. The susceptibility $\chi$, near $T = 0$ but above the

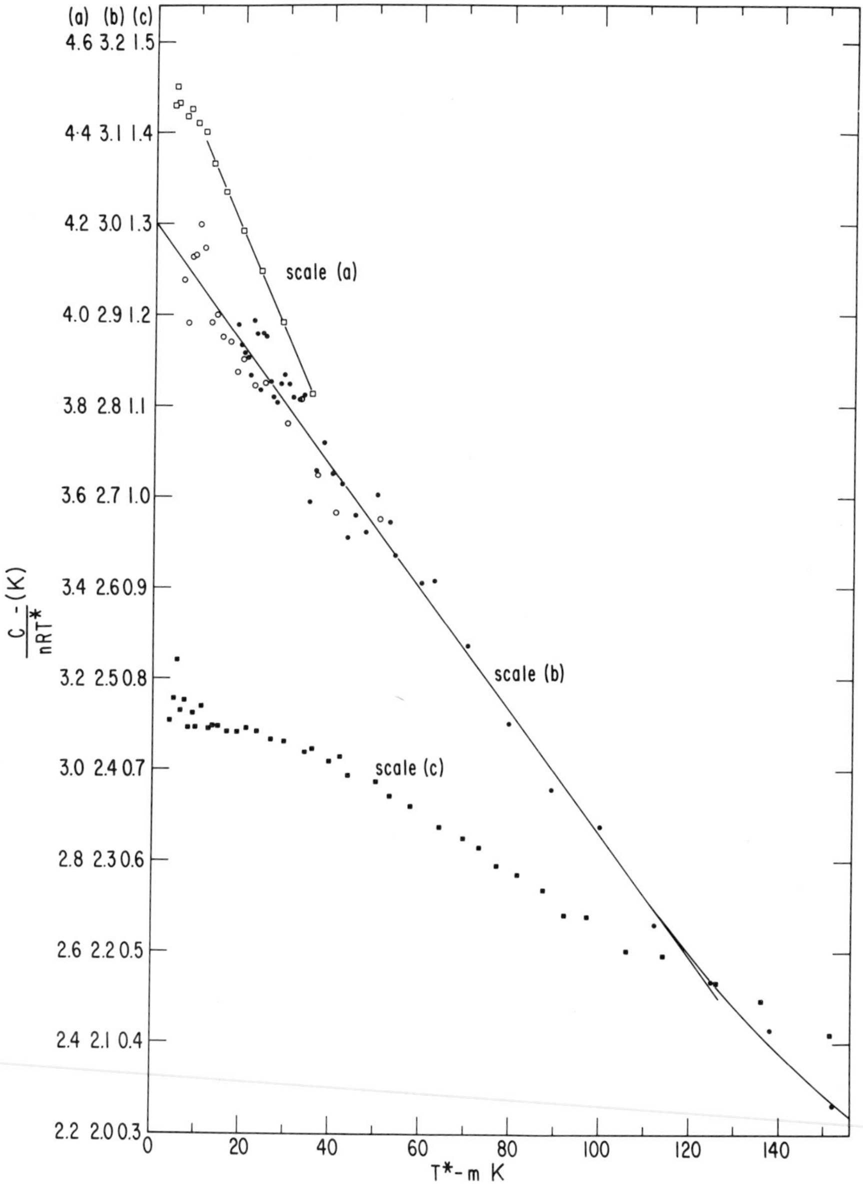


FIG. I.1. Molar specific heat of pure He$^3$ divided by $RT^*$ for pure He$^3$ at 0·24 Atm (●) and at 0·28 Atm (○)—scale (b); for pure He$^3$ at 27·0 Atm (□)—scale (a); and, for comparison, for a 5% solution of He$^3$ in superfluid He$^4$ with respect to the total number of moles $n$ of solution at saturated vapour pressure (■)—scale (c). The CMN magnetic temperature $T^*$, and the Kelvin temperatures are the same within one or two tenths mK. (After Wheatley, Ref. 7.)

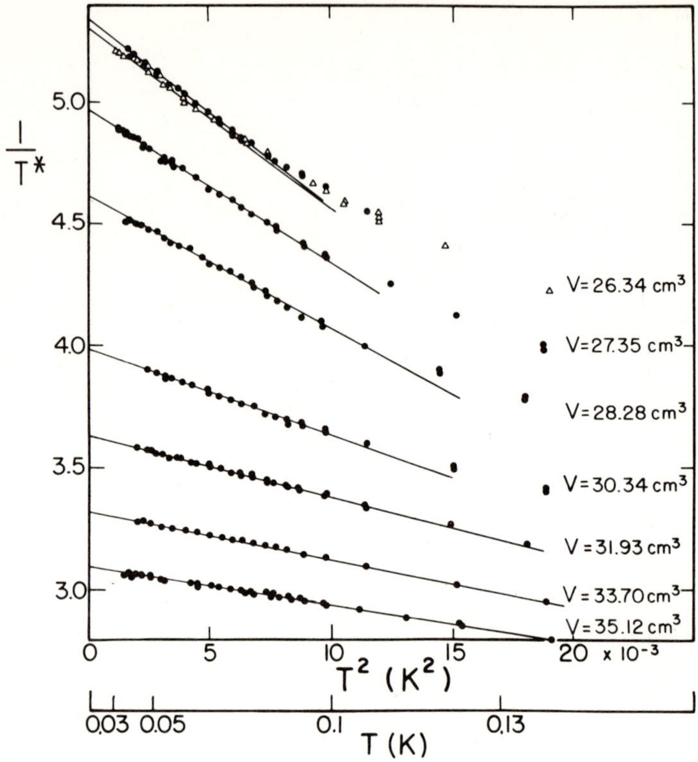


FIG. I.2. The effective magnetic temperature $T^*$ of pure liquid $He^3$ defined as $C/\chi$, where $C$ is the Curie constant for the nuclear magnetism and $\chi$ is the nuclear susceptibility, for various *molar* volumes as function of $T^2$. The value of $T^*$ extrapolated to $T^2 = 0$ is the quantity used in Eqn (I.10). (After Ramm, Pedroni, Thompson and Meyer, Ref. 4.)

temperature of the superfluid transition, is of the form

$$\chi = \chi_{\mathrm{N}} - aT^2 \tag{I.1}$$

where $\chi_N$ and $a$ are functions of pressure. The limiting low temperature value $\chi_N$ of the susceptibility is one of the most interesting properties of the normal Fermi liquid. It may be written in the form

$$\chi_N = \frac{C}{T^*}, \tag{I.3}$$

where $C$ is the Curie constant ($C = N\gamma^2\hbar^2/4k$, where $N$ is the number of $He^3$ atoms per unit volume, $\gamma$ is the gyromagnetic ratio, $\hbar$ is Planck's constant divided by $2\pi$, and $k$ is Boltzmann's constant) and $T^*$ is an effective magnetic temperature of the degenerate Fermi system. The measured values of $T^*$ are 10 to 20 times smaller, the factor being greater the higher the pressure, than would have been expected for an ideal Fermi gas of the same density. This has been, in part, the basis for the so-called paramagnon model to describe the properties of liquid $He^3$.

The specific heat, the susceptibility, and the first sound velocity (which we have yet to discuss) are used to evaluate parameters to describe the fluid in Landau's theory of the Fermi liquid.[5] According to this theory the liquid is comprised of quasiparticles of momentum $\mathbf{p}$ and spin $\boldsymbol{\sigma}$ and has energy per unit volume given by

$$E = \sum_{\mathbf{p},\boldsymbol{\sigma}} \varepsilon_0(\mathbf{p},\boldsymbol{\sigma})\delta n(\mathbf{p},\boldsymbol{\sigma}) + \tfrac{1}{2}\sum_{\substack{\mathbf{p},\mathbf{p}'\\ \boldsymbol{\sigma}\,\boldsymbol{\sigma}'}} f(\mathbf{p},\boldsymbol{\sigma};\mathbf{p}',\boldsymbol{\sigma}')\delta n(\mathbf{p}',\boldsymbol{\sigma}'), \tag{I.4}$$

where $\varepsilon_0(\mathbf{p},\boldsymbol{\sigma})$ is the energy of a quasiparticle singly excited, $\delta n$ is an incremental change in the distribution function, and $f$ is a function, fundamental to Landau's theory, which describes how the particles interact. The energy of any given quasiparticle excitation thus depends on how all the other excitations are distributed. For example, referring to Fig. I.3 where we imagine

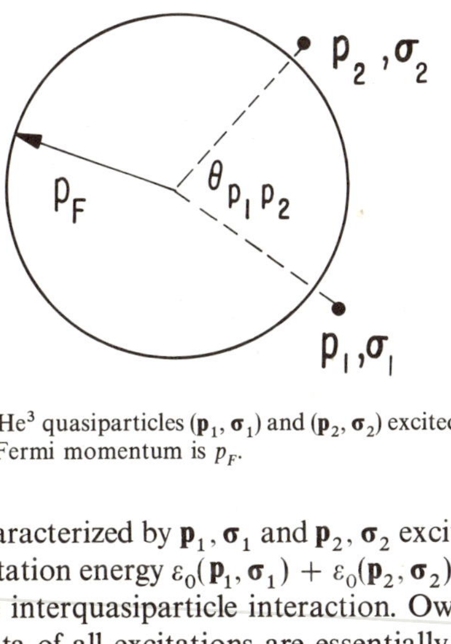


FIG. I.3. Illustration of two $He^3$ quasiparticles $(\mathbf{p}_1, \boldsymbol{\sigma}_1)$ and $(\mathbf{p}_2, \boldsymbol{\sigma}_2)$ excited above the Fermi sphere in momentum space. The Fermi momentum is $p_F$.

two quasiparticles characterized by $\mathbf{p}_1, \boldsymbol{\sigma}_1$ and $\mathbf{p}_2, \boldsymbol{\sigma}_2$ excited above the Fermi sea, we expect an excitation energy $\varepsilon_0(\mathbf{p}_1,\boldsymbol{\sigma}_1) + \varepsilon_0(\mathbf{p}_2,\boldsymbol{\sigma}_2) + f(\mathbf{p}_1,\boldsymbol{\sigma}_1;\mathbf{p}_2,\boldsymbol{\sigma}_2)$, where $f$ describes the interquasiparticle interaction. Owing to the low temperatures the momenta of all excitations are essentially the Fermi momentum $p_F$ in magnitude, so the interaction function $f$ in (I.4) depends only on the angle $\theta_{pp'}$. The function may then be written in reduced form

$$2N(0)f(\mathbf{p},\mathbf{p}',\boldsymbol{\sigma},\boldsymbol{\sigma}') = \sum_{l=0}^{\infty} (F_l + Z_l\boldsymbol{\sigma}\cdot\boldsymbol{\sigma}')\, P_l(\cos\theta_{pp'}), \tag{I.5}$$

where $N(0)$ is the density of states per unit energy and volume for spins of one sign at the Fermi surface, $P_l$ are the Legendre polynomials, $F_l$ and $Z_l$ and phenomenological parameters, and $\boldsymbol{\sigma}$, $\boldsymbol{\sigma}'$ are the spin angular momenta divided by $\hbar$. (According to this notation the spin angular momentum of a single quasiparticle is $\hbar\boldsymbol{\sigma}$, *not* $\frac{1}{2}\hbar\boldsymbol{\sigma}$.) To date most experiments are described in terms of $F_0$, $F_1$, and $Z_0$. A particularly simple case to understand is the static effect of a magnetic field, where the term in $Z_0$ contributes a "molecular field." Thus from (I.4) and (I.5), in a magnetic field $\mathbf{H}_0$ a given quasiparticle $(\mathbf{p}, \boldsymbol{\sigma})$ will have magnetic energy

$$-\gamma\hbar\boldsymbol{\sigma}\cdot\mathbf{H}_0 + \frac{Z_0}{2N(0)}\boldsymbol{\sigma}\cdot\sum_{\mathbf{p}',\boldsymbol{\sigma}'}\boldsymbol{\sigma}'\,\delta n(\mathbf{p}',\boldsymbol{\sigma}') = -\gamma\hbar\boldsymbol{\sigma}\cdot\left(\mathbf{H}_0 - \frac{Z_0}{2\gamma\hbar N(0)}\mathbf{S}\right), \tag{I.6}$$

where $\mathbf{S} = \Sigma_{\mathbf{p}',\boldsymbol{\sigma}'}\,\boldsymbol{\sigma}'\,\delta n(\mathbf{p}',\boldsymbol{\sigma}')$ is $(\hbar)^{-1}$ times the spin angular momentum per unit volume. The factor 1/2 in front of the interaction term is not appropriate when we consider the energy of a single quasiparticle interacting with the rest. Thus the interaction parameter $Z_0$ leads to a molecular field $-Z_0\mathbf{S}/2\gamma\hbar N(0)$. Following Leggett[6] the interaction energy in (I.4) can be expressed

$$E_{\text{interaction}} = \frac{1}{4N(0)}\left[F_0(\delta N)^2 + Z_0(\mathbf{S})^2 + \frac{F_1}{p_F^2}(\mathbf{Q})^2 + \ldots\right] \tag{I.7}$$

where

$$(\delta N)^2 = \sum_{\substack{\mathbf{p},\mathbf{p}'\\ \boldsymbol{\sigma},\boldsymbol{\sigma}'}} \delta n(\mathbf{p},\boldsymbol{\sigma})\,\delta n(\mathbf{p}',\boldsymbol{\sigma}')$$

so that $\delta N$ is the change in the number of particles per unit volume, $\mathbf{S}$ is as given above, and $\mathbf{Q} = \Sigma_{\mathbf{p},\boldsymbol{\sigma}}\,\mathbf{p}\,\delta n(\mathbf{p},\boldsymbol{\sigma})$ is the linear momentum per unit volume. $F_1$ will act as a "molecular field" parameter for flow just as $Z_0$ does for magnetism.

The parameters $F_0$, $Z_0$, and $F_1$ are obtained from experiment by the formulas:

$$1 + \tfrac{1}{3}F_1 = \frac{m^*}{m} = \frac{1}{k}\frac{\hbar^2}{m}\left(\frac{3}{\pi}N\right)^{2/3}\left(\frac{C}{nRT}\right) \tag{I.8}$$

$$\frac{1 + F_0}{1 + \frac{1}{3}F_1} = \frac{c_1^2}{\frac{1}{3}(p_F/m)^2} \tag{I.9}$$

$$\frac{1 + \frac{1}{4}Z_0}{1 + \frac{1}{3}F_1} = \frac{\frac{3}{2}kT^*}{p_F^2/2m}, \tag{I.10}$$

where $p_F = \hbar(3\pi^2 N)^{1/3}$, $N$ is the measured number density, $m$ is the mass of a $He^3$ atom while $m^*$ is the effective mass, $C/nRT$ relates to measured heat capacity, $c_1$ is the measured velocity of ordinary hydrodynamic sound, and

$T^*$ is the measured magnetic temperature in the limit $T \to 0$ as given in (I.3). Some values of these parameters are given in Appendix B.

The effective mass $m^*$ and the atomic mass $m$ may be related as in (I.8) by considering a uniform motion of the fluid at some velocity $\mathbf{v}$ so that the centre of the Fermi sphere is translated $m\mathbf{v}$ as shown in Fig. I.4. This has the effect of changing the distribution function, adding particles to the right and holes to the left. As a consequence of (I.4) and (I.7) the energy of the fluid is then

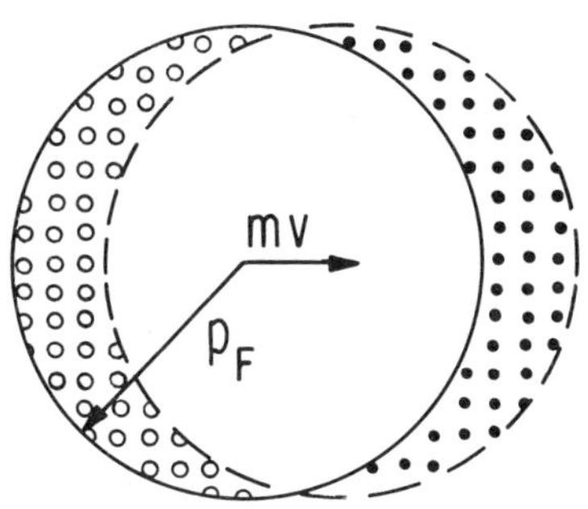


FIG. I.4. Gas of He$^3$ quasiparticles in momentum space uniformly translating with velocity $\mathbf{v}$ as imagined in relating $F_1$ and $m^*/m$. The ● represent filled states and the ○ holes. The Fermi momentum is $p_F$.

$$\sum_{\mathbf{p},\sigma} [\varepsilon_0(\mathbf{p},\boldsymbol{\sigma}) - \mathbf{v}\cdot\mathbf{p}]\,\delta n(\mathbf{p},\boldsymbol{\sigma}) + \tfrac{1}{2}\frac{F_1}{2N(0)p_F^2}\mathbf{Q}^2$$

$$= \sum_{\mathbf{p},\sigma} \varepsilon_0(\mathbf{p},\boldsymbol{\sigma})\,\delta n(\mathbf{p},\boldsymbol{\sigma}) - \mathbf{v}\cdot\mathbf{Q} + \tfrac{1}{2}\frac{F_1}{2N(0)p_F^2}\mathbf{Q}^2 \tag{I.11}$$

where $\mathbf{Q}$ is the momentum density and where the energy of a single excitation in the moving system has been replaced by $\varepsilon_0(\mathbf{p},\boldsymbol{\sigma}) - \mathbf{v}\cdot\mathbf{p}$, both $\varepsilon_0$ and $\mathbf{p}$ referring to the lab system.† In (I.11) $\varepsilon_0$ may be thought of as the energy of a free particle of mass $m^*$. The last two terms may be combined to give a "molecular field" approximation to an effective velocity $\mathbf{v}_{\text{eff}}$ for computing the energy of a single excitation of mass $m^*$ in the moving fluid:

$$-\mathbf{v}\cdot\mathbf{Q} + \frac{F_1}{2N(0)p_F^2}\mathbf{Q}^2 = -\left(\mathbf{v} - \frac{F_1}{2N(0)p_F^2}\mathbf{Q}\right)\cdot\mathbf{Q} = -\mathbf{v}_{\text{eff}}\cdot\mathbf{Q}, \tag{I.12}$$

so that

$$\mathbf{v}_{\text{eff}} = \mathbf{v} - \frac{F_1}{2N(0)p_F^2}\mathbf{Q}. \tag{I.13}$$

† See L. D. Landau and E. M. Lifshitz, "Statistical Physics", Pergamon Press, London (1958), p. 204.

The factor $\frac{1}{2}$ in the $\mathbf{Q}^2$ term of (I.11) is omitted since it came from a sum of energies of pairs and we are looking for the average effect of all other excitations on a given excitation. Now we can use this "molecular field" result to compute the momentum density. Equate the momentum density of quasi-particles of effective mass $m^*$, number density $N$, and velocity $\mathbf{v}_{\text{eff}}$ to that of particles of mass $m$, number density $N$, and velocity $\mathbf{v}$. The result is

$$Nm^*\mathbf{v}_{\text{eff}} = Nm^*\left(\mathbf{v} - \frac{F_1}{2N(0)p_F^2}\mathbf{Q}\right) = \mathbf{Q} = Nm\mathbf{v}, \tag{I.14}$$

or

$$m^*\left(\mathbf{v} - \frac{NmF_1}{2N(0)p_F^2}\mathbf{v}\right) = m\mathbf{v}. \tag{I.15}$$

Now by using the equation

$$2N(0)p_F^2 = 3Nm^* \tag{I.16}$$

(I.15) can be reduced to

$$\frac{m^*}{m} = 1 + \tfrac{1}{3}F_1. \tag{I.17}$$

So the parameter $F_1$ may be obtained from $m^*/m$, which is obtained in turn from the specific heat and number density (I.8).

Equation (I.10) for the magnetic temperature $T^*$ follows in a similar manner to the above on setting the magnetization $\gamma\hbar\mathbf{S}$ equal to the product of the "bare" susceptibility, $\frac{1}{2}\gamma^2\hbar^2N(0)$, and the effective field $\mathbf{H}_0 - Z_0\mathbf{S}/2\gamma\hbar N(0)$ displayed in the relation (I.6). One solves $\chi = \gamma\hbar S/H_0$ and uses the definition $\chi = N\gamma^2\hbar^2/4kT^*$.

The relationship between $F_0$, $F_1$ and the velocity of sound may be obtained as follows. At the absolute zero a change in $\mu$, the chemical potential per particle, is given by

$$\delta\mu = \frac{1}{N}\delta P \tag{I.18}$$

where the pressure $P$ depends on number density $N$. Equation (I.18) may be rewritten, using the mass density $\rho = Nm$,

$$\delta\mu = \frac{m}{N}\frac{dP}{d\rho}\delta N = \frac{m}{N}c_1^2\delta N \tag{I.19}$$

where we have substituted the well known result $c_1^2 = dP/d\rho$ for the square of the velocity of hydrodynamic sound. Hence to compute $c_1^2$ we need to compute the change in the energy of a particle on the Fermi surface on changing the number density by $\delta N$ as illustrated in Fig. I.5. Following (I.4) and (I.7) the energy of the fluid is

$$\sum_{\mathbf{p},\sigma} \varepsilon_0(\mathbf{p}, \boldsymbol{\sigma})\, \delta n(\mathbf{p}, \boldsymbol{\sigma}) + \tfrac{1}{2}\frac{F_0}{2N(0)}(\delta N)^2,$$

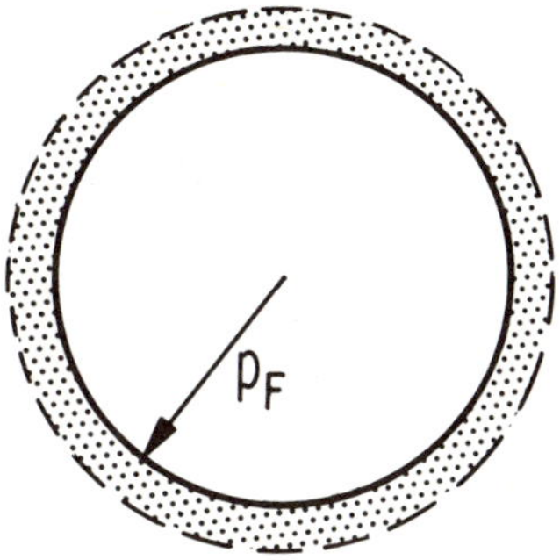

FIG. I.5. Changes of the Fermi sphere of radius $p_F$ on increasing the number density. The ● refer to additional filled states imagined in relating $F_0$ to the velocity of sound.

where $\varepsilon_0 = p^2/2m^*$. Hence the change of energy of a single particle at the Fermi surface is

$$\delta\mu = \delta\left(\frac{p_F^2}{2m^*}\right) + \frac{F_0}{2N(0)}\delta N. \tag{I.20}$$

Using the result that $p_F^2 \propto N^{2/3}$ this may be rewirtten

$$\delta\mu = \tfrac{1}{3}\frac{p_F^2}{m^*}\frac{\delta N}{N} + \frac{F_0}{2N(0)}\delta N. \tag{I.21}$$

This may be simplified using (I.16) to find

$$\frac{\delta\mu}{\delta N} = \tfrac{1}{3}\frac{p_F^2}{m^* N}(1 + F_0), \tag{I.22}$$

from which follows via (I.19) that

$$c_1^2 = \tfrac{1}{3}\frac{p_F^2}{m^* m}(1 + F_0). \tag{I.23}$$

Then using (I.17) we find

$$c_1^2 = \tfrac{1}{3}\frac{p_F^2}{m^2}\frac{1 + F_0}{1 + \frac{1}{3}F_1} \tag{I.23a}$$

as indicated by (I.9).

Measurements of the thermal, magnetic, and sonic properties of liquid $He^3$ are of great importance; but, with the possible exception of zero sound, they are not as spectacular as the transport properties. Here we will discuss the thermal conductivity and viscosity, omitting spin diffusion since at this writing no experiments on this phenomenon have been performed on the superfluid state. We might note parenthetically, however, that the probable existence of supercurrents of magnetization as well as diffusive modes will make such experiments very interesting indeed.

The coefficients of viscosity $\eta$ and of thermal conductivity $\kappa$ have the form[7]

$$\eta = \tfrac{1}{5}\rho v_F^2\left(\frac{m^*}{m}\right)\tau_\eta \tag{I.24}$$

and

$$\kappa = \tfrac{1}{3}C_V v_F^2 \tau_\kappa, \tag{I.25}$$

where $\rho$ is mass density, $C_V$ is specific heat per unit volume, $v_F$ is Fermi velocity, $\tau_\eta$ is a relaxation time for viscosity, and $\tau_\kappa$ is a relaxation time for thermal conductivity. Since the $\tau$'s involve different kinds of averages of collisions they are different. The characteristically large transport properties of liquid $He^3$ at very low temperatures comes from the temperature dependence of the relaxation time $\tau$. Collisions between quasiparticles must not only conserve momentum and energy but also obey Pauli's exclusion principle as well. This severely restricts the number of possible collisions. Consider Fig. I.6 in which a thermally excited quasiparticle in state $\mathbf{p}_1$ scatters against the quasiparticle in state $\mathbf{p}_2$ with final states $\mathbf{p}'_1$ and $\mathbf{p}'_2$. Conservation of energy and momentum in the collision requires that $\varepsilon_1 + \varepsilon_2 = \varepsilon'_1 + \varepsilon'_2$ and $\mathbf{p}_1 + \mathbf{p}_2 = \mathbf{p}'_1 + \mathbf{p}'_2$. Now since the final states $\mathbf{p}'_1$ and $\mathbf{p}'_2$ must be unoccupied the state $\mathbf{p}_2$ must lie within energy $kT$ of the Fermi surface to

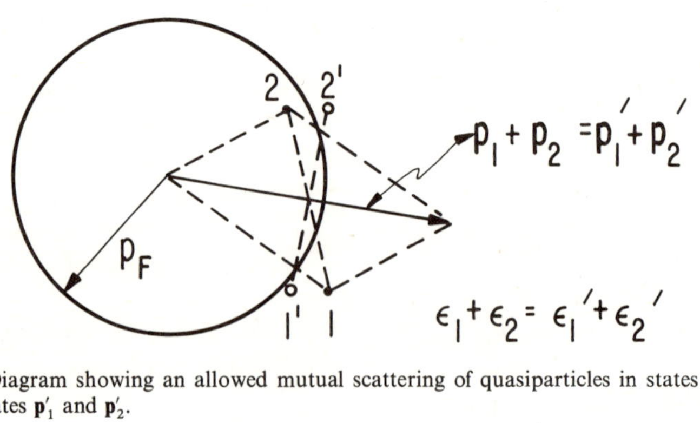


FIG. I.6. Diagram showing an allowed mutual scattering of quasiparticles in states $\mathbf{p}_1$ and $\mathbf{p}_2$ to final states $\mathbf{p}'_1$ and $\mathbf{p}'_2$.

conserve energy. Furthermore, of all final states satisfying energy and momentum conservation only those unoccupied states such as $\mathbf{p}'_1$ (and then, by conservation of momentum, $\mathbf{p}'_2$) within $kT$ of the Fermi surface are possible. Hence the number of collisions possible using conservation of energy and momentum alone is reduced because of the Pauli principle by a factor estimated by $(kT/\varepsilon_F)^2$, where $\varepsilon_F$ is the Fermi energy. Consequently the probability per unit time for scattering around the Fermi surface is proportional to $T^2$ and the collision times $\tau$ are proportional to $T^{-2}$. Thus in the limit $T \to 0$ the viscosity coefficient should increase with decreasing $T$ as $T^{-2}$ while the thermal conductivity coefficient, owing to the limiting linear dependence of $C_V$ on $T$, should increase with decreasing $T$ as $T^{-1}$.

Some experimental results of Alvesalo *et al.*[8] for the viscosity coefficient at melting pressure as determined by the damping of a vibrating wire are shown in Fig. I.7 both as $\ln \eta$ versus $\ln T$ and as a plot of $(\eta T^2)^{-1}$ versus $T$. The form $\eta T^2 + a_\eta - b_\eta T$, where $a_\eta$ and $b_\eta$ are functions of pressure, gives a good low temperature representation of the data. This form was originally suggested by Rice[9] on the basis of the so-called paramagnon model. What is particularly interesting is that at the temperature of the superfluid transition (2·6 mK at melting pressure[10] and 0·9 mK at saturated vapour pressure[11]) the viscosity is 0·13 poise at melting pressure and 2·3 poise at SVP, enormous values compared with the 20 micropoise viscosity of liquid $He^4$ just above the $\lambda$-point. These high viscosities are reflecting the very long free path of the quasiparticle excitations in the liquid ($v_F \tau_\eta \sim 2$–$4 \times 10^{-3}$ cm mK$^2$/$T^2$ in the pressure range above 10 Bar). They furthermore profoundly affect the design of experiments to study flow properties: it is not necessary to resort to a tightly packed powder to make a superleak

Thermal conduction in the liquid is also very high. An example of experimental data[7] is shown in Fig. I.8, where we find the data represented by the expression $1/(\kappa T) = a_\kappa - b_\kappa T$, where $a_\kappa$ and $b_\kappa$ are functions of pressure, and $T^* + \Delta$ is to be interpreted as $T$. This form is also expected on the basis of the paramagnon theory. Extensive measurements of $\kappa$ have not been made at higher pressures, but from measurements near the superfluid transition[10] it is known that $a_\kappa$ has increased with pressure by about a factor three near melting pressure compared with its value at SVP. Again, the values of the thermal conductivity are very high just above the temperature of the superfluid transition: about 4 erg/sec cm mK near melting pressure and nearly 40 erg/sec cm mK at SVP. To put these in perspective note that the thermal conductivity of electrical purity copper at 1 mK is about 10 erg/sec cm mK; so $He^3$ is, comparatively, a very good conductor of heat indeed at these low temperatures. Since the specific heat of the liquid is low, the thermal diffusivity is high; and liquid $He^3$ tends rapidly to come into internal thermal equilibrium.

Also of great interest for the properties of the Fermi liquid is the propagation of collisionless sound, or zero sound as it was called by Landau[12] in the paper in which he proposed its existence. If for sound of angular frequency $\omega$ we have the condition $\omega\tau \ll 1$, where $\tau$ is the collision time, there will be plenty of time for the distribution of particles to conform to that for local thermal equilibrium. The sound will propagate with a velocity determined by a thermodynamic formula $[c_1^2 = (\partial P/\partial \rho)_S]$ and with an attenuation due to viscosity. The attenuation coefficient is proportional to $\omega^2\eta$. and hence *decreases* as $T$ increases. However, in a Fermi liquid with $\tau \propto T^{-2}$, at some low enough temperature we may have $\omega\tau \gg 1$. Landau showed that in this limit sound can propagate without the need for quasiparticle collisions to maintain thermal equilibrium. To get some understanding of the nature

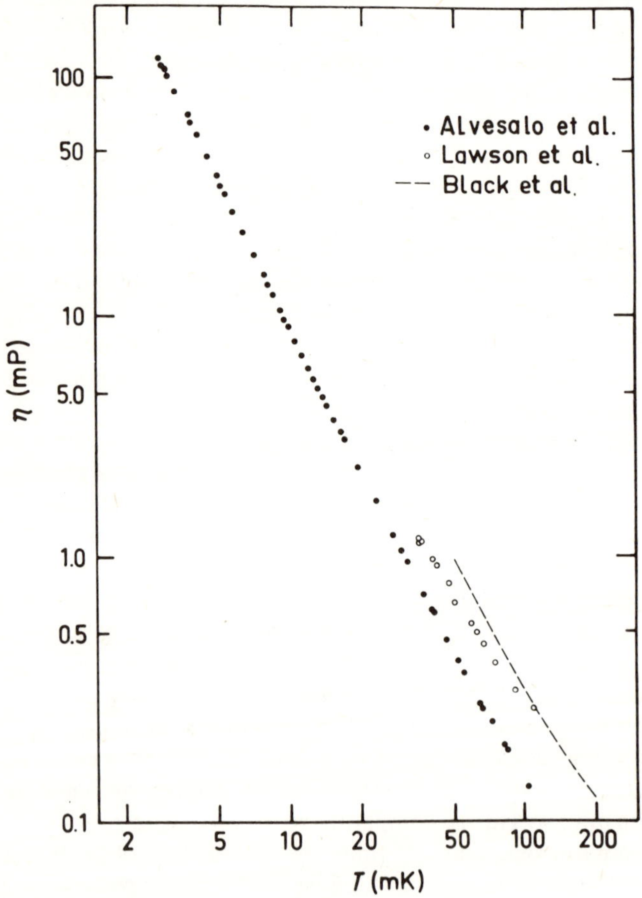


FIG. I.7. (a) The viscosity $\eta$ of pure liquid He$^3$ at melting pressure.

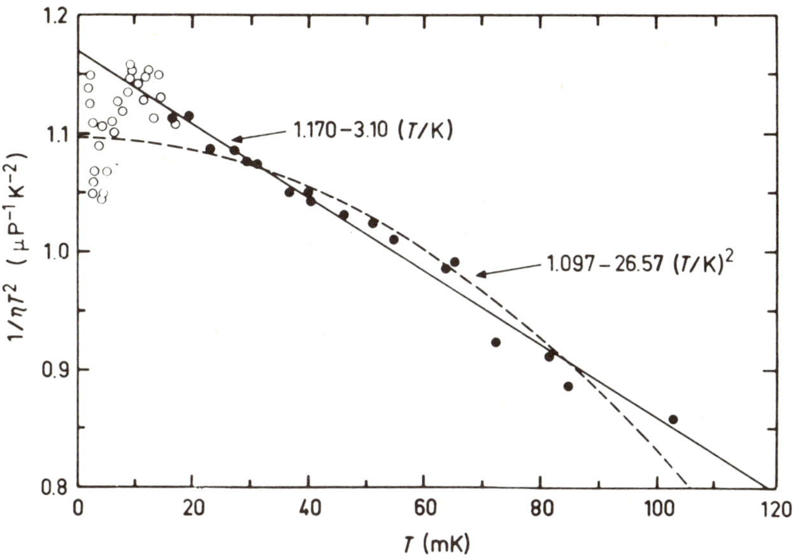


FIG. I.7. (b) Plot of $(\eta T^2)^{-1}$ versus $T$ for pure He$^3$ at melting pressure. The straight line (solid) is the form predicted by the paramagnon theory. (After Alvesalo, Collan *et al.*, Ref. 8).

of zero sound imagine that the Fermi surface is changed in some way from its equilibrium spherical shape centered on $p = 0$, as in Fig. I.9, and that this change varies in a wavelike way with space and time. The distorted Fermi surface need not be spherical since collisions are not frequent enough to maintain sphericity. To be concrete let this distortion be characterized by the changed distribution function

$$\delta n(\mathbf{p}, \mathbf{r}, t) = \left(-\frac{\partial n_0}{\partial \varepsilon}\Delta\varepsilon\right)v(\theta)\exp\left[i(\mathbf{k}\cdot\mathbf{r} - \omega t)\right], \tag{I.26}$$

where $\Delta\varepsilon$ is a constant energy change included for dimensional purposes, $\partial n_0/\partial\varepsilon$ is the derivative of the equilibrium distribution function with respect to energy and thus confines the change of the distribution function to the region near the Fermi surface, and $v$ is a function of the angle $\theta$ between $\mathbf{p}$ and the wave propagation vector $\mathbf{k}$ which will have to be determined. To simplify the thinking we will further assume that in the expression for the energy of a quasiparticle, from (I.4), the phenomenological function is a constant $f$. Then

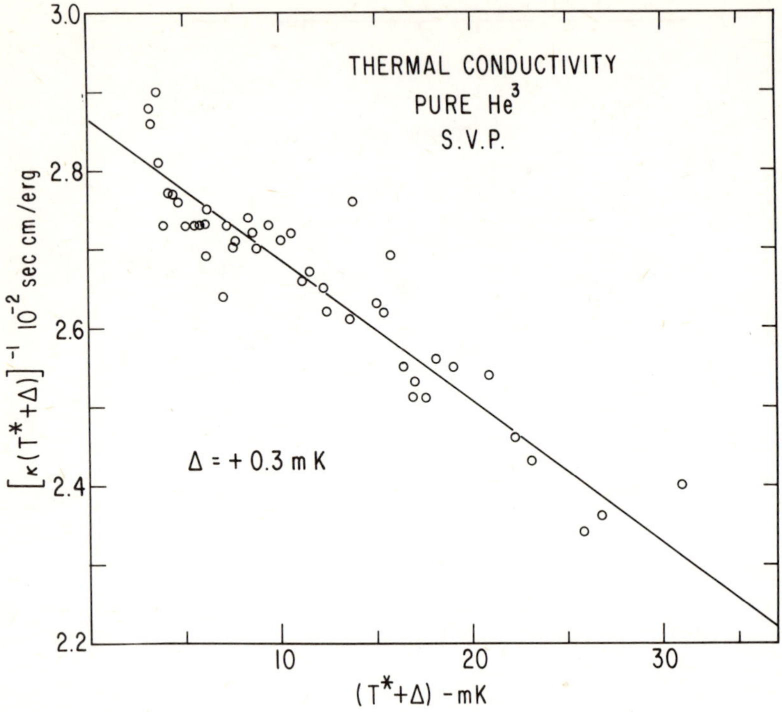


FIG. I.8. Plot of $[\kappa(T^* + \Delta)]^{-1}$ versus $(T^* + \Delta)$ for pure $He^3$ at saturated vapour pressure. $\kappa$ is the thermal conductivity and $T^* + \Delta$ is approximately equal to the Kelvin temperature. (After Wheatley. Ref. 7.)

$$\begin{aligned}
\varepsilon &= \frac{p^2}{2m^*} + \sum_{\mathbf{p}'} f\,\delta n(\mathbf{p}') \\
&= \frac{p^2}{2m^*} + \int 2N(0)\,\mathrm{d}\varepsilon\left(-f\frac{\partial n_0}{\partial \varepsilon}\Delta\varepsilon\right)\int \frac{\mathrm{d}\Omega'}{4\pi}\,\nu(\theta')\exp\left[i(\mathbf{k}\cdot\mathbf{r} - \omega t)\right] \\
&= \frac{p^2}{2m^*} + \left[F_0\Delta\varepsilon\int\frac{\mathrm{d}\Omega'}{4\pi}\,\nu(\theta')\right]\exp\left[i(\mathbf{k}\cdot\mathbf{r} - \omega t)\right] \qquad \text{(I.27)}
\end{aligned}$$

where (I.26) and the definition of $F_0$ have been used. It is the "Fermi liquid" term in (I.27) which provides the spatially dependent energy needed to propagate the sound wave.

Both the form of the function $\nu$ and the velocity of zero sound may now be obtained from the kinetic equation assuming that the distribution function is $n = n_0 + \delta n(\mathbf{p}, \mathbf{r}, t)$. The kinetic equation is

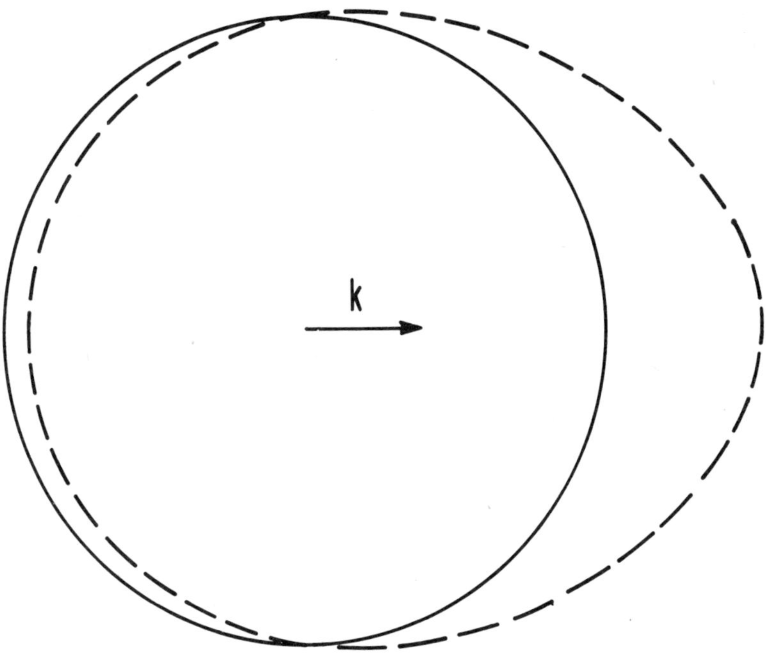


FIG. I.9. Shape of the distorted Fermi surface (shown dashed) as compared with the undistorted Fermi surface (shown solid) for a wave of zero sound characterized by $c_0/v_F = 1{\cdot}4$ propagating with wave vector **k**, as computed using Eqn (I.31). In liquid $He^3$ $c_0/v_F$ varies from 3·5 at zero pressure to 14 at melting pressure, and the distorted Fermi surface is nearly spherical.

$$\frac{\partial n}{\partial t} + \frac{\partial \varepsilon}{\partial \mathbf{p}}\cdot\frac{\partial n}{\partial \mathbf{r}} - \frac{\partial \varepsilon}{\partial \mathbf{r}}\cdot\frac{\partial n}{\partial \mathbf{p}} = I(n) = 0, \tag{I.28}$$

where the collision integral $I(n)$ is taken to be zero. In substituting into this equation note that $\partial\varepsilon/\partial\mathbf{p} = \mathbf{v}$, the particle group velocity; and

$$\partial n/\partial \mathbf{p} \simeq \partial n_0/\partial \mathbf{p} = (\partial n_0/\partial \varepsilon)\mathbf{v}.$$

Then, using (I.27), we find that (I.28) can be written

$$(\mathbf{k}\cdot\mathbf{v} - \omega)\delta n = \mathbf{k}\cdot\mathbf{v}\frac{\partial n_0}{\partial \varepsilon}\left[F_0\Delta\varepsilon\int\frac{\mathrm{d}\Omega'}{4\pi}\nu(\theta')\right]\exp\left[i(\mathbf{k}\cdot\mathbf{r} - \omega t)\right]. \tag{I.29}$$

Substitute $\delta n$ from (I.26) into the left side of (I.29) to obtain

$$(\mathbf{k}\cdot\mathbf{v} - \omega)\nu(\theta) = -\mathbf{k}\cdot\mathbf{v}F_0\int\frac{\mathrm{d}\Omega'}{4\pi}\nu(\theta'). \tag{I.30}$$

Finally, noting that $\theta$ is the angle between **k** and **v**, that $c_0 = \omega/k$ ($c_0$ is the velocity of zero sound), and that the magnitude of $\mathbf{v}$ is $v_F$ we have

$$\nu(\theta) = \frac{\cos\theta}{(c_0/v_F) - \cos\theta} F_0 \int \frac{d\Omega'}{4\pi} \nu(\theta'). \tag{I.31}$$

This gives the form of the distortion of the Fermi surface caused by the wave. For $c_0/v_F \gg 1$ as is actually the case in $He^3$, the distortion is essentially proportional to $\cos\theta$ as in a regular sound wave; but for $c_0/v_F \simeq 1$ there is a pronounced "nose" in the direction of propagation of the wave. An equation for $c_0/v_F$ can be obtained by integrating (I.31) with respect to $d\Omega/4\pi$:

$$\int \frac{d\Omega}{4\pi} \frac{\cos\theta}{s_0 - \cos\theta} = \frac{1}{F_0}, \tag{I.32}$$

where $s_0 \equiv c_0/v_F$. The integral gives

$$\frac{s_0}{2} \ln\left(\frac{s_0 + 1}{s_0 - 1}\right) - 1 = \frac{1}{F_0} \tag{I.33}$$

as an implicit equation for $s_0$. If the $F_1$ parameter is also taken into account then the implicit equation for $s_0$ is

$$\frac{s_0}{2} \ln\left(\frac{s_0 + 1}{s_0 - 1}\right) - 1 = \frac{1}{F_0 + s_0^2[F_1/(1 + F_1/3)]}. \tag{I.34}$$

For large $F_0$ we find the interesting result

$$\frac{c_0^2 - c_1^2}{c_1^2} = \frac{4}{5} \frac{1}{1 + F_0}, \tag{I.35}$$

so that the difference in the velocities of zero and first sound is small for large $F_0$.

Assuming that the quantum of sonic energy is much less than $2\pi kT$ a wave of zero sound is attenuated by collisions between quasiparticles.(12) The attenuation coefficient is proportional to the quasiparticle collision rate and is thus proportional to $T^2$.

Experimental evidence(13) for the behaviour suggested in the preceding is shown in Fig. I.10, where both the attenuation of sound and the velocity of sound are shown for two frequencies as a function of temperature for liquid $He^3$ at low pressure. At low temperatures the attenuation is frequency independent but $T^2$ dependent. At higher temperatures the attenuation is proportional to $\omega^2/T^2$ as for classical viscous attenuation. In the vicinity of $\omega\tau = 1$ there is a maximum attenuation. The relative sound velocity changes in the vicinity of the attenuation maximum by an amount consistent with (I.35). This is an important confirmation of Landau's theoretical concepts.

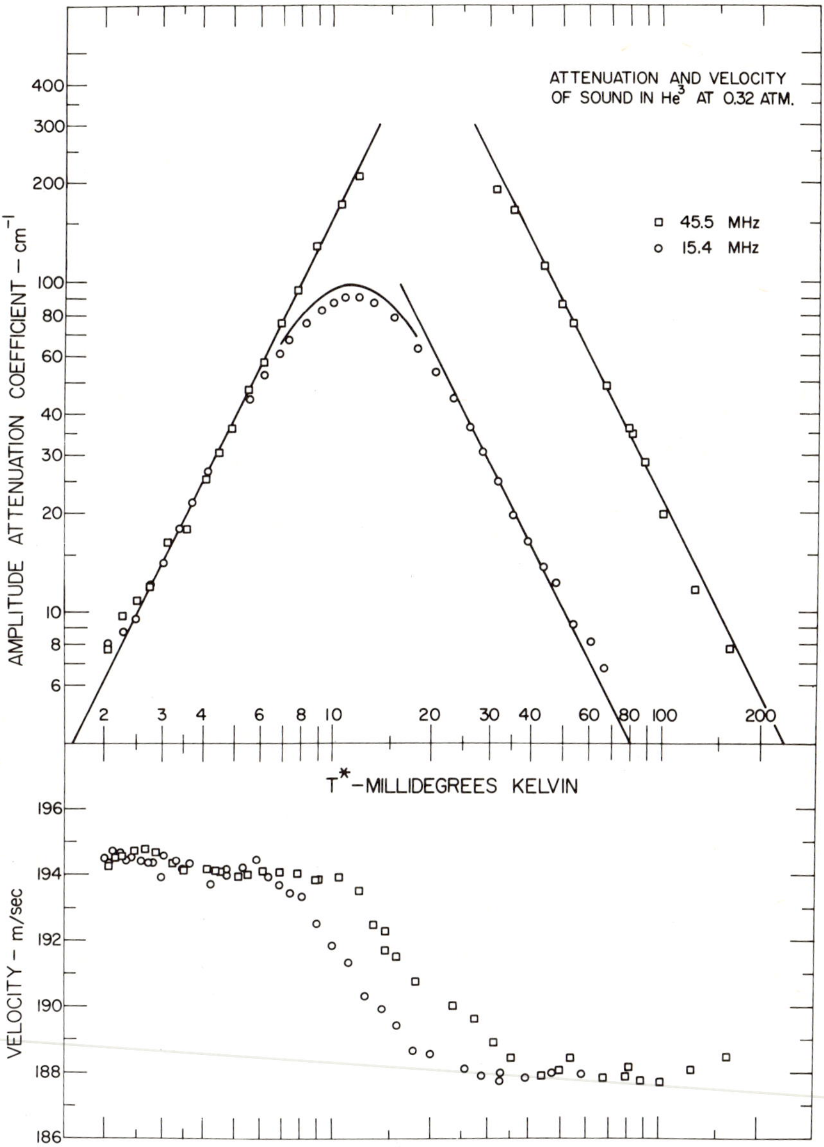


FIG. I.10. Amplitude attenuation coefficient and sound propagation velocity versus CMN magnetic temperature $T^*$ in pure liquid He$^3$ at 0·32 Atm and for frequencies of 15·4 and 45·5 MHz. (After Abel *et al.*, Ref. 13.)

Collisionless sound in $He^3$ (zero sound) has also been useful in studying the superfluid state of $He^3$ although a detailed theoretical understanding of the results has not yet been achieved. An example of experimental data[(14)] is given in Fig. I.11, where the attenuation relative to that of normal zero sound and the velocity relative to that of first sound are plotted vs. a reduced temperature $T/T_c$, $T_c$ being the temperature of the transition to the superfluid state—about 2·6 mK in this case. At the critical temperature $T_c$ a sudden change occurs from the frequency independent attenuation above the transition. The additional attenuation is thought to arise from absorption of

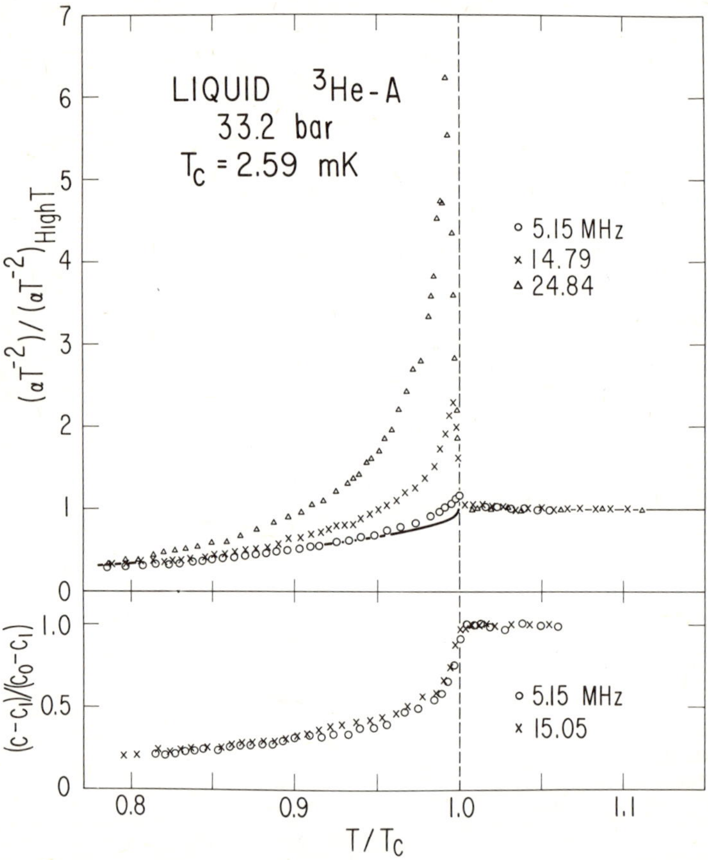


FIG. I.11. The reduced amplitude attenuation coefficient $(\alpha T^{-2})/(\alpha T^{-2})_{\text{High } T}$ and the reduced velocity $(c - c_1)/(c_0 - c_1)$ as a function of reduced temperature $T/T_c$, where $T_c$ is the temperature of the second order transition, for collisionless sound (zero sound) propagating at several frequencies in $He^3$–A and normal Fermi liquid at a pressure of 33·2 Bar. (After Paulson *et al.*, Ref. 14.)

phonons by correlated pairs of quasiparticles, the peak in attenuation occurring when the quantum of sound energy is equal to an energy gap. There is a corresponding decrease in the sound velocity. At low temperatures it appears that the collisionless sound velocity approaches that of ordinary sound.

## II. THERMODYNAMIC PROPERTIES OF SUPERFLUID HE$^3$

So far we have only an imperfect knowledge of the properties of the superfluid state; most experiments still seek just a qualitative notion of what the superfluid is like. Furthermore few experiments have been designed to do a good job of measurement since the effects to be measured have been unknown at the outset. Once into the experiment new effects have been "measured," but usually not very well. The technology is moderately difficult and the experiments long (typically months in our laboratory). It is not a simple matter to change geometry or configuration. For example, in the spring of 1973 we performed experiments on heat flow in superfluid He$^3$ which we interpreted in terms of the hydrodynamic flow of heat. (These experiments were analogous to those of Allen and Jones[(15)] when they discovered the fountain effect in He$^4$ below the $\lambda$-point in 1938.) But proof of our conclusions could only be made by changing the diameter of the tube through which the heat flows. Now we can say, over a year later, that the experiment with a changed geometry has just been performed and the original conclusions upheld—proving the point. Progress is slow but nevertheless very interesting. With this apology for the rather spotty nature of results available thus far, we shall begin by discussing what is known of the phase diagram.

The phase diagram[(16)] is shown in Fig. II.1. The liquid is confined to pressures below 34·4 Bar in the temperature range shown. Actually, the melting pressure is changing in this temperature range in a highly significant way, but the effects are too small to be seen on the scale of the diagram. The line $T_c$ on the diagram is the locus in zero magnetic field of a second order phase transition in Ehrenfest's sense: there is a discontinuity, but not a divergence, in the specific heat.[(17)] In a magnetic field the line $T_c$ is split into two transitions (and possibly three), though each seems to be second order also.[(18)] We have usually identified $T_c$ in each experiment by the specific heat discontinuity, but there are other effects which may give a sharper identification experimentally. One is the abrupt change in ultrasonic attenuation as shown in Fig. I.11. A second is a weak increase[(19)] (a few tenths percent in a few hundred gauss) in static magnetization at $T_c$. A third is a change in wire damping as observed in Helsinki.[(8)] So far one of the most precise indicators of $T_c$ that we know is the time constant for response of *spurious* mag-

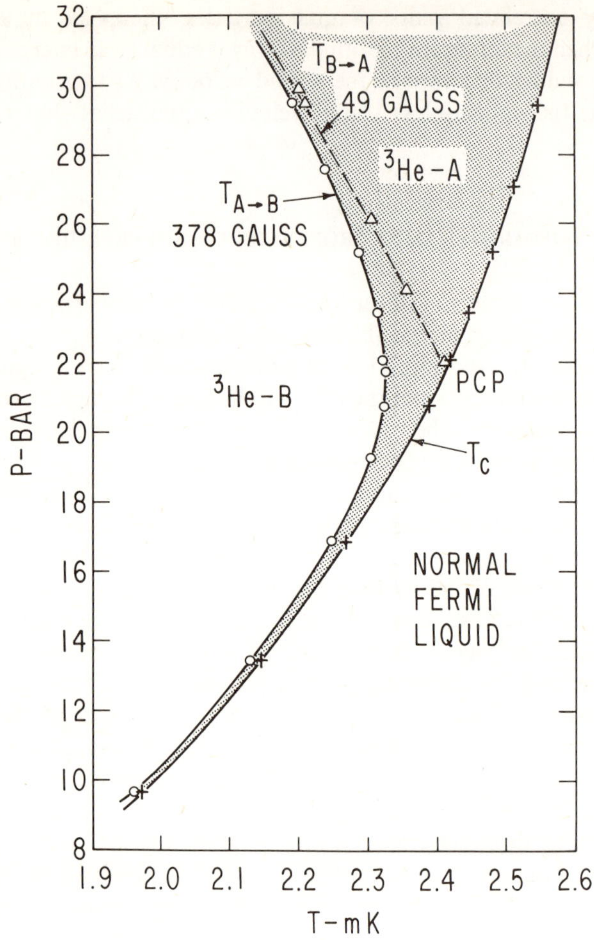


FIG. II.1. Phase diagram of liquid $He^3$ in a limited range of temperature and pressure in both 49 and 378 gauss. The transition line $T_c$ is split by a small amount, too small to indicate on this diagram for these fields, in a magnetic field. (After Paulson *et al.*, Ref. 16.)

netization to a sudden change of field in our static magnetization apparatus. This time constant suddenly decreases by a factor of order two as we warm through $T_c$. Yet neither the source of the magnetism observed nor the reason for the sudden change of time constant is understood An example of a determination of the location of $T_c$ by specific heat is shown in Fig. II.2 in which the rate of change of magnetic temperature $T^*$ is plotted as a function of $T^*$ for a cell containing both CMN and $He^3$. We believe that the width of the transition is caused by thermal inhomogeneities rather than by intrinsic

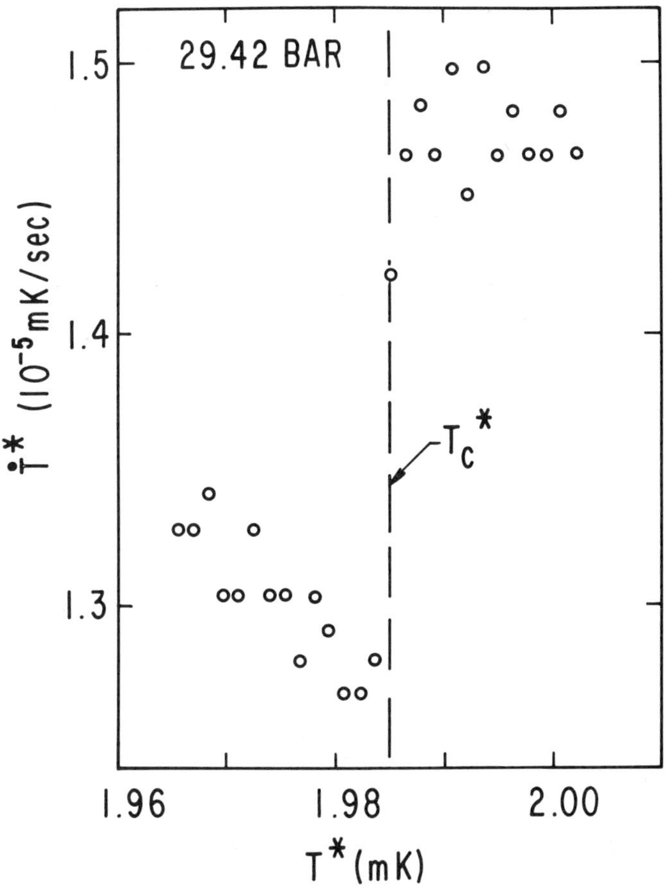


FIG. II.2. Plot of the time rate of change of CMN magnetic temperature $T^*$ versus the magnetic temperature near the second order transition temperature $T_c^*$ for a cell containing a mixture of CMN powder and liquid He$^3$. (Unpublished measurements of Webb, Kleinberg, and Wheatley.)

processes. Although the He$^3$ specific heat changes by a factor of 2·5–3 at the transition as shown in Fig. II.3 the heat capacity of the CMN cooling salt and thermometer is so large that the effect of the He$^3$ heat capacity discontinuity is quite small.

The line $T_{AB}$ in Fig. II.1 is profoundly influenced by a magnetic field, particularly near the point on the $T_c$ line labelled PCP. The line $T_{AB}$ was first discovered[20] as the locus of thermal resistance transitions from a low temperature fluid with linear heat flow to one readily exhibiting nonlinearities and hysteresis. Somewhat later we found[21] that the line $T_{AB}$ was the locus of discontinuities in the static nuclear magnetization at the transition from a

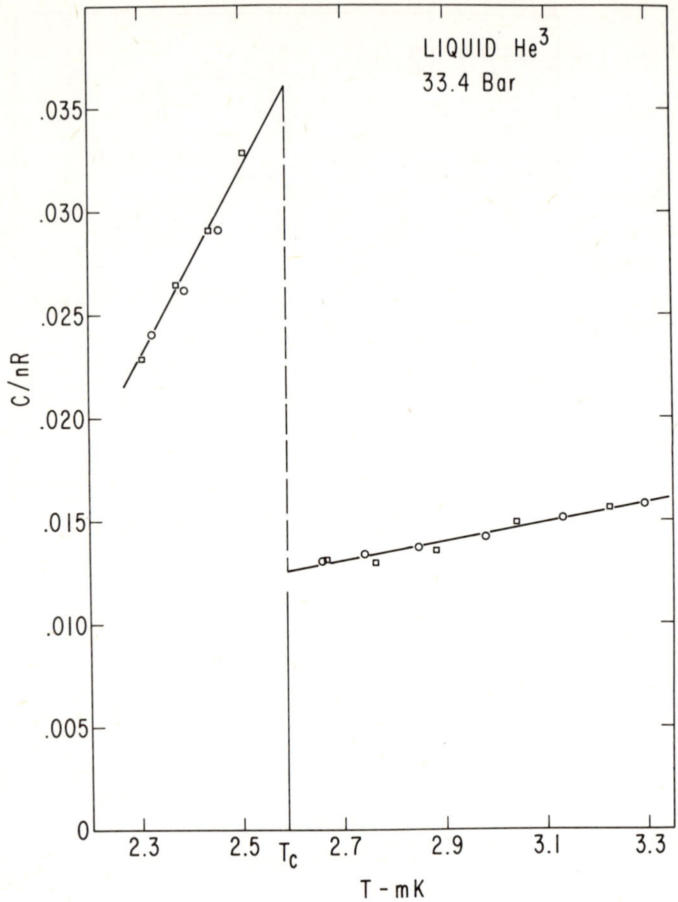


FIG. II.3. Molar specific heat relative to the gas constant $R$ for liquid $He^3$ at 33·4 Bar near the second order transition. (After Webb *et al.*, Ref. 17.)

low temperature phase called $He^3$–B whose static magnetism is temperature dependent to a higher temperature phase called $He^3$–A whose static magnetism is rather precisely (few tenths percent) constant at a value slightly greater than that of the normal liquid.(19) In our initial work(21) in a measuring field of 49 gauss the discontinuity in magnetization seemed to approach zero on the $T_c$ line at a pressure very close to that estimated for the thermal resistance discontinuities. On the basis of the magnetization discontinuities the transition line $T_{AB}$ was taken to be a line of first order transitions terminating in a critical point on the $T_c$ line. Owing to the known effect(18) of a magnetic field on the transition $T_c$ we expected that this critical point might be at the confluence of several lines or surfaces in a pressure–temperature-field space, so we labelled the point PCP for polycritical point.† The nature of the PCP and its precise position are still not known owing both to the profound effect which a magnetic field has on the $T_{AB}$ line near the PCP and to a supercooling effect in low fields which allows the A liquid to be observed,[22] for a field as small as 5 gauss, several tenths of a Bar below a possible pressure of the PCP at 21·2 Bar. Actually neither the form of the $T_{AB}$ line near the $T_c$ line nor the precise pressure of the PCP are known at present. There also appears to be a sharp bend in the $T_c$ line in the pressure region about 21·2 Bar, but this requires further study. At present we can only say that the region of the supposed PCP should be very interesting indeed, though it will take a fine hand and low fields to study it. That a few gauss acting on nuclear magnetism has such a significant effect is indeed remarkable, and this is a clue to the thermal properties of the superfluid states which we shall discuss later.

The effect of a 378 gauss field on the line $T_{AB}$ is also shown on Fig. II.1 .For this field the A phase is interposed between the normal liquid and the B liquid over the full pressure range available to us. The dependence of the susceptibility discontinuity at $T_{AB}$ on pressure for several fields is shown in Fig. II.4. Note how the discontinuity is deflected away from zero in the 60 gauss data near the PCP pressure.

The $T_{AB}$ line, reflecting a first order transition, should exhibit a latent heat. However, the latent heat is small and we have seen it only recently,[23] when it coincided exactly with the magnetic transition. One problem that we may have with respect to observation of latent heat is that the A → B transition may be broadened or distorted in some way for $He^3$ in the interstices of the powdered CMN cooling salt. The effect of confining $He^3$ to small pores does have a profound effect on magnetic properties.[24] This subject has however scarcely been studied.

Thermodynamic properties of the superfluid which have been studied are the specific heat,[17, 25, 26] the static magnetization,[19] and the superfluid density.[27] Of these, only the static magnetization has been measured in bulk $He^3$ off the melting curve. Both the specific heat and the superfluid density at pressures below melting pressure have been measured for $He^3$ in the interstices of packed CMN powder. The specific heat has been measured[17] only near to $T_c$ and in $He^3$–A, but when compared at a pressure near melting pressure to a measurement effectively in bulk at melting pressure[26] rather good agreement is obtained. We conclude that for the pore sizes presently involved the *thermal* properties of $He^3$–A have not been grossly altered by the CMN grains with which the $He^3$ has been mixed. Obtaining the $He^3$ heat capacity in the presence of a large background heat capacity of CMN

† We are indebted to Dr. John C. Wheeler for suggesting this name.

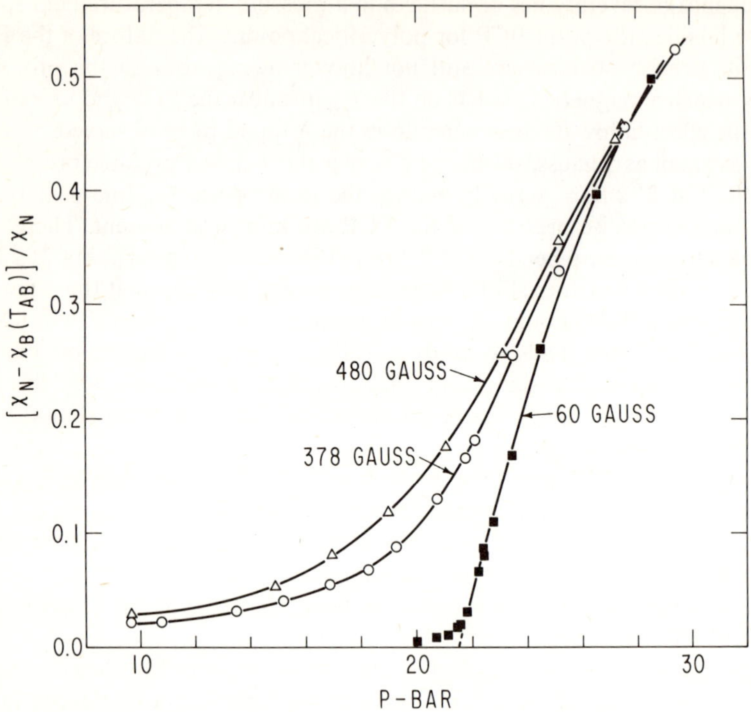


FIG. II.4. Nuclear magnetic susceptibility discontinuity $\chi_N - \chi_B(T_{AB})$ at the A → B transition relative to the normal susceptibility $\chi_N$ as a function of pressure. The 480 and 378 gauss points are thermodynamic A → B transitions while the 60 gauss points are nonthermodynamic B → A transitions. The different curves have different calibrations and hence can be shifted slightly with respect to one another. (After Paulson *et al.*, Ref. 16.)

is quite difficult, but it can be done by taking differences between the heat capacity at pressure $P$ and that at a low pressure. The heat capacity is measured on the magnetic temperature scale of the CMN refrigerant, but to be useful the results have to be expressed on a Kelvin temperature scale. This has been achieved by requiring the heat capacity to be linear in $T$, the Kelvin temperature, in the normal liquid above $T_c$. The resulting heat capacity is shown in Fig. II.3. The ratio $C_</C_>$ of heat capacities just below to just above $T_c$ can be obtained without recourse to the temperature scale. It varies from about 2·5 at 23 Bar, somewhat above the PCP, to 2·8 at 34 Bar near melting pressure. Specific heat measurements have not been made at lower pressures with lower $T_c$ owing to excessive CMN heat capacity, but rough indications suggest that $C_</C_>$ may rise again below the PCP. Careful heat capacity measurements in bulk liquid, both normal and superfluid, are badly needed over the full pressure and temperature range. However, many technical and thermometric problems need to be solved before this goal can be achieved.

Static nuclear magnetization measurements have provided a great deal of insight into the superfluid state. Although we cannot digress to discuss experimental matters, static nuclear magnetism measurements are technically very interesting owing to the extremely small size of the nuclear paramagnetic susceptibility near $T = 0$. This susceptibility is about $10^{-7}$ near melting pressure, decreasing with decreasing pressure, and measurements to about a tenth percent precision are both possible and useful. The static magnetic properties of He$^3$–A and He$^3$–B are quite different. The magnetization in the A phase is temperature independent to within 0·2–0·3%. In the B phase the magnetization depends on temperature, decreasing rather rapidly below $T_c$. When reduced susceptibility $\chi_B/\chi_N$ is plotted versus reduced temperature as in Fig. II.5 for a static field of 378 gauss and a variety of pressures one sees first of all that is is nearly pressure independent. Next, it would appear that $\chi_B/\chi_N$ would extrapolate, at least approximately, to unity at $T/T_c = 1$. This is a remarkable observation since $T_c$ refers to the boundary between He$^3$–A and normal fluid while the reduced susceptibility plotted is that of He$^3$–B. It appears that He$^3$–B may have the same critical temperature as He$^3$–A, not only in zero field at the pressure of the PCP but also over the full pressure range. The quantitative basis for this assumption is strongest below the PCP.

Also shown in Fig. II.5 as an inset is the magnetic feature at the transition from He$^3$–A to normal fluid (A → N transition). This feature has been interpreted by Patton[28] as comprising in part an average effect due to the splitting of the transition $T_c$ in a magnetic field[18] and in part a manifestation of fluctuations of the superconducting order parameter above $T_c$ which lead to a broadening of the transition above $T_c$. By comparison with the measurements he finds a low temperature coherence length of about 70Å.

We can use (1) our assumption that $T_c$ is the same for both He$^3$–A and He$^3$–B, (2) the displacement of the transition $T_{AB}$ below $T_c$ in zero field, and (3) the profound effect of a magnetic field on the AB transition to deduce a relationship between the thermal properties of the A and B liquids by means of thermodynamics. It should be intuitively clear that at the PCP their specific heats must be the same and that a small magnetic field could not possibly have a profound effect unless the difference of the specific heats of He$^3$–A and He$^3$–B was also small.

A schematic representation of the simplest notion of how the entropies of the two phases depend on temperature in zero field is shown in Fig. II.6a for $P > P_{PCP}$ and in Fig. II.6b for $P < P_{PCP}$. They have the same entropy at $T_c$. Consider first the case where $P > P_{PCP}$. The A phase is stable in the region from $T_{AB}$ to $T_c$. Owing to the first order character of the AB transition

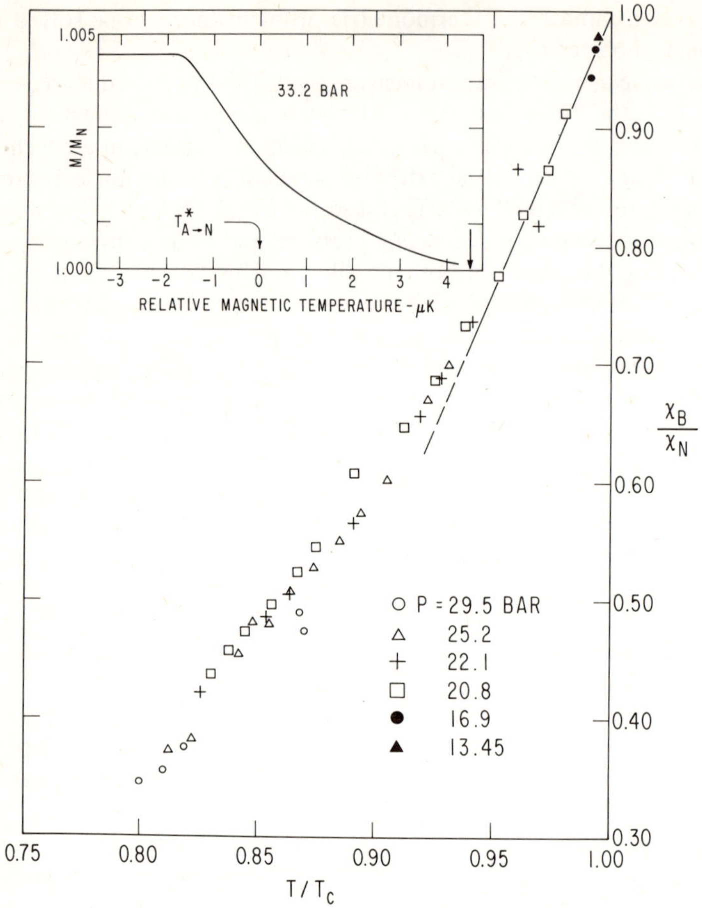


FIG. II.5. Nuclear magnetic susceptibility in He$^3$–B, $\chi_B$, relative to the normal susceptibility $\chi_N$ as a function of reduced temperature $T/T_c$ for various pressures as observed in a field of 378 gauss. The inset shows the magnetization relative to that in normal liquid in the vicinity of $T_c$ for a pressure of 33·2 Bar and a field of 378 gauss. (After Paulson *et al.*, Ref. 19.)

$S_A > S_B$ at $T_{AB}$. Furthermore $F_A(T_{AB}) = F_B(T_{AB})$ and $F_A(T_c) = F_B(T_c)$, where $F$ is the free energy. Thus we have

$$F_A(T_c) - F_A(T_{AB}) = F_B(T_c) - F_B(T_{AB}). \tag{II.1}$$

In zero field we have $dF = -SdT$ so that, from (II.1)

$$\int_{T_{AB}}^{T_c} S_A \, dT = \int_{T_{AB}}^{T_c} S_B \, dT, \tag{II.2}$$

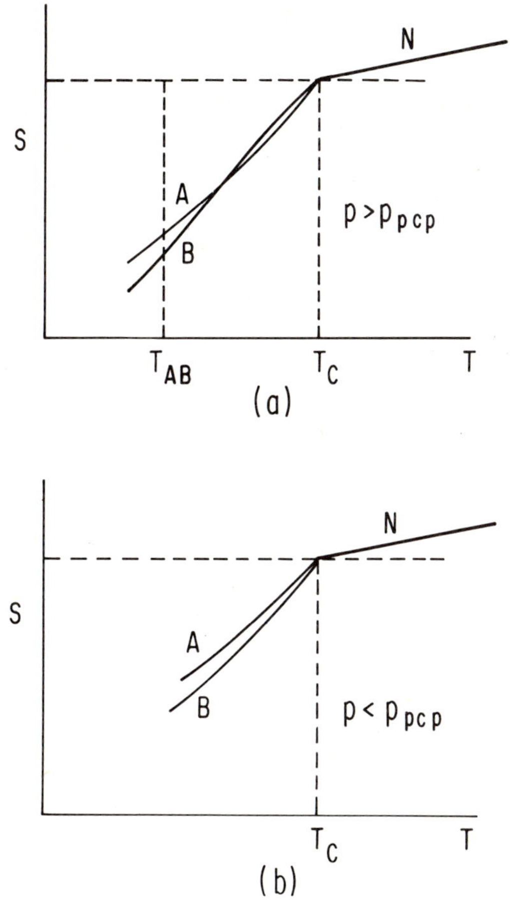


FIG. II.6. Schematic assumed dependence of entropy on temperature in liquid He$^3$ near the second order transition at $T_c$ for (a) pressures greater than that of the PCP and temperatures down to below $T_{AB}$, and for (b) pressures below that of the PCP.

or the average entropies of the A and B phases must be the same between these two temperatures. Thus while $S_A > S_B$ at $T_{AB}$, the two must cross qualitatively as shown so that their average values are the same over the $T_{AB} - T_c$ interval. We thus expect $(dS_A/dT)_{T_c} > (dS_B/dT)_{T_c}$ or $C_A(T_c) > C_B(T_c)$. We thus qualitatively expect that $C_A - C_B$ as function of temperature might look like the upper curve in Fig. II.7. As the pressure decreases $T_{AB}$ approaches $T_c$, so the entropy crossover point in Fig. II.6a approaches $T_c$. Finally at the PCP the A and B entropy curves have the same slope at $T_c$, $[C_A(T_c) = C_B(T_c)$ at the PCP] while at lower pressures $S_B < S_A$ at all temperatures below $T_c$.

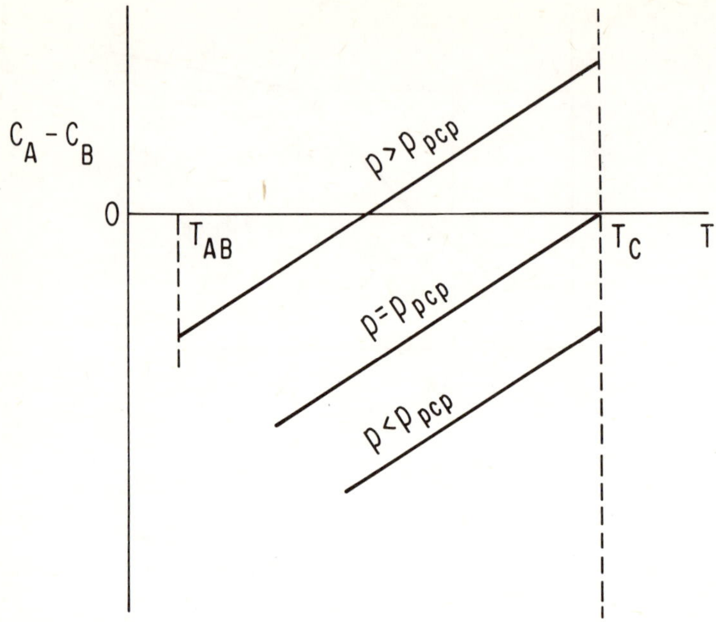


FIG. II.7. Simple possible dependence of the difference $C_A - C_B$ of the specific heats of the A and B phases on temperature for pressures above, at, and below that of the PCP.

Possible dependences of $C_A - C_B$ on temperature for $P = P_{PCP}$ and for $P < P_{PCP}$ are also shown on Fig. II.7. In view of the limited data available we made the simple assumption

$$C_A - C_B = \Delta C(P) + \alpha t \tag{II.3}$$

where $t \equiv (T - T_c)/T_c < 0$. All pressure dependence is put into the specific heat difference at $T_c$, and the temperature dependence is assumed to be the same at all pressures. This gives an easy form to fit, but no doubt loses accuracy particularly as one moves away from the PCP. This form also avoids exotic temperature dependences which are not required by the data in our present state of knowledge. With the form for $C_A - C_B$ assumed in (II.3) and the known magnetic properties of $He^3$–B and $He^3$–A a magnetic field will always act to interpose $He^3$–A between normal liquid and $He^3$–B at any pressure below the PCP, as we shall see below.

Substituting the assumed form (II.2) into (II.4) with the further observation that $S_A(T_c) = S_B(T_c) = S_N(T_c)$ we find to lowest order for $H_0 = 0$ and $P > P_{PCP}$

$$\Delta C(P) = \frac{-\frac{1}{3}\alpha t_{AB}}{1 - \frac{1}{3}t_{AB}}, \tag{II.4}$$

where

$$t_{AB} \equiv \left(\frac{T_{AB} - T_c}{T_c}\right).$$

Note that since in zero field $t_{AB} = 0$ at the PCP, $\Delta C(P_{PCP}) = 0$ also as expected.

The parameter $\alpha$ may be evaluated by the known effect of a magnetic field on the phase diagram. In a magnetic field the change in free energy is $dF = -SdT - MdH_0$, where $M$ is the sample magnetic moment and $H_0$ is the externally applied field. For all practical purposes density is constant as temperature changes so we can imagine $S$ and $M$ as referring to a unit volume of He$^3$. Then in our equations $S$ refers to entropy per unit volume and $M$ to magnetization. Empirically, we have to sufficient accuracy near $T_c$

$$M_A(T, P, H_0) = \chi_A(T, P)H_0 = \chi_N(P)H_0, \tag{II.5}$$

and

$$M_B(T, P, H_0) = \chi_B(T, P)H_0 = \chi_N(P)[1 + 4{\cdot}7t]H_0, \tag{II.6}$$

where $\chi_N$ is normal susceptibility and where the reduced susceptibility is taken to be pressure independent with a temperature dependence given by data on Fig. II.5. Recall $t \equiv (T - T_c)/T_c \leqslant 0$.

Since

$$F_A(T_{AB}, H_0) = F_B(T_{AB}, H_0) \quad \text{and} \quad F_A(T_c, H_0) = F_B(T_c, H_0)$$

while $S_A(T_c, H_0) = S_B(T_c, H_0)$ we have for a fixed pressure

$$\int_{T_{AB}}^{T_c} [S_A(T) - S_B(T)]\,dT = \tfrac{1}{2}[\chi_A - \chi_B]H_0^2. \tag{II.7}$$

Putting in (II.3), (II.5) and (II.6) we then find to lowest order in $t_{AB}$

$$\Delta C(P) = \frac{(4{\cdot}7\chi_N(P)H_0^2/T_c) - \alpha t_{AB}^2/3}{t_{AB}}. \tag{II.8}$$

The parameter $\alpha$ is obtained at the pressure of the PCP where $\Delta C(P) = 0$. Then $\Delta C(P)$ may be evaluated using data from the phase diagram. In this connection we have used both values of $t_{AB}$ determined thermometrically and values obtained by the equation

$$t_{AB} = -\frac{1}{4{\cdot}7}\,\frac{\chi_N(P) - \chi_B(P, T_{AB})}{\chi_N(P)} \tag{II.9}$$

obtained from (II.6) (or Fig. II.5). The latter values of $t_{AB}$ are more precise. The results for $\Delta C(P)$ are given in Fig. II.8. Values of $\Delta C(P)$ determined from the $T_{AB}$ line in $H_0 = 0$ above the PCP reasonably extrapolate from those determined for $P < P_{PCP}$ for magnetic data. The total specific heat of the superfluid itself is about $80\,erg/cm^3$ mK near the PCP so we see that $\Delta C$ is just a few percent at most of the total specific heat. The thermal properties of the two phases are very similar indeed The susceptibility information in (II.5) and (II.6) together with the assumption (II.3) and its numerical evaluation as on Fig. II.8 are adequate to explain the entire phase diagram shown on Fig. II.1.

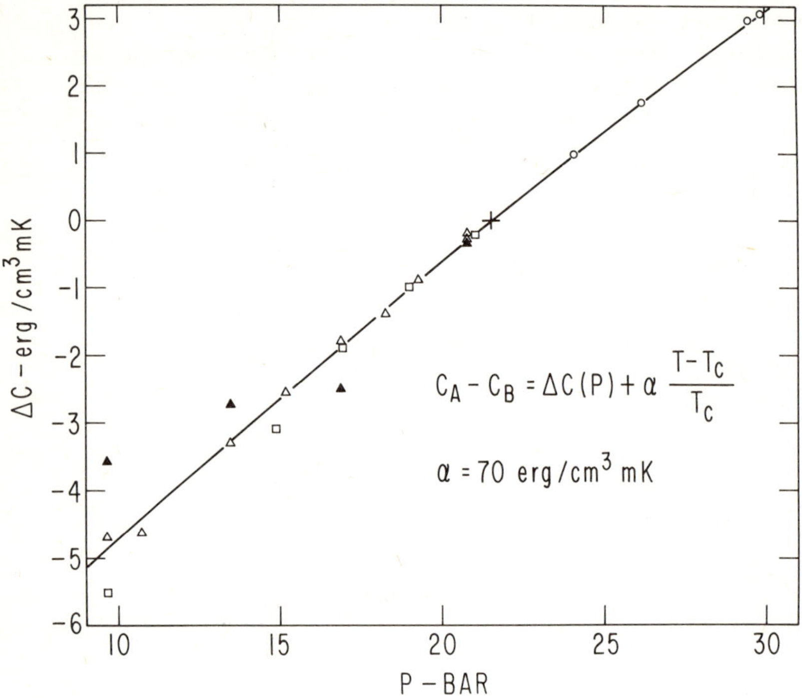


FIG. II.8. Difference in specific heat per unit volume of $He^3$–A and $He^3$–B at $T_c$ as derived from data for $(1 - T_{AB}/T_c)$: (○) 49 gauss, thermometric; (▲) 378 gauss, thermometric; (△) 378 gauss, susceptibility; (□) 480 gauss, susceptibility; +, assumed PCP. (After Paulson *et al.*, Ref. 16.)

Finally, for the form of specific heat difference given in (II.3) and for $P < P_{PCP}$ the free energy difference between the A and B phases *in zero field* increases as $t^2$, or quadratically, $F_A$ being greater than $F_B$. However, since the A phase is *more magnetic* than the B phase, the magnetism of the B phase, decreasing linearly with decreasing temperature, the magnetic free energy of the A phase is less than that of the B phase by an amount which is *linearly* dependent on temperature difference (and quadratically dependent on $H_0$).

Hence the magnetic effect always wins out to make $F_A$ less than $F_B$ for $T$ very near $T_c$ and the transition from normal liquid is first into $He^3$–A, as shown in Fig. II.1.

Next let us consider the question of superfluid density. We have evidently been able to steer clear of theoretical bias regarding superfluid properties up to this point, but it will be helpful to introduce some concepts now. In the case of liquid $He^4$ below the $\lambda$-point experiments can be understood on the basis of a two fluid model in which there are two interpenetrating fluids: one has entropy and viscosity and behaves like a normal fluid—it is thus called normal fluid; the other has zero entropy and flows with a velocity field having zero curl so that it acts as if it had zero viscosity—it is called superfluid. The superfluid has density $\rho_s$ and the normal fluid density $\rho_n$. That the superfluid has zero curl is thought to derive from the idea that it is the manifestation of a single quantum state of macroscopic occupation which, in a state of uniform flow at velocity $\mathbf{v}_s$, is described by a phase factor $\phi = (m_4\mathbf{v}_s \cdot \mathbf{r} - \mu t)/\hbar$, where $m_4$ is the $He^4$ atomic mass, $\mathbf{r}$ is position, $\mu$ is the chemical potential (the energy to add a particle at constant $S$ and $V$) and $t$ is time. Thus we have $\mathbf{v}_s = (\hbar/m)\nabla\phi$, so that curl $\mathbf{v}_s = 0$. Naturally a key question regarding the new phases of $He^3$ has been whether or not they are superfluid. The question is really not regarding an enhanced fluidity below the transition but rather whether in liquid $He^3$ there is also a macroscopically occupied single quantum state whose flow is curl free because the superfluid velocity is the gradient of the phase of a macroscopic wave function. Now I believe that in liquid $He^3$ there is already evidence that we have not just a two-fluid model but a multi-fluid model with interpenetrating normal fluid and several weakly coupled superfluids. The latter is a consequence of the nuclear spin of the $He^3$.

Putting aside for the moment the notion of interpenetrating superfluids, which is manifested when the spin degrees of freedom are important, one may ask what type of experiment can test the question of whether or not there is superfluidity at all. At Helsinki an enhanced fluidity was demonstrated[(29)] while in our laboratory heat was shown to flow hydrodynamically.[(20)] However, an experimental proof of superfluidity to be generally accepted amounts to the observation of a phenomenon which does not occur at all unless superfluidity exists. In the present case the observation of the propagation of fourth sound in liquid $He^3$ in analogy to the observations in liquid $He^4$ by Rudnick and Shapiro[(30)] gave the proof of superfluidity. Fourth sound is a sound wave propagating through superfluid in a superleak, a superleak being a channel with such a high impedance for the viscous flow of normal fluid that the normal fluid is effectively clamped.

To see what sound propagation through a superleak, if it exists at all, can tell us about superfluid density let us write down in analogy to $He^4$ the phase $\phi$ of the macroscopic wavefunction related to the superflow of $He^3$:

$$\phi = (2m_3\mathbf{v}_s\cdot\mathbf{r} - 2\mu t)/\hbar \tag{II.10}$$

Here we have used the basic idea of Pitaevskii,[31] of Brueckner, Soda, Anderson, and Morel[32] of Emery and Sessler,[33] and of Thouless[34] that the new phases of $He^3$ consist of correlated pairs of $He^3$-atoms as in the BCS theory of superconductivity. Thus we have

$$\mathbf{v}_s = \frac{\hbar}{2m_3}\,\text{grad}\,\phi, \tag{II.11}$$

so that curl $\mathbf{v}_s = 0$, the condition we are supposing is necessary for superfluidity. The equation of motion for $\mathbf{v}_s$ is then obtained by taking the partial derivative of $\mathbf{v}_s$ with respect to time in (II.11) and noting from (II.10) that $\partial\phi/\partial t = -2\mu/\hbar$. Thus we find

$$\frac{\partial\mathbf{v}_s}{\partial t} = -\text{grad}\,\mu/m_3 \tag{II.12}$$

where $\mu/m_3$ is the chemical potential per unit mass of $He^3$. For fourth sound propagation in $He^3$ the thermal term in grad $\mu/m_3$ is negligible compared to the mechanical term so that we have

$$\frac{\partial v_s}{\partial t} = -\frac{1}{\rho}\,\text{grad}\,P, \tag{II.13}$$

where $\rho$ is mass density and $P$ is pressure.

An interesting comparison here is to apply Newton's law to one dimensional motion of fluid of density $\rho$ in a tube of cross-sectional area $A$. For the fluid between $x$ and $x + \Delta x$ we have

$$-\left(\frac{\partial P}{\partial x}\right)A\Delta x = \rho A\Delta x\frac{\partial x_x}{\partial t}$$

or

$$\frac{\partial v_x}{\partial t} = -\frac{1}{\rho}\frac{\partial P}{\partial x} \tag{II.14}$$

which is the same equation as (II.13) except that the actual velocity has been substituted for the superfluid velocity.

Let us return now to sound propagation through a superleak. Since normal fluid does not flow there is no entropy flow. The equation of continuity is

$$\frac{\partial\rho}{\partial t} + \text{div}\,\rho_s\mathbf{v}_s = 0, \tag{II.15}$$

since $\rho_n\mathbf{v}_n = 0$. If now we assume that $\rho = \rho_0 + \delta\rho$ and both $\delta\rho$ and $v_s$ vary as exp $i\omega(t - x/c_4)$, (II.13) and (II.15) become

$$\omega v_s = \frac{\omega}{c_4}\frac{1}{\rho}\left(\frac{\partial P}{\partial \rho}\right)\delta\rho \tag{II.16}$$

and

$$\omega \delta\rho = \frac{\omega}{c_4}\rho_s v_s. \tag{II.17}$$

These are solved for the velocity of fourth sound by

$$c_4^2 = \frac{\rho_s}{\rho}\left(\frac{\partial P}{\partial \rho}\right) = \frac{\rho_s}{\rho}c_1^2, \tag{II.18}$$

using $c_1^2 = (\partial P/\partial \rho)$. Looking back, we see that whereas the acceleration equation is the same for the superfluid as for an ordinary fluid the particle current is less by a factor $\rho_s/\rho$ for motion of superfluid only, and this leads to a reduced sound velocity.

In an actual fourth sound measurement the value of $c_4$ may turn out to be less than that calculated from (II.18) owing to the effect of refraction in the superleak itself. This effect could be measured in the case of our He$^3$ experiments by measuring fourth sound velocity in superfluid He$^4$ at such such low temperatures that $\rho_s/\rho = 1$.

The experiments on propagation of fourth sound in He$^3$ did give positive results, so that we have conclusive proof of the existence of superfluidity. The fourth sound velocity was obtained by measuring the frequency of the first longitudinal resonance in a cylinder packed with CMN powder, which acted as refrigerant and thermometer as well as superleak. Results at various pressures[27] on a magnetic temperature scale are given in Fig. II.9. At least for the higher pressures we would now expect, on the basis of the observed magnetic temperatures dependence of the squares of parallel ringing frequencies to be discuss later, that $\rho_s/\rho$ would be linear in $T^*$ near $T_c^*$. Actually there is some upward curvature. (The very pronounced curvature at the lowest temperatures is an effect of the magnetic temperature scale.) Furthermore there is no evidence in $\rho_s/\rho$ of the A $\rightarrow$ B transition, although the data are obtained rather near to the PCP where the thermal and probably superfluid density differences are expected to be small. This is a good place to emphasize that caution must be exercised in using weak coupling predictions on the basis of specific models to discuss relatively small differences between the A and B phases since after all the weak coupling theory does not predict the phase diagram.

Pore size may have an effect on the results both with respect to a possible broadening or alteration of the B $\rightarrow$ A transition and with respect to the

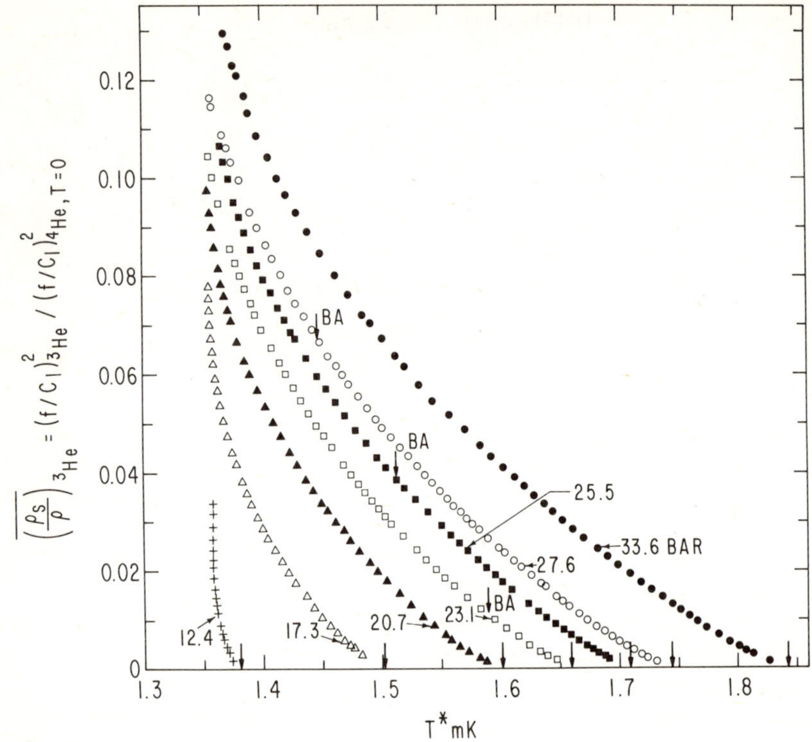


FIG. II.9. Experimental values of $\bar{\rho}_s/\rho$ from fourth sound as a function of CMN magnetic temperature for various pressures. The corresponding second order transition temperatures $T_c^*$ are indicated by arrows on the horizontal axis. The expected locations of $T_{AB}^*$ are shown by arrows on the appropriate curves. The PCP is near 21·2 Bar. (After Kojima *et al.*, Ref. 27.)

small upward curvature near $T_c$ of $\rho_s/\rho$ as plotted against $T^*$. Possible depairing at the CMN surfaces also would have an important effect, which we discuss later, with the likely result on the basis again of weak coupling theory applied to specific models that the effective superfluid density would not be expected to change across the B → A transition. Such questions can only be decided experimentally with an improved geometry.

A plot of $\bar{\rho}_s/\rho$ for a pressure of 33·6 Bar ($He^3$-A) on a reduced temperature difference scale based on provisional absolute temperature is shown in Fig. II.10. The data approximately obey the law $\bar{\rho}_s/\rho \propto (1 - T/T_c)^\alpha$ with $\alpha \simeq 1{\cdot}1$ for the case shown. Data from the other pressures shown on Fig. II.9 when plotted this way give similar results (little pressure effect) with $\alpha$ in the range 1·1 to 1·3. The solid line on Fig. II.10 is based on a comparison with susceptibility data and will be discussed later.

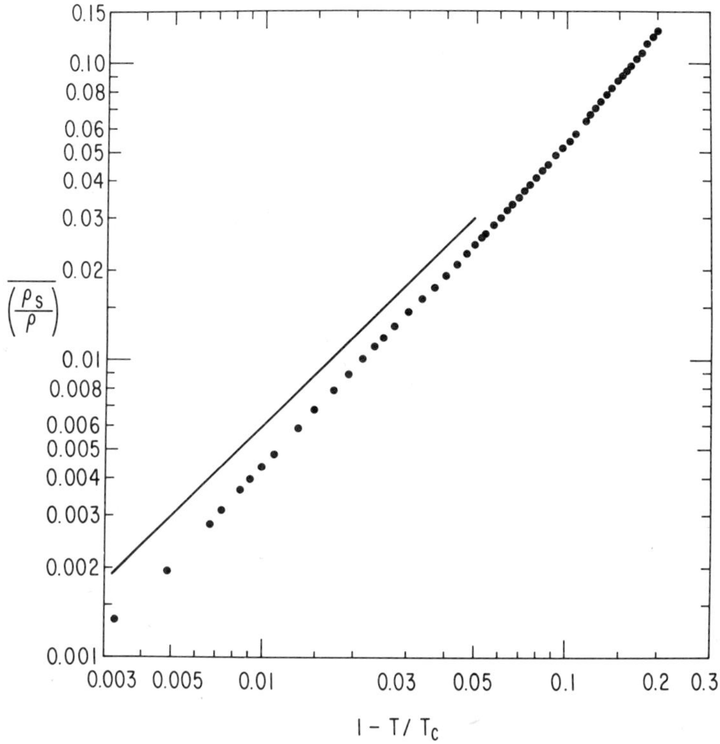


FIG. II.10. Logarithmic plot of $\bar{\rho}_s/\rho$ against $(1 - T/T_c)$ for He$^3$–A at a pressure of 33·6 Bar. The solid line is $\bar{\rho}_s/\rho = 0{\cdot}60\,(1 - T/T_c)$ obtained from susceptibility measurements as suggested by (II.47) and following material in the text. (In part from Kojima *et al.*, Ref. 27.)

Now it is believed that if He$^3$ is a BCS-like superfluid it may have an anisotropic superfluid density in the sense that the proportionality between momentum density and drift velocity will be dependent on some internal parameter of the liquid.[35–37] However, before we get into this possibility in detail we want to show how measurements of superfluid density and magnetic susceptibility might be related to one another.† Our remarks will be in a context of pairing theories of the superfluid where the pairs can either be in non-magnetic singlet states or triplet states. For liquid He$^3$ it is likely that both the A and B phases are triplet states. For triplet pairing we have three possible interpenetrating superfluids in which the pairs are in $\uparrow\uparrow$, $\downarrow\downarrow$, and $(1/\sqrt{2})(\uparrow\downarrow + \downarrow\uparrow)$ states with respect to some axis of quantization. (Following convention, $\uparrow$ refers to the direction of magnetic moment, which happens to be opposite that of spin since He$^3$ has a negative moment. We will ignore

† We are indebted to Professor A. J. Fetter for his help in this area.

the sign of the moment.) The $\uparrow\uparrow$ and $\downarrow\downarrow$ superfluids are magnetic, but the $(1/\sqrt{2})\,(\uparrow\downarrow + \uparrow\downarrow)$ superfluid is not magnetic. We will refer to it as the non-magnetic triplet superfluid or the $\uparrow\downarrow$ superfluid.

Following Leggett[(6)] we will first look at susceptibility and superfluid density for a weakly coupled system and then put in the effects of the strong Fermi liquid interactions in the form of "molecular fields". The $\uparrow\uparrow$ and $\downarrow\downarrow$ superfluids, presumed in equal amounts in zero field, will have a susceptibility equal to that of the normal fluid. Quasiparticles excited out of the non-magnetic triplet state will lead to a magnetization

$$M_w = \frac{\gamma\hbar}{2}\sum_p \left[f(E_{p\uparrow}) - f(E_{p\downarrow})\right] \tag{II.19}$$

where the subscript $w$ refers to weak coupling, $f$ is the Fermi function, and

$$E_{p\uparrow} = E_p - \frac{\gamma\hbar}{2}H_0,$$

and (II.20)

$$E_{p\downarrow} = E_p + \frac{\gamma\hbar}{2}H_0,$$

with $E_p = [\varepsilon_p^2 + \Delta^2(\hat{\mathbf{n}})]^{\frac{1}{2}}$. The energy gap $\Delta(\hat{\mathbf{n}})$ depends in general on the pairing axis $\hat{\mathbf{n}} \equiv \mathbf{p}/|\mathbf{p}|$, and $\varepsilon_p$ is kinetic energy measured with respect to the fermi energy. In our notation $\gamma$ is the absolute value of the gyromagnetic ratio of the $He^3$ nuclei, so $\gamma\hbar/2$ is the component of the magnetic moment in the direction of the field. In the limit of $H_0 \to 0$ (II.19) can thus be written

$$M_w = \left[\frac{\gamma^2\hbar^2}{2}\sum_p -\frac{\partial f}{\partial E_p}\right] H_0. \tag{II.21}$$

Defining

$$g(T/T_c) \equiv \frac{1}{N(0)}\sum_p -\frac{\partial f}{\partial E_p}, \tag{II.22}$$

we have

$$\chi_{\uparrow\downarrow w} = \frac{\gamma^2\hbar^2}{2} N(0) g(T/T_c), \tag{II.23}$$

where $g(1) = 1$. If now a fraction $x$ of the pairs are in the $\uparrow\downarrow$ state the weak coupling susceptibility will be given by

$$\chi_w = \frac{\gamma^2\hbar^2}{2} N(0)[(1-x) + x g(T/T_c)]. \tag{II.24}$$

To the extent that a molecular field constant over the Fermi surface is accurate in describing the magnetization we can write, following (I.6) (recall that $\hbar\mathbf{S}$ is the spin angular momentum per unit volume)

$$\mathbf{M} = \gamma\hbar\mathbf{S} = \chi_w\left(\mathbf{H}_0 - \frac{Z_0}{2\gamma\hbar N(0)}\mathbf{S}\right) = \chi\mathbf{H}_0. \qquad \text{(II.25)}$$

Eliminating $\mathbf{S}$ and $\mathbf{H}_0$ we find

$$\chi = \frac{\chi_w}{1 + \dfrac{Z_0\chi_w}{2\gamma^2\hbar^2 N(0)}} \qquad \text{(II.26)}$$

If we substitute

$$\chi_N = \frac{\gamma^2\hbar^2}{2} N(0)/(1 + Z_0/4) \qquad \text{(II.27)}$$

and (II.24) into (II.26) we find the equation

$$\frac{\chi}{\chi_N} = \frac{(1 + Z_0/4)[(1 - x) + xg(T/T_c)]}{1 + Z_0/4[(1 - x) + xg(T/T_c)]}. \qquad \text{(II.28)}$$

Near $T = T_c$ we have

$$\frac{d(\chi/\chi_N)}{d(T/T_c)} = \frac{xg'(T/T_c)}{1 + Z_0/4}, \qquad \text{(II.29)}$$

while if $g(0) = 0$ we have

$$\frac{\chi(T = 0)}{\chi_N} = \frac{(1 + Z_0/4)(1 - x)}{1 + Z_0/4(1 - x)}. \qquad \text{(II.30)}$$

Formula (II.28) was derived by Leggett[6] in the context of $l = 0$ pairing. It may hold near $T_c$ for He$^3$ but it is not known how accurate it is at lower temperatures. Since $(1 + Z_0/4)$ in the pressure range of interest here is about 1/4, $\chi_N$ from (II.27) is greater by a factor of 4 than would have been expected without it. But as normal quasiparticles are condensed into correlated pairs the molecular field which led to the factor 4 enhancement of $\chi$ is reduced and consequently $\chi$ is reduced. That is why in (II.29) the susceptibility reduction below $T_c$ due to pairing in nonmagnetic states is enhanced by the factor $(1 + Z_0/4)^{-1}$.

Susceptibility and superfluid (or normal fluid) density are related to one another. Let us calculate the *normal* fluid density first for a weakly interacting gas[35] and then attempt to correct for the strong Fermi liquid interactions by means of a "molecular field" as in (I.13). Suppose the superfluid is at rest and the normal fluid is moving with velocity $v$. Then, treating the quasiparticle excitations out of the three superfluids (↑↑, ↓↓, and ↑↓) alike, the momentum density is

$$\mathbf{Q} = 2\sum_{\mathbf{p}} \mathbf{p} f(E_p - \mathbf{p}\cdot\mathbf{v}), \tag{II.31}$$

where the factor 2 comes from the sum over spin states, $f$ is the Fermi function, and $E_p = [\varepsilon_p^2 + \Delta^2(\hat{n})]^{\frac{1}{2}}$. Writing $f(E_p - \mathbf{p}\cdot\mathbf{v}) = f(E_p) - \mathbf{p}\cdot\mathbf{v} \times \partial f/\partial E_p$ we have

$$\mathbf{Q} = \overleftrightarrow{\rho}_{n_w}\mathbf{v} = \left[2\sum_p \mathbf{pp}\left(-\frac{\partial f}{\partial E_p}\right)\right]\cdot\mathbf{v} \tag{II.32}$$

giving for the normal fluid density $\overleftrightarrow{\rho}_{n_w}$ in a weakly coupled system the tensor

$$\overleftrightarrow{\rho}_{n_w} = 2\sum_p \mathbf{pp}\left(-\frac{\partial f}{\partial E_p}\right). \tag{II.33}$$

This tensor has principal components

$$\begin{aligned}(\rho_{n_w})_{xx} &= 2p_F^2 \sum_p \sin^2\theta\cos^2\varphi\left(-\frac{\partial f}{\partial E_p}\right)\\ (\rho_{n_w})_{yy} &= 2p_F^2 \sum_p \sin^2\theta\sin^2\varphi\left(-\frac{\partial f}{\partial E_p}\right)\\ (\rho_{n_w})_{zz} &= 2p_F^2 \sum_p \cos^2\theta\left(-\frac{\partial f}{\partial E_p}\right),\end{aligned} \tag{II.34}$$

where $\theta$ and $\varphi$ are polar and azimuthal angles of $\mathbf{p}$ with respect to any domain axis. If now we define

$$h_{\parallel}(T/T_c) \equiv \frac{3}{N(0)}\sum_p \cos^2\theta\left(-\frac{\partial f}{\partial E_p}\right)$$

and (II.35)

$$h_{\perp}(T/T_c) \equiv \frac{3}{N(0)}\sum_p \sin^2\theta\left(-\frac{\partial f}{\partial E_p}\right)$$

then if there is no $\varphi$ dependence of the sums in (II.34) the three tensor components of $\overleftrightarrow{\rho}_n$ are

$$(\rho_{n_w})_{zz} = \tfrac{2}{3}p_F^2\, N(0)\, h_{\parallel}(T/T_c)$$

and (II.36)

$$(\rho_{n_w})_{xx} = (\rho_{n_w})_{yy} = \tfrac{1}{3}p_F^2 N(0) h_{\perp}(T/T_c).$$

To take into account the effect of the strongly interacting Fermi liquid we will again assume that the same "molecular velocity field" $\mathbf{v}_{\text{eff}}$ is valid over the whole Fermi surface and write, using (I.13)

$$\mathbf{Q} = \vec{\vec{\rho}}_{n_w}\left(\mathbf{v} - \frac{F_1}{2N(0)p_F^2}\mathbf{Q}\right) = \rho_n \mathbf{v} \tag{II.37}$$

Eliminating **Q** and **v** we have for the case where these are along principal axes we have

$$\frac{(\rho_n)_{ii}}{\rho} = \frac{(\rho_{n_w}/\rho)_{ii}}{1 + \dfrac{F_1/3}{\frac{2}{3}N(0)p_F^2/\rho}\left(\dfrac{\rho_{n_w}}{\rho}\right)_{ii}} \tag{II.38}$$

Then using both the equation

$$\tfrac{2}{3}p_F^2 N(0) = Nm^* = \rho\frac{m^*}{m} = \rho(1 + F_1/3), \tag{II.39}$$

from (I.16), (I.17) and (II.36) we find

$$\left(\frac{\rho_n}{\rho}\right)_{zz} = \frac{(1 + \frac{1}{3}F_1)h_{\|}(T/T_c)}{1 + \frac{1}{3}F_1 h_{\|}(T/T_c)}$$

and (II.39)

$$\left(\frac{\rho_n}{\rho}\right)_{xx} = \left(\frac{\rho_n}{\rho}\right)_{yy} = \frac{(1 + \frac{1}{3}F_1)h_{\perp}(T/T_c)}{1 + \frac{1}{3}F_1 h_{\perp}(T/T_c)}.$$

Now let us use the simplest possible assumptions based on weak coupling theory to compare susceptibility and superfluid density results near $T_c$. For $He^3$-$A$ we assume the ABM state[38] where, for $(1 - T/T_c) \ll 1$,

$$(h_{\|})_{\text{ABM}} = 1 - (1 - T/T_c),$$
$$(h_{\perp})_{\text{ABM}} = 1 - 2(1 - T/T_c), \tag{II.41}$$

and

$$\chi(T/T_c)_{\text{ABM}} = \chi_N.$$

For $He^3$-$B$ we will assume the Balian–Werthamer state[39] even though our susceptibility data as presently interpreted do not quantitatively support this assumption. For the BW state, again for $(1 - T/T_c) \ll 1$ we have

$$g_{\text{BW}} = (h_{\|})_{\text{BW}} = (h_{\perp})_{\text{BW}} = 1 - 2(1 - T/T_c), \tag{II.42}$$

and $x = \frac{1}{3}$ in (II.28). Now as we shall see below it is likely that in a fourth sound measurement using a confined geometry it is $(\rho_s/\rho)_{\perp}$ which is sensed. Hence, approximating (II.40) near $T_c$ we might expect, using (II.41) and (II.42),

$$(\rho_{n\perp}/\rho)_{\mathrm{ABM}} = (\rho_n/\rho)_{\mathrm{BW}} = 1 - \frac{2(1 - T/T_c)}{1 + F_1/3}, \tag{II.43}$$

or

$$(\rho_{s\perp}/\rho)_{\mathrm{ABM}} = (\rho_s/\rho)_{\mathrm{BW}} = \frac{2}{1 + F_1/3}(1 - T/T_c). \tag{II.44}$$

This result has also been emphasized recently by Cómbescot.[40] Approximating (II.28) near $T_c$ and using (II.42) we get

$$\left(\frac{\chi}{\chi_N}\right)_{\mathrm{BW}} = 1 - x\frac{2(1 - T/T_c)}{1 + Z_0/4}. \tag{II.45}$$

From the appendix, for a pressure of 33·6 Bar we expect $1 + F_1/3 = 6{\cdot}1$ and $1 + Z_0/4 = 0{\cdot}26$. We thus would calculate near $T_c$ that $\mathrm{d}(\rho_s/\rho)/\mathrm{d}(1 - T/T_c) = 0{\cdot}33$ and, for $He^3$–B, $\mathrm{d}(\chi/\chi_N)/\mathrm{d}(T/T_c) = 2{\cdot}6$, where we have used $x = \frac{1}{3}$. Experimentally the density derivative is close to 0·5 and the susceptibility derivative to 4·7, so the quantitative agreement is poor. However, (II.44) and (II.45) can be combined to give a result which may have greater validity:

$$\frac{\mathrm{d}(\rho_s/\rho)/\mathrm{d}(1 - T/T_c)}{\mathrm{d}(\chi_B/\chi_N)/\mathrm{d}(T/T_c)} = \frac{1}{x}\frac{1 + Z_0/4}{1 + F_1/3}. \tag{II.46}$$

Using (I.10) this may be written

$$\frac{\mathrm{d}(\rho_s/\rho)/\mathrm{d}(1 - T/T_c)}{\mathrm{d}(\chi_B/\chi_N)/\mathrm{d}(T/T_c)} = \frac{1}{x}\frac{\frac{3}{2}kT^*}{p_F^2/2m}. \tag{II.47}$$

In addition to $x$ only well known magnetic and molar volume data are needed to evaluate this quantity. We can use (II.47) to "predict" the superfluid density results from the static susceptibility measurements. Again using $x = \frac{1}{3}$ and with $\mathrm{d}(\chi_B/\chi_N)/\mathrm{d}(T/T_c) = 4{\cdot}7$ we thus might expect at 33·6 Bar from (II.47) that $\rho_s/\rho = 0{\cdot}60(1 - T/T_c)$. This equation, shown as a solid line on Fig. II.10 is rather close to the experimental results. Further experiments are needed to investigate whether or not this is a fortuitous result. At the present time there are substantial questions being raised regarding the basic experimental facts. Resonance measurements[41, 42] suggest that the static magnetization changes observed by us are too large while measurements of the damping and resonant frequency shift of a vibrating wire[8, 43] suggest that $\rho_s/\rho$ as determined by fourth sound is too small. It will be interesting to see as time goes on how these uncertainties are resolved.

As of present writing no measurements have been published demonstrating an anisotropy of flow as reflecting a density tensor with unequal principal components. However, there has been considerable theoretical work[35, 36, 44, 45] which serves to illuminate the possibilities. We will give a short discussion based on this theoretical work and oriented toward experiments on propagation of fourth sound.

Let us begin by discussing some concepts regarding the ordering in $He^3$–A. It is generally believed that $He^3$–A reflects the so-called ABM pairing state proposed by Anderson and Brinkman.[38] For this state the superfluid consists of two weakly interacting components corresponding to pairing in $\uparrow\uparrow$ and in $\downarrow\downarrow$ states. The ordering is described by a vector order parameter **d** in spin space which in a strong field $B\hat{z}$ lies in the plane perpendicular to **B**. The components of **d** are related to the order parameters $d_{\uparrow\uparrow}$ and $d_{\downarrow\downarrow}$ for the two components of the superfluid by the equation

$$d_{\uparrow\uparrow} = d^*_{\downarrow\downarrow} = d_y + id_x. \tag{II.48}$$

This order parameter **d** is a function of the direction $\hat{\mathbf{n}}$ of pairing, and for the ABM state in equilibrium **d** is given by

$$\mathbf{d}_{\text{equil}} = \sqrt{\tfrac{3}{2}}\hat{\mathbf{y}}(n_x + in_z) \tag{II.49}$$

where $n_x$ and $n_z$ are the $x$- and $z$-components of $\hat{\mathbf{n}}$. The numerical coefficient normalizes $|d|^2$ to unity over a sphere. **d** is related to the energy gap $\Delta(\hat{\mathbf{n}})$ by multiplication so that

$$\Delta(\hat{\mathbf{n}})_{\text{equil}} = \sqrt{\tfrac{3}{2}}\Delta\hat{\mathbf{y}}(n_x + in_z). \tag{II.50}$$

The orbital part of the order parameter function, $n_x + in_z$, is a $Y_1^1$ with respect to the $\hat{\mathbf{y}}$ direction as axis. The pairing then has associated with it a direction **l** which is the axis of this spherical harmonic.

In the absence of a coherent dipolar interaction amongst the spins the vectors **d** and **l** could have any relative orientation. However, if the dipolar interaction is taken into account[46–48] there is an energy coupling **l** and **d** given by

$$\Delta F_D = -\tfrac{1}{2}\chi_N B_D^2 \cos^2\theta_{\mathbf{l},\mathbf{d}} \tag{II.51}$$

where $\theta_{\mathbf{l},\mathbf{d}}$ is the angle between **l** and **d** and

$$B_D^2 = \frac{\Omega^2(T/T_c)}{\gamma^2}. \tag{II.52}$$

The angular frequency $\Omega$ can be measured for $He^3$–A. This will be discussed in Lecture III, but for the moment we will regard $B_D^2$ as an empirical quantity.

In large enough magnetic field the vector **d** will lie in a plane perpendicular to the field direction, and (II.51) tells us that in bulk quiescent fluid **d** and **l** will be coaxial. However, this may not necessarily occur since **l** can also be oriented both by flow and by boundaries. Let us first consider flow as an orienting means since it is intuitively the simplest to consider. We have already seen, (II.40) and (II.41), that the density tensor is expected to be anisotropic for the ABM state with the superfluid density half as great along the axis **l** as perpendizular to that axis. This anisotropy reflects the fact, from (II.50), that superfluid density is proportional to $|\Delta|^2$ and for the ABM state $|\Delta(\hat{\mathbf{n}})|^2 = \Delta^2 \sin^2 \theta$, where $\theta$ is the angle between $\hat{\mathbf{n}}$ and **l**. In the presence of a superfluid velocity field $\mathbf{v}_s$ the free energy $\frac{1}{2}\rho_s v_s^2$ is minimized by having **l** point along the direction of flow.

The direction of **l** can also be influenced by boundaries. This is because scattering off a boundary tends to depair the correlated particles. As a result it is energetically favourable for correlated pairs to have large components of momentum parallel to the boundaries: $\hat{\mathbf{n}}$ for pairs is most favourable when it is parallel to the boundary or **l** is perpendicular to the boundary. *Within the bulk liquid* the effect of the boundaries is still felt since energy is required to

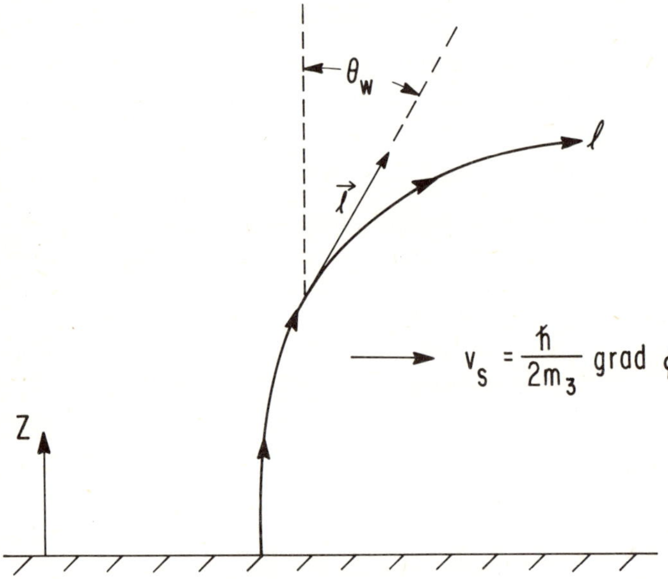


FIG. II.11. Diagram illustrating the influence in zero field of boundaries and flow on the orientation of the vector **l** in He$^3$–A. In this case there is a superfluid flow parallel to the boundary.

bend the direction of **l** in space. For a plane boundary as in Fig. II.11 the free energy due to flow and to the bending of **l** is, in the notation of Refs. 36, and 37,

$$\Delta F_w = \tfrac{1}{2}\Delta^2\Big[(K_L + K_T)\,\text{grad}\,\varphi)^2 + (K_T - K_L)(\mathbf{l}\cdot\text{grad}\,\varphi)^2 + (K_L\cos^2\theta_w + K_T\sin^2\theta_w)\left(\frac{d\theta_w}{dz}\right)^2\Big], \tag{II.53}$$

where

$$K_L = 3K_T = \tfrac{9}{5}N(0)\xi_s^2 \tag{II.54}$$

and

$$\xi_s^2 = \tfrac{7}{8}\zeta(3)\left(\frac{\hbar^2 v_F^2}{6\pi^2 k^2 T_c^2}\right) \tag{II.55}$$

is the square of an $s$-wave coherence length.

Near to $T_c$ the thermal free energy difference between the superfluid and normal states is given by

$$F_S - F_N = \tfrac{1}{2}N(0)\Delta^2 t \tag{II.56}$$

where $t = (T - T_c)/T_c$ and

$$\Delta^2 = -\frac{(kT_c)^2}{\beta}t. \tag{II.57}$$

For this free energy difference, we find by differentiating (II.57) twice

$$C_S - C_N = N(0)k^2T_c/\beta \tag{II.58}$$

where

$$C_N = \frac{2\pi^2}{3}N(0)k^2T_c. \tag{II.59}$$

From this we find

$$\beta^{-1} = \frac{2\pi^2}{3}\left(\frac{C_S - C_N}{C_N}\right)_{T=T_c},$$

so that, using (II.57), we have

$$\Delta^2 = -\frac{2\pi^2}{3}\left(\frac{C_S - C_N}{C_N}\right)_{T=T_c}(kT_c)^2 t \tag{II.60}$$

From (II.54), (II.55) and the known values of $(C_S - C_N)/C_N$ one can evaluate all the parameters in (II.53).

Now let us consider a hypothetical orientation problem for an experiment in which it is desired to measure superfluid density in a direction parallel to a plane surface as in Fig. II.12. We will take $\mathbf{v}_s$ to be small enough that it does

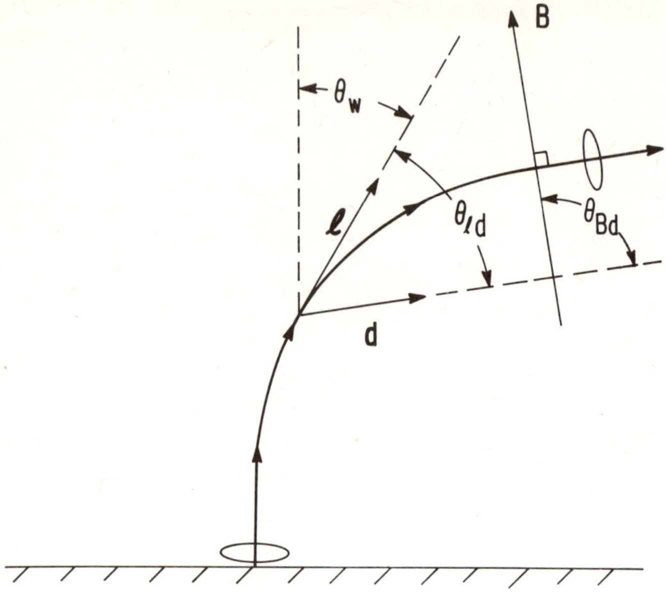


FIG. II.12. Diagram illustrating the influence of boundaries and a magnetic field on orientation of the vector **l** in He$^3$–A.

not affect the orientation. Thus terms in grad $\varphi$ in (II.53) will be neglected. In the absence of a magnetic field depairing at the wall will dominate, **l** will be perpendicular to the wall, **d** will be along **l**, and the measured $\rho_s$, $\rho_{s\perp}$ in (II.44), will be relatively large. Now apply a magnetic field which is primarily perpendicular to the surface. The **d** vector will try to swing perpendicular to the field. If the magnetic field energy

$$\Delta F_B = -\tfrac{1}{2}\chi_N B^2 \sin^2 \theta_{\mathbf{B},\mathbf{d}}, \tag{II.61}$$

where $\theta_{\mathrm{B},\mathbf{d}}$ is the angle between **B** and **d**, is small compared to the dipolar energy $\Delta F_D$ which in turn is small compared to $\Delta F_w$, then the **d** vector cannot turn perpendicular to **B** and **l** will remain perpendicular to the wall. But if **B** is large enough ($B^2 \gg B_D^2$) then **d** will turn perpendicular to **B** and the dipolar energy will try to turn **l** to make the minimum angle with respect to **d**. The purpose of tilting **B** away from perpendicular to the wall is to eliminate the degeneracy in the direction of **l**. Thus at some distance from the wall **l** will have a large component parallel to $\mathbf{v}_s$ and the measured superfluid density should be $\rho_{s\parallel}$; or, from (II.41) smaller than $\rho_{s\perp}$ by a factor two.

We can work out the magnitude of some of these energies for a pressure of 33 Bar, where

$$v_F = 3{\cdot}05 \times 10^3 \text{ cm/sec}$$

$$N(0) = 1{\cdot}23 \times 10^{38} \,(\text{erg cm}^3)^{-1}$$

$$T_c = 2{\cdot}59 \text{ mK}$$

$$B_D^2 = (74 \text{ gauss})^2 |t|$$

$$\xi_s = 1{\cdot}2 \times 10^2 \text{ Å}$$

$$(C_S - C_N)/C_N = 1{\cdot}8$$

$$K_L \Delta^2 = 4{\cdot}8 \times 10^{-10} |t| \text{ erg/cm}$$

$$K_T \Delta^2 = 1{\cdot}6 \times 10^{-10} |t| \text{ erg/cm}$$

$$\chi_N = 1{\cdot}1 \times 10^{-7}$$

$$\chi_N B_D^2 = 6{\cdot}0 \times 10^{-4} |t| \text{ erg/cm}^3.$$

Thus for $|t| = 0{\cdot}16$ we require $\mathbf{B} \gg 30$ gauss to orient the **d** vector perpendicular to **B** while for **l** to be oriented significantly by **B** it is necessary to move out into the bulk an approximate distance $d$ such that

$$K_L \Delta^2 \left(\frac{\pi}{2d}\right)^2 \approx \chi_N B_D^2 \qquad \text{(II.62)}$$

which corresponds to $d \approx 10^{-3}$ cm. According to this estimate the walls have a very significant macroscopic effect on any orientation or anisotropy experiment.

## III. Dynamic Properties of the Superfluid

In this Lecture we will consider mainly dynamic properties of the nuclear spin system and flow properties of the liquid. Since this is not a general review we shall discuss here only two experiments which have been performed in our laboratory in the last year, one on the dynamics of nuclear magnetism oscillating parallel to itself and the other on the flow of heat. There are a number of other beautiful experiments bearing on the same concepts which we shall not have the time to discuss in detail.

### 1. Parallel Spin Dynamics

The experiments which we shall describe here were motivated in concept by theoretical ideas of Leggett[46, 47, 48] regarding the possible development in a nonequilibrium state of a torque on the nuclear spin system parallel to the magnetization which would allow the magnetization to change parallel to itself without any flow of heat. Our method of experimental realization of Leggett's idea was motivated by our[21] observation that the $He^3$ nuclear magnetization could readily be detected using an RF-biased SQUID system as a wideband, non-overloading detector of flux with, however, a low dynamic range.[49]

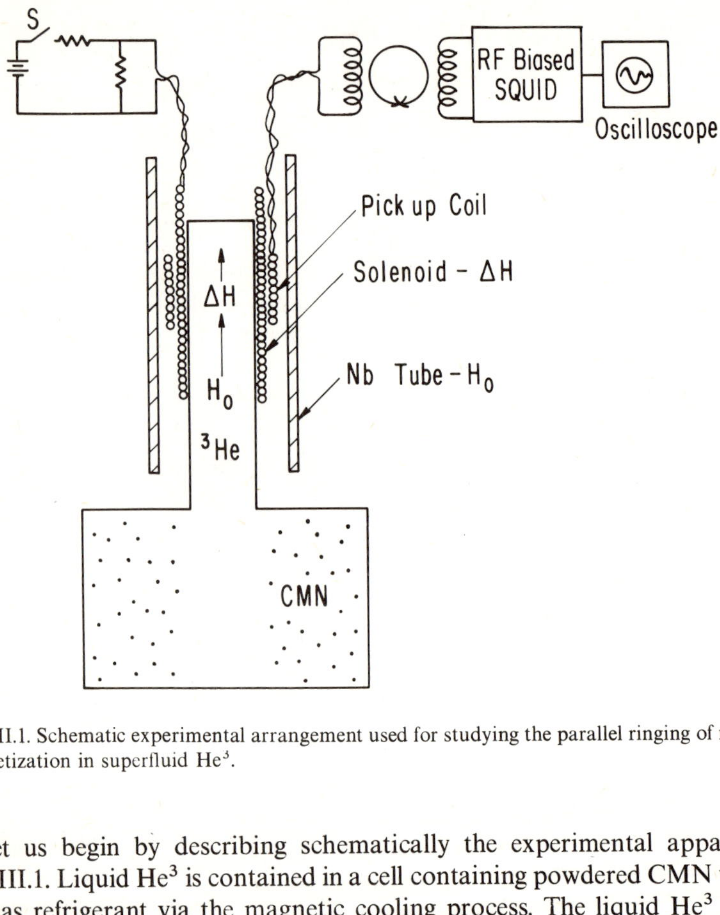


FIG. III.1. Schematic experimental arrangement used for studying the parallel ringing of nuclear magnetization in superfluid $He^3$.

Let us begin by describing schematically the experimental apparatus, Fig. III.1. Liquid $He^3$ is contained in a cell containing powdered CMN which acts as refrigerant via the magnetic cooling process. The liquid $He^3$ being studied is contained in a tower above the main cell. Around the $He^3$ in the tower is a small uniformly wound Nb wire solenoid which can produce an incremental field $\mathbf{\Delta H}$ parallel to the tower axis, a pick-up coil of Nb wire which is part of a completely superconducting flux transfer circuit, and a niobium tube in which we trap a stable field $\mathbf{H}_0$ parallel to the tower axis. This tube also serves to shield the pick-up coil from changing external fields.

In the experiment the switch $S$ is closed and the incremental field $\mathbf{\Delta H}$ is applied. Then at some later time the switch is opened, $\mathbf{\Delta H}$ falls to zero in a few microseconds, the non-overloading flux detector system clears, and we detect the changing nuclear magnetism. These actions are shown schematically in Fig. III.2, where $t_0$ is the time at which $\mathbf{\Delta H}$ is turned off. What normally

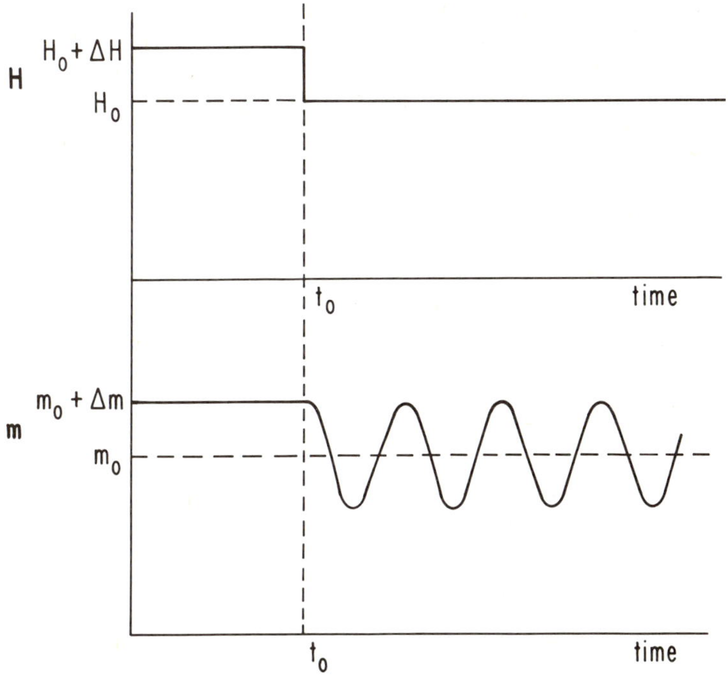


FIG. III.2. Idealized dependence of magnetic field and parallel magnetization on time in a parallel ringing experiment.

would be expected in such an experiment is an exponential decrease in magnetization with a time constant $T_1$, the spin–lattice relaxation time. The magnetization parallel to the field would change only to the extent that heat flows between the spins and their surrounding. But this is not what is observed in the superfluid. Rather the magnetization oscillates—or rings—with a frequency which depends on temperature and approaches zero at $T_c$.

To get some concept of why the ringing occurs we will assume that He$^3$–A is a manifestation of the so-called ABM state[38] (for Anderson, Brinkman and Morel). For this state in presence of field $B\hat{\mathbf{z}}$ the normalized equilibrium vector order parameter may be written (Lecture II)

$$\mathbf{d}(\hat{\mathbf{n}}) = \sqrt{\tfrac{3}{2}}\hat{\mathbf{x}}(n_y + in_z), \tag{III.1}$$

where $n_y$ and $n_z$ are, respectively, the components of $\hat{\mathbf{n}}$ along the $\hat{\mathbf{y}}$ and $\hat{\mathbf{z}}$ directions. The vector $\mathbf{l}$ introduced in Lecture II is along $\hat{\mathbf{x}}$ and $\mathbf{d}$ has the same direction for all pairs $\hat{\mathbf{n}}$. However, if the $He^3$–A is suddenly subjected to an incremental field change $\Delta H\hat{\mathbf{z}}$ then Leggett showed that the vector $\mathbf{d}$ precesses about $\hat{\mathbf{z}}$ with angular frequency

$$\dot{\theta} = \gamma(\Delta H - \gamma \Delta S_z/\chi), \tag{III.2}$$

where $\gamma$ is the gyromagnetic ratio, $\chi$ is the susceptibility, and $\Delta S_z$ is the resultant change of the spin angular momentum per unit volume along $\hat{\mathbf{z}}$. Note that when the change of magnetization $\gamma \Delta S_z$ is equal to $\chi \Delta H$, then $\dot{\theta} = 0$. The relationships of various vectors could be as shown in Fig. III.3. For times sufficiently short compared with the ringing period the orbital variables are not expected to change. This means in particular that $\mathbf{l}$ does not change. But as we indicated in (II.51), the dipolar energy per unit volume $E_D$ depends on

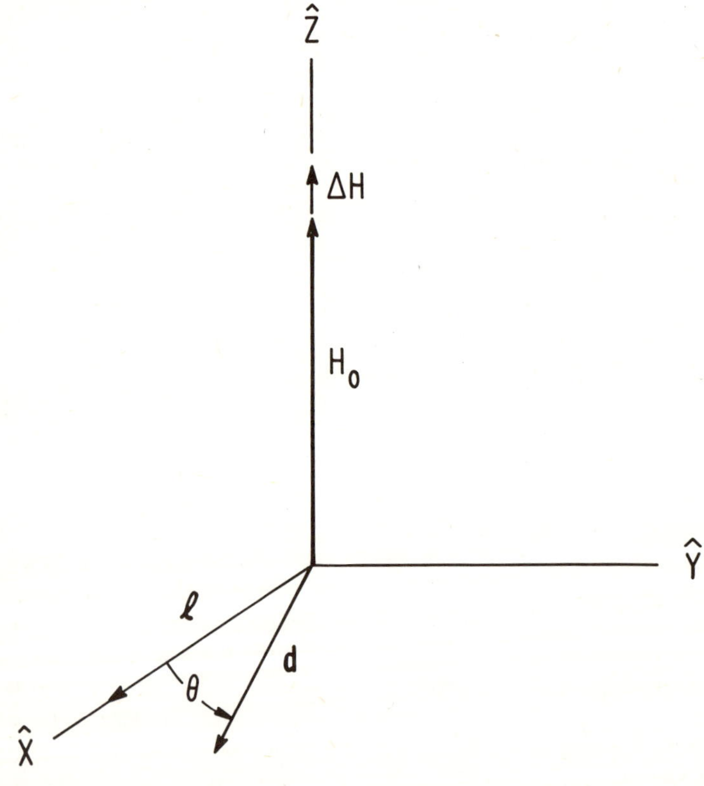


FIG. III.3. Diagram illustrating the relative orientation of the various vectors involved in understanding the parallel ringing experiment in $He^3$–A.

angle $\theta$ as

$$E_D = E_0 - \tfrac{1}{2}\lambda \cos^2 \theta, \qquad \text{(III.3)}$$

where $E_0$ does not depend on $\theta$ and $\lambda$ depends on temperature and pressure, as shown in Fig. III.4a. In the vicinity of $\theta = 0$ or $\pi$ this form of $E_D$ leads to a torque per unit volume in the $\hat{\mathbf{z}}$ direction on the spins equal to $\lambda\theta$, as shown in Fig. III.4b. The time rate of change of the angular momentum

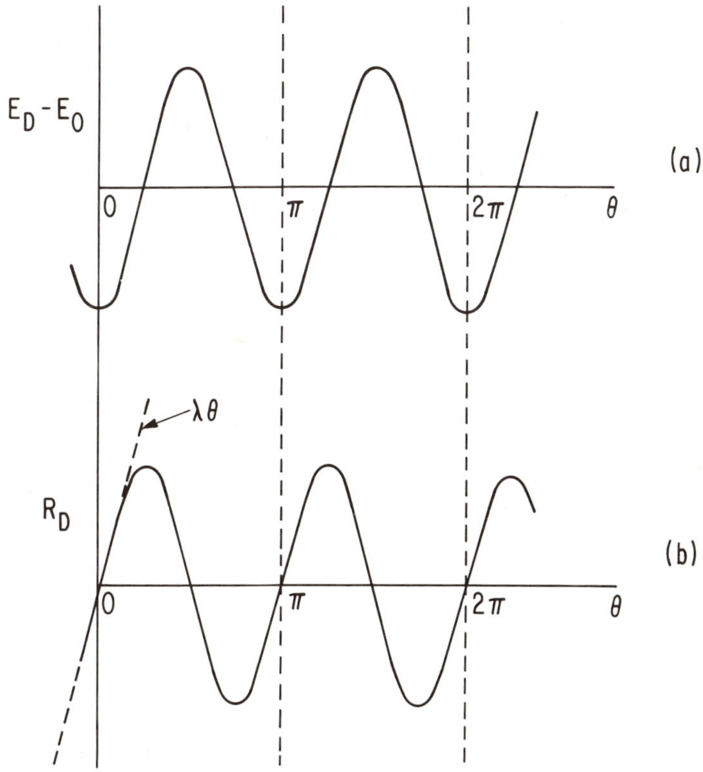


FIG. III.4. (a) Dipolar energy and (b) dipolar torque along $S_z$ for He$^3$–A as a function of the angle $\theta$ between **d** and **l**.

per unit volume is given by the equation

$$\frac{\mathrm{d}}{\mathrm{d}t} \Delta S_z = R_D \simeq \lambda\theta. \qquad \text{(III.4)}$$

Now taking into account (III.2) we have

$$\frac{d^2}{dt^2}\Delta S_z = \lambda\dot{\theta} = \lambda\gamma\left(\Delta H - \frac{\gamma\Delta S_z}{\chi}\right), \tag{III.5}$$

or

$$\frac{d^2}{dt^2}\Delta S_z + \frac{\gamma^2\lambda}{\chi}\Delta S_z = \gamma\lambda\Delta H. \tag{III.6}$$

This equation is solved by the magnetization $\gamma\Delta S_z$ varying harmonically with time at angular frequency

$$\Omega^2(T) = \frac{\gamma^2\lambda}{\chi} \tag{III.7}$$

and with amplitude $\chi\Delta H$. The solution should be valid for $\theta_{\text{max}}$ sufficiently small. The oscillation takes place at constant entropy and reflects an interchange of spin and orbital angular momentum dictated by the mutual torques exerted by the spin and orbital systems on one another by a *coherent* dipolar interaction.

The order of magnitude and temperature dependence of the parameter $\lambda$ may be estimated by the following arguments. For $He^3$–A in the equal spin pairing ABM state, if one spin of a correlated pair is ↑ (or ↓) then the second is ↓ (or ↑) also. In the normal Fermi liquid, where the spins are not correlated in pairs, there is a constant relative flipping of spins which precludes any significant coherent spin energy. However, in the superfluid we may expect a dipolar interaction energy per unit volume of order

$$\lambda \approx N\cdot\frac{\gamma^2\hbar^2}{a^3}\cdot\left(\frac{\Delta}{\varepsilon_F}\right)^2, \tag{III.8}$$

where the first factor is the number of atoms/volume, the second is the energy of interaction of a single pair ($a$ is a nearest neighbour distance), and the last is the joint probability that both members of a pair are in the correlated state. Here $\Delta$ is the energy gap function and $\varepsilon_F$ is the Fermi energy. Thus near $T_c$ we estimate $\lambda$ very crudely by

$$\lambda \approx \left(\frac{8\pi^2}{3}\frac{C_<}{C_>}\right)(\gamma\hbar N(0)kT_c)^2(1 - T/T_c) \tag{III.9}$$

where we have put

$$Na^3 \simeq 1, \qquad N/\varepsilon_F \simeq 2N(0), \qquad \Delta^2 \simeq \left(\frac{2\pi^2}{3}\frac{C_<}{C_>}\right)(kT_c)^2(1 - T/T_c).$$

For $He^3$ near the melting curve, we have $\gamma = 2{\cdot}04 \times 10^4$ (gauss sec)$^{-1}$, $N(0) \simeq 1{\cdot}2 \times 10^{38}$ (erg cm$^3$)$^{-1}$, $T_c = 2{\cdot}6 \times 10^{-3}$ K, and $C_</C_> = 2{\cdot}8$, so that

$\lambda \simeq 6 \times 10^{-5}\ (1 - T/T_c)$ erg/cm$^3$. With $\chi = 1{\cdot}1 \times 10^{-7}$ we find from (III.7) $(\Omega(T)/2\pi)^2 \approx (0{\cdot}8 \times 10^2\ \text{kHz})^2(1 - T/T_c)$. Leggett estimated $\lambda$ carefully and found $(\Omega(T)/2\pi)^2 \simeq (2{\cdot}5 \times 10^2\ \text{kHz})^2(1 - T(T_c)$ for the ABM state. The experimental value is $(2{\cdot}4 \times 10^2\ \text{kHz})^2(1 - T/T_c)$.

Actual experimental observations[23] of the ringing frequencies in $He^3$–A are shown in Fig. III.5. The ringing responds to field changes, not to the

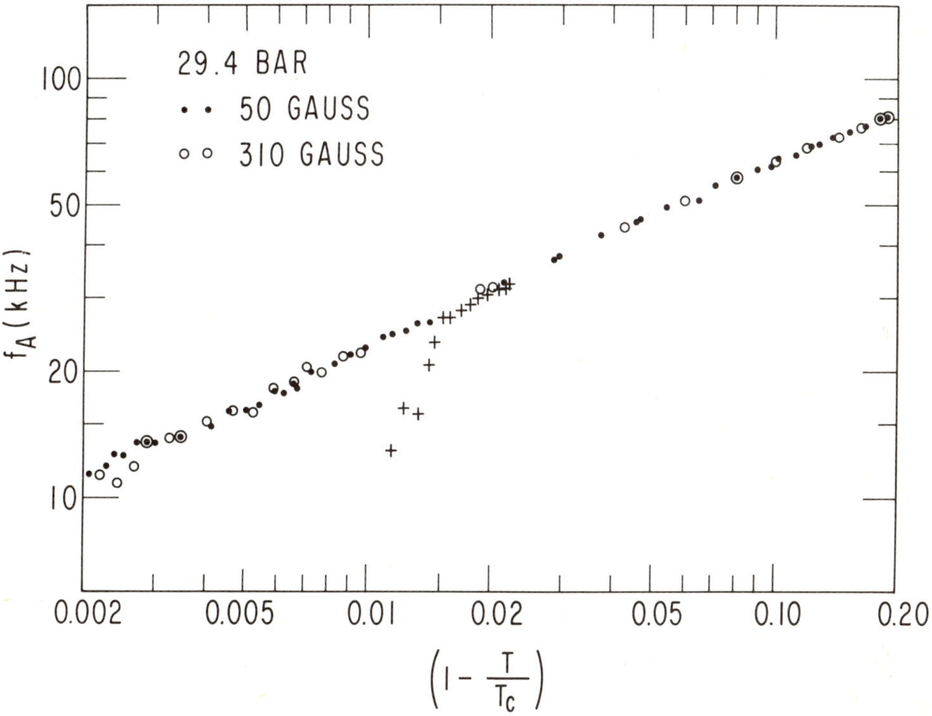


FIG. III.5. Observed parallel ringing frequencies in $He^3$–A plotted logarithmically against $(1 - T/T_c)$ for a pressure of 29·4 Bar and fields of 50 and 310 gauss. The (+) symbols refer to a case where $\Delta H$ was too large at certain frequencies so that the ringing frequency was reduced by nonlinearities. (After Webb *et al.*, Ref. 23).

steady field, so the frequency should be accurately field independent. This is observed. The temperature dependence also is the expected $(1 - T/T_c)$ near $T_c$. The ringing response to a step change in field in the time domain has as its analog parallel resonance in the frequency domain as in the experiments of Osheroff and Brinkman[50] and of Bozler *et al.*[51] Our results are in excellent quantitative agreement with those of Ref. 50.

There is excellent agreement between theoretical concepts and experimental spin dynamics in the case of $He^3$–A including the equality of parallel resonance frequencies and transverse resonance frequency shifts.[50, 51]

Our transient ringing method has some other interesting features which further illuminate the theoretical concepts. Consider the full dependence of $R_D$ on $\theta$ in Fig. III.4(b). If $\Delta H$ is small enough the sequence of events is as follows, following (III.2) and (III.4). After the rapid turnoff of $\Delta H$ the $d$-vector starts to precess at $\dot{\theta} = \gamma \Delta H$. As $\theta$ increases a torque develops to decrease $S_z$. As $S_z$ changes $\dot{\theta}$ decreases, finally reaching $\dot{\theta} = 0$ when $\gamma \Delta S_z = \chi \Delta H$. However, at this point $\theta$ itself is not zero; rather $\theta$ still has the sign to decrease $S_z$, so although $\theta$ now decreases the torque is still such as to decrease $S_z$. Thus $S_z$ continues to decrease, reaching a minimum when $\theta = 0$ again. The next half cycle then ensues with $\theta$ on the opposite side of equilibrium. But the dipolar torque is not unlimited. If $\Delta H$ is too large, then the angle $\theta$ will be carried beyond the region where the torque is linear in $\theta$. The system will appear less "stiff," and the ringing frequency will drop. This is shown by the (+) symbols on Fig. III.5. If $\Delta H$ is made large enough then the ringing is extinguished altogether. If $\Delta H$ is made very large then $\Delta S_z$ never has a chance to respond fully to the rapidly rotating $d$-vector. Rather, examination of Fig. III.4(b) shows that the spins are subject to a doubly periodic torque at frequency $2(\gamma \Delta H)$. The magnetization then oscillates at a "driven" frequency $2\gamma \Delta H$. The detailed theory of this has been worked out by Maki and Tsuneto.[52]

An example of results[53] for this non-linear effect is given in Fig. III.6 along with a solid theoretical curve. It is possible that the extinction point of the "theoretical" curve should be farther to the left with respect to the experimental points since both $\Delta H$ and $f_A$ are subject to error. From examination of Fig. III.6 we conclude that the observed frequency $f_{obs}$ for $\Delta H$ incremental field change stays closer to the low $\Delta H$ value, $f_A$, than expected by the theory —as if the linear region were larger than anticipated. The ringing signals for $\gamma \Delta H / 2\pi f_A > 1$ are of poorer quality and rather rapidly disappear in noise as the higher frequency branch develops.

The preceding experiments are illuminated further by an idea introduced by Leggett[48] and further developed by Maki and Tsuneto.[52] The two ESP states in $He^3$–A may be conceived as two interpenetrating superfluids with order parameters $d_{\uparrow\uparrow}$ and $d_{\downarrow\downarrow}$. Following Leggett, for $\mathbf{d}$ making angle $\theta$ with $\mathbf{l}$ these order parameters are $d_{\uparrow\uparrow} = d^*_{\downarrow\downarrow} = id(\hat{\mathbf{n}})\, e^{i\theta}$. The phase difference $\varphi$ between the $\uparrow\uparrow$ and $\downarrow\downarrow$ states is just $2\theta$, so the energy $E_0$ in (III.3) may be written

$$E_D = E_0 - \tfrac{1}{4}\lambda - \tfrac{1}{4}\lambda \cos \varphi. \qquad \text{(III.10)}$$

One may then think of two independent superfluids weakly coupled together by an energy dependent on the cosine of their phase difference as in the

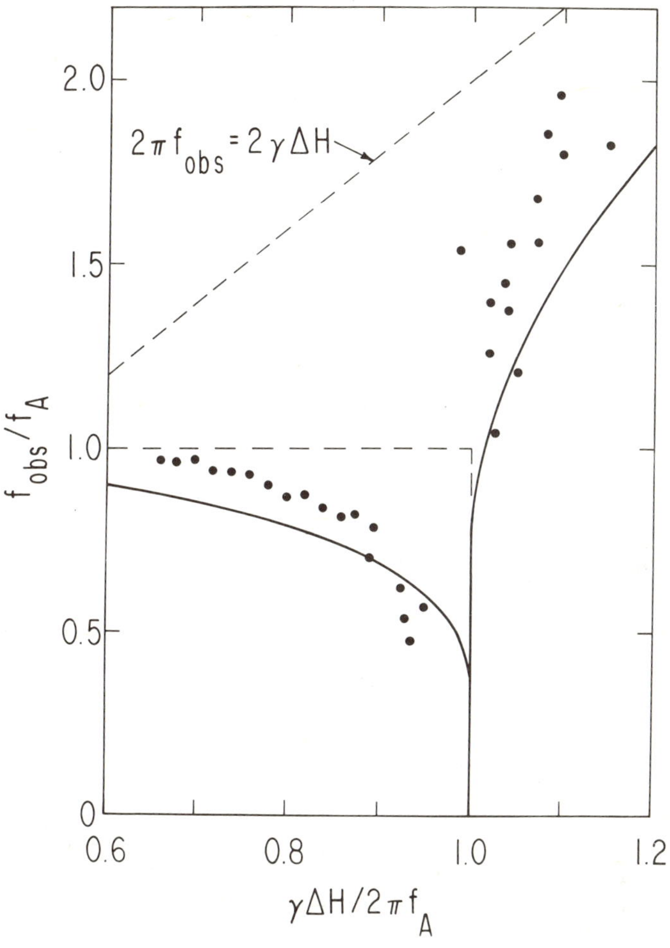


FIG. III.6. Nonlinear ringing phenomena in He$^3$–A $f_{obs}/f_A$ is the ratio of the observed frequency $f_{obs}$ to that $f_A$ measured in the linear regime with small $\Delta H$ as a function of $\gamma\Delta H/2\pi$, the maximum precessional frequency of the **d** vector, divided by $f_A$. The solid line is calculated from the theory by Maki and Tsuneto, Ref. 52. (After Webb *et al.*, Ref. 23.)

Josephson effect. Let us explore this idea further, following Maki and Tsuneto but ignoring the effective spin interactions among the particles. Apart from constant terms, at a given temperature the energy may be written

$$\mathscr{H} = \Sigma(\mu_\uparrow \delta n_\uparrow + \mu_\downarrow \delta n_\downarrow) - \frac{\lambda}{4}\cos\varphi, \tag{III.11}$$

where $\mu_\uparrow$ and $\mu_\downarrow$ are the chemical potentials for the $\uparrow$ and $\downarrow$ spins, $\delta n_\uparrow$ and

$\delta n_\downarrow$ refer to changes in the distribution of ↑ and ↓ spins, and the phase term comes from (III.10). When $\delta n$ pairs of correlated particles are transferred between the ↑↑ distribution and the ↓↓ distribution the corresponding difference in the number in the numbers of ↑ and ↓ spins is related to $\delta n$ by

$$\delta n = \tfrac{1}{4}(\delta n_\uparrow - \delta n_\downarrow). \tag{III.12}$$

In terms of $n$ the equations of motion are

$$\dot{n} = \frac{1}{\hbar}\frac{\partial \mathscr{H}}{\partial \varphi}, \tag{III.13}$$

and

$$\hbar\dot{\varphi} = -\frac{\partial \mathscr{H}}{\partial n}. \tag{III.14}$$

Application of the first equation to (III.11) gives

$$\dot{n} = \frac{\lambda}{4\hbar}\sin\varphi. \tag{III.15}$$

If now we imagine a change in which $\delta n$ pairs are transferred from the ↓↓ to the ↑↑ superfluids at constant $\varphi$ we have

$$\delta\mathscr{H} = \mu_\uparrow \delta n_\uparrow + \mu_\downarrow \delta n_\downarrow = 2(\mu_\uparrow - \mu_\downarrow)\delta n, \tag{III.16}$$

where we have used (III.12) together with $\delta n_\uparrow + \delta n_\downarrow = 0$. Thus, from (III.14) we find

$$\hbar\dot{\varphi} = -2(\mu_\uparrow - \mu_\downarrow). \tag{III.17}$$

Before proceeding further consider Fig. III.7 in which we show the Fermi functions $f_\uparrow$ and $f_\downarrow$ for the ↑ and ↓ spins plotted as a function of $\varepsilon = p^2/2m \pm (\gamma\hbar/2)H_0$, where the ↑ spins have the (−) sign and the ↓ spins the (+) sign. The distributions shown are valid for the instant after the field $H_0$ has been applied when $\mu_\uparrow - \mu_\downarrow = -\gamma\hbar H_0$. From (III.17), $\varphi$ will start to increase at rate $2\gamma\hbar H_0/\hbar$ while $n$ will then start to increase at rate $\lambda\dot{\varphi}/4\hbar$. But as ↓↓ pairs are changed to ↑↑ pairs the difference in chemical potential increases, and by an amount $4/N(0)$ for each pair transferred. Thus the chemical potential difference after $n$ pairs are transferred is

$$\mu_\uparrow - \mu_\downarrow = -\gamma\hbar H_0 + 4n/N(0). \tag{III.18}$$

Equation (III.17) can then be written

$$\hbar\dot{\varphi} = 2\gamma\hbar H_0 - 8n/N(0). \tag{III.19}$$

In the equilibrium case we have $\dot{\varphi} = 0$ so that $n_{\text{equil}} = \frac{1}{4}\gamma\hbar N(0)H_0$. The resul-

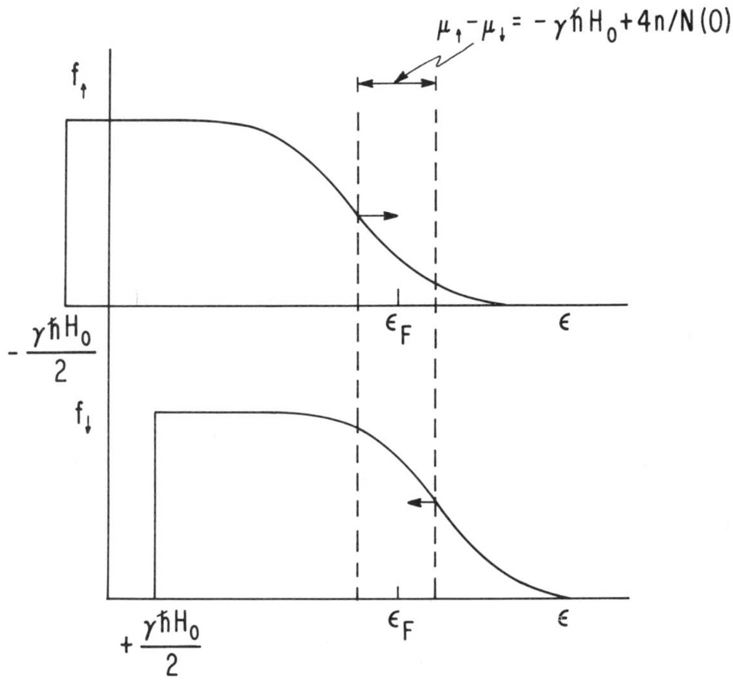


FIG. III.7. Diagram illustrating the distribution functions $f_\uparrow$ and $f_\downarrow$ and the relative chemical potentials $\mu_\uparrow$ and $\mu_\downarrow$ for ↑ and ↓ spins as function of quasiparticle energy when a field $H_0$ is suddenly applied to He$^3$.

tant magnetic moment per unit volume in equilibrium is

$$M = 4\cdot\frac{\gamma\hbar}{2}\cdot n_{\text{equil}} = \tfrac{1}{2}\gamma^2\hbar^2 N(0)H_0, \tag{III.20}$$

or

$$\chi = \tfrac{1}{2}\gamma^2\hbar^2 N(0), \tag{III.21}$$

the susceptibility expected for the non-interacting gas which we have assumed. Equation (III.19) can then be written

$$\tfrac{1}{2}\dot{\varphi} = \gamma H_0 - 4n/\hbar N(0). \tag{III.22}$$

Now if we use $n = S/2\hbar$ and $N(0) = 2\chi/\gamma^2\hbar^2$ (from (III.21)) in (III.22) we find

$$\dot{\theta} = \tfrac{1}{2}\dot{\varphi} = \gamma(H_0 - \gamma S/\chi), \tag{III.23}$$

in agreement with (III.2), apart from the use of incremental field and spin changes. So far as that is concerned all the preceding equations hold for incremental changes of $H_0$ and $n$, so the results will not depend on the actual value of a steady field $H_0$.

Let us next make the assumption that the field change $H_0$ is small enough that $\varphi$ always remains small and $\sin\varphi \simeq \varphi$. Then taking the time derivative of (III.15) and substituting (III.19) we get

$$\ddot{n} = \frac{\lambda}{4\hbar}\left[2\gamma H_0 - 8n/\hbar N(0)\right]. \tag{III.24}$$

Rewriting this equation using (III.21) to eliminate $N(0)$ we then find

$$\ddot{n} + \frac{\gamma^2\lambda}{\chi} n = \frac{\lambda\gamma}{2\hbar} H_0. \tag{III.25}$$

This is the same equation, effectively, as (III.6). Thus this approach gives the same results as our earlier one, but it does put these results in a new and interesting perspective. We have an effect which can be interpreted as analogous to the Josephson effect in superconductors. Actual experimental observation then suggests that the concept may be good that $He^3$–A is a two-component superfluid in which the two components interact weakly with one another. Actually this concept was first given experimental and theoretical support by the observation[(54)] of the splitting of the second order transition in a magnetic field and its interpretation[(55)] as the formation of first an ↑↑ superfluid and then both ↑↑ and ↓↓ superfluids as the temperature decreased through $T_c$. It would appear that parallel ringing should not occur unless both ↑↑ and ↓↓ superfluids are present. In considering flow and other properties in superfluid $He^3$ one must take into account not only the relative motion of superfluid and normal fluid but also the relative motion or action of the individual components of the superfluid.

Parallel ringing and resonance phenomena have also been observed in $He^3$–B, but the behaviour is more complicated. In our experiments[(23)] ringing is consistently observed only very near to $T_c$, and then at a frequency about 1·9 times that observed in the A phase. This is shown in Fig. III.8. Our initial thought on the reason for the difference in degradation of ringing in $He^3$–B compared with that observed in $He^3$–A is related to the observation that in $He^3$–B the ringing signals rapidly improve as the temperature warms through an "onset" temperature near $T_c$. This is not unlike the onset phenomena found in heat flow experiments for comparable thermal, as compared to magnetic, chemical potential differences in the superfluid. In $He^3$–A it appears relatively easy to quench superflow over a wide temperature range, whereas in $He^3$–B this is effective only very near $T_c$. We have thus hypothesized that spatial variations in $\Delta H$ in our apparatus have led to degradation by mixing via magnetization supercurrents in $He^3$–B except very near to

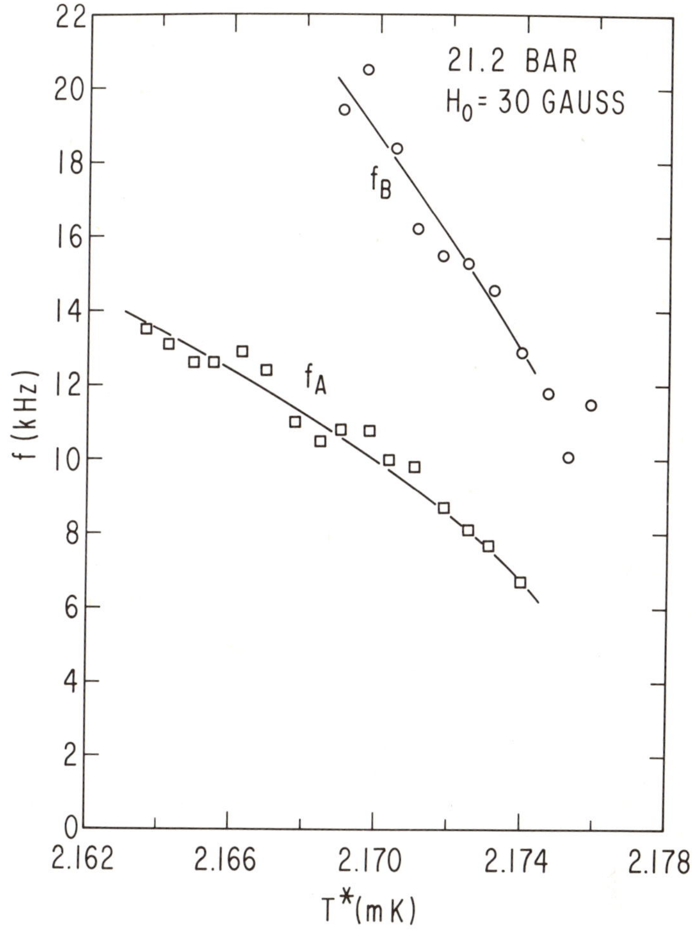


FIG. III.8. Ringing frequencies $f_A$ and $f_B$ in He$^3$–A and He$^3$–B at 21·2 Bar in a steady field of 30 gauss as a function of CMN magnetic temperature near $T_c$. Observations of the different phases at the same temperature was made possible by the judicious use of supercooling and superheating phenomena. (After Webb *et al.*, Ref. 23.)

$T_c$. The existence of counterflowing ↑↑ and ↓↓ components[52, 56] should grossly affect experiments on spin transfer in liquid He$^3$ below $T_c$.

## 2. HEAT FLOW

Our experiments on heat flow in liquid He$^3$ were motivated initially by the qualitative question of whether or not He$^3$ was a superfluid below $T_c$. The first experiments[20] were consistent with the interpretation of heat flow in both the A and B fluids below $T_c$ as being hydrodynamic rather than diffusive, and subsequent experiments[57] have shown that the effective *conductivity* does have the dependence on geometry expected for hydrodynamic heat flow. However, as a proof of superfluidity in $He^3$ we feel they are less positive than the observation of fourth sound as described in Lecture II. But as part of a quantitative experiment the heat flow measurements have shown for example that the viscosity of the normal component does drop rapidly just below $T_c$, as has been observed recently at melting pressure by Alvesalo *et al.*[43]; and they also have provided the first quantitative information on critical velocities in the superfluid. In the case of $He^4$ the hydrodynamic heat flow possible below the $\lambda$-point is one of the most dramatic properties of the superfluid state.[58, 59] Although heat flow can be understood similarly in $He^3$ below $T_c$, the effect of superfluidity is much less profound, as we shall understand presently.

Hydrodynamic heat flow may be understood, following London and Zilsel[60] and Gorter and Mellink,[61] on the basis of concepts familiar from $He^4$. For present purposes we will assume that all spin components of the superfluid behave similarly and that there is a definite $\rho_s$, as in $He^4$. This may in fact not be the case. Furthermore the orientation of domains (for example the direction of the orbital orientation vector $\mathbf{l}$ in $He^3$–A) may depend on the flow velocity itself as discussed in Lecture II. "Critical" or "onset" velocities may reflect reorientation of the domains. These reservations will have to be kept in mind in interpreting the results. Thus we assume a superfluid of density $\rho_s$, zero entropy, and curl free velocity field $\mathbf{v}_s$ and a normal fluid of density $\rho_n$ and velocity field $\mathbf{v}_n$ which carries all the entropy.

Consider a tube of diameter $d$ and length $l \gg d$ with one end closed and fitted with a heater (and thermometer) and the other end opening to a large reservoir at temperature $T$ and pressure $P$ as shown in Fig. III.9(a). When heat is applied, the superfluid will be accelerated down the chemical potential gradient due to $\Delta T$ (II.12) leading to a stationary state once a fountain pressure $\Delta P$ has been developed. Such a stationary state is characterized by grad $\mu = 0$ or

$$\Delta P = S\Delta T, \tag{III.26}$$

where $S$ is the entropy per unit volume. Under the action of the pressure difference $\Delta P$, normal fluid will move down the tube with a volume rate of flow given for small $\Delta P$ by the laminar flow formula

$$\dot{V}_n = \frac{\pi d^4}{128 l \eta} \Delta P. \tag{III.27}$$

where $\eta$ is the viscosity coefficient. The superfluid then moves with curl-free

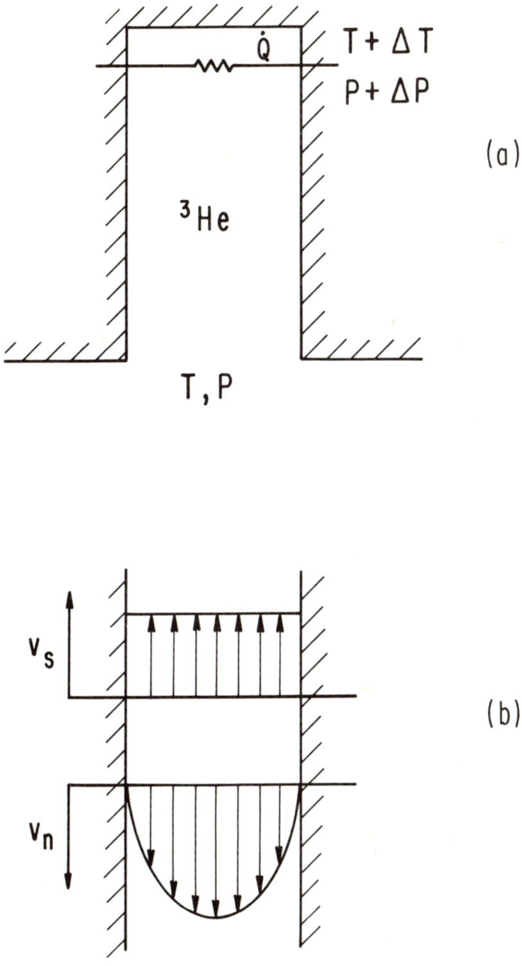


FIG. III.9. (a) Schematic arrangement for a heat flow measurement in superfluid He$^3$ and (b) the corresponding velocity profiles $v_s$ and $v_n$ assumed to hold across the heat flow tube cross-section for superfluid and normal fluid.

motion up the tube such that

$$\rho_s \mathbf{v}_s + \rho_n \bar{\mathbf{v}}_n = 0, \tag{III.28}$$

where $\bar{\mathbf{v}}_n$ is the average normal fluid velocity over the cross section of the tube. The velocity profiles for the normal and superfluid velocities are shown in Fig. III.9(b). Since the normal fluid carries entropy $S$ per unit volume the flow of normal fluid constitutes a hydrodynamic flow of heat $\dot{Q}_h$ down the tube equal to

$$\dot{Q}_h = TS\dot{V}_n. \tag{III.29}$$

Expressing $\dot{V}_n$ as in (III.27) and $(\Delta P)$ as in (III.26) this may then be written

$$\dot{Q}_h = \frac{TS^2d^2}{32\eta} \cdot \frac{\pi d^2/4}{l} \Delta T. \tag{III.30}$$

This gives an effective hydrodynamic thermal conductivity

$$\kappa_h = \frac{TS^2d^2}{32\eta}. \tag{III.31}$$

The same formula holds for superfluid $He^4$. The hydrodynamic conductivity depends on geometry, so it is of course not a true thermal conductivity. However it is exactly this geometrical dependence which is characteristic of hydrodynamic heat flow. The hydrodynamic conductivity is much reduced in $He^3$ from what it is in $He^4$ for the same geometry. In the first place $T$ is about a factor $10^3$ less while $\eta$ is greater by a factor of $10^4$–$10^5$, depending on pressure. Thus the ratio of $\kappa_h$ in $He^3$ just below $T_c$ to that in $He^3$ just below $T_\lambda$ is $10^{-7}$–$10^{-8}$ of the ratio of the squares of the entropies at these two temperatures. Since the entropy of $He^4$ near $T_\lambda$ is orders of magnitude greater than that of $He^3$ near $T_c$ the hydrodynamic conductivity in $He^3$ at $T_c$ is extremely small, infinitesimal, compared with that in $He^4$ near $T_\lambda$ for the same geometry. What this means experimentally is that for the effect to be dominant in $He^3$ the tube diameter must be much larger and its length much smaller than what is required to produce a large effect in $He^4$ near $T_\lambda$. There is the additional property that while the ordinary diffusive thermal conductivity of $He^4$ near $T_\lambda$ is quite small, the diffusive conductivity is quite large in $He^3$ near $T_c$. Thus if hydrodynamic heat flow is to dominate diffusive heat flow in superfluid $He^3$ near $T_c$ then the tube diameter $d$ in (III.31) must be made as large as possible.

Now it is known from the work of Alvesalo *et al.*[(43)] that the effective viscosity in $He^3$–A drops rapidly just below $T_c$. It is reasonable to assume that other diffusive type transport phenomena such as diffusive thermal conductivity or spin diffusion would likewise drop sharply below $T_c$. Thus for a constant heat flow down the tube of Fig. III.9(a) we would expect that if heat were transferred diffusively the temperature difference across the column would *drop* on warming through $T_c$ if only diffusive conductivity were effective, while in a suitable geometry the temperature difference would *increase* on warming through $T_c$ if heat flowed hydrodynamically at temperatures below $T_c$. What actually happens is shown in Fig. III.10 taken from work by Johnson *et al.*[(57)] This figure needs some detailed explanation since it is made up of almost raw experimental data obtained with an apparatus like

Fig. III.9(a) having a tube diameter of 3 mm. The temperatures $T_2^*$ and $T_1^*$ refer to the magnetic temperatures of, respectively, a small CMN thermometer at the top of the column and a large CMN thermometer also acting as thermal reservoir in the space below the column. Owing to shape and un-unknown factors these two thermometers do not read the same when they are at the same temperature. That is most of the reason why, when the externally applied $\dot{Q}$ is zero, the difference $T_2^* - T_1^*$ is not zero. This difference is, however, highly reproducible and only weakly dependent on temperature, so if one can get accustomed to a slowly changing baseline the difference in $T_2^* - T_1^*$ due to $\dot{Q}$ reflects directly the temperature difference due to heat flow.

It is obvious from Fig. III.10 that for a fixed $\dot{Q}$ the temperature difference across the column rises dramatically on warming through $T_c$, so there really is little doubt that heat flows hydrodynamically below $T_c$ and not just diffusively. This is a major qualitative result. A proof of hydrodynamic heat flow consists of showing that the hydrodynamic conductivity has the $d^2$ dependence of (III.31). A glance at Fig. III.10 shows that the level of heat flow that we need to produce the effect shown is very small. Indeed, it is not possible to turn off the heat flow down the column, and we have concluded that a residual heat flow of $0{\cdot}73 \times 10^{-3}$ erg/sec flows down the column even when the external $\dot{Q}$ is zero. It is this small flow that produces the characteristic temperature jump near $T_c^*$ for $\dot{Q} = 0$.

Measurement of thermal conductivity requires the precise measurement of two closely adjacent temperatures. We must measure magnetic temperatures, and we can measure precisely the difference $\Delta(T_2^* - T_1^*)$ produced by applying a heat flow $\dot{Q}$. We can define a thermal *resistance* on a magnetic scale by $R^* \equiv \Delta(T_2^* - T_1^*)/\dot{Q}$. but to find a thermal resistance that can be interpreted we need accurate values for $dT/dT^*$ where $T$ is absolute temperature. We simply do not have $dT/dT^*$ with sufficient accuracy. Until the problem of absolute thermometry is solved our most reliable alternative is to base our temperatures on some well-known property of the Fermi liquid, such as the low pressure thermal conductivity [see Lecture I, Fig. I.8] as measured by Abel *et al.*[62] So what is required is to measure the thermal resistance of low pressure normal Fermi liquid in the same apparatus using the same magnetic temperature scale. For the present measurements we chose a low pressure of 32 mm-Hg and obtained at this pressure thermal resistance $R_{32}^*$ as a function of magnetic temperature over the full range of our higher pressure measurements. If now for the same average temperature in the heat flow tube we form the ratio $R^*/R_{32}^*$ we can deduce thermal conductivities $\kappa$ via the formula

$$\kappa = (R^*/R_{32}^*)^{-1}\kappa_{32}(T), \tag{III.32}$$

where $\kappa_{32}(T)$ is the low pressure thermal conductivity extrapolated from

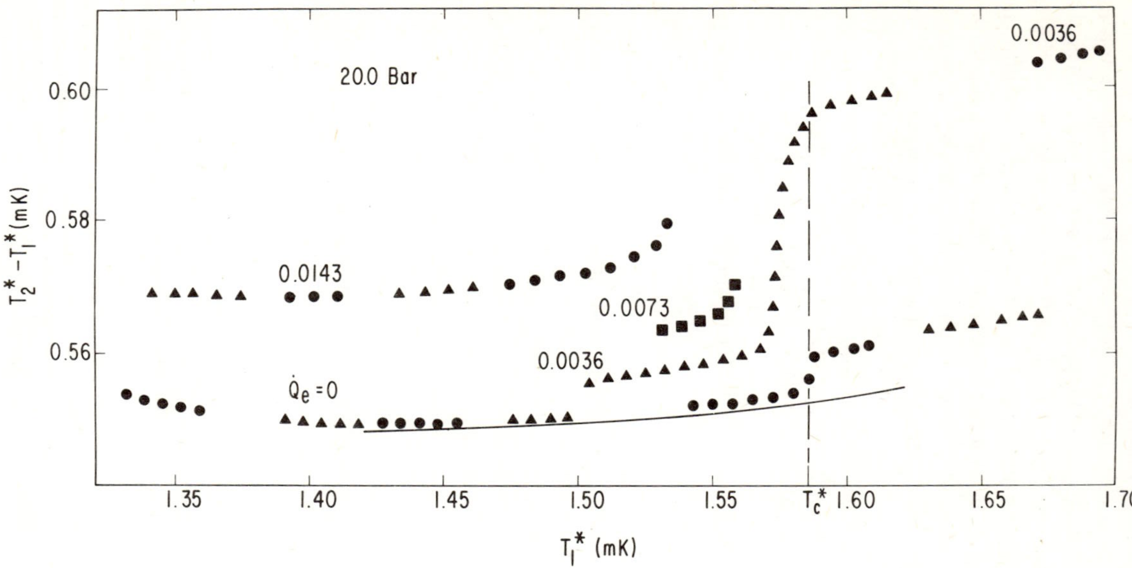


FIG. III.10. The indicated magnetic temperature difference $T_2^* - T_1^*$ across heat flow column as a function of the magnetic temperature $T_1^*$ of the larger (reservoir) thermometer. $T_c^*$ is the temperature of the centre of the thermal second order transition from $He^3$–B to normal liquid and $He^3$ pressure is 20·0 Bar. The different symbols correspond to different runs and the numbers refer to externally applied $\dot{Q}$ in erg/sec. All points at the bottom refer to zero applied $\dot{Q}$. The solid line is the empirically deduced dependence of $T_2^* - T_1^*$ on $T_1^*$ for zero total $\dot{Q}$, the effect of the residual heat flow down the column of $0{\cdot}73 \times 10^{-3}$ erg/sec having been accounted for.

data of Ref. 62. Using a provisional absolute scale (see appendix) based on zero sound attenuation (again a Fermi liquid scale) we can readily find $T/T_c$ from magnetic temperature and pressure data, so $\kappa_{32}(T)$ can be evaluated.

The above does seem to be a bit too much, especially for the student primarily interested in theory, but it does help to illustrate the present state of experimental science in the millidegree range.

Experimental results from Ref. 57 and from Greytak *et al.*[20] for pressures of 20·0 Bar (all B phase) and 29·6 Bar (both B and A phases and the transition B → A between the two) are shown in Fig. III.11, where $R^*/R^*_{32}$ is plotted

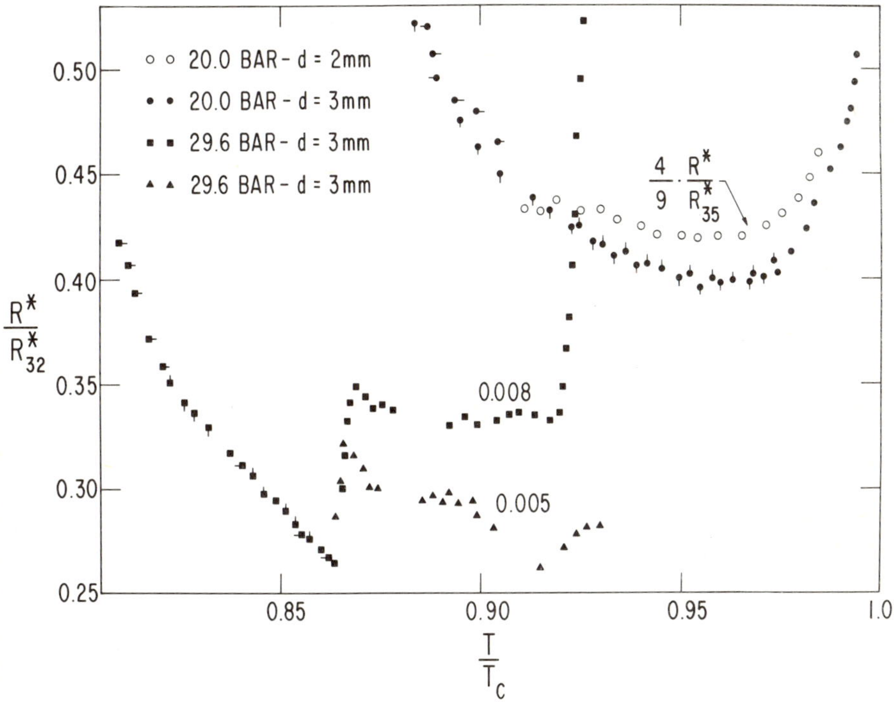


FIG. III.11. Ratio $R^*/R^*_{32}$ as a function of $T/T_c$ for He$^3$–B at 20·0 Bar [(●), 3 mm diam tube; (○), 2 mm diam tube] and at 29·6 Bar [(■), 3 mm diam tube] and for He$^3$–A at 29·6 Bar in a 3 mm diam tube [(■), $\dot{Q}_e$ = 0·008 erg/sec; (▲), $\dot{Q}_e$ = 0·005 erg/sec]. The flags on the points for He$^3$–B refer to the following $\dot{Q}_e$ in $10^{-3}$ erg/sec. Solid circles. to right, 28·6; up, 20·0; down, 14·3; to left, 7·1; no flag, 3·6. Solid squares: to right, 19·9; down, 10·0; up, 9·0; no flag, 8·0; left, 3·5.

as a function of $T/T_c$. These data show the principal characteristics of the two phases. Consider first the results for the B phase at 20 Bar in the upper right section of the figure. Data are available both for the 2mm diameter tube of Ref. 20 and the 3 mm diameter tube of Ref. 57. For *pure* hydrodynamic conductivity, according to (III.31) the thermal *resistance* for the 2 mm diameter tube should be $\frac{9}{4}$ as great as that for the 3 mm diameter tube; so we have multiplied $R^*/R^*_{35}$ for the 2 mm tube data by $\frac{4}{9}$ to make the points directly comparable. Results for the two tubes, so compared, are the same within the experimental accuracy. We regard this comparison as proof of hydrodynamic conductivity in $He^3$. There is the additional conclusion that diffusive conductivity does not contribute significantly to $R^*/R^*_{32}$, so it must drop rapidly below $T_c$ as might be expected from the behaviour of the viscosity.

At 29.6 Bar in the B phase, the thermal resistance is independent of $\dot{Q}$, but it is evident from Fig. III.11 for $\dot{Q} = 0{\cdot}008$ and $0{\cdot}005$ erg/sec that following the transition to the A phase the thermal resistance becomes dependent on $\dot{Q}$. The dependence of $R^*/R^*_{32}$ on $\dot{Q}$ at constant $T/T_c$ is displayed in Fig. III.12 where it is seen to have the form $R^*/R^*_{32} = a + b\dot{Q}$. This behaviour is observed unless some critical flow phenomenon has occurred. An example of the latter is also evident in Fig. III.11 for $\dot{Q} = 0{\cdot}008$ erg/sec at $T/T_c \approx 0{\cdot}92$, where the thermal resistance suddenly increases.

If thermal resistance at a fixed temperature depends on $\dot{Q}$ this may also be expressed by saying that the effective viscosity depends on the relative velocity $v$ of superfluid with respect to normal fluid. That is because thermal resistance at a fixed $T$ is proportional to $\eta$ (or better an effective viscosity $\bar{\eta}$) according to (III.31), while the velocity $v$ is given in terms of $\dot{Q}$ by the equation

$$v = \frac{\dot{Q}}{(\rho_s/\rho)\cdot TS\cdot\frac{1}{4}\pi d^2}. \tag{III.33}$$

This is immediately derived by combining (III.28) and (III.29) with $v = |\mathbf{v}_s - \bar{\mathbf{v}}_n|$ and $\rho_s + \bar{\rho}_n = \rho$. Now in an ordinary fluid if $\bar{\eta}$ depended on $v$ as in Fig. III.13 we would be inclined to postulate the existence of some dissipation means other than that of the viscosity in laminar flow. But in $He^3$–A, as discussed in Section II, it is possible that the fluid is anisotropic and that, following de Gennes and Rainer,[45] flow tends to order the orbital orientation vector **l** parallel to the tube axis rather than perpendicular to it. Furthermore, if the fluid is anisotropic then the density is probably a tensor quantity and the theory of our measurement is probably inadequate. Thus one is led to speculate that the effective $\eta$ depends on average orientation which in turn depends on flow rate. The concept is subject to test since geometry can be expected to influence strongly the $v$ dependence.

It was also possible in Refs. 20 and 57 to get some information on critical flow velocities. In the B phase it was found that nonlinear thermal resistance just started to occur for $v = 0{\cdot}4$–$0{\cdot}5$ cm/sec in a 3 mm tube. This is not very far from the value of critical velocity in $He^4$ for a similar size tube. In the A

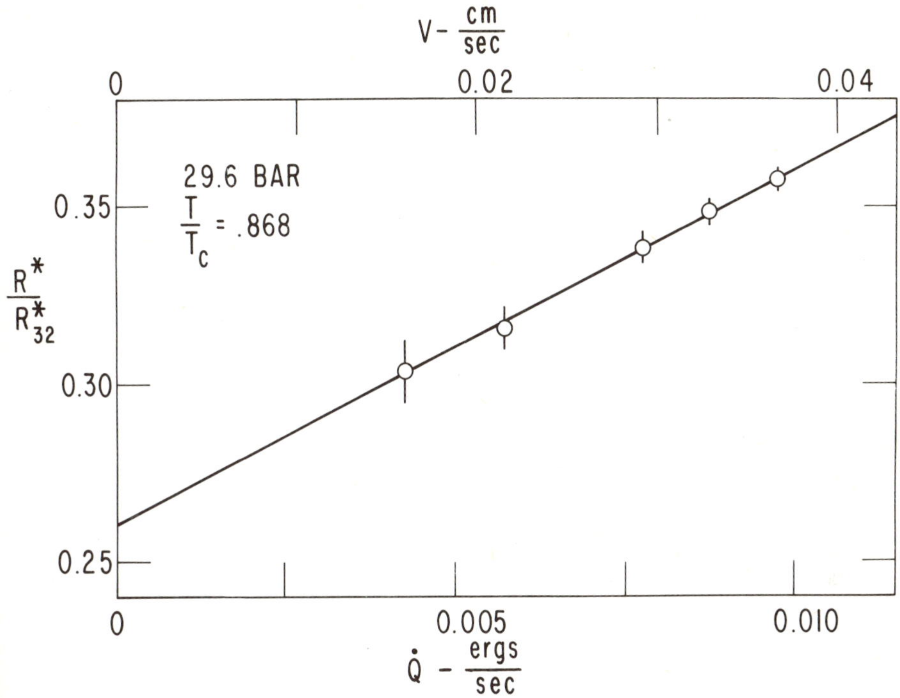


FIG. III.12. Dependence of $R^*/R^*_{32}$ on total $\dot{Q}$ in He$^3$–A at fixed $T/T_c = 0{\cdot}87$ and a pressure of 29·6 Bar. The flags on the points show the effect on $R^*/R^*_{32}$ of a $10^{-4}$ mK change in $\Delta(T_2^* - T_1^*)$. Also shown horizontally is the relative velocity between superfluid and normal fluid as calculated from Eqn (III.33).

phase there were a number of types of critical flow phenomena. An example of the most reproducible type is shown near $T/T_c = 0{\cdot}92$ for $\dot{Q} = 0{\cdot}008$ erg/sec. The onset velocity for this phenomenon was about 0·08 cm/sec for low $\dot{Q}$ near $T_c$ and 0·045 cm/sec for higher $\dot{Q}$ farther from $T_c$. But there were other more subtle phenomena with lower onset velocities. From time to time in a nonreproducible fashion while studying the reproducible nonlinear resistance of Fig. III.12 we found that, after turning $\dot{Q}$ to zero temporarily to establish baseline, on reapplication of the same $\dot{Q}$ the resistance would jump to a much higher value than expected. The onset velocities for such phenomena are as low as 0·02 cm/sec. Hence, critical flow in He$^3$–A is quite complicated. Unfortunately the experimental arrangement used in Refs. 20 and 57 is very poorly suited for critical velocity measurements.

## Appendix A

## Temperature Scales

An absolute temperature scale is essential to progress in the temperature region of the experiments described in this paper. Without such a scale it is not possible to compare experiments done in different laboratories or indeed within the same laboratory on different occasions. Comparison with theoretical concepts and calculations is likewise not possible, on a quantitative basis, without such a scale. In this appendix we present scales in current use.

### (a) *La Jolla*

This scale has been described by Wheatley.[(10)] It is based in part on unpublished noise thermometry measurements of Paulson *et al.*[(63)] of the absolute temperature of the "A" feature on the pressurization curve in a compressional cooling cell. It is assumed that this feature coincides with the intersection of the ($P_c$, $T_c$) second order line with the melting curve. A temperature of 2·6 mK was found for this feature, so for purposes of obtaining temperature differences precisely we took the ($P_c$, $T_c$) coordinates at melting pressure to be 34·36 Bar, 2·6000 mK.

Our temperature scale is based on a provisional determination of the ($P_c$, $T_c$) curve from 9·5 Bar to melting pressure which was obtained by assuming that in the normal Fermi liquid the attenuation of zero sound is proportional to the square of the absolute temperature.[(64, 13)] If the temperature of the "A" feature is different from 2·6000 mK then the whole scale would then be multiplied by a constant factor equal to ($T_A$/2·6000 mK). In Table I we give the coordinates ($P_c$, $T_c$) currently in use in our laboratory.

### (b) *Cornell melting pressure scale*

This scale, based on work by Halperin, Rasmussen, Archie, and Richardson,[(65)] has not yet been described in detail. However, one can surmise from a paper by Halperin, Archie, Rasmussen, Buhrman, and Richardson[(66)] that it is based on Clapeyron's equation in the form

$$T\frac{\mathrm{d}P}{\mathrm{d}T} = \frac{L}{\Delta V}, \tag{A.1}$$

where $T$ is absolute temperature, $P$ is pressure, $L$ is the latent heat of transformation from liquid to solid $He^3$ and $\Delta V$ is the corresponding change of volume. In their work $T(\mathrm{d}P/\mathrm{d}T)$ is measured along the melting curve by introducing a known amount of heat $L$ and measuring the volume change $\Delta V$

TABLE I. Coordinates ($P_c$, $T_c$) of second order line in zero field (La Jolla)

| $P$ (Bar) | $T_{sound}$ (mK) | $P$ (Bar) | $T_{sound}$ (mK) |
|---|---|---|---|
| 9·5 | 1·9646 | 22·5 | 2·4267 |
| 10·0 | 1·9917 | 23·0 | 2·4364 |
| 10·5 | 2·0171 | 23·5 | 2·4462 |
| 11·0 | 2·0411 | 24·0 | 2·4562 |
| 11·5 | 2·0636 | 24·5 | 2·4658 |
| 12·0 | 2·0853 | 25·0 | 2·4748 |
| 12·5 | 2·1069 | 25·5 | 2·4836 |
| 13·0 | 2·1272 | 26·0 | 2·4925 |
| 13·5 | 2·1471 | 26·5 | 2·5008 |
| 14·0 | 2·1671 | 27·0 | 2·5088 |
| 14·5 | 2·1864 | 27·5 | 2·5168 |
| 15·0 | 2·2051 | 28·0 | 2·5245 |
| 15·5 | 2·2201 | 28·5 | 2·5320 |
| 16·0 | 2·2395 | 29·0 | 2·5392 |
| 16·5 | 2·2564 | 29·5 | 2·5463 |
| 17·0 | 2·2730 | 30·0 | 2·5532 |
| 17·5 | 2·2893 | 30·5 | 2·5599 |
| 18·0 | 2·3054 | 31·0 | 2·5660 |
| 18·5 | 2·3210 | 31·5 | 2·5720 |
| 19·0 | 2·3365 | 32·0 | 2·5775 |
| 19·5 | 2·3517 | 32·5 | 2·5828 |
| 20·0 | 2·3665 | 33·0 | 2·5878 |
| 20·5 | 2·3808 | 33·5 | 2·5921 |
| 21·0 | 2·3930 | 34·0 | 2·5970 |
| 21·5 | 2·4045 | 34·36 | 2·6000 |
| 22·0 | 2·4161 | | |

needed to maintain the pressure and hence the temperature strictly constant. The ratio of $T$ to that for the "A" feature, for example, is then found by numerical integration to be

$$\frac{T}{T_A} = \exp\left[\int_{P_A}^{P} \left(T\frac{dP}{dT}\right)^{-1} dP\right]. \tag{A.2}$$

They find that the reduced temperature difference from $T_A$ is then given by

$$\frac{T_A - T}{T_A} = 1{\cdot}00 \times 10^{-2}\left(\frac{P - P_A}{1\ \text{mBar}}\right) + 2{\cdot}1 \times 10^{-5}\left(\frac{P - P_A}{1\ \text{mBar}}\right)^2. \tag{A.3}$$

The pressure difference $P_{B'} - P_A$ is 20·04 mBar for zero external field according to Osheroff's measurements[67] and 19·90 mBar according to Halperin's measurements.[68] If we take $T_A = 2{\cdot}6000$ mK as assumed in the La Jolla scale then for the average of these measurements we flnd, using Eqn (A.3), that $T_{B'} = 2{\cdot}059$ mK in zero field.

TABLE II. Experimental parameters for normal Fermi liquid $He^3$ as $T \to 0$.

| | $P$ (bar) | | | | | | | | | | | | |
|---|---|---|---|---|---|---|---|---|---|---|---|---|---|
| | 0 | 3 | 6 | 9 | 12 | 15 | 18 | 21 | 24 | 27 | 30 | 33 | 34·36 |
| $\frac{V}{n}$ (cm$^3$/mol) | 36·84 | 33·87 | 32·07 | 30·76 | 29·71 | 28·86 | 28·13 | 27·56 | 27·06 | 26·58 | 26·14 | 25·71 | 25·54 |
| $\frac{C}{nRT}$ (K$^{-1}$) | 3·00 | 3·22 | 3·43 | 3·61 | 3·78 | 3·95 | 4·10 | 4·24 | 4·37 | 4·51 | 4·64 | 4·78 | 4·85 |
| $c_1$ (m/s) | 182·9 | 227·5 | 259·7 | 285·9 | 308·0 | 327·1 | 345·0 | 360·5 | 375·1 | 389·3 | 403·0 | 415·9 | 421·7 |
| $T^*$ (K) | 0·359 | 0·305 | 0·277 | 0·256 | 0·238 | 0·224 | 0·212 | 0·205 | 0·198 | 0·191 | 0·185 | 0·179 | 0·177 |
| $\eta T^2$ (poise mK$^2$) | 1·834 | 1·73 | 1·63 | 1·54 | 1·46 | 1·38 | 1·30 | 1·22 | 1·14 | 1·06 | 0·99 | 0·92 | 0·88 |
| $\kappa T$ (erg/sec cm) | 34·8 | 29·7 | 25·8 | 22·7 | 20·3 | 18·4 | 16·7 | 15·3 | 14·0 | 12·9 | 11·9 | 11·0 | 10·7 |

## APPENDIX B

## PROPERTIES OF THE NORMAL FERMI LIQUID

In this appendix we give provisional properties of the normal Fermi liquid in the vicinity of $T = 0$. Smoothed and interpolated values of the principal experimental parameters as a function of pressure are given in Table II. Formulas for the computation of derived properties are given in Table III, and the corresponding derived quantities are given in Table IV.

For molar volumes we have used the tabulation of Wheatley[69] and the measurements of Grilly.[70] The most doubtful quantity at intermediate pressures is the limiting value of $(C/nRT)$, which is based on the measurements of Abel *et al.*[1] at 0·28 and 27 atm and the higher temperature data at intermediate pressures of Anderson *et al.*[2] New measurements over the full pressure range are badly needed. For the velocity of first sound we have

TABLE III. Formulas for the computation of derived Fermi liquid parameters.

$$m = 5{\cdot}009\times 10^{-24}\ \text{gm} \qquad \rho = \frac{3{\cdot}016\ \text{gm/mol}}{(V/n)}$$

$$2N(O) = \frac{(C/nRT)}{k(\pi^2/3)(V/N)} = \frac{(C/nRT)}{7{\cdot}5421\ V/n + 10^{-40}\ \text{erg/K mol}}$$

$$p_F = h\left(\frac{3}{8\pi}\frac{N}{V}\right)^{\frac{1}{3}} = 2{\cdot}7551\times 10^{-19}\,\frac{\text{gm cm}}{\text{sec}}\left(\frac{1\ \text{cm}^3}{\text{mol}}\right)^{\frac{1}{3}}(V/n)^{-\frac{1}{3}}$$

$$\frac{m^*}{m} = \frac{h^3}{8\pi m}\frac{2N(O)}{p_F} = 2{\cdot}3110\times 10^{-57}\,\frac{2N(O)}{p_F}\frac{\text{erg}^3\ \text{sec}^3}{\text{gm}}$$

$$v_F = \frac{p_F}{(m^*/m)m} = \frac{p_F}{m^*/m}\frac{1}{5{\cdot}009\times 10^{-24}\ \text{gm}}$$

$$F_0 = \frac{3m^2c_1^2}{p_F^2}\frac{m^*}{m} - 1 = \left[7{\cdot}5270\times 10^{-47}\ \text{gm}^2\,\frac{c_1^2}{p_F^2}\left(\frac{m^*}{m}\right)\right] - 1$$

$$F_1 = 3\left(\frac{m^*}{m} - 1\right)$$

$$Z_0 = 4\left[\frac{3kT^*(m^*/m)m}{p_F^2} - 1\right] = 4\left[2{\cdot}0746\times 10^{-39}\,\frac{\text{erg gm}}{\text{K}}\frac{T^*(m^*/m)}{p_F^2} - 1\right]$$

$$\tau_\eta T^2 = \frac{\eta T^2}{\frac{1}{5}\rho v_F^2(m^*/m)} = \frac{(\eta T)(V/n)}{0{\cdot}6032\ \text{gm/mol}\ v_F^2(m^*/m)}$$

$$\tau_\kappa T^2 = \frac{\kappa T}{\frac{1}{3}(C_V/T)v_F^2} = \frac{(\kappa T)(V/n)}{\frac{1}{3}R(C/nRT)v_F^2} = \frac{(\kappa T)(V/n)}{2{\cdot}7714\times 10^7\ \text{erg/mol K}\ (C/nRT)v_F^2}$$

$$\chi_N = \frac{1{\cdot}362\times 10^{-8}\ \text{K}}{T^*}\frac{37{\cdot}0\ \text{cm}^3/\text{mol}}{(V/n)}$$

TABLE IV. Derived quantities for the normal Fermi liquid.

| | $P$ (bar) | | | | | | | | | | | | |
|---|---|---|---|---|---|---|---|---|---|---|---|---|---|
| | 0 | 3 | 6 | 9 | 12 | 15 | 18 | 21 | 24 | 27 | 30 | 33 | 34·36 |
| $2N(O)$ [$10^{38}$ (erg cm$^3$)$^{-1}$] | 1·08 | 1·26 | 1·42 | 1·56 | 1·69 | 1·82 | 1·93 | 2·04 | 2·14 | 2·25 | 2·35 | 2·46 | 2·52 |
| $p_F$ ($10^{-20}$ g cm/s) | 8·28 | 8·52 | 8·67 | 8·79 | 8·89 | 8·98 | 9·06 | 9·12 | 9·18 | 9·23 | 9·28 | 9·34 | 9·36 |
| $v_F$ ($10^3$ cm/s) | 5·48 | 4·97 | 4·58 | 4·29 | 4·05 | 3·84 | 3·67 | 3·52 | 3·40 | 3·27 | 3·16 | 3·05 | 3·00 |
| $m^*/m$ | 3·01 | 3·42 | 3·78 | 4·09 | 4·38 | 4·67 | 4·93 | 5·17 | 5·39 | 5·63 | 5·86 | 6·10 | 6·22 |
| $F_0$ | 10·07 | 17·39 | 24·51 | 31·55 | 38·58 | 45·62 | 52·83 | 59·78 | 66·81 | 74·38 | 82·13 | 90·17 | 94·13 |
| $F_1$ | 6·04 | 7·27 | 8·34 | 9·27 | 10·15 | 11·01 | 11·79 | 12·51 | 13·18 | 13·90 | 14·58 | 15·31 | 15·66 |
| $z_0$ | −2·690 | −2·805 | −2·845 | −2·876 | −2·902 | −2·923 | −2·940 | −2·943 | −2·949 | −2·953 | −2·955 | −2·958 | −2·955 |
| $z_1$ | −2·68 | −2·33 | −2·28 | −2·21 | −2·13 | −2·06 | −2·00 | −2·05 | −2·06 | −2·09 | −2·13 | −2·16 | −2·22 |
| $\tau_n T^2$ ($10^{-6}$ s mK$^2$) | 1·24 | 1·15 | 1·09 | 1·00 | 0·96 | 0·96 | 0·91 | 0·87 | 0·82 | 0·78 | 0·73 | 9·69 | 0·66 |
| $\tau_\kappa T^2$ ($10^{-6}$ s mK$^2$) | 0·51 | 0·46 | 0·41 | 0·38 | 0·35 | 0·33 | 0·31 | 0·29 | 0·27 | 0·26 | 0·24 | 0·23 | 0·22 |
| $\chi_N$ ($10^{-8}$) | 3·81 | 4·88 | 5·67 | 6·38 | 7·99 | 7·75 | 8·38 | 8·92 | 9·42 | 9·91 | 10·42 | 10·93 | 11·17 |

used the recent unpublished low temperature measurements of Paulson *et al.*,[71] early low temperature velocity measurements of Abel *et al.*,[72] and more recent measurements of Abraham *et al.*[73] For the magnetic temperature $T^*$ we have used the limiting value as $T \to 0$ of $C/\chi$, where $C$ is the curie constant and $\chi$ the susceptibility, as given by Ramm *et al.*[4] The limiting $T \to 0$ values of $\eta T^2$ were obtained from low pressure (Bertinat *et al.*[74] and melting pressure (Alvesalo *et al.*[8]) wire damping measurements and from the ultrasonic attenuation measurements of Paulson *et al.*[72] with heavy weight placed on the wire damping measurements. The limiting $T \to 0$ values of $\kappa T$ are based on the zero pressure measurement of Abel *et al.*[62] and the measurements of thermal resistance ratios under pressure of Greytak *et al.*[75] Significant parts of the Tables are based on tabulations worked out by Paulson.[76] Derivation of Fermi liquid parameters from experimental data, including a means of computing $Z_1$, is discussed by Wheatley.[69]

## Acknowledgements

I take this opportunity to thank E. E. Pichelmann, who prepared most of the figures and helped in preparation of the Tables; and R. T. Johnson, R. L. Kleinberg, and R. A. Webb, who gave me their critical appraisal of the text. I thank them and, as well T. J. Greytak, H. Kojima and D. N. Paulson for their contributions to the experimental work on which these lectures are largely based. My physical insight into the nature of the experiments has been greatly helped by discussions and correspondence with W. F. Brinkman, A. L. Fetter, A. J. Leggett, K. Maki, B. R. Patton, and S. Putterman. I am grateful to O. V. Lounasmaa for sending me results, prior to their publication, on the flow properties of $He^3$ based on the damping and resonance frequency of a vibrating wire.

## References

1. W. R. Abel, A. C. Anderson, W. C. Black and J. C. Wheatley. *Phys. Rev.* **147**, 111 (1966).
2. A. C. Anderson, W. Reese and J. C. Wheatley. *Phys. Rev.* **130**, 495 (1963).
3. R. A. Webb, R. P. Giffard and J. C. Wheatley. *J. Low Temp. Phys.* **13**, 383 (1973).
4. H. Ramm, P. Pedroni, J. R. Thompson and H. Meyer. *J. Low Temp. Phys.* **2** 539 (1970).
5. L. D. Landau. *Zh. Eksperim. i Teor. Fiz.* **30**, 1058 (1956); [translation: *Sov. Phys. JETP* 3, 920 (1957)].
6. A. J. Leggett. *Phys. Rev.* **140A**, 1869 (1965).
7. J. C. Wheatley. *In* "Progress in Low Temperature Physics" Vol. VI (C. J. Gorter, Ed.), North-Holland Publishing Co., Amsterdam (1970), p. 77.

8. T. A. Alvesalo, H. K. Collan, M. T. Loponen, O. V. Lounasmaa and M. C. Veuro. *J. Low Temp. Phys.* (to be published).
9. M. J. Rice. *Phys. Rev.* **159**, 153 (1967).
10. J. C. Wheatley. *Physica* **69**, 218 (1973).
11. A. I. Ahonen, M. T. Haikala, M. Krusius and O. V. Lounasmaa (to be published).
12. L. D. Landau. *Zh. Eksperim. i Teor. Fiz.* **32**, 59 (1957); [*Sov. Phys. JETP* **5**, 101 (1957)].
13. W. R. Abel, A. C. Anderson and J. C. Wheatley. *Phys. Rev. Lett.* **17**, 74 (1966).
14. D. N. Paulson, R. T. Johnson and J. C. Wheatley. *Phys. Rev. Lett.* **30**, 829 (1973).
15. J. F. Allen and J. Jones. *Nature* **141**, 243 (1938).
16. D. N. Paulson, H. Kojima and J. C. Wheatley. *Phys. Rev. Lett.* **32**, 1098 (1974).
17. R. A. Webb, T. J. Greytak, R. T. Johnson and J. C. Wheatley. *Phys. Rev. Lett.* **30**, 210 (1973).
18. W. J. Gully, D. D. Osheroff, D. T. Lawson, R. C. Richardson and D. M. Lee. *Phys. Rev.* **A8**, 1633 (1973).
19. D. N. Paulson, H. Kojima and J. C. Wheatley. *Phys. Letters* **47A**, 457 (1974).
20. T. J. Greytak, R. T. Johnson, D. N. Paulson and J. C. Wheatley. *Phys. Rev. Lett.* **31**, 452 (1973).
21. D. N. Paulson, R. T. Johnson and J. C. Wheatley. *Phys. Rev. Lett.* **31**, 746 (1973).
22. R. L. Kleinberg, D. N. Paulson, R. A. Webb and J. C. Wheatley. *J. Low Temp. Phys.* (to be published).
23. R. A. Webb, R. L. Kleinberg and J. C. Wheatley. *Phys. Rev. Lett.* **33**, 145 (1974).
24. A. I. Ahonen, M. T. Haikala and M. Krusius. *Phys. Lett.* **47A**, 215 (1974).
25. Yu. D. Anufriyev, T. A. Alvesalo, H. K. Collan, N. T. Opheim and P. Wennerström. *Phys. Letters* **43A**, 175 (1973).
26. W. P. Halperin, R. A. Buhrman and R. C. Richardson. *Bull. Am. Phys. Soc.* **18**, 642 (1973).
27. H. Kojima, D. N. Paulson and J. C. Wheatley. *Phys. Rev. Lett.* **32**, 141 (1974).
28. B. R. Patton. *Phys. Letters* **47A**, 459 (1974).
29. T. A. Alvesalo, Yu. D. Anufriyev, H. K. Collan, O. V. Lounasmaa and P. Wennerström. *Phys. Rev. Lett.* **30**, 962 (1973).
30. K. A. Shapiro and I. Rudnick. *Phys. Rev.* **137A**, 1383 (1965).
31. L. P. Pitaevskii. *Zh. Eksperim. i Teor. Fiz.* **37**, 1794 (1959) [translation: *Soviet Phys. JETP* **10**, 1267 (1960)].
32. K. A. Brueckner, T. Soda, P. W. Anderson and P. Morel. *Phys. Rev.* **118**, 1442 (1960).
33. V. J. Emery and A. M. Sessler. *Phys. Rev.* **119**, 43 (1960).
34. D. J. Thouless. *Ann. Phys.* **10**, 553 (1960).
35. W. M. Saslow. *Phys. Rev. Letts* **31**, 870 (1973).
36. V. Ambegaokar, P. G. de Gennes and D. Rainer. *Phys. Rev.* **A9**, 2676 (1974).
37. A. E. Glassgold and A. M. Sessler. *Nuovo Cimento* **19**, 723 (1961).
38. P. W. Anderson and W. F. Brinkman. *Phys. Rev. Letts* **30**, 1108 (1973).
39. R. Balian and N. R. Werthamer. *Phys. Rev.* **131**, 1553 (1963).
40. R. Combescot, preprint.
41. D. D. Osheroff and W. F. Brinkman, *Phys. Rev. Letts* **32**, 584 (1974).
42. W. F. Brinkman, H. Smith, D. D. Osheroff and E. I. Blount. Preprint.
43. T. A. Alvesalo, H. K. Collan, M. T. Loponen and M. C. Veuro. *Phys. Rev. Letts* **32**, 981 (1974).
44. P. G. de Gennes. *Phys. Letters* **44A**, 271 (1973).
45. P. G. de Gennes and D. Rainer. *Phys. Letters* **46A**, 429 (1974).

46. A. J. Leggett. *Phys. Rev. Letts* **31**, 352 (1973).
47. A. J. Leggett. *J. Phys. C, Solid State Physics* **6**, 3187 (1973).
48. A. J. Leggett. Preprint.
49. R. P. Giffard, R. A. Webb and J. C. Wheatley. *J. Low Temp. Phys.* **6**, 533 (1972).
50. D. D. Osheroff and W. F. Brinkman. *Phys. Rev. Letts* **32**, 584 (1974).
51. H. M. Bozler, M. E. R. Bernier, W. J. Gully, R. C. Richardson and D. M. Lee. *Phys. Rev. Letts* **32**, 875 (1974).
52. K. Maki and T. Tsuneto. *Prog. Theor. Phys.* (in press).
53. R. A. Webb, R. L. Kleinberg and J. C. Wheatley. *Phys. Letters* (in press).
54. D. T. Lawson, W. J. Gully, S. Goldstein, R. C. Richardson and D. M. Lee. *Phys. Rev. Letts* **30**, 541 (1973).
55. V. Ambegaokar and N. D. Mermin. *Phys. Rev. Letts* **30**, 81 (1973).
56. M. Vuorio. *J. Phys. C, Solid State Physics* **7**, L5 (1974).
57. R. T. Johnson, R. L. Kleinberg, R. A. Webb and J. C. Wheatley. *J. Low Temp. Phys.* (to be published).
58. D. F. Brewer and D. O. Edwards. *Proc. Roy. Soc.* **A251**, 247 (1959).
59. D. F. Brewer and D. O. Edwards. *Phil. Mag.* **6**, 775 (1961).
60. F. London and P. R. Zilsel. *Phys. Rev.* **74**, 1148 (1948).
61. C. J. Gorter and J. H. Mellink. *Physica* **15**, 285 (1949).
62. W. R. Abel, R. T. Johnson, J. C. Wheatley and W. Zimmermann. *Phys. Rev. Letts* **18**, 737 (1967).
63 D. N. Paulson, R. T. Johnson and J. C. Wheatley. (Unpublished).
64. A. A. Abrikosov and I. M. Khalatnikov. *Zh. Eksperim. i Teor. Fiz.* **33**, 110 (1957); [*Sov. Phys. JETP* **6**, 84 (1958)].
65. W. P. Halperin, F. Rasmussen, C. N. Archie and R. C. Richardson. (Unpublished).
66. W. P. Halperin, C. N. Archie, F. B. Rasmussen, R. A. Buhrman and R. C. Richardson. *Phys. Rev. Letts* **32**, 927 (1974).
67. D. D. Osheroff. Private communication.
68. W. P. Halperin. Private communication to Osheroff.
69. J. C. Wheatley. *In* "Quantum Fluids" (D. F. Brewer, Ed.), North-Holland Publishing Co., Amsterdam (1966), p. 183.
70. E. R. Grilly. *J. Low Temp. Phys.* **4**, 615 (1971).
71. D. N. Paulson, R. T. Johnson and J. C. Wheatley. (Unpublished).
72. W. R. Abel, A. C. Anderson and J. C. Wheatley. *Phys. Rev. Letts* **7**, 299 (1961).
73. B. M. Abraham, D. Chung, Y. Eckstein, J. B. Ketterson and P. R. Roach. *J. Low Temp. Phys.* **6**, 521 (1972).
74. M. P. Bertinat, D. S. Betts, D. F. Brewer, G. J. Butterworth. Preprint.
75. T. J. Greytak, R. T. Johnson, D. N. Paulson and J. C. Wheatley. (Unpublished).
76. D. N. Paulson. Thesis, University of California (San Deigo), 1974. (Unpublished).

# 8. THEORY OF ANISOTROPIC SUPERFLUIDITY IN $He^3$

P. W. Anderson

*Bell Laboratories, Murray Hill, New Jersey 07974, USA*
*also*
*Cavendish Laboratory, Cambridge, England*
*and*

W. F. Brinkman

*Bell Laboratories, Holmdel, New Jersey 07733, USA*
*also*
*H. C. Ørsted Institute, Copenhagen, Denmark*

## I. Historical Introduction

### 1. PREFACE

From a microscopic point of view helium is the simplest of all condensed substances: the atoms may be treated as structureless particles (except for the convenient nuclear spin of $\frac{1}{2}$ carried by $He^3$) interacting via an interatomic potential which is quite accurately known. Nonetheless both solid and liquid phases of both isotopes exhibit properties which seem as complex as those of any inorganic system, and which in all cases are beyond the present capacity of many-body theory to calculate quantitatively, and in many cases to understand except in qualitative terms. Of all these varied phenomena, perhaps the most challenging intellectually and most complex phenomenologically are the recently discovered anisotropic superfluid phases of $He^3$. These phases are superfluid liquid crystals exhibiting ferromagnetism; their nuclear magnetic resonance behaviour has its closest analogue in antiferromagnetism; they exhibit the totally unique phenomenon of orbital supercurrents, as well as various new kinds of structural singularity, such as the disgyrations of de Gennes and the point defects proposed by Brinkman. It is the task of the theorist to show how Nature can construct, using the basic principle of broken symmetry, all this richness of phenomenology out of such a totally simple microscopic Hamiltonian.

Why should the physicist go to the very considerable trouble necessary to study in detail this minimicrocosm, all of which occurs well below 3 mK? Partly, of course, for the challenge: "it's there". This study stretches to their absolute limits his intellectual and experimental resources. More, perhaps, the motivation must be the expansion of the human consciousness. But one can by no means assume that at least the theoretical lessons learned will be of no value elsewhere in science. The understanding of superconductivity introduced new perspectives into the theory of the elementary particles, for instance, and we have no way of knowing whether this far more complex case of broken symmetry may not be of even greater value. Certainly, as well, advances in our capability of understanding Fermi systems can be important both to nuclear and to electronic physics.

## 2. EARLY (PRE-1972) HISTORY

In spite of uncertainties and difficulties which remain even today, the fact that a fair amount of still useful theoretical work was done on this system in the years 1959–1970, while the experimental discovery dates from 1972, correctly suggests a remarkable triumph of theoretical many-body physics. Theory, in the absence of experiment, often went off down blind alleys;† but also many of the most important phenomena were first suggested from the theoretical side.

Almost immediately after the BCS theory[(1)] several authors proposed anisotropic generalizations of the BCS phenomena. (Thouless[(2)] and Fisher[(3)] may well not be a complete list.) But it was not until the general realization that $He^3$ is a genuine Fermi liquid, which realization took place around 1958–1960, that these generalizations were taken seriously. Pitaevskii[(4)] was perhaps the first to propose an anisotropic BCS state for $He^3$, followed very shortly by Brueckner, Soda, Anderson and Morel[(5)] and by Emery and Sessler.[(6)]

Cooper, Mills and Sessler[(7)] had previously attempted unsuccessfully to find a conventional BCS transition in $He^3$. The reason for their failure was the nature of the $He^3$ interparticle potential (see Fig. 1). The repulsive hard core cannot be renormalized out completely enough, apparently, so that the $l = 0$ isotropic interaction potential at the Fermi surface is almost cer-

† We will try not to refer to papers which seem to have made no contribution to the advancement of the subject. We assure the reader that any impression of steady progress in the positive direction which he obtains is to some extent a result of selection.

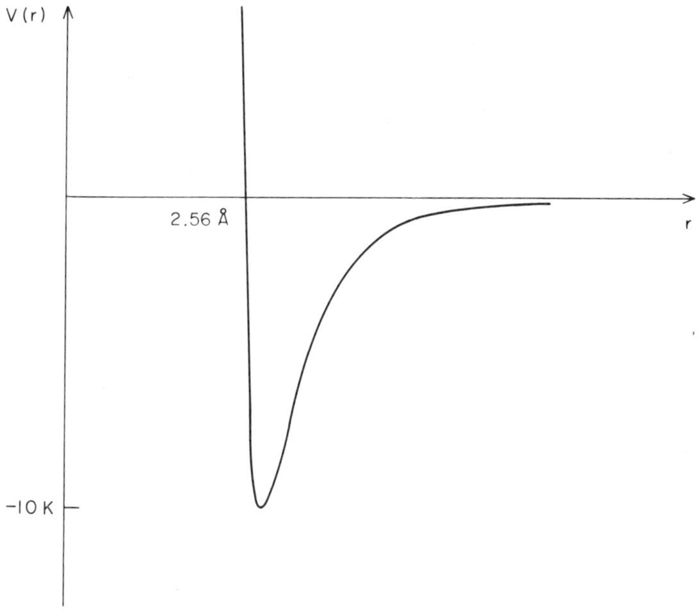


FIG. 1. A plot of the interparticle potential as a function of separation.

tainly repulsive and can lead to no pair condensation. The reason for trying $l \neq 0$ was in order to avoid this hard core. All of these calculations really amounted to rather naively introducing the bare interatomic potential

$$V(\mathbf{k}, \mathbf{k}') = \int \exp\left[i(\mathbf{k} - \mathbf{k}')\cdot\mathbf{r}\right] \mathrm{d}^3 r \, V(r) \tag{I.1}$$

into the straightforward BCS transition temperature equation

$$\Delta(\mathbf{k}) = \sum_{\mathbf{k}'} \frac{V(\mathbf{k}, \mathbf{k}')\Delta(\mathbf{k}')}{\varepsilon(\mathbf{k}')} \tanh \beta_c \varepsilon(\mathbf{k}'). \tag{I.2}$$

Then one looks for a solution

$$\Delta(\mathbf{k}) = \Delta_0(\varepsilon_{\mathbf{k}}) Y^l_m(\hat{\mathbf{k}}) \tag{I.3}$$

having the symmetry of a spherical harmonic of order $l$. Since (I.2) is linear in $\Delta$ and $V$ is rotationally symmetric, $V$ may be separated out into its spherical harmonic components

$$V(\mathbf{k}, \mathbf{k}') = \sum_l \overline{V}_l \tag{I.4}$$

where

$$\begin{aligned}\bar{V}_l &= V_l(k, k')P_l(\hat{\mathbf{k}}\cdot\hat{\mathbf{k}})(2l+1)\\ &= V_l(k, k')\sum_m Y^l_m(\hat{\mathbf{k}}')\,Y^l_{-m}(\hat{\mathbf{k}}')\,(4\pi).\end{aligned}\tag{I.5}$$

Inserting (I.4) and (I.5) into (I.2) separates that equation into independent equations for each spherical harmonic $l$. Since the coefficients $V_l(k, k')$ are spherical harmonic averages themselves,

$$V_l(k, k') = \langle V(\mathbf{k}, \mathbf{k}')P_l(\hat{\mathbf{k}}\cdot\hat{\mathbf{k}}')\rangle,$$

the hard core, which is represented by a positive constant rather uniformly distributed in $k$-space, will tend to average out for higher $l$'s. Indeed, for $l = 1$ a solution was problematic, while for $l = 2$ and 3, $V_l(k, k')$ was definitely attractive (negative).

Such a procedure was clearly too naive, since, as became increasingly clear as the Fermi-liquid data on $He^3$ came in (see Abel *et al.*[(8)] for instance) the interactions are so strong that all the properties are renormalized by factors considerably greater than unity. That is, the specific heat, $\gamma T$, at the melting pressure is approximately six times that of the free Fermi gas, the susceptibility about 24 times, etc. We would expect interactions to be renormalized—a generalized screening—by factors of the same order, and with $T_c = \beta_c^{-1}$ depending exponentially on the coupling constant according to the standard formula $T_c \propto \exp[-N(0)V_l(k_F, k_F)]^{-1}$, the $T_c$ values were merely wild guesses. Nonetheless this parameter is the most vital one to experimentalists and the point that the range of possible values included experimentally accessible ones was important to make.

The first papers seriously attempting a description of what the resulting state might be like were by Anderson and Morel.[(9, 10)] The first of these called attention to two aspects of these phases which have turned out to be of importance. These were, first, that a number of inequivalent solutions belonging to a single $l$ existed, and that these different phases had to be tested for relative stability: different spherical harmonics led to inequivalent solutions. Second, they observed that a favourable type of solution in many cases might be complex, exhibiting internal orbital currents: as they named it, "orbital ferromagnetism", or aligned orbital angular momentum.

The reasoning behind these statements becomes clear if we look at the BCS gap equation below $T_c$—say, at absolute zero. As Anderson and Morel showed, so long as the coupling is reasonably weak (i.e. $\Delta$ considerably less than $E_F$) different $l$'s continue to separate to a good approximation, so we may write with good accuracy (we omit spin considerations for the time being)

$$\Delta_l(\mathbf{k}) = -\sum_{\mathbf{k}'}\frac{V_l(k, k')P_l(\hat{\mathbf{k}}\cdot\hat{\mathbf{k}}')\Delta_l(\mathbf{k}')}{E(\mathbf{k}')},$$

where

$$E^2(\mathbf{k}') = |\Delta(\mathbf{k}')|^2 + \varepsilon_{\mathbf{k}'}^2.$$

Because of the dependence of $E$ on $\Delta$, this is a nonlinear equation, and therefore different $Y_m^l$'s will give different solutions. For instance, for $l = 1$ either $\Delta \propto \cos\theta$ or $\Delta \propto \sin\theta e^{i\varphi}$ will be solutions, but they have different energies. It is easy to see intuitively that this simple weak-coupling theory favours the most uniformly distributed $|\Delta|^2$ possible. Thus in fact of the two solutions above $\sin\theta e^{i\varphi} = \hat{k}_x + i\hat{k}_y$ is favoured: the complex one. While the various solutions are inequivalent, they must all have the same $T_c$, since the equation for $T_c$ is a linear one.

Anderson and Morel borrowed from the BCS paper a device which is of great value in visualizing these states. This is the idea of the extra correlation or "correlation hump" caused by the condensation. This "correlation hump" is defined as

$$\Delta\langle nn(\mathbf{r})\rangle = \langle n(\mathbf{0})n(\mathbf{r})\rangle - \langle n\rangle^2.$$

$\Delta\langle nn\rangle$ may be calculated from a quality $b(\mathbf{r})$ which takes the formal place of a wave function of condensed pairs and is given by

$$b(\mathbf{r}) = \int d^3k \exp(i\mathbf{k}\cdot\mathbf{r})\frac{\Delta(\mathbf{k})}{E(\mathbf{k})}.$$

Then

$$\Delta\langle nn\rangle = |b(\mathbf{r})|^2.$$

Clearly, if $\Delta(\mathbf{k})$ has a certain spherical harmonic form $\Delta = \Delta(k)Y^l(\hat{\mathbf{k}})$, the "wave function" $b$ will very nearly be proportional to the same harmonic.

One could imagine a gas of $(He^3)_2$ molecules, in an orbital state of given $L$, undergoing a Bose condensation. Such a Bose-condensed gas of molecules would also have a "correlation hump" since each atom would be correlated with its molecular partner. The total magnitude of that correlation hump would be almost precisely unity:

$$\int [\langle n(\mathbf{0})n(\mathbf{r})\rangle - n^2]_{B-E\text{ gas}}\, d^3r \simeq 1 - na^3,$$

where $n$ is the density of molecules and $a$ the molecular diameter. In contrast, the "correlation hump" of a BCS condensed Fermi gas is very small:

$$\int d^3r[\langle n(\mathbf{0})n(\mathbf{r})\rangle - n^2] \simeq \frac{\Delta}{E_F} \ll 1.$$

This is the basic difference between the two cases: in this sense they belong to the same continuum but are opposite extreme cases. It is often useful in thinking about the system to keep the molecular picture in mind.

If a particular solution $\Delta(\mathbf{k})$ is a complex spherical harmonic such as $\hat{k}_x + i\hat{k}_y$, the resulting $b$ will be complex and, in this case, the corresponding molecular state could have a net angular momentum $M_L^z = +1$. In precisely the same way the correlation hump will have a number-current correlation calculated, again, as though $b$ were a wave function:

$$\langle n(\mathbf{0})\mathbf{j}(\mathbf{r})\rangle = \frac{\hbar}{m}(b^*\nabla b - b\nabla b^*).$$

In the given case, this will lead to a net orbital angular momentum of the sample, but, as Anderson and Morel showed, one may have $\langle n\mathbf{j}\rangle$ correlation without net angular momentum.

A word should be said about spin: just as for the molecule, a symmetric orbital function implies a spin singlet, an antisymmetric one a triplet. For even $l$, then, the spin behaviour will be just as in BCS, but for odd $l$ there will be triplet pairing. Anderson and Morel considered only parallel spin pairing, ignoring the third ($\uparrow\downarrow + \downarrow\uparrow$) component of the triplet. This error unfortunately has a qualitative effect on some results, as we shall see. All of the theory we have so far discussed assumes a single spin wave function, either up and down spins independently paired parallel, and $l$ odd, or BCS type singlet pairing as one always has for $l$ even. The magnetic behaviour in either case is simple: for odd $l$ the spin susceptibility of independently paired up and down spin systems is unchanged, since the relative Fermi levels may shift at will. For even $l$ a curve similar to Yosida's[11] as calculated for BCS is correct, giving a $\chi$ which vanishes as $T \to 0$ because shifting the Fermi surfaces relatively breaks up the $\uparrow\downarrow$ pairs.

In their final paper Anderson and Morel worked out a number of results with some care. Unfortunately, the greatest effort went into the case $l = 2$ which, at the time, was considered to be the most favourable.

They made the point that the higher $l$ states would be superfluid even though the energy gap vanishes at several points on the Fermi surface. This was the first example of "gapless superfluidity" belying the Landau criterion for superfluidity and showing that the real criterion is the existence of the anomalous self-energy $\Delta$. They gave the presently used formula for superfluid density in weak coupling and observed that in many cases this would be anisotropic. A final remark which is relevant to the modern theory is that anisotropic BCS states would be very sensitive to all forms of scattering—impurities, walls, etc., since all such would be "pair-breaking", in modern terms, like magnetic scattering in superconductors.

A few years later Balian and Werthamer[12] gave the first proper treatment of all spin components and worked out what is still the accepted formalism of the weak-coupling theory. Since they were concerned not with liquid He$^3$, primarily, but (despite developing the remark about scattering made by Anderson and Morel) to try to explain discrepancies in the spin susceptibility of normal superconductors in terms of an $s = 1$ state, they concentrated on the $l = 1$, $s = 1$ case, fortunately for the later development. (The singlet case is not modified, of course.) Since we will have occasion to use the BW (Balian–Werthamer) formalism at length in the body of this review, we shall not present it in detail here. Suffice it to say that upon including the third $(\uparrow\downarrow + \downarrow\uparrow)$ pairing component, one has available the possibility of using each of the three $l = 1$ spherical harmonics $m_l = \pm 1, 0$ (or $k_x, k_y, k_z$) with a different $s = 1$ spin state. One thereby can derive a pseudoisotropic, essentially real state with properties very like the standard $l = 0$ BCS state. In weak coupling this is clearly preferable to the complex AM (Anderson–Morel) state and apparently scotches the chance for orbital ferromagnetism in real systems. The BW formalism for the spin susceptibility shows that this state follows an intermediate course, extrapolating to $\chi_S/\chi_N = \frac{2}{3}$ at $T = 0$.

A final major early advance in discussing the properties of these hypothetical states was made by Leggett(13, 14) in 1965. Leggett observed that even if the transition temperature were so low as to validate the weak-coupling assumption so far made, at the very least Landau Fermi-liquid-parameter corrections would have to be made to the properties—such as $\chi_S/\chi_N$, $\rho_S$, etc.—of the anisotropic state. His results again will be used here repeatedly and thus will be treated in more detail later. Therefore we quote only the most striking ones: that the exchange correction to the spin susceptibility would bring the BW value down to $\frac{1}{3}$ or so at $T = 0$, and that the temperature dependence of the superfluid density is drastically changed because of large mass enhancement.

In the meantime, a new and fruitful line of discussion of the problem of the transition temperature was opening up. Emery(15) appears to have been the first to remark that magnetic renormalization effects might very appreciably favour odd-$l$, triplet states *vis-a-vis* even-$l$, singlet ones. The quantitative working out of this idea stemmed from the application by Berk and Schrieffer(16) of spin-fluctuation theory to the problem of superconducting transition temperatures.

The spin-fluctuation theory applies to Fermi systems of the type known as "nearly magnetic" in which the renormalization of the spin susceptibility by exchange effects is large (Pd, Pt, Ni among the metals; and, as we have observed, He$^3$). In such materials the spectrum of single-particle pairs has an enhanced low-frequency, long-wavelength region which can be thought of as a spectrum of a non-propagating type of collective excitation, the "spin fluctuation" or paramagnon. Berk and Schrieffer showed that the exchange of such fluctuations leads to a repulsive effective potential between the members of a singlet pair, and thus depresses the superconducting $T_c$. Although quantitatively this theory is never reliable, this and another qualitative result—that exchange of spin fluctuations enhances the linear specific heat $\gamma T$ and leads to thermal anomalies—are well borne out by experiment. One can think of the depression of $T_c$ in another way, which is equally valid and gives one a good feeling for the physics: that the strong reduction which singlet BCS causes in the spin susceptibility makes the propagation of paramagnons more difficult, and reduces the paramagnon free energy. This interference of the two disfavours BCS superconductivity. Applied to $He^3$ by Brinkman and Engelsberg,[17] even quantitatively the specific heat theory was not a total failure.

Layzer and Fay[18] seem to have been the first to attempt a calculation of the effect of paramagnons on the $T_c$ for anisotropic BCS states in $He^3$. One sees immediately that the effect must be very large, and indeed in this and later papers the major theoretical difficulty remains simply that. Not only is the even-$l$, $s = 0$, $T_c$ very severely depressed, one finds, qualitatively because of the very different spin susceptibility of the $s = 1$ states, a strong enhancement of $T_c$ for them. While quantitatively the effect on $T_c$ cannot be calculated with any confidence at all even yet, Layzer and Fay's work strongly indicated that if anisotropic superfluidity were found, odd $l$ was the correct choice. Incidentally, it is still not clear why it has so far not been discovered in paramagnon metals such as Pt and Pd as well.

## 3. POST-EXPERIMENT HISTORY—LATE 1972 TO PRESENT

Here we will give only a quick sketch of the developments as an introduction to the remainder of this chapter. Once correctly interpreted, the experiment of Osheroff *et al.*[19, 20] called out strongly to all those aware of the theoretical background for an interpretation as anisotropic BCS superfluidity. A low $T_c/E_F$, second-order phase transition, and approximately correct specific heat jump suggested BCS. The existence of more than one low phase, the A and the B, with what had to be equal or similar $T_c$'s, is, as we have seen, a characteristic of these states. The magnetic susceptibility behaviour could be relatively easily fitted into the AM–BW structure, and militated against any alternative possibility such as some spin-density wave or liquid-crystal state. Finally, the shifts of the NMR frequency in the A phase required, rigorously, an anisotropic $\langle n(\mathbf{0})n(\mathbf{r})\rangle$ correlation.

The first to fit some of these results into the anisotropic BCS picture correctly was Leggett.[21, 22, 23] Leggett's theory of the resonance shifts has been remarkably successful in interpreting both the original data and subsequent measurements which have played the major role in identifying the A and B phases.

To briefly recapitulate the early experiments: these revealed a second-order phase transition along the melting curve at about 2·6 mK, into a phase which had unchanged nuclear susceptibility but a large NMR shift obeying

$$\omega_R^2 = (\gamma H)^2 + \Delta\omega_T^2(T).$$

At a lower $T$, about 2·0 mK, there was a first-order transition into a state with $\frac{1}{2}$ to $\frac{1}{3}$ of the nuclear susceptibility.

Leggett showed that any axial, odd-$l$ phase such as those postulated by Anderson and Morel in 1960 would show such a shift. His theory will be discussed in detail in our section on resonance effects. His early work was marred by two slight difficulties: an incorrect computation of the shift in the BW phase, which is correctly zero; with this, and an understandable unwillingness to accept the stability of an AM type phase, he hesitated to identify the phases correctly.

On the basis of a less accurate theory of shifts Anderson[24, 25] reproduced Leggett's frequencies in the high-field regime and correctly identified the A phase as an $l = 1$, AM phase, the B phase as BW. Where they differ Anderson's theory is wrong, except possibly very near to $T_c$.

To continue with the resonance story, Leggett's full paper in 1973 correctly calculated shifts for all possible $l = 1, s = 1$ phases, predicted a "longitudinal resonance" which has now been observed at the correct frequency in both phases (Osheroff and Brinkman,[26] Osheroff[27]), and produced a complete theory of orientation effects by dipolar interactions. These theories have been among the most useful results in phenomenologically identifying the phases.

A second confirmation of the triplet BCS nature of the transitions came from the small, but characteristic, effect of magnetic fields on the second-order "A" transition (Ambegaokar and Mermin;[28] Varma and Werthamer[29]). In a magnetic field there is a linear displacement of the two Fermi levels and an accompanying change $\pm(\mathrm{d}^2n/\mathrm{d}\varepsilon^2)\mu H$ in the densities of states of both up and down spins. Thus $T_c$ for a pure ↑↑ pairing, say, is slightly, and linearly, raised, leading to a small region of the so-called "A$_1$" phase. The theory does not give the correct, observed effect in detail, but the presence of this effect is a clear signal of a triplet BCS phase.

Early calculations of the attenuation of ultrasound (Patton,[30] Wölfle[31,32]), when compared with that measured by Richardson and Lee[33] confirmed, to a considerable extent, Anderson's assignment of the phases.

Although all the experimental data (magnetic susceptibility, resonance, ultrasound) pointed in the direction of the AM ↔ A, BW ↔ B assignment, it was clearly in contradiction with microscopic BCS theory and one was very reluctant to accept it whole-heartedly. The reason for this inversion of the stability of the two phases was found by Anderson and Brinkman[34] in a feedback effect of the susceptibility difference upon the spin-fluctuation interactions. The feedback calculation gives rise to a coupling between up and down spin pairs which favours parallel orbital angular momentum which is a specialization of the AM state; therefore this phase is now called the ABM (Anderson–Brinkman–Morel) state.

The many experimental consequences of this idea, as worked out in subsequent papers (Brinkman and Anderson;[35,36] Brinkman *et al.*;[37] Kuroda[38,39]) have been confirmed in considerable detail, especially by the series of phase diagrams and thermodynamic measurements carried out by Wheatley *et al.*[40,41,42,43] and the NMR measurements in the $A_1A_2$ region by Osheroff and Anderson.[44]

The superfluidity properties have not been as accessible experimentally, but have attracted a great deal of theoretical attention. The formal calculation of superfluid density was carried out by Saslow.[45] demonstrating the anisotropy of $\rho_s$, and Legget's considerations on Fermi-liquid renormalization were applied to the anisotropic $\rho_s$. Unfortunately, the viscosity measurements of Alvesalo *et al.*,[46,47] the anomalous heat conductivity measurements of Wheatley's group[41] and the fourth-sound measurements of Wheatley's group[48] and the Cornell groups,[49] while adding up to a very conclusive demonstration of superfluidity, do not settle many quantitative questions.

A remarkable new phenomenon has been predicted theoretically by de Gennes.[50,51,52] This is the phenomenon of "orbital supercurrents", resulting from cross-coupling of superflow and the angular properties. These effects are intrinsically dependent upon the complex wave function and their observations will present direct proof of its existence. The original speculations of de Gennes have been modified and completed by Wölfle[53] and Blount, Brinkman and Anderson.[54]

Another area which has been fertile for theoretical discussion but has not had direct experimental confirmation is that of collective modes of motion. Pre-1972 speculations were, on the whole, found to be infertile, and it is only from the detailed equations of motion of Anderson[24,25] and Leggett,[22,23,55] the detailed calculations of Maki and Ebisawa,[56] Combescot[57,58] and Brinkman and Smith,[59] and the current equations of de Gennes[50,51] and Wölfle[53] that a coherent picture has arisen. Anderson's "orbit waves" are apparently to be expected at quite low frequencies in the ABM–A phase, while spin waves (Brinkman and Smith;[59] Maki and Tsuneto;[60] Combescot[57,58]) will be present in both phases.

Finally, de Gennes[50,51] has opened the discussion of "textures" (in the liquid crystal sense), singularities and surface orientation effects. Brinkman, Smith, Osheroff and Blount[61] have given a first computation of resonance effects in the BW–B phase in terms of textures and the weak orientation effects in the B phase suggested by Leggett.[21,22,23,55]

At the present moment one may summarize the situation as follows: The theoretical structure based on Leggett's equations of motion and the spin-fluctuation stabilization of the ABM state has had a series of remarkable successes in correlating the static and NMR properties of these phases. The theoretical suggestions of de Gennes and others on the dynamical side are adequate to foreshadow a much greater richness of phenomena than demonstrated either by $He^4$II or by superconductors, and thus perhaps it is not surprising that on that side theory and experiment, while both very active, are yet to be reconciled in detail.

## II. PROPERTIES OF $HE^3$ IN THE NORMAL STATE

The properties of $He^3$ as a normal Fermi liquid have recently been reviewed by Baym and Pethick.[62] Here we shall only briefly summarize the features which are relevant to our considerations of the superfluid phases. Using Landau's theory of a normal Fermi liquid the low temperature properties of $He^3$ can be parameterized in terms of the interaction function between quasiparticles defined as

$$f_{\alpha\beta}(\mathbf{k}, \mathbf{k}') \equiv \frac{\delta^2 E}{\delta n_{\mathbf{k}\alpha} \delta n_{\mathbf{k}\beta}} \tag{II.1}$$

where $E$ is the total energy of the Fermi liquid and $n_{\mathbf{k}\alpha}$ is the occupation number of the quasiparticle state with momentum $\mathbf{k}$ and spin $\alpha$. Since in most experiments quasiparticles are always excited near the Fermi surface, $|\mathbf{k}| \simeq |\mathbf{k}'| \simeq k_F$ and the $f_{\alpha\beta}(\mathbf{k}, \mathbf{k}')$ depends only on the angle $\theta$ between $\mathbf{k}$ and $\mathbf{k}'$. Therefore $f$ can be expanded in Legendre polynomials. Using the spin symmetry of the function one is left with two sets of parameters

$$F_l^s \equiv \frac{(2l+1)}{2} N(0) \int_{-1}^{1} \mathrm{d}\cos\theta\, P_l(\cos\theta)(f_{\uparrow\uparrow}(\theta) + f_{\uparrow\downarrow}(\theta)) \tag{II.2}$$

$$F_l^a \equiv \frac{(2l+1)}{2} N(0) \int_{-1}^{1} \mathrm{d}\cos\theta\, P_l(\cos\theta)(f_{\uparrow\downarrow}(\theta) - f_{\uparrow\downarrow}(\theta)). \tag{II.3}$$

In principle, one should be able to determine all of these parameters from experiment and thus give a complete characterization of the Fermi liquid. In practice, one can only determine the values of $F_0^s$, $F_1^s$, $F_0^a$ and crude estimates of $F_1^a$ and it turns out that even if we knew all the Fermi-liquid parameters we would not have the essential information necessary to determine $T_c$ and other properties of the superfluid and normal states. However, considerable information on the nature of the state of normal $He^3$ is available in the values of the known parameters. The experimental values[63] of $F_0^s$ are large and positive, varying from 10·8 at the saturated vapour pressure to

$\approx 100$ near the melting curve. The values of $F_1^s$, which determines the effective mass, i.e. $m^*/m = 1 + F_1^s/3$, of the quasiparticles, vary from 6·3 at the s.v.p. to $\approx 15$ on the melting curve. Finally, $F_0^a$ is determined from the so-called Stoner enhancement factor defined from the ratio of the measured susceptibility to the non-interacting Pauli susceptibility

$$\frac{\chi}{\chi_P} = \frac{m^*/m}{(1 + F_0^a)}. \tag{II.4}$$

It is found that $F_0^a$ is negative and only slightly pressure dependent, varying from $\approx -0{\cdot}67$ at s.v.p. to $-0{\cdot}74$ near the melting curve. The large positive values of $F_0^s$ imply that those particle–hole pair excitations which are of the density-fluctuation type are strongly coupled together into the higher energy zero-sound mode, while the negative value of $F_0^a$ implies that the low energy particle–hole pair excitations of the spin-fluctuation type are strongly enhanced. Therefore, the dynamic spin susceptibility $\chi(q, \omega)$ will be dominated by the low energy excitations while the density response function will be dominated by the zero-sound collective mode. These facts led Doniach and Engelsberg[(64)] to show that deviations of the specific heat of $He^3$ from the simple linear temperature dependence given by the Landau theory could be explained in terms of the low-frequency long-wavelength spin fluctuations. In particular, they showed that the singular nature of the fluctuations at long wavelengths and low energies gives rise to an enhancement of the effective mass and to a $T^3 \ln T$ correction to the specific heat whose coefficient is sufficiently large to explain the observed anomalous behaviour.

Shortly thereafter, Rice[(65, 66)] showed that the spin-fluctuation model also leads to corrections to the transport coefficients that differed by only one power of temperature from the low temperature limiting forms. A $T^3 \ln T$ correction was also found in the temperature dependence of the susceptibility.[(67)] These results were all discovered by studying a simple model with a contact interaction $I$ which was assumed to be determined from the susceptibility enhancement $\chi/\chi_p = (1 - IN(0))^{-1}$, where $N(0)$ is the bare-particle density of states at the Fermi surface $mk_F/2\pi^2$. More recently Pethick,[(68)] Emery,[(69)] Amit, Kane, and Wagner[(70)] and others have shown that spin-fluctuation-model results can be placed into the context of Fermi-liquid theory and that such corrections are a general property of any Fermi liquid, not just almost ferromagnetic liquids for which the Stoner enhancement is large.

In the spin-fluctuation model one uses the parameter $\bar{I} \equiv IN(0)$ to describe the entire susceptibility enhancement including the effective mass corrections, i.e.

$$\frac{\chi}{\chi_P} \equiv \frac{1}{1 - \bar{I}} \equiv \frac{m^*/m}{(1 + F_0^a)}. \tag{II.5}$$

Thus in this model $He^3$ "looks like" an almost ferromagnetic system in that $\bar{I}$ becomes closer to one as the pressure is increased. However, from a Fermi-liquid point of view, as we already mentioned, $F_0^a$ is almost independent of pressure and the variation of $m^*/m$ gives rise to the pressure dependence of $\chi/\chi_P$. Therefore, one would not say that $He^3$ is almost ferromagnetic. In fact, this variation of $F_0^s$, $F_1^s$, and $F_0^a$ as a function of pressure is qualitatively what one expects for a Fermi system approaching localization in the Mott(71) sense. Indeed, $He^3$ does localize at 33 atm when it becomes a solid. To see this we compare the behaviour of $He^3$ with the Gutzwiller(72) results on the Hubbard model with one particle per site. In this model $I$ is the intrasite repulsive interaction between the particles. Let us assume that increasing pressure on $He^3$ is equivalent to increasing the repulsive interaction in the model. Brinkman and Rice(73, 74) showed that the Gutzwiller trial-wave-function approach to this problem gives rise to a critical value of the interaction $I$ at which the system becomes localized, each particle occupying a single site. As $I(P)$ approaches the critical value the Fermi liquid has the following properties

$$\frac{m^*}{m} = \left\{1 - \left[\frac{I(P)}{I_c}\right]^2\right\}^{-1}, \tag{II.6}$$

$$F_0^a = IN(0)\frac{1 + I/2I_c}{(1 + I/I_c)^2} \tag{II.7}$$

and

$$F_0^s = IN(0)\left(2 + \frac{I}{I_c}\right)^{-1}\left(1 - \frac{I}{I_c}\right)^{-2}. \tag{II.8}$$

Here $I_c$ is the critical interaction strength at which localization sets in. Varying $I(P)$ to fit the $m^*/m$ versus pressure data we find that $F_0^a$ is almost constant, while $F_0^s$ varies roughly as the square of the mass. These results qualitatively agree with the experimental facts, so that the properties of $He^3$ liquid are undoubtedly related to its incipient localization rather than to its being almost ferromagnetic. However, this does not mean that the spin fluctuations which characterize the behaviour of $\chi(q, \omega)$ are unimportant, but indicates rather that they are a consequence of the incipient localization. We will find that these fluctuations are very important in understanding the stability of the superfluid $A$ phase.

Since the Landau parameters offer a means of parameterizing a normal Fermi liquid, one would like to be able to estimate their values from first-principle calculations. In practice, such work has not proved very successful. The most recent work is that of Babu and Brown,(75) who attempt to calculate the interaction function as a self-consistency problem in the virtual excitation of quasiparticle–hole pairs. Although the approach is interesting it does not appear to give good quantitative results. In the final analysis it is found necessary to introduce the spin-fluctuation concepts. This problem has also been attacked using variational wavefunction techniques by Feenberg[(76)] and his collaborators with slightly more quantitative success. Older work using the Brueckner $K$-matrix approach[(77)] has been shown by several authors to lead to poor results for the effective interaction. However, it has been shown by Østgaard[(78)] that by including the so-called rearrangement terms one obtains reasonable estimates of the Landau parameters from the Brueckner scheme. These calculations are simply pointed out because their sophistication makes one appreciate the difficulty of calculating the effective quasiparticle interaction.

## III. Higher Angular Momentum Pairs and Estimates of $T_c$

The superfluid transition of a Fermi system can be determined by examining the stability of the normal state with respect to the formation of bound states of pairs of quasiparticles with total momentum zero. The transition temperature is determined by a divergence in the scattering amplitude of two quasiparticles at the Fermi surface.[(6)] The scattering amplitudes satisfy an integral obtained by summing the particle–particle ladder diagram shown in Fig. 2.

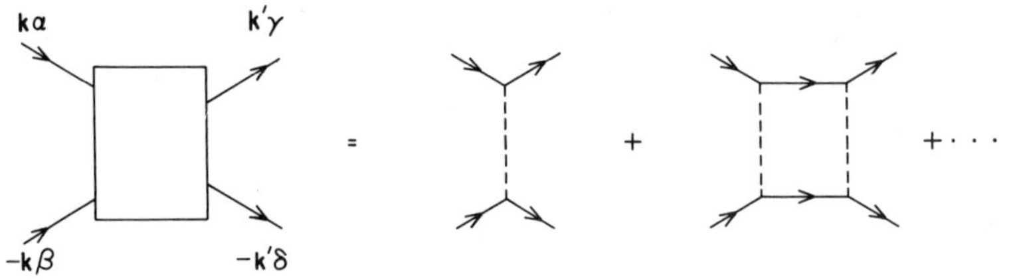


FIG. 2. The ladder diagrams that must be summed to calculate the stability of the normal state with respect to the creation of bound Cooper pairs.

Since each quasiparticle has spin $\frac{1}{2}$, the scattering states can be divided into states with total spin angular momentum 0 or 1. The two spin channels satisfy separate integral equations

$$A_\sigma(\mathbf{k}, \mathbf{k}') = v_\sigma(\mathbf{k}, \mathbf{k}') - \int \frac{d^3k''}{(2\pi)^3} v_\sigma(\mathbf{k}, \mathbf{k}'') \frac{(1 - 2f_{k''})}{2\varepsilon_{k''}} A_\sigma(\mathbf{k}'', \mathbf{k}'). \qquad \text{(III.1)}$$

We have assumed that the total momentum and energy of the pair is zero. The $f_k$ is the occupation number of the quasiparticle state $k$ of energy $\varepsilon_k$. Energy is measured relative to the Fermi momentum. The variable $\sigma$ indicates total spin (=0,1). The $v_\sigma(\mathbf{k}, \mathbf{k}')$ is the matrix element of the interaction potential. For a system invariant under spin rotations

$$v_1(\mathbf{k}, \mathbf{k}') = \tfrac{1}{2}\langle \mathbf{k}\uparrow, -\mathbf{k}\uparrow | v | \mathbf{k}'\uparrow, -\mathbf{k}'\uparrow \rangle, \tag{III.2}$$

where

$$|\mathbf{k}\sigma, -\mathbf{k}\sigma'\rangle = a^\dagger_{\mathbf{k}\sigma} a^\dagger_{-\mathbf{k}\sigma} |\psi\rangle$$

and $|\psi\rangle$ is the state with no quasiparticles.

Since the scattering of an up spin with a down spin must be half the sum of the triplet and singlet matrix elements we have

$$v_0(\mathbf{k}, \mathbf{k}') = \tfrac{1}{2}[2\langle \mathbf{k}\uparrow, -\mathbf{k}\downarrow | v | \mathbf{k}'\uparrow, -\mathbf{k}'\downarrow \rangle - \langle \mathbf{k}\uparrow, -\mathbf{k}\uparrow | v | \mathbf{k}'\uparrow, -\mathbf{k}'\uparrow \rangle]. \tag{III.3}$$

It is easy to see that $v_1(\mathbf{k}, \mathbf{k}')$ is an odd function of $\mathbf{k}$ and $\mathbf{k}'$, while $v_0$ is an even function. This in turn implies that $A_0$ and $A_1$ are even and odd respectively. The early papers on He$^3$ concentrated on the solution of (III.1) when $v(\mathbf{k}, \mathbf{k}')$ is chosen to be the bare interaction potential between two He$^3$ atoms.

To do this we write

$$v(\mathbf{k}, \mathbf{k}') = v(k, k', \theta) \tag{III.4}$$

$$= \sum_{l=0}^{\infty} (2l + 1) v_l(k, k') P_l(\cos\theta), \tag{III.5}$$

where $\theta$ is the angle between $\mathbf{k}$ and $\mathbf{k}'$ and where the $P_l$ are the Legendre polynomials. By making the same expansion for $A_\sigma(\mathbf{k}, \mathbf{k}')$ we obtain an integral equation for states of given orbital angular momentum

$$A_l(k, k') = v_l(k, k') - \int_0^\infty \frac{k''^2 dk''}{(2\pi^2)} v_l(k, k'') \frac{(1 - 2f_{k''})}{2\varepsilon_{k''}} A_l(k'', k'). \tag{III.6}$$

We use a single index $l$, since its value implies the spin value; i.e. odd $l$ implies spin triplet and even $l$ spin singlet. This equation is usually solved in two steps by introducing an artificial cut-off $\omega_c$ and first solving (III.6) with the integration over $\mathbf{k}''$ restricted to be outside the region of momentum space defined by $|\varepsilon_{k''}| < \omega_c$. If the solution to the resulting equation is defined to be $K_l(k, k')$ then one can show that

$$A_l(k, k') = K_l(k, k') - \int_{|\varepsilon_k| < \omega_c} \frac{k''^2 dk''}{(2\pi^2)} K_l(k, k'') \frac{(1 - 2f_{k''})}{2\varepsilon_{k''}} A_l(k'', k'). \tag{III.7}$$

By doing this one treats the problem of the hard-core repulsion in solving for $K_l$ and then studies (III.7) for the superfluid transition temperature. By a proper choice of $\omega_c$ it is possible to make $K_l$ relatively independent of $k$ and $k'$. In this case the solution to (III.7) is

$$A_l = \frac{K_l}{\left\{1 + K_l N(0) \int_0^{\omega_c} \frac{d\varepsilon}{\varepsilon} \tanh\left(\frac{\beta\varepsilon}{2}\right)\right\}}. \tag{III.8}$$

The transition temperature is thus given by the vanishing of the denominator,

$$1 = -K_l N(0) \int_0^{\omega_c} \frac{d\varepsilon}{\varepsilon} \tanh\left(\frac{\beta\varepsilon}{2}\right), \tag{III.9}$$

or defining

$$\lambda_l = -N(0)K_l \tag{III.10}$$

for negative $K_l$ we have

$$T_c^l = 1{\cdot}14\omega_c \mathrm{e}^{-1/\lambda_l}. \tag{III.11}$$

Although somewhat different procedures have been developed in order to treat separately the hard core, the principle behind them is the same as in the above treatment. Estimates by Brueckner *et al.*[(5)] and by Emery and Sessler[(6)] based on the 6–12 potential of de Boer and other suitable bare potentials give the result that the $l = 2$ *d*-wave pairs should become unstable at temperatures in the region of 0·01 K although the latter number was very uncertain. The *p*-wave pairs were found to have a small $K_l$ whose sign was uncertain.

Pitaevskii[(4)] in his early work on $l \neq 0$ pairing examined carefully the consequences of the long-range nature of the van der Waals force and was able to show that $v_l$ was necessarily attractive for sufficiently large $l$ even if one includes many-body polarization effects. This problem was re-examined by Gor'kov and Pitaevskii[(79)] who attempted to extrapolate the large $l$ results to $l \approx 2{\cdot}3$. Recently Bookbinder[(80)] showed that solutions for the function $K_l$, which is the Brueckner $K$-matrix, do not approach the large $l$ limit until $l > 20$, so that in practice the exact result of Pitaevskii, although interesting, is not very useful.

Since liquid $He^3$ is not a low-density gas one cannot expect the estimates made from the two-particle $K$-matrix to be very accurate. This point was made clear in 1964 when Emery[(15)] investigated the use of the scattering amplitude calculated from the bare interaction as a means of determining the Landau parameters of the normal state. As mentioned in the previous section, he showed that this approximation leads to a poor description of these parameters and argued that it is quite important to include the effect of other $He^3$ atoms on the effective interaction. He thus noted that the values of the Landau parameters suggested that *p*-wave pairing would be most favourable. Since it has not yet proven possible to calculate this amplitude from first principles we proceed in a phenomenological fashion.

The scattering amplitude needed is that between two quasiparticles, so that $v_l$ is the four-point vertex irreducible with respect to particle–particle scattering. We can gain some insight into its structure by considering the spin-fluctuation model, as was first done for singlet $s$-wave pairing by Berk and Schrieffer(16) and for triplet pairing by Layzer and Fay.(81)

In this model one assumes that the spin fluctuations are much like phonons in the case of strong-coupling superconductivity. The interaction between two quasiparticles is assumed to be due to the attraction of one quasiparticle by the spin-polarization cloud of the other. Clearly the force will be attractive only if the spins of the two quasiparticles are aligned. To see this first consider a spin-dependent interaction of the general form(82)

$$v = \tfrac{1}{2} \sum_{\substack{\mathbf{kk}' \\ \mathbf{q}}} [\Gamma_\rho(\mathbf{q})\delta_{\alpha\beta}\delta_{\gamma\delta} + \Gamma_s(\mathbf{q})\boldsymbol{\sigma}_{\alpha\beta}\cdot\boldsymbol{\sigma}_{\gamma\delta}]\, a^\dagger_{\mathbf{k}+\mathbf{q}\alpha} a^\dagger_{\mathbf{k}'-\mathbf{q}\gamma} a_{\mathbf{k}'\delta} a_{\mathbf{k}\beta}. \tag{III.12}$$

Taking the matrix elements in (III.2) and (III.3) we find that

$$v_1(\mathbf{k}, \mathbf{k}') = \tfrac{1}{2}[\Gamma_\rho(\mathbf{k} - \mathbf{k}') + \Gamma_s(\mathbf{k} - \mathbf{k}') - (\mathbf{k} \to -\mathbf{k})] \tag{III.13}$$

and

$$v_0(\mathbf{k}, \mathbf{k}') = \tfrac{1}{2}[\Gamma_\rho(\mathbf{k} - \mathbf{k}') - 3\Gamma_s(\mathbf{k} - \mathbf{k}') + (\mathbf{k} \to -\mathbf{k})]. \tag{III.14}$$

If $\Gamma_s$ is negative, it enhances triplet or odd-$l$ pairing while suppressing singlet even-$l$ pairing. In the spin-fluctuation model using the contact interaction $I$ the Γ's are obtained from the summation of the diagrams shown in Fig. 3.

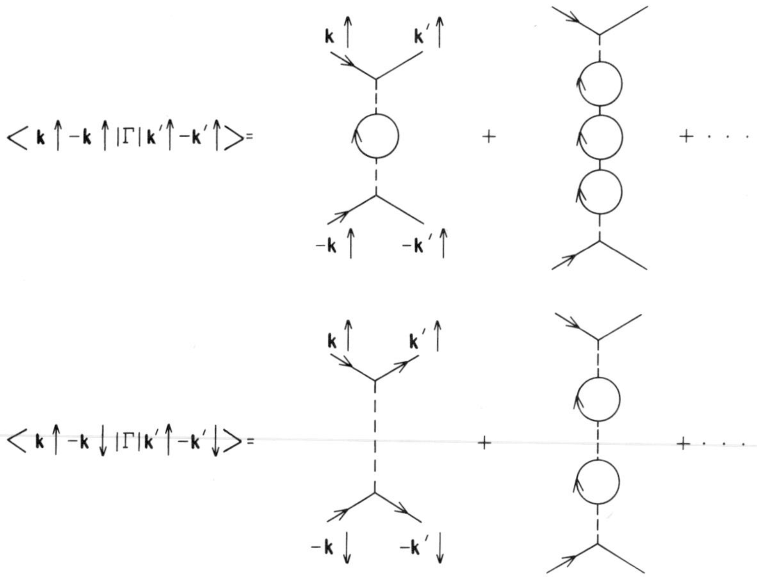


FIG. 3. The diagrams that must be summed to obtain the spin-fluctuation contributions to the effective interaction $\Gamma_s$. The bubble represents $\chi^0(q, \omega)$.

These diagrams can be summed to give

$$\langle \mathbf{k}\uparrow, -\mathbf{k}\uparrow|v|\mathbf{k}'\uparrow, -\mathbf{k}'\uparrow\rangle = -\frac{I^2\chi^0(\mathbf{k}-\mathbf{k}',0)}{\{1-[I\chi^0(\mathbf{k}-\mathbf{k}',0)]^2\}} + (\mathbf{k}\to-\mathbf{k}), \quad \text{(III.15)}$$

$$\langle \mathbf{k}\uparrow, -\mathbf{k}\downarrow|v|\mathbf{k}'\uparrow, -\mathbf{k}'\downarrow\rangle = +\frac{I}{\{1-[I\chi^0(\mathbf{k}-\mathbf{k}',0)]^2\}}. \quad \text{(III.16)}$$

The $\chi^0$ is the noninteracting susceptibility

$$\chi^0(\mathbf{q},\omega) \equiv \int \frac{d^3k}{(2\pi)^3}\frac{(f_{\mathbf{k}+\mathbf{q}}-f_{\mathbf{k}})}{\omega-\varepsilon_{\mathbf{k}+\mathbf{q}}+\varepsilon_{\mathbf{k}}}. \quad \text{(III.17)}$$

Equating (III.15, 16) to (III.13, 14) via (III.2, 3) we find

$$\Gamma_\rho(\mathbf{k}-\mathbf{k}') = \tfrac{1}{2}\frac{I}{[1+I\chi^0(\mathbf{k}-\mathbf{k}',0)]}, \quad \text{(III.18)}$$

$$\Gamma_s(\mathbf{k}-\mathbf{k}') = -\tfrac{1}{2}\frac{I}{[1-I\chi^0(\mathbf{k}-\mathbf{k}',0)]}. \quad \text{(III.19)}$$

The $\Gamma_\rho$ describes the exchange of density fluctuations in the model. Since these fluctuations are high energy fluctuations we will ignore them for the present discussion. It should be remarked that the diagram in Fig. 4 must be considered if one desires to examine all the matrix elements. We have avoided it by our separation (III.3). To see the low-frequency fluctuation in

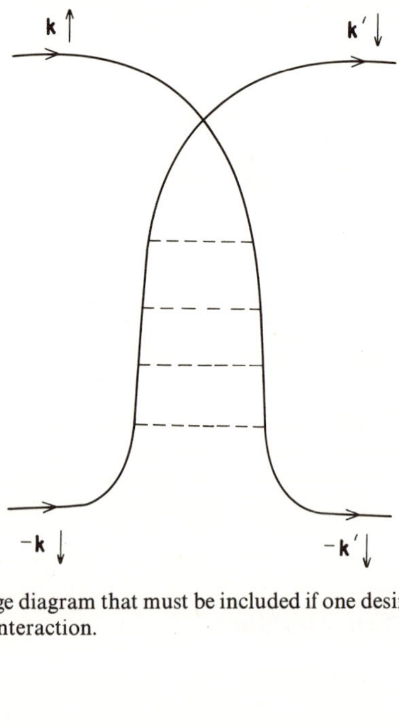


FIG. 4. The particle-exchange diagram that must be included if one desires to obtain the complete symmetry of the effective interaction.

$\Gamma_s$ we calculate the noninteracting susceptibility $\chi^0$ in the long-wavelength limit

$$\chi^0(q, \omega) = N(0)\left(1 - \frac{q^2}{12k_F^2} + \frac{i\pi}{2}\frac{\omega}{qv_F}\right). \tag{III.20}$$

Inserting this expression into (III.19) and taking the imaginary part, a peak is found in $\text{Im}\,\Gamma_s$ at $\omega_q = (1 - \bar{I})(2/\pi)v_F q$ where $\bar{I} = IN(0)$. If $\bar{I}$ is close to one, $\omega_q$ is considerably less than $v_F q$. The situation is illustrated in Fig. 5. The

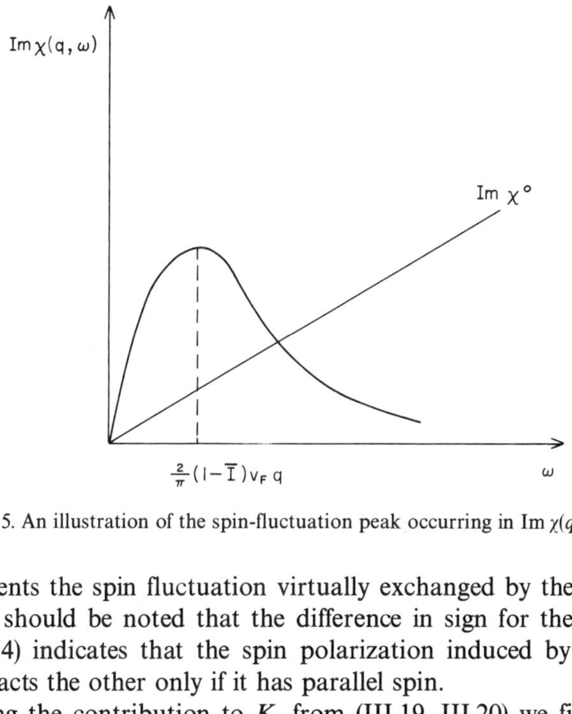


FIG. 5. An illustration of the spin-fluctuation peak occurring in Im $\chi(q, \omega)$.

peak represents the spin fluctuation virtually exchanged by the two quasi-particles. It should be noted that the difference in sign for the $\Gamma_s$ term in (III.13, III.14) indicates that the spin polarization induced by one quasi-particle attracts the other only if it has parallel spin.

Calculating the contribution to $K_l$ from (III.19, III.20) we find that the spin fluctuations give rise to a contribution

$$\lambda_l^{\text{SF}} \simeq \frac{-3}{2}\ln\left(\frac{1 - 2/3\bar{I}}{1 - \bar{I}}\right) \tag{III.21}$$

for triplet pairing. For even-$l$ pairing $\lambda_l^{\text{SF}}$ is $-3$ times this result. (III.21) holds only for relatively small $l$ since we have set $P_l(\cos\theta) = 1$ in doing the integrals. If we use $\bar{I} = 0{\cdot}75$ we obtain a number close to one. This is presumably cancelled to a large extent by the repulsive density-fluctuation term. What is essential here is that the calculation illustrates the importance of the spin fluctuations. Since one expects the result (III.21) to be most applicable for $p$-wave pairing one expects that the $p$-wave state will be most stable. Layzer and Fay[18] have attempted to calculate $T_c^1$ by using the sum of the spin-fluctuation interaction and the bare interaction for $v(\mathbf{k}, \mathbf{k}')$. In this way they obtained rather high values of $T_c^1 \approx 0{\cdot}1$ K. Recently, Layzer and Fay[83] have attempted to improve on their early work by examining the strong coupling equations used by Berk and Schrieffer for the $l = 0$ case. They find that the quasiparticle renormalization effects included in these equations tend to strongly lower $T_c^1$ and that its pressure dependence is the opposite of experiment. They then suggest a scaling of the renormalization factors which appears to give better results. The validity of the latter procedure is not presently clear.

## IV. Weak-Coupling BCS and Model Theories of the Anisotropic Superfluid State

### 1. BALIAN–WERTHAMER FORMALISM

#### (a) *General formalism*

In the Introduction we have reviewed the primitive Anderson–Morel formalism indicating the general nature of the anisotropic superfluid state. A much more complete, general formalism, still within the framework of BCS weak-coupling theory, was given by Balian and Werthamer,[12] which is particularly valuable for the specific $l = 1,\ s = 1$ case of interest in $He^3$. This formalism underlies all further work in the field, even where one must abandon some of the specific weak-coupling assumptions on which it is based.

The model to which this formalism applies is a Fermi liquid with interactions which are either weak or act only over a relatively small region of phase space near the Fermi surface or both:

$$\mathscr{H} = \sum_{\mathbf{k}\sigma} \varepsilon_{\mathbf{k}} n_{\mathbf{k}\sigma} + \sum_{\substack{\mathbf{k}\mathbf{k}'\mathbf{q} \\ (|\varepsilon_{\mathbf{k}}|, |\varepsilon_{\mathbf{k}'}| < \omega_s)}} \sum_{\sigma\sigma'} V_{\mathbf{k}\mathbf{k}'} a^\dagger_{\mathbf{k}\sigma} a^\dagger_{\mathbf{k}'\sigma'} a_{\mathbf{k}'-\mathbf{q}\sigma'} a_{\mathbf{k}+\mathbf{q}\sigma} \tag{IV.1}$$

where $\omega_s$, the cutoff energy, is $\ll E_F$ and we measure the kinetic energy $\varepsilon_{\mathbf{k}}$ from the Fermi surface. We shall see that (IV.1) is far less restrictive than it appears. To a great extent, the BW formalism, like the BCS theory, can serve as a *model* even when it is not a valid microscopic *theory*. We will discuss the degree to which these restrictions may be removed, and the types of modifications which result, in later sections. The basic restriction, however, which makes BCS-type theories workable, cannot be abandoned: $T_c/E_F \ll 1$. This restriction ensures among other things that the part of the interaction which is responsible for the pairing condensation is in some real sense small (since $T_c/E_F \approx e^{-1/\lambda}$, where $\lambda$ is the coupling constant). More importantly, the pairing interaction, and the pairing phenomenon, does not cause major changes in the basic correlations among the particles, because pairing only involves a few particles $\approx T_c/E_F$ near the Fermi surface. The bulk of the binding energy, the correlation function behaviour at most energies and wavelengths, and the major part of the Fermi-liquid renormalizations, and thus the normal quasiparticle self-energies, are not much affected.

With this restriction, and only with this, we can treat the pairing phenomenon as an independent additive contribution. The truncated hamiltonian formalism used by Bardeen, Cooper and Schrieffer and by Balian and Werthamer is then a perfectly satisfactory model, but we will use the more modern idea of introducing an anomalous self-energy which connects particle and hole excitations, and may be derived self-consistently from the existence of an anomalous amplitude or "pair amplitude"

$$\tfrac{1}{2}x^{\mathbf{k}}_{\sigma\sigma'} = \langle a_{\mathbf{k}\sigma} a_{-\mathbf{k}\sigma'} \rangle. \tag{IV.2}$$

Let us start the self-consistency process with this anomalous amplitude, which as you can see is already assumed to be constant in space, i.e. we for the present treat only pairs with equal and opposite momenta.

In the presence of an anomalous amplitude such as (IV.2) the fermions will experience an anomalous self energy

$$\Delta^{\mathbf{k}}_{\sigma\sigma'} = -\tfrac{1}{2}\sum_{\mathbf{k}'} V_{\mathbf{k}\mathbf{k}'} x^{\mathbf{k}}_{\sigma\sigma'} \tag{IV.3}$$

which will have the effect of mixing particles and holes in the equations of motion, by simple calculation from (IV.1):

$$\frac{\hbar}{i}\frac{\partial}{\partial t} a^{\dagger}_{\mathbf{k}\sigma} = [\mathscr{H}, a^{\dagger}_{\mathbf{k}\sigma}] = \varepsilon_{\mathbf{k}\sigma} a^{\dagger}_{\mathbf{k}\sigma} + \sum_{\sigma'} \Delta^{\mathbf{k}}_{\sigma\sigma'} a_{-\mathbf{k}\sigma'} + \ldots \tag{IV.4}$$

The + ... in (IV.4) are the many-body corrections, which may be subsumed into many-body corrections to the ordinary and anomalous self-energies.[84, 85] The anomalous term in (IV.4) requires us, in searching for elementary excitations, to make a canonical transformation involving both spins and both particles and holes:

$$a_{\mathbf{k}\sigma} = \sum_{\sigma'} u^{\mathbf{k}}_{\sigma\sigma'} \alpha_{\mathbf{k}\sigma'} + v^{\mathbf{k}}_{\sigma\sigma'} \alpha^{\dagger}_{-\mathbf{k}\sigma'} \tag{IV.5}$$

where the $\alpha$'s are the quasiparticle elementary excitations. Balian and Werthamer introduce a convenient 4-component matrix notation (following Eliashberg[86])

$$\hat{a}^{\mathbf{k}} \equiv \begin{pmatrix} a_{\mathbf{k}\uparrow} \\ a_{\mathbf{k}\downarrow} \\ a^{\dagger}_{-\mathbf{k}\uparrow} \\ a^{\dagger}_{-\mathbf{k}\downarrow} \end{pmatrix} \tag{IV.6}$$

and (IV.5) becomes

$$\hat{a}^{\mathbf{k}} = \begin{pmatrix} \hat{u}^{\mathbf{k}} & \hat{v}^{\mathbf{k}} \\ \hat{v}^{-\mathbf{k}*} & \hat{u}^{-\mathbf{k}*} \end{pmatrix} \hat{\alpha}^{\mathbf{k}} = \hat{U}^{\mathbf{k}} \hat{\alpha}^{\mathbf{k}}. \tag{IV.7}$$

$\hat{u}$ and $\hat{v}$ are $2 \times 2$ matrices, $\hat{U}$ is a $4 \times 4$ matrix. For (IV.7) to be canonical, $\hat{U}$ must be unitary. Clearly in this theory (as opposed to BCS) we cannot allow $\hat{U}^{\mathbf{k}}$ and $\hat{U}^{-\mathbf{k}}$ to be independent, and $\hat{U}$ satisfies the symmetry relationship

$$\hat{U}^{\mathbf{k}} = \begin{pmatrix} 0 & 1 \\ 1 & 0 \end{pmatrix} \hat{U}^{-\mathbf{k}*} \begin{pmatrix} 0 & 1 \\ 1 & 0 \end{pmatrix}. \tag{IV.8}$$

The equation of motion (IV.4) may be written in 4-component notation as well:

$$\frac{\hbar}{i} \frac{\partial (\hat{a}^{\mathbf{k}})^{\dagger}}{\partial t} = \hat{\mathscr{E}}^{\mathbf{k}} (\hat{a}^{\mathbf{k}})^{\dagger} \tag{IV.9}$$

where

$$\hat{\mathscr{E}}^{\mathbf{k}} = \begin{pmatrix} \varepsilon_{\mathbf{k}} & \hat{\Delta}_{\mathbf{k}} \\ -\hat{\Delta}^{*}_{-\mathbf{k}} & -\varepsilon_{\mathbf{k}} \end{pmatrix} = \begin{pmatrix} \varepsilon_{\mathbf{k}} & \hat{\Delta}_{\mathbf{k}} \\ (\hat{\Delta}_{\mathbf{k}})^{\dagger} & -\varepsilon_{\mathbf{k}} \end{pmatrix}, \tag{IV.10}$$

$\hat{\Delta}_{\mathbf{k}}$ is a $2 \times 2$ matrix in spin space given by the components (IV.3). The symmetry relation for $\hat{\Delta}$ is, by the anticommutation relations,

$$\hat{\Delta}_{\mathbf{k}} = -\tilde{\hat{\Delta}}_{-\mathbf{k}} \tag{IV.11}$$

as is true for $\hat{x}$, and this has been used in the second equality of (IV.10). What we need to do is to diagonalize the equation of motion with the quasiparticle transformation $\hat{U}$ of (IV.7):

$$\hat{E}^{\mathbf{k}} = \hat{U}^{-1} \hat{\mathscr{E}}^{\mathbf{k}} \hat{U}, \tag{IV.12}$$

where $\hat{E}$ is a diagonal matrix:

$$\hat{E}^{\mathbf{k}} = \begin{pmatrix} E_{\mathbf{k}+} & & & 0 \\ & E_{\mathbf{k}-} & & \\ & & E_{-\mathbf{k}+} & \\ 0 & & & -E_{-\mathbf{k}-} \end{pmatrix}. \tag{IV.13}$$

When we have done that, the $\alpha$'s will be eigen-excitations whose thermal averages obey

$$\langle \alpha_{\mathbf{k}\sigma}\alpha^{\dagger}_{\mathbf{k}\sigma}\rangle = 1 - f(E_{\mathbf{k}\sigma}/kT), \tag{IV.14}$$

$f$ being the Fermi function. This can be rewritten as a matrix equation

$$\langle \alpha^{\mathbf{k}}\alpha^{\mathbf{k}\dagger}\rangle = \begin{pmatrix} 1 - f(E_{\mathbf{k}+}) & & & 0 \\ & 1 - f(E_{\mathbf{k}-}) & & \\ & & f(E_{-\mathbf{k}+}) & \\ 0 & & & f(E_{-\mathbf{k}-}) \end{pmatrix}$$

$$= \tfrac{1}{2}(1 + \tanh \tfrac{1}{2}\beta \hat{E}^{\mathbf{k}}). \tag{IV.15}$$

From $\langle \alpha\alpha^{\dagger}\rangle$ we can, by transforming back to the $a$'s, calculate the mean values of the real, bare-particle operators $a$. We write these in matrix form as well, defining matrices $\hat{w}$ and $\hat{x}$ by

$$\langle a^{\mathbf{k}}a^{\mathbf{k}\dagger}\rangle = \frac{1}{2}\begin{pmatrix} 1 + \hat{w}^{\mathbf{k}} & \hat{x}^{\mathbf{k}} \\ -\hat{x}^{-\mathbf{k}} & 1 - \hat{w}^{-\mathbf{k}} \end{pmatrix} \equiv \tfrac{1}{2}(1 + \hat{W}^{\mathbf{k}}). \tag{IV.16}$$

Here $\hat{x}$ is the matrix of anomalous amplitudes we started from, (IV.2), and the elements of the $2 \times 2$ matrix $\hat{w}$ are defined by

$$\langle a_{\mathbf{k}\sigma}a^{\dagger}_{\mathbf{k}\sigma'}\rangle = \tfrac{1}{2}(\delta_{\sigma\sigma'} + \hat{w}^{\mathbf{k}}_{\sigma\sigma'}). \tag{IV.17}$$

Initially we defined the transformation $\hat{U}$ to carry the $a$'s into the $\alpha$'s and vice versa; thus

$$\langle \hat{a}^{\mathbf{k}}\hat{a}^{\mathbf{k}\dagger}\rangle = \hat{U}^{\mathbf{k}}\langle \alpha^{\mathbf{k}}\alpha^{\mathbf{k}\dagger}\rangle(\hat{U}^{\mathbf{k}})^{-1} \tag{IV.18}$$

or, using (IV.15),

$$\begin{aligned} \hat{W}^{\mathbf{k}} &= \hat{U}^{\mathbf{k}}(\tanh \tfrac{1}{2}\beta\hat{E}^{\mathbf{k}})(\hat{U}^{\mathbf{k}})^{-1} \\ &= \tanh \tfrac{1}{2}\beta\hat{\mathscr{E}}^{\mathbf{k}}. \end{aligned} \tag{IV.19}$$

Since $\hat{x}^{\mathbf{k}}$ is determined by $\hat{W}^{\mathbf{k}}$, and in turn $\hat{\Delta}^{\mathbf{k}}$ is defined from $\hat{x}^{\mathbf{k}}$ by (IF.3), this gives us the essential self-consistency relationship. (IV.19) with (IV.3) is the gap equation in $4 \times 4$ matrix form.

Balian and Werthamer immediately specialize to a particular form of $\hat{\Delta}$ in order to simplify this equation. This is not necessary as yet, if we observe that the matrix

$$(\hat{\mathscr{E}}^{\mathbf{k}})^2 = (\hat{\varepsilon}_{\mathbf{k}} + \hat{\Delta}_{\mathbf{k}})^2$$

is block diagonal:

$$\hat{\mathscr{E}}^2 = \begin{pmatrix} (\varepsilon_{\mathbf{k}}^2 \hat{1} + \hat{\Delta}_{\mathbf{k}}^\dagger \hat{\Delta}_{\mathbf{k}}) & 0 \\ 0 & (\varepsilon_{\mathbf{k}}^2 \hat{1} + \hat{\Delta}_{\mathbf{k}} \hat{\Delta}_{\mathbf{k}}^\dagger) \end{pmatrix}. \tag{IV.20}$$

Thus the eigenvalues always appear in $\pm$ pairs, corresponding to quasiparticle and quasihole. Any function of $\hat{\varepsilon}^2$ will have the same diagonal form, as will

$$F(\hat{\mathscr{E}}^2) = \frac{1}{\hat{\mathscr{E}}} \tanh \tfrac{1}{2}\beta\hat{\mathscr{E}} \tag{IV.21}$$

which may be expanded as a power series in $\hat{\mathscr{E}}^2$. This allows us to break up (IV.19) into two $2 \times 2$ equations

$$\begin{aligned} \hat{w}^{\mathbf{k}} &= \varepsilon_{\mathbf{k}} F[(\hat{\mathscr{E}}^{\mathbf{k}})^2] \\ &= \varepsilon_{\mathbf{k}} \left[ \frac{1}{(\varepsilon_{\mathbf{k}}^2 + \hat{\Delta}^\dagger \hat{\Delta})^{\frac{1}{2}}} \tanh \tfrac{1}{2}\beta(\varepsilon_{\mathbf{k}}^2 + \hat{\Delta}^\dagger \hat{\Delta})^{\frac{1}{2}} \right] \\ \hat{x}^{\mathbf{k}} &= \hat{\Delta}_{\mathbf{k}} F(\varepsilon_{\mathbf{k}}^2 + \hat{\Delta}_{\mathbf{k}}^\dagger \hat{\Delta}_{\mathbf{k}}), \end{aligned} \tag{IV.22}$$

where in (IV.22) $F$ signifies the $2 \times 2$ block of $F$ in which the $\hat{\Delta}^\dagger\hat{\Delta}$ order is taken. From now on except where noted all matrices will be $2 \times 2$ rather than $4 \times 4$, with spin indices only.

It is useful to note the theorem that

$$\tilde{\hat{x}}^{\mathbf{k}} = \tilde{\hat{\Delta}}_{\mathbf{k}} F(\varepsilon_{\mathbf{k}}^2 + \tilde{\hat{\Delta}}_{\mathbf{k}}^\dagger \tilde{\hat{\Delta}}_{\mathbf{k}}). \tag{IV.23}$$

That is, (IV.22) retains the transposition symmetry of $\hat{\Delta}$. Thus the full gap equation, which we may now write, does as well, since $V$ is symmetric in its spin indices:

$$\begin{aligned} \Delta_{\sigma\sigma'}^{\mathbf{k}} &= (-\tfrac{1}{2}) \sum_{\mathbf{k}'} V_{\mathbf{k}\mathbf{k}'} [\hat{\Delta}_{\mathbf{k}'} F(\varepsilon_{\mathbf{k}'}^2 + \hat{\Delta}_{\mathbf{k}'}^\dagger \hat{\Delta}_{\mathbf{k}'})]_{\sigma\sigma'} \\ &= (-\tfrac{1}{2}) \sum_{\mathbf{k}'} V_{\mathbf{k}\mathbf{k}'} [\hat{\Delta}_{\mathbf{k}} \hat{\mathscr{E}}_{\mathbf{k}'}^{-1} \tanh \tfrac{1}{2}\beta \hat{\mathscr{E}}_{\mathbf{k}'}]_{\sigma\sigma'} \end{aligned} \tag{IV.24}$$

where

$$\hat{\mathscr{E}}_{\mathbf{k}}^2 = \varepsilon_{\mathbf{k}}^2 + \hat{\Delta}_{\mathbf{k}}^\dagger \hat{\Delta}_{\mathbf{k}}. \tag{IV.25}$$

The first of the equations (IV.22) simply determines $\hat{w}$ given $\hat{\Delta}$; (IV.24) is the basic equation of the system, or "gap equation".

### (b) *Singlet–triplet separation and spin rotation symmetry*

Now we can separate out the singlet and triplet parts of (IV.24). The symmetry (IV.23) shows us that symmetry in interchange of $\sigma$ and $\sigma'$ is a good quantum number: a symmetric $\hat{\Delta}$ on the right-hand side leads self-consistently to a symmetric one on the left, and vice versa (though we see no proof that mixed spins *cannot* occur!). Thus we have the two cases

$$\hat{\Delta} = -\tilde{\hat{\Delta}} = \begin{pmatrix} 0 & d_0 \\ -d_0 & 0 \end{pmatrix}$$

(Singlet)

and

$$\hat{\Delta} = \tilde{\hat{\Delta}} = \begin{pmatrix} d_v + id_u & -id_w \\ -id_w & d_v - id_u \end{pmatrix}. \tag{IV.26}$$

(Triplet)

We have chosen the matrix elements in (IV.26) to conform to the BW convention

$$\hat{\Delta} = \sum_{i=0}^{3} \sigma_i \sigma_v d_i, \tag{IV.27}$$

where $\sigma_u$, $\sigma_v$, $\sigma_w$ are Pauli spin matrices along a set of axes $u$, $v$, $w$ in spin space, and $\sigma_0$ is $-i$ times the unit matrix. $d_u$, $d_v$ and $d_w$ are amplitudes of the pairs in the $m_s = 0$ quantum state along the corresponding axis, e.g. $d_w$ has the symmetry of a wave-function $w$ belonging to $s = 1$, where we have quantized our spins along $\hat{w}$.

The rotational properties of $\hat{\Delta}$ are peculiar, but the $d$'s transform just like $u$, $v$ and $w$ under rotations of the spin axes. The unitary transformation on particle operators $a_{\mathbf{k}\sigma}$ which rotates the spin system through the Eulerian angles $\theta$, $\varphi$, $\psi$ is given by [87]

$$U(\theta, \varphi, \psi) = \begin{pmatrix} \cos[\theta/2]\exp[-\tfrac{1}{2}i(\varphi + \psi)] & -\sin[\theta/2]\exp[\tfrac{1}{2}(\psi - \varphi)] \\ \sin[\theta/2]\exp[\tfrac{1}{2}i(\varphi - \psi)] & \cos[\theta/2]\exp[\tfrac{1}{2}i(\psi + \varphi)] \end{pmatrix}.$$

The gap matrix $\hat{\Delta}$, by its definition, transforms not unitarily but according to

$$R(\hat{\Delta}_{\mathbf{k}}) = \hat{U}(R)\hat{\Delta}_{\mathbf{k}}\tilde{\hat{U}}(R), \tag{IV.28}$$

which in turn gives a conventional orthogonal rotation transformation to the $d$-vector:

$$R(d_u, d_v, d_w) = \hat{O}(R)\mathbf{d}.$$

The unusual rotation transformation leads to some counterintuitive results—such as, for instance, that it is the antisymmetric part of $\hat{\Delta}$ which is spin rotation-invariant. Because of the further symmetry of the commutation relationship (IV.11), (IV.26) implies that (as one would expect from Fermi statistics) the singlet corresponds to a $\hat{\Delta}(k)$ even in $k$, the triplet to an odd one.

The two sets of states see essentially different potentials

$$V^S_{\mathbf{kk}'} = \tfrac{1}{2}(V_{\mathbf{k}+\mathbf{k}'} + V_{\mathbf{k}-\mathbf{k}'})$$
$$V^T_{\mathbf{kk}'} = \tfrac{1}{2}(V_{\mathbf{k}+\mathbf{k}'} - V_{\mathbf{k}-\mathbf{k}'}). \tag{IV.29}$$

In actual fact, the real potential in $He^3$ is a result of complicated many-body screening and renormalization effects, and has a spin dependence which is not necessarily derivable from a real potential $V_{\mathbf{kk}'}$. It is then convenient to give the potential an explicit spin dependence: replace the potential in (IV.1) by

$$\tfrac{1}{2}\sum_{\mathbf{kk}'}\sum_{\sigma_i} a^\dagger_{\mathbf{k}\sigma_1} a^\dagger_{-\mathbf{k}\sigma_2} a_{-\mathbf{k}\sigma_3} a_{\mathbf{k}\sigma_4} V^{\sigma_1\sigma_2;\sigma_3\sigma_4}_{\mathbf{kk}'}. \tag{IV.30}$$

Here, if the system is isotropic in spin space we may write

$$V = V_1\delta(\sigma_1, \sigma_4)\delta(\sigma_2, \sigma_3) + V_2(\sigma)_{\sigma_1\sigma_4}\cdot(\sigma')_{\sigma_2\sigma_3}$$
$$= V_1 + V_2\sigma\cdot\sigma'. \tag{IV.31}$$

The triplet sees the potential $V_1 + V_2$, the singlet $V_1 - 3V_2$. We shall see that in the anisotropic BCS theory of $He^3$, (IV.31) is not quite adequate and we have to use the general expression (IV.30). The gap equation now reads

$$\Delta^{\sigma_1\sigma_2}_{\mathbf{k}} = \sum_{\sigma_3\sigma_4}\sum_{\mathbf{k}'} V^{\sigma_3\sigma_4;\sigma_1\sigma_2}_{\mathbf{kk}'} \chi^{\sigma_3\sigma_4}_{\mathbf{k}'} \tag{IV.32}$$

but the property of $V$ that

$$V^{\alpha\beta;\gamma\delta} = V^{\gamma\delta;\alpha\beta}$$

still ensures the singlet–triplet separation.

### (c) *Orbital angular momentum separation*

The gap equation may be further separated, to a very good approximation, according to orbital angular momentum $l$. At $T_c$ the energy $\Delta$ is very small and

$$F(\hat{\mathscr{E}}^2_{\mathbf{k}}) = \frac{1}{\varepsilon_{\mathbf{k}}}\tanh\frac{\beta_c\varepsilon_{\mathbf{k}}}{2}, \tag{IV.33}$$

which is independent of $\Delta$ and isotropic in $\hat{\mathbf{k}}$. The gap equation (IV.24) then becomes a linear eigenvalue equation for the critical temperature $\beta_c$. Since the interaction $V$ is isotropic, we may make the standard spherical harmonic separation of it as given in the Introduction

$$V_{\mathbf{kk}'} = \sum_l \overline{V}_l$$
$$\overline{V}_l = V_l(k, k')P_l(\hat{\mathbf{k}}\cdot\hat{\mathbf{k}}')(2l + 1)$$
$$= \overline{V}_l(k, k')(4\pi)\sum_m Y^m_l(\hat{\mathbf{k}})Y^{-m}_l(\hat{\mathbf{k}}'). \tag{IV.34}$$

If we assume a spherical harmonic solution of (IV.24)

$$\hat{\Delta}_{\mathbf{k}} = \hat{\Delta}_l(k) Y_l(\hat{\mathbf{k}}),$$

where $Y_l$ is any linear combination of spherical harmonics of order $l$, the orthogonality of spherical harmonics of different $l$ will eliminate all terms but $Y_l$,

$$\int d\Omega_{\hat{\mathbf{k}}} Y_l(\hat{\mathbf{k}}) P_{l'}(\hat{\mathbf{k}} \cdot \hat{\mathbf{k}}') = \delta_{ll'} Y_l(\hat{\mathbf{k}}') 4\pi/(2l+1), \qquad \text{(IV.35)}$$

and in fact will reproduce $Y_l$ on the left-hand side. Thus all spherical harmonics of order $l$ are solutions at the appropriate $T_c$ for $l$, and all therefore have the same $T_c$.

This is given by the standard BCS theory as in Section III:

$$\Delta_0 Y^l(\hat{\mathbf{k}}) = \frac{V_l}{2} \sum_{\mathbf{k}'} (2l+1) P_l(\hat{\mathbf{k}} \cdot \hat{\mathbf{k}}') \Delta_0 Y^l(\hat{\mathbf{k}}') \varepsilon_{\mathbf{k}}^{-1} \tanh \frac{\beta_c \varepsilon_{\mathbf{k}}}{2}$$

or

$$1 = N(0) V_l \int_0^{\omega_s} \frac{d\varepsilon_{\mathbf{k}}}{\varepsilon_{\mathbf{k}}} \tanh \frac{\beta_c \varepsilon_{\mathbf{k}}}{2}$$

$$T_c \simeq 1{\cdot}14 \omega_s \exp\left(-\frac{1}{N(0) V_l}\right),$$

where $V_l \equiv V_l(k_F, k_F)$.

Below $T_c$ $F(\hat{\mathscr{E}}^2)$ is no longer independent of $\Delta$ or isotropic in $k$; but at most energies (if $T_c$ is small) $\varepsilon_{\mathbf{k}} \gg \Delta$, and it is approximately so. Since the $T_c$'s for different $l$'s will in general be very different, the amount of coupling between, say, $l = 1$ and 3, or 0 and 2, is quite small—Anderson and Morel estimated a few percent at most. Thus, as they showed, it is an excellent approximation to neglect $l$–$l'$ coupling, and to retain only $V_l$ at all temperatures in the gap equation.

### (d) *Special cases: Singlet $l = 2$ and free energy minimization*

Only two cases have been explored in any detail in the literature: $l = 1$ and $l = 2$. There has been some discussion of $l = 3$ but no definitive treatment. It is now rather certain that $l = 1$ describes He$^3$, so that detailed treatments of general $l$ are pointless; but it may be interesting, for orientation, to summarize $l = 2$.

In the singlet case

$$\hat{\Delta}_{\mathbf{k}} \hat{\Delta}_{\mathbf{k}}^{\dagger} = \hat{\Delta}_{\mathbf{k}}^{\dagger} \hat{\Delta}_{\mathbf{k}} = \begin{pmatrix} |d_0|^2 & 0 \\ 0 & |d_0|^2 \end{pmatrix}$$

$$E_{\mathbf{k}}^2 = \varepsilon_{\mathbf{k}}^2 + |d_0|^2 \qquad \text{(IV.36)}$$

and matrix notation is irrelevant. The gap equation becomes precisely the BCS equation, but with an $l = 2$ interaction

$$d_0(\hat{\mathbf{k}}) = V_2 \sum_{\mathbf{k}'} \frac{(2l+1)}{2} P_2(\hat{\mathbf{k}}\cdot\hat{\mathbf{k}}') d_0(\hat{\mathbf{k}}') \frac{\tanh \frac{1}{2}\beta E_{\mathbf{k}'}}{E_{\mathbf{k}'}}. \tag{IV.37}$$

Anderson and Morel showed that symmetry considerations controlled, to a great extent, the allowable solutions of (IV.37). A good way to form a solution is to take a subgroup $S$ of the rotation group $R$, such as $D_\infty$ or the cubic group $T$. If we can find an $l = 2$ spherical harmonic such that $|d_0|^2 = |Y_l|^2$ belongs to the identity representation of $S$, $d_0$ will automatically be a solution, since the kernel $FP_2$ will be invariant under $S$. This may be done in two obvious ways for $l = 2$: (1) $|Y^l_m|^2$ for any $m$ belongs to the identity of $D_\infty$, so $Y^l_m$, $m = 0, 1, 2$ is a solution. (2) We choose (with $\hat{\mathbf{k}} \equiv \hat{k}_x\hat{\mathbf{x}} + \hat{k}_y\hat{\mathbf{y}} + \hat{k}_z\hat{\mathbf{z}}$)

$$\begin{aligned} d^r_0 &= \hat{k}_z^2 - \tfrac{1}{2}(\hat{k}_x^2 + \hat{k}_y^2) \\ d^i_0 &= \frac{\sqrt{3}}{2}(\hat{k}_x^2 - \hat{k}_y^2); \end{aligned} \tag{IV.38}$$

i.e. $d^r_0$ and $d^i_0$ are the two members of the 2-dimensional $l = 2$ representation of the cubic group.

To choose among these solutions (and any others: there appears, for instance, to be an unfavourable one based on the orthorhombic group) we must compute the total free energy of each solution and choose the lowest. Within the weak-coupling approximation, the following is the appropriate procedure, since we can assume that the total modifications of background contributions to the free energy, such as collective excitations, are negligible (primarily because the response functions are modified only for $\hbar\omega \lesssim kT_c$, $q \lesssim (1/\xi_0) = k_F(\Delta/E_F)$, which is a tiny portion of phase space).

We assume that for each direction $\hat{\mathbf{k}}$ the gap matrix $\hat{\Delta}_{\mathbf{k}}$ is independent of $|\mathbf{k}|$, and as a function of $\hat{\mathbf{k}}$ represents an independent variable. The free-energy calculation is best done following the method of Eilenberger[88] which we shall need in later work as well. The scheme is to try to find a best single-quasiparticle hamiltonian

$$\mathscr{h} = \sum_{\mathbf{k}\sigma} \varepsilon_{\mathbf{k}} n_{\mathbf{k}\alpha} - \tfrac{1}{4}\sum_{\mathbf{k}} \mathrm{Tr}\,(\hat{\Delta}_{\mathbf{k}}\hat{\mathscr{X}}^\dagger_{\mathbf{k}} + \mathrm{H.C.}) \tag{IV.39}$$

where $\hat{\Delta}_{\mathbf{k}}$ is a given trial value of the energy gap to be optimized as a function of $\hat{\mathbf{k}}$. In more general terms, $\hat{\Delta}$ can be thought of as a trial self-energy and the scheme fits within the more general Luttinger–Ward[89] technique.

We calculate the trial free energy $g$ given $\mathscr{h}$ (IV.39)

$$g = -kT \ln \mathrm{Tr}\, e^{-\beta \mathscr{h}} \tag{IV.40}$$

(averages are to be taken with the trial hamiltonian $\mathscr{h}$) and correct $g$ by subtracting the mean value of the artificial interaction and replacing it with the real one:

$$F = g + \tfrac{1}{4}\langle \mathrm{Tr} \sum_{\mathbf{k}} (\hat{\Delta}_{\mathbf{k}} \hat{x}^{\dagger}_{\mathbf{k}} + \mathrm{H.C.})\rangle - V_l \sum_{\mathbf{kk}'} P_l(\hat{\mathbf{k}} \cdot \hat{\mathbf{k}}') \langle \hat{x}^{\dagger}_{\mathbf{k}} \hat{x}_{\mathbf{k}'} \rangle \frac{2l+1}{2}. \qquad \text{(IV.41)}$$

Let us add and subtract the quantity

$$\frac{1}{2(2l+1)} \frac{N(0)}{V_l} \int d\Omega_{\hat{k}} \int d\Omega_{\hat{k}'} P_l(\hat{\mathbf{k}} \cdot \hat{\mathbf{k}}') \, \mathrm{Tr}\,[\hat{\Delta}(\hat{\mathbf{k}}) \hat{\Delta}^{\dagger}(\hat{\mathbf{k}})]$$

$$= \frac{N(0)}{(2l+1)V_l} \int d\Omega_{\hat{k}} \, \mathrm{Tr}(\hat{\Delta}^{\dagger}(\hat{\mathbf{k}}) \hat{\Delta}(\hat{\mathbf{k}}))$$

and we obtain

$$F = g + \frac{1}{(2l+1)V_l} \int d\Omega_{\hat{k}} \, \mathrm{Tr}\, \hat{\Delta}^{\dagger}(\hat{\mathbf{k}}) \hat{\Delta}(\hat{\mathbf{k}})$$

$$- \frac{\mathrm{Tr}}{2(2l+1)V_l} \int d\Omega_{\hat{k}} \left( \hat{\Delta}(\hat{\mathbf{k}}) - \frac{(2l+1)}{2} V_l \sum_{\mathbf{k}'} P_l(\hat{\mathbf{k}} \cdot \hat{\mathbf{k}}') \hat{x}_{\mathbf{k}'} \right)$$

$$\times \left( \hat{\Delta}^{\dagger}(\hat{\mathbf{k}}) - \frac{(2l+1)}{2} V_l \sum_{\mathbf{k}'} P_l(\hat{\mathbf{k}} \cdot \hat{\mathbf{k}}') \hat{x}^{\dagger}_{\mathbf{k}'} \right). \qquad \text{(IV.42)}$$

$F$ is now to be minimized by differentiating with respect to $\hat{\Delta}(\hat{\mathbf{k}})$ functionally. By the optimal properties of the free energy, this must give us the correct gap equation (IV.24), by virtue of which the last term above vanishes. Since this term is negative definite, it is equally valid to use as the free energy for minimization the far simpler expression obtained by removing this term, which must be greater than or equal to $F$:

$$F' = g + \frac{1}{(2l+1)V_l} \mathrm{Tr} \int d\Omega_{\hat{k}} \hat{\Delta}^{\dagger}_{\hat{k}} \hat{\Delta}_{\hat{k}}. \qquad \text{(IV.43)}$$

This is the form which corresponds to the free energy used in the generalized Ginsburg–Landau theories (see also Werthamer,[90]) and which results from perturbation theory.

Now $g$ may be calculated directly from

$$g = \langle h \rangle - T\langle S \rangle,$$

where contributions from each $\mathbf{k}$ add in each term, and $S_k$ is given by Balian and Werthamer:

$$\langle S \rangle = -\sum_{\mathbf{k}} \mathrm{Tr}\,\{\tfrac{1}{2}(1 \pm \hat{W}^{\mathbf{k}}) \ln \tfrac{1}{2}(1 \pm \hat{W}^{\mathbf{k}})\}. \qquad \text{(IV.44)}$$

The result for $g$ takes on especially simple forms in two limits, although of course it may be integrated out in general. In the limit $T \to 0$, TS vanishes and one need only do $\langle h \rangle$. We will follow the procedure of integrating radially over $\varepsilon_{\mathbf{k}}$ and leaving the angular integrations to be done later.

$$\langle h_k \rangle = 2N(0) \int_0^{\omega_s} d\varepsilon_{\mathbf{k}} \operatorname{Tr} \hat{w}^{\mathbf{k}} - \frac{N(0)}{4} \int_{-\omega_s}^{\omega_s} d\varepsilon_{\mathbf{k}} \operatorname{Tr} \langle \hat{\Delta}_{\mathbf{k}} \hat{x}_{\mathbf{k}}^{\dagger} + \text{H.C.} \rangle. \qquad \text{(IV.45)}$$

The somewhat arbitrary cutoff $\omega_s$ must be introduced.[10] At $T = 0$, $\hat{x}_{\mathbf{k}} = (E_{\mathbf{k}})^{-1} \hat{\Delta}_{\hat{\mathbf{k}}}$ and the integral on the right is

$$\frac{\Delta_{\hat{\mathbf{k}}}^{\dagger} \hat{\Delta}_{\hat{\mathbf{k}}}}{(\varepsilon_{\mathbf{k}}^2 + \Delta_{\hat{\mathbf{k}}}^{\dagger} \Delta_{\hat{\mathbf{k}}})^{\frac{1}{2}}}.$$

Thus all quantities are functions of $\Delta_{\hat{\mathbf{k}}}^{\dagger} \Delta_{\hat{\mathbf{k}}}$ and may be simultaneously diagonalized; all traces become, then, simply sums over the two eigenvalues of this quantity. We give the result in the appropriate weak-coupling limit $\Delta \ll \omega_s$:

$$\langle h_k \rangle = \frac{N(0)}{4} \operatorname{Tr} \left( -\hat{\Delta}_{\hat{\mathbf{k}}} \hat{\Delta}_{\hat{\mathbf{k}}}^{\dagger} \ln \frac{2\omega_s}{\Delta\Delta^{\dagger}} - \hat{\Delta}\hat{\Delta}^{\dagger} \right) \qquad \text{(IV.46)}$$

so that the total "free energy" (IV.43) becomes

$$F_0 = \frac{N(0)}{4} \int \frac{d\Omega_{\hat{\mathbf{k}}}}{4\pi} \operatorname{Tr} \left[ -\hat{\Delta}\hat{\Delta}^{\dagger} \ln \frac{2\omega_s}{\Delta\Delta^{\dagger}} - \hat{\Delta}\hat{\Delta}^{\dagger} \left( 1 - \frac{4}{(2l+1)V_l} \right) \right]. \qquad \text{(IV.47)}$$

Inserting the gap equation, which we obtain by differentiating with respect to $\hat{\Delta}\hat{\Delta}^{\dagger}$, this becomes the expression first given by Anderson and Morel:

$$E_0 = -\frac{N(0)}{8\pi} \int d\Omega_{\hat{\mathbf{k}}} \operatorname{Tr} \hat{\Delta}_{\hat{\mathbf{k}}} \hat{\Delta}_{\hat{\mathbf{k}}}^{\dagger}. \qquad \text{(IV.48)}$$

The ground-state energy is proportional to the total normalization of the gap.

In the "Ginzburg–Landau" limit $T \to T_c$ we can expand all quantities in power series in the gap parameter. The coefficients may be deduced from the corresponding coefficients in a power-series expansion of the gap equation (IV.24), if we can determine the normalization. This may be done by observing that the entropy, which is proportional to $\tanh(\beta\mathscr{E}/2) - \tanh(\beta\varepsilon/2)$, converges exponentially to zero at large $\varepsilon_{\mathbf{k}}$, so that the logarithmic terms must all come from (IV.45), and can be shown to be of the form

$$(F(T))_{\text{logarithmic}} \simeq -\frac{N(0)}{2} \int_{-\omega_s}^{\omega_s} \varepsilon^2 \, d\varepsilon \operatorname{Tr} \frac{\hat{\Delta}\hat{\Delta}^{\dagger} \tanh(\beta E/2)}{E^3}$$

$$= -\frac{N(0)}{2} \operatorname{Tr} \hat{\Delta}\hat{\Delta}^{\dagger} \left[ \ln \frac{\omega_s}{T} + \text{const.} \right]; \qquad \text{(IV.49)}$$

and since the remainder of the integrals are all convergent they can only depend on $(\Delta\Delta^\dagger)/T^2$, so that we have

$$F(T) = -\frac{N(0)}{2}\int\frac{d\Omega_{\hat{\mathbf{k}}}}{4\pi}\operatorname{Tr}\left\{\hat{\Delta}_{\mathbf{k}}\hat{\Delta}_{\mathbf{k}}^\dagger \ln T_c/T + T^2 f\left(\frac{\hat{\Delta}\hat{\Delta}^\dagger}{T^2}\right)\right\} \tag{IV.50}$$

where, since any quadratic term can be included in $\ln T_c$, $f$ must start with fourth-order terms. Explicitly, from well-known expansions (see, e.g. Ambegaokar and Mermin,(28) Anderson,(25) Brinkman, Serene and Anderson,(37) Kuroda(38, 39)) we have

$$F = -N(0)\int\frac{d\Omega_{\hat{\mathbf{k}}}}{4\pi}\frac{1}{2}\operatorname{Tr}\left\{\hat{\Delta}_{\mathbf{k}}\hat{\Delta}_{\mathbf{k}}^\dagger \ln T_c/T + \frac{(\hat{\Delta}\hat{\Delta}^\dagger)^2}{T^2}\frac{7\zeta(3)}{8\pi^2} - \frac{(\hat{\Delta}\hat{\Delta}^\dagger)^3}{T^4}(0{\cdot}007735) + \ldots\right\}. \tag{IV.51}$$

Again, if we stop at fourth order and use the gap equation

$$\frac{\partial F}{\partial(\Delta\Delta^\dagger)} = 0,$$

the fourth-order term becomes $-\frac{1}{2}$ times the second-order one and we have

$$F(T \simeq T_c) \simeq -\frac{N(0)}{4}\ln\frac{T_c}{T}\int\frac{d\Omega_{\hat{\mathbf{k}}}}{4\pi}\operatorname{Tr}\hat{\Delta}_{\hat{\mathbf{k}}}\hat{\Delta}_{\hat{\mathbf{k}}}^\dagger. \tag{IV.52}$$

Again, the free energy is proportional to the normalization of $\Delta^2$.

Specialized to the singlet case, the question is then for which state the gap eqn (IV.37) allows the largest $\langle d_0^2\rangle$. The general rule may be understood by realizing that $(\tanh \beta E_k/2)/E_k$ is a concave function of $E^2$ and hence of $d_0^2$.

Integration over $\varepsilon$ cannot remove this property: the curvature will remain upwards. Multiplying (IV.37) on the left by $d_0(k)$, and using the normalization of $P_2(\hat{\mathbf{k}}\cdot\hat{\mathbf{k}}')$, we recognize that it is in an equation of the form

$$\langle d_0^2(k)\rangle = \int d\Omega_{\hat{\mathbf{k}}} d_0^2(k')\Psi(d_0^2(k')) = \langle d_0^2\Psi(d_0^2)\rangle$$

where

$$\frac{d^2\Psi(x)}{dx^2} > 0 \tag{IV.53}$$

everywhere. This means that more uniform solutions are favoured, because the right-hand side will allow a bigger $\langle\Psi\rangle$. We see easily from the free-energy expressions (IV.47) and (IV.51) that the function $\Psi$ near $T = 0$ is essentially $\propto -\ln|d_0^2|$, while near $T_c$ it must clearly be $|d_0|^2$. The solution

(IV.38) is favoured near $T = 0$, while as Mermin and Stare[91] have pointed out the $Y_1^2$ and $Y_2^2$ solutions have the same $\langle d_0^4 \rangle$ and are *a priori* acceptable at $T_c$. It seems likely, however, that the more uniform "cubic" solution will be favoured at all $T < T_c$ and is the only stable solution in the weak-coupling limit.

(e) *Special cases: Triplet* $l = 1$

(i) *Nonunitary case.* The $l = 2$ case is only of academic interest. What occurs in the $l = 1$, $s = 1$ case which almost certainly represents $He^3$?

The eigenvalue spectrum is given by diagonalizing (IV.20) or (IV.25) with (IV.26)

$$\hat{\Delta}_{\mathbf{k}}^{\dagger}\hat{\Delta}_{\mathbf{k}} =$$
$$\begin{pmatrix} |d_u|^2 + |d_v|^2 + |d_w|^2 + i(d_u d_v^* - d_v d_u^*) & -i(d_v d_w^* - d_w d_v^*) + d_w d_u^* - d_u d_w^* \\ i(d_v d_w^* - d_w d_v^*) + d_u d_w^* - d_w d_u^* & |d_u|^2 + |d_v|^2 + |d_w|^2 - i(d_u d_v^* - d_v d_u^*) \end{pmatrix} \tag{IV.54}$$

i.e.

$$(E_{\mathbf{k}}^2 - \varepsilon_{\mathbf{k}}^2 - |\mathbf{d}|^2)^2 = |i\mathbf{d}^* \times \mathbf{d}|^2. \tag{IV.55}$$

The eigenvalues of $\hat{\Delta}\hat{\Delta}^{\dagger}$ are thus

$$|\mathbf{d}|^2 \pm |i\mathbf{d}^* \times \mathbf{d}|. \tag{IV.56}$$

The quantity $\mathbf{d}^* \times \mathbf{d}$ represents a kind of spin angular momentum in the sense that in the pictorial description in terms of $(He^3)_2$ molecules, it would give the molecular spin moment. We can see this, for instance, from (IV.26) by setting

$$d_v = id_u = d$$
$$d_w = 0$$

which gives a $\Delta$ only for $\uparrow\uparrow$ pairing, $m_s = 1$, while in this case $i\mathbf{d} \times \mathbf{d}^* = +2|d|^2\hat{w}$. Then general spin moments follow from the fact that $\mathbf{d}$ transforms like a vector under rotations. In the presence of $(i\mathbf{d}^* \times \mathbf{d})$, time-reversal symmetry is broken and the Kramers degeneracy removed. $(i\mathbf{d}^* \times \mathbf{d})$ does not, however, imply the presence of a total spin angular momentum, both because it may average out over the Fermi surface, since $d_u$, $d_v$ and $d_w$ are functions of $\hat{\mathbf{k}}$, and because moments can and do also arise from displacements of the Fermi surface. The correlation-hump picture mentioned in the Introduction clarifies this point: the existence of a $(\mathbf{d}^* \times \mathbf{d})_w$ implies a spin-polarized correlation hump $\langle n\uparrow(\mathbf{0})n\uparrow(\mathbf{r})\rangle \neq \langle n\downarrow(\mathbf{0})n\downarrow(\mathbf{r})\rangle$ but that does not imply an average $\langle n\uparrow - n\downarrow\rangle$. In fact, in the most obvious physical phase where $\mathbf{d}^*\times\mathbf{d}$ exists, the $A_1$ phase which occurs in a magnetic field near $T_c$, there is a moment because of Fermi-surface displacement and also a tiny $\mathbf{d}^* \times \mathbf{d}$ moment resulting from hole-particle disymmetry (see (IV.98) following).

Calculating the free energy of solutions with $\mathbf{d}^* \times \mathbf{d}$ finite is complicated even near $T_c$ or 0. These are called "non-unitary" because, as (IV.54) and (IV.55) show, in their absence $\hat{\Delta}^\dagger\hat{\Delta}$ is proportional to a unit matrix and $E_\mathbf{k}$ is given by the standard BCS gap expression

$$\begin{aligned} E_\mathbf{k}^2 &= \varepsilon_\mathbf{k}^2 + |d_u|^2 + |d_v|^2 + |d_w|^2 \\ &= \varepsilon_\mathbf{k}^2 + |\mathbf{d}|^2. \qquad (\text{"unitary"}) \end{aligned} \tag{IV.57}$$

Our general free-energy formalism clearly applies even in the non-unitary cases, and tells us that we must minimize the variation of $\hat{\Delta}\hat{\Delta}^\dagger$ *including* the variation caused by $\mathbf{d}^* \times \mathbf{d}$. That is, if we call the two eigenvalues of $\hat{\Delta}$ (IV.56). $\Delta_1$ and $\Delta_2$, we want to minimize

$$T \to 0 \qquad \frac{\langle -|\Delta_1|^2 \ln|\Delta_1|^2 - |\Delta_2|^2 \ln|\Delta_2|^2\rangle}{\langle|\Delta_1|^2 + |\Delta_2|^2\rangle} \tag{IV.58}$$

or

$$T \to T_c \qquad \frac{\langle|\Delta_1|^4 + |\Delta_2|^4\rangle}{\langle|\Delta_1|^2 + |\Delta_2|^2\rangle^2}.$$

A non-unitary solution can make $\Delta$ more uniform by making $|\mathbf{d}|^2$ more uniform: Re $\mathbf{d}$ can be made up of a different set of functions from Im $\mathbf{d}$. This is not, however, possible for $l = 1$, since Re $\mathbf{d}$ can saturate the set of only 3 independent $l = 1$ functions.

For $l = 3$ a non-unitary solution will improve $|\mathbf{d}|^2$ since there are 7 functions to choose from, but in actual fact [contrary to a previous remark (Anderson[24, 25])] this does not seem to compensate for the extra variation $|\mathbf{d}^* \times \mathbf{d}|^2$.

Thus non-unitary solutions occur only in the presence of magnetic fields or in boundaries. The one well-known case is the A phase near $T_c$ in a large magnetic field, which will be discussed as an interesting NMR phenomenon.

(ii) *Unitary Solutions: BW versus AM*. For the discussion of $l = 1$ solutions it is very convenient to introduce a "bivector" notation in which we expand the spin-space vector $\mathbf{d}$ in orbit-space components:[91, 35, 36]

$$d_u(\hat{\mathbf{k}}) = \sum_\alpha d_{\alpha u}\hat{k}_\alpha \text{ etc.} \tag{IV.59}$$

$d_{\alpha i}$ transforms like a vector in orbit space as regards its first component, in spin space as regards its second and thus like a tensor under full rotation. Except for the weak dipolar interactions the energy is invariant under both types of transformation.

The general principle given for $l = 2$ allows us to construct at least three solutions. That of Balian and Werthamer[12] comes from using the full rotation group as our "subgroup", under which $k_x$, $k_y$ and $k_z$ determine a 3-dimensional representation, which we may use for $d_u$, $d_v$ and $d_w$. The BW solution is then given by all spin and space rotations of

$$\text{BW}: \; d_{\alpha i} = d \begin{pmatrix} 1 & 0 & 0 \\ 0 & 1 & 0 \\ 0 & 0 & 1 \end{pmatrix}. \tag{IV.60}$$

A second subgroup is the axial group, and we may construct solutions based on either $Y_0^1$ or $Y_{\pm 1}^1$ relative to a given axis $z$. The solution based on $Y_0^1 = \hat{k}_z$ is the most non-uniform possible function, and, although it could be favoured for extreme spin fluctuations, it does not occur and we will not further discuss it. Anderson and Morel pointed out that we can use $Y_{\pm 1}^1 = \hat{k}_x \pm i\hat{k}_y$ and pair up spins and down spins independently. In terms of the gap matrix $\hat{\Delta}$ this leads to

$$\hat{\Delta} = \Delta_0 \begin{pmatrix} \hat{k}_x + i\hat{k}_y & 0 \\ 0 & \hat{k}_{x'} + i\hat{k}_{y'} \end{pmatrix}, \tag{IV.61}$$

where we observe that in weak coupling the $\uparrow\uparrow$ and $\downarrow\downarrow$ pairs may be totally decoupled and so we can allow the $\uparrow\uparrow$ and $\downarrow\downarrow$ orbital directions to be different. In fact, there are two explicitly unitary and axially symmetric solutions. The first, chosen more or less by chance by Anderson and Morel, is the parallel one now known as the ABM solution, which in terms of **d** is

$$\text{ABM}: \; \mathbf{d} = \begin{pmatrix} d & 0 & 0 \\ id & 0 & 0 \\ 0 & 0 & 0 \end{pmatrix} \begin{matrix} x \\ y; \\ z \end{matrix} \qquad \begin{matrix} u & v & w \end{matrix} \tag{IV.62}$$

and the second is the "real" solution suggested by Balian and Werthamer, in which $y' \to -y$:

$$\text{A}': \mathbf{d} = \begin{pmatrix} d & 0 & 0 \\ 0 & d & 0 \\ 0 & 0 & 0 \end{pmatrix}. \tag{IV.63}$$

This solution is very likely to be the limit of BW near a vessel wall, where the component of $k$ perpendicular to the wall will be severely depaired. Although ABM and A′ have the same energy in weak coupling, their response functions are very different, as we shall see, and as a consequence their higher-order free energies can be very different. (IV.63) never occurs in equilibrium.[91]

In weak-coupling theory there was never any doubt that the BW solution (IV.60) is the minimum one, as Balian and Werthamer proved. The gap parameter is given by $\overline{(\mathbf{d})^2} = \overline{d^2}$, which is totally uniform and cannot be improved on under any criterion. In fact $\overline{d^2_{\mathrm{ABM}}} = 0{\cdot}88\, \overline{d^2_{\mathrm{BW}}}$ at $T = 0$, $\overline{d^2_{\mathrm{ABM}}} = \frac{5}{6}\overline{d^2_{\mathrm{BW}}}$ as $T \to T_c$. Most of the thermal properties of BW are identical with those of BCS. It was thus rather a puzzle that the "A" stable state first discovered could clearly not be BW, experimentally.

In attempting to discuss this problem, Mermin and Stare[91] and Brinkman and Anderson[35, 36] developed a general scheme for the Ginzburg–Landau regime near $T_c$, which has been of value in later work. In the absence of a theory of the phases they investigated by group theory what quantities might enter the free energy (IV.50) or (IV.51) to second and fourth order, taking into account rotation symmetry for both spins and orbits. Clearly to second order only the invariant

$$
\begin{aligned}
I_2 &= \mathrm{Tr}\int \frac{d\Omega_{\hat{\mathbf{k}}}}{4\pi}\hat{\Delta}_{\hat{\mathbf{k}}}\hat{\Delta}^{\dagger}_{\hat{\mathbf{k}}} \\
&= d^*_{\alpha i} d_{\alpha i} \qquad \text{(IV.64)}
\end{aligned}
$$

(using the usual summation convention for repeated indices) can enter, but in fourth order there are five invariants:

$$
\begin{aligned}
A_1 &= \Big|\sum_{\alpha, i} (d_{\alpha i})^2\Big|^2; \qquad & A_2 &= d_{\alpha i} d_{\alpha j} d^*_{\beta i} d^*_{\beta j}; \\
A_3 &= d^*_{\alpha i} d^*_{\beta i} d_{\alpha j} d_{\beta j}; \qquad & A_4 &= \Big(\sum_{\alpha i} |d_{\alpha i}|^2\Big)^2 = (I_2)^2; \\
A_5 &= d^*_{\alpha i} d^*_{\beta j} d_{\alpha j} d_{\beta i}. & & \qquad \text{(IV.65)}
\end{aligned}
$$

They proposed that the coefficients of these five invariants be considered as arbitrary parameters of the theory:

$$
F = \alpha I_2 + \sum_i a_i A_i \qquad \text{(IV.66)}
$$

and observed that the correct weak-coupling form of the coefficients is

$$
\mathrm{BW}: a_2 = a_3 = a_4 = -a_5 = -2a_1 \qquad \text{(IV.67)}
$$

where $a_1 = (N(0)/T_c^2)7\zeta(3)/120\pi^2$ (Ambegaokar and Mermin[28]). In a general strong-coupling theory $\alpha$ remains the same but the $a_i$'s become parameters to be calculated, as we shall see in Section V.

## 2. STATIC AND LOW-FREQUENCY PROPERTIES OF ANISOTROPIC SUPERFLUID MODEL PHASES

In this subsection we will discuss various low-frequency and static phenomena most of which can be calculated reasonably directly from old-fashioned BCS theory and do not require the apparatus of temperature perturbation theory or Landau–Ginzburg–Gor'kov methods, and on the whole we will not indulge in these sophisticated techniques. We are essentially looking at the ways in which the simple properties of the anisotropic superfluid differ from those of a classic BCS superconductor. It is a special feature of $He^3$ that because it is a rather strongly coupled Fermi liquid, many of these results will be modified by Landau Fermi-liquid factors. This aspect of the problem we reserve for the next section, where we will see that the BCS calculation is a necessary preliminary to that one.

### (a) *Correlations*

As we remarked in the Introduction, many of the properties follow, and more can be understood, from a consideration of the additional correlations which are introduced by the existence of the anomalous amplitudes $x^{\mathbf{k}}_{\sigma\sigma'}$. Except for negligible corrections to the Wigner exchange hole, these follow directly by simply factorizing the appropriate two-particle correlation into the relevant $x$'s. For instance,

$$\begin{aligned}\Delta(\langle n_\sigma(\mathbf{r})n_{\sigma'}(\mathbf{r}')\rangle &= \langle\Psi^\dagger_\sigma(\mathbf{r})\Psi^\dagger_{\sigma'}(\mathbf{r}')\rangle\langle\Psi_\sigma(\mathbf{r}')\Psi_\sigma(\mathbf{r})\rangle \\ &= b^\dagger_{\sigma\sigma'}(\mathbf{r}-\mathbf{r}')b_{\sigma'\sigma}(\mathbf{r}'-\mathbf{r}) \qquad \text{(IV.68)}\end{aligned}$$

where

$$\begin{aligned}b_{\sigma\sigma'}(\mathbf{r}-\mathbf{r}') &= \sum_{\mathbf{k}}\exp\left[i\mathbf{k}\cdot(\mathbf{r}-\mathbf{r}')\right]\langle c_{\mathbf{k}\sigma}c_{-\mathbf{k}\sigma'}\rangle \\ &= \tfrac{1}{2}\sum_{\mathbf{k}}\exp\left[i\mathbf{k}\cdot(\mathbf{r}-\mathbf{r}')\right]x^{\mathbf{k}}_{\sigma\sigma'}. \qquad \text{(IV.69)}\end{aligned}$$

Anderson and Morel, using methods suggested by the BCS paper[1], computed this quantity for absolute zero and for $k_F r \gg 1$ in terms of the gap matrix $\hat{\Delta}$, and found

$$\hat{b}(r) = \frac{N_0\cos k_F r}{ik_F r}K_0\left(\frac{r|\hat{\Delta}(\hat{\mathbf{r}})|}{\hbar v_F}\right)\hat{\Delta}(\hat{\mathbf{r}}) \qquad \text{(IV.70)}$$

(here it is unnecessary to go beyond the unitary case). By $\hat{\Delta}(\hat{\mathbf{r}})$ we mean $\hat{\Delta}(\hat{\mathbf{k}})$ with $\hat{\mathbf{k}}\|\hat{\mathbf{r}}$. The integration technique used by Anderson and Morel and by Bardeen, Cooper and Schrieffer is actually even easier in the opposite limit† $T \to T_c$ and gives us

$$\hat{b}(r) = \hat{\Delta}(\hat{\mathbf{r}}) \frac{N_0 \cos k_F r}{ik_F r} \ln \coth\left(\frac{\pi k T_c}{2\hbar v_F}\right) r$$

$$= \hat{\Delta}(\hat{\mathbf{r}}) \frac{2N_0 \cos k_F r}{ik_F r} \ln \coth\left(\frac{r}{3{\cdot}5\zeta_0}\right), \qquad \text{(IV.71)}$$

defining the average coherence length $\xi_0$ in terms of $T_c$. The correlation hump is an oscillating function with exponentially decaying amplitude at large $r$, with a range of the order of the coherence length, $\approx 10^3$ Å. Its total amplitude may be calculated by integrating over $r$, or, what is equivalent because of completeness, by summing over $k$:

$$\int \Delta\langle n_\sigma(\mathbf{0}) n_{\sigma'}(\mathbf{r})\rangle \, \mathrm{d}^3 r = n_c = \tfrac{1}{4}\sum_{\mathbf{k}} |x^{\mathbf{k}}_{\sigma\sigma'}|^2. \qquad \text{(IV.72)}$$

As calculated by Anderson and Morel for temperatures near absolute zero, this is

$$n_c(T=0) = \tfrac{1}{4}\sum_{\mathbf{k}} \frac{|\Delta^{\sigma\sigma'}_{\mathbf{k}}|^2}{E^2_{\mathbf{k}}}$$

$$= \frac{\pi N(0) |\Delta^{\sigma\sigma'}_{\mathbf{k}}|^2}{4} \int \frac{\mathrm{d}\Omega}{4\pi} (\mathrm{Tr}\, \Delta_{\mathbf{k}} \Delta^\dagger_{\mathbf{k}})^{-\frac{1}{2}}$$

$$\approx \frac{\pi N(0)}{4} \Delta. \qquad \text{(IV.73)}$$

For temperatures near $T_c$ the result is

$$n_c(T_c) = |\Delta^{\sigma\sigma'}_{\mathbf{k}}|^2 \frac{N(0)}{4} \int \frac{\mathrm{d}\varepsilon}{\varepsilon^2} \tanh^2 \frac{\beta\varepsilon}{2}$$

$$= |\Delta^{\sigma\sigma'}_{\mathbf{k}}|^2 \frac{N(0)}{T_c} \frac{7}{\pi^2} \zeta(3). \qquad \text{(IV.74)}$$

Note that the result is of order $\Delta/E_F$ or about $10^{-3}$ in He$^3$.

Among effects which follow directly from the correlation hump itself, is first, of course, the condensation energy

$$\int V(\mathbf{r} - \mathbf{r}') |b(\mathbf{r} - \mathbf{r}')|^2 \, \mathrm{d}^3 r \, \mathrm{d}^3 r',$$

† The relevant integral is[92]

$$\int_0^\infty \cos \frac{r\varepsilon}{\hbar v_F} \frac{\mathrm{d}\varepsilon}{\varepsilon} \tanh \frac{\beta\varepsilon}{2} = \ln \coth \frac{\pi r}{2\hbar v_F \beta}.$$

which we have computed in the last section. The anisotropic spin–position correlation leads to a non-zero nuclear spin–spin dipolar interaction, which is vital to the discussion of orientation effects. Another effect which may eventually be experimentally important is the effect of this correlation on dielectric constants and index of refraction.(93) This comes about through the modification of local field corrections. From the magnitude of the magnetic dipolar orientation energies (see later) we find that $\langle 1/r^3 \rangle \approx 10^{20}\,\text{cm}^{-3}$, so that we can expect a change in the local field correction $\Delta L \approx 1/nr^3$, i.e. of order $10^{-3}$. This will lead to an anisotropy of the dielectric constant $\varepsilon$ of order $(\varepsilon - 1)^2 \Delta L \approx 10^{-6}$. This will give a barely observable birefringenece and an appreciable orienting effect in an electric field of order $10^4$–$10^5$ volt $\text{cm}^{-1}$, in the ABM state, at which field the effect is comparable to the dipolar interaction.

In addition to the static density and spin correlations, Anderson and Morel pointed out the existence of density–current correlations $\langle n\mathbf{j} \rangle$, and Balian and Werthamer generalized this to spin–current correlations in the pseudoisotropic BW state.

Again, in the most direct model theory such correlations may be computed by factorizing the product of one-particle operators which defines it:

$$\langle n_\sigma(\mathbf{r})\mathbf{j}_{\sigma'}(\mathbf{r}') \rangle = \frac{i\hbar}{2m}\big[\langle \Psi_\sigma^\dagger(\mathbf{r})\nabla\Psi_{\sigma'}^\dagger(\mathbf{r}')\rangle\langle \Psi_{\sigma'}(\mathbf{r}')\Psi_\sigma(\mathbf{r})\rangle - \langle \Psi_\sigma^\dagger(\mathbf{r})\Psi_{\sigma'}^\dagger(\mathbf{r}')\rangle\langle \nabla\Psi_{\sigma'}(\mathbf{r}')\Psi_\sigma(\mathbf{r})\rangle\big]. \tag{IV.75}$$

Obviously this too follows the general rule that $b$ may be manipulated as though it were the wave function of the pairs:

$$\langle n_\sigma(\mathbf{0})\mathbf{j}_\sigma(\mathbf{r}) \rangle = \frac{i\hbar}{2m}\langle \nabla b^\dagger(\mathbf{r})b(\mathbf{r}) - b^\dagger(\mathbf{r})\nabla b(\mathbf{r})\rangle. \tag{IV.76}$$

Since in (IV.76) the $r$-dependent factors have constant phase, the current correlation is entirely azimuthal and determined by the phases of the components of $\hat{\Delta}$. In the ABM state,

$$\hat{\Delta}^\dagger(\hat{\mathbf{r}}) = \hat{\Delta}^\dagger\, e^{i\psi}$$

where $\psi$ is the angle in the $x$–$y$ plane: as a result

$$\langle n\mathbf{j} \rangle = \hat{\boldsymbol{\Psi}}\frac{\hbar}{mr}\langle nn \rangle, \tag{IV.77}$$

or the density–density correlation rotates with a net angular momentum $\hbar$ about the $z$ axis. The total angular momentum of this motion is, then, $\int \mathbf{r} \times \langle n\mathbf{j} \rangle$ which is equal to $\hbar$ times the total number of correlated particles. This, from (IV.73) or (IV.74), gives a total orbital angular momentum in the correlation hump of order $\hbar\Delta/E_F \approx 10^{-3}\hbar$.

This total angular momentum must, as Anderson and Morel emphasized, imply a net orbital angular momentum of the sample as a whole. As in the conventional discussions of diamagnetism or ferromagnetism, in a uniform sample the correlation currents at any interior point will average to zero, but at a surface there will be a net surface current. Equally, if the state is slowly-varying but not uniform in space, we can expect a net current related to this orbital angular momentum density:

$$\mathbf{J}_L = \nabla \times \mathbf{L}. \tag{IV.78}$$

The existence of $\langle L \rangle$ is indirectly fairly important to a number of physical phenomena such as the orbital supercurrents, and directly to the possible existence of "orbit waves" with a $k^2$ dispersion relation. $\mathbf{J}_L$ should appear as part of the general supercurrent structure,[49, 50, 53] which we will describe later in this article. Relative to other supercurrent terms this first appears in order $\Delta^2/E_F^2$. This is one order smaller than given by Anderson and Morel, the difference being the result of a cancellation due to hole–particle symmetry.[94] For this reason orbit waves may not have quite as low a frequency as has been proposed.[24, 25, 53]

A convenient formal expression for $\langle L \rangle$ has been suggested by de Gennes.[50, 51] We write the gap function in terms of the $d$ bivector $d_{\alpha i}$ (IV.59)

$$d_i(\hat{\mathbf{k}}) = d_{\alpha i}\hat{k}_\alpha$$

and observed that the $\alpha$ component of the density–current correlation (IV.76) may be written, using (IV.70) or (IV.71), as

$$\langle n_\sigma \mathbf{j}_{\sigma'} \rangle^\alpha = \frac{\hbar}{2m}\frac{f(r)}{\rho}(\sigma_v\sigma_i)_{\sigma\sigma'}(\sigma_j\sigma_v)_{\sigma'\sigma}(d_{i\alpha}d^*_{j\beta} - d^*_{i\beta}d_{j\alpha})x_\beta,$$

where $f(r)$ is the numerical $r$-dependent factor of $b$ in (IV.70) or (IV.71) and

$$\int r^2\,\mathrm{d}r \int \mathrm{d}\Omega(m(\hat{\mathbf{r}} \times \langle n\mathbf{j} \rangle)_\alpha) = \langle L \rangle_\alpha$$

$$\int r^2\,\mathrm{d}r \int \frac{\hbar f(r)}{\rho}(\sigma_v\sigma_i)_{\sigma\sigma'}(\sigma_j\sigma_v)_{\sigma'\sigma}\varepsilon_{\alpha\beta\gamma}(d_{i\alpha}d_{i\beta} - d_{i\beta}d_{i\alpha})$$

$$= \int \frac{\hbar}{6}\frac{f(r)}{\rho} r^2\,\mathrm{d}r\,(\text{spin factors}) \times (\mathbf{d} \times \mathbf{d}^*)_\alpha; \tag{IV.79}$$

i.e. when integrated over $r$ the total orbital angular momentum is proportional to $\mathbf{d} \times \mathbf{d}^*$, where in this case the cross-product refers to the orbital, not spin, components of the $d$ bivector. As in the case of the spin, $\mathbf{d}^* \times \mathbf{d}$ is the nominal angular momentum of the pairs, and the total real angular momentum has a rather complex relation to this quantity which is not yet wholly understood. For the ABM state, both spins have the same orbital shape of $d$; but in the pseudoisotropic state, as Balian and Werthamer indeed pointed out, or in the A′ state, there is an orbital angular momentum for each spin direction but it cancels out on the average when summed over spins.

### (b) *Susceptibility and orientation effects*

The nuclear spin susceptibility plays a great role in many phenomena in $He^3$ and will be derived here in some qualitative detail. First, however, we would like to clarify the confusing remark which one finds in the literature that this susceptibility is not "really" anisotropic. This is no more meaningful than it would be to say that the optical or elastic properties of crystalline quartz are "really" isotropic. The phases of $He^3$ are broken symmetry states[95] for which the only practical method of discussion is to assume that the spin and orbital axis systems, as well as the complex phase of the order parameter, are macroscopic variables of state, equivalent in nature to the orientation and position of a crystal (and in fact, in resonance phenomena the orbital axis system plays precisely the mathematical role of the crystal axes in antiferromagnetic resonance). Thus we calculate in every case under the assumption of a fixed numerical value and orientation of the $d_{\alpha i}$ bivector. Under some circumstances the $d$ bivector will orient itself so as to optimize the nuclear susceptibility in a magnetic field, and $\chi$ will appear isotropic; but experimentally this is by no means always the case.

Here we calculate the susceptibilities purely within the weak-coupling theory, assuming non-interacting quasiparticles. As we have remarked, these must, and will in the next subsection, be corrected severely for Fermi-liquid interactions. There is clearly a question whether that is the only correction which need be applied. Again, it seems certain that the universal justification, that pairing simply modifies too small a region of phase space to be significant, is adequate to guarantee that.

This being the case, the calculation of $\omega$, $k = 0$ susceptibilities becomes almost trivial (the far more complicated problem of general $q$ and $\omega$ is carried through in the spin-fluctuation section). The total susceptibility will be the sum of those for each $k$, and therefore, of course, also for each direction $\hat{\mathbf{k}}$. In the sub-hamiltonian $h_k$ for each $\mathbf{k}$ we insert the effect of the magnetic field on the one-particle energy $\varepsilon_{\mathbf{k}}$

$$\varepsilon_{\mathbf{k},\sigma} \rightarrow \varepsilon_{\mathbf{k}} + \mu\boldsymbol{\sigma}\cdot\mathbf{H} \qquad \text{(IV.80)}$$

and recalculate the eigenvalues $E_{\mathbf{k}}$, and from them we can calculate the susceptibilities and add them up.

Even this need not be done in general. Since each of these susceptibilities is a tensor, we can pick the three principal axes of that tensor and then rotate to calculate a general direction.

For any unitary odd-$l$ state, and for any Fermi-surface direction $\hat{\mathbf{k}}$, $d_i(\hat{\mathbf{k}})$ is a vector whose components all have the same phase, and thus defines a real vector $\mathbf{d}_i$ in spin space. Clearly the susceptibility axis must be $\mathbf{d}_i$. For this direction, the $\mathscr{E}_{\mathbf{k}}$ tensor becomes

$$\mathscr{E}'_{\mathbf{k}} = \begin{pmatrix} \varepsilon_{\mathbf{k}} + \mu H & 0 & 0 & -i|\mathbf{d}| \\ 0 & \varepsilon_{\mathbf{k}} - \mu H & -i|\mathbf{d}| & 0 \\ 0 & i|\mathbf{d}| & -\varepsilon_{\mathbf{k}} - \mu H & 0 \\ i|\mathbf{d}| & 0 & 0 & -\varepsilon_{\mathbf{k}} + \mu H \end{pmatrix}.$$

This separates into two sub-matrices in which the effect of the magnetic field is identical to that in BCS theory itself: essentially the pairing is between opposite spins quantized along the magnetic field, and the susceptibility in that case was calculated by Yosida[(11)]:

$$\chi_{\|} = \chi_N Y(\beta|\mathbf{d}|) \tag{IV.81}$$

where $Y$ is the "Yosida function"

$$Y = \tfrac{1}{2}\beta \int_0^\infty d\varepsilon \, \mathrm{sech}^2 \left[\tfrac{1}{2}\beta(\varepsilon^2 + d^2)^{\frac{1}{2}}\right]. \tag{IV.82}$$

On the other hand, we note that when $\mathbf{H} \perp \mathbf{d}$ the $\Delta$ matrix is diagonal, and we have strictly the "equal-spin pairing" (ESP) case of Anderson and Morel in which

$$\chi_{\perp} = \chi_N, \tag{IV.83}$$

since the $\uparrow$ and $\downarrow$ spin Fermi surfaces may still move perfectly independently of each other in the magnetic field. (Incidentally, we may remark on the other two cases of interest. Even $l$: all pairing is opposite-spin and (IV.81) is correct in general. Odd $l$, non-unitary, is complicated in general (an expression is given by Balian and Werthamer[(12)]), but in fact the only important case is the high-field regime near $T_c$ in the "A" phase, which is purely ESP, and (IV.83) is correct.)

Now $\chi$ of the system as a whole depends on the relationship of the $\mathbf{d}$ vectors for different $\hat{\mathbf{k}}$ directions, i.e. on $d_{\alpha i}$. Since the Yosida function (IV.82) depends in a complicated way on the magnitude of $\mathbf{d}$, this is a little complicated except in the two limits $T \to 0$ ($Y(0) = 0$) and $T \to T_c$, where

$$Y \simeq 1 - \tfrac{7}{8}\frac{\zeta(3)d^2}{\pi^2 T^2} \cdots$$

As $T \to 0$

$$\frac{\chi_{ij}(T=0)}{\chi_N} = \delta_{ij} - d_{\alpha i} d_{\alpha j} \tag{IV.84}$$

whereas as $T \to T_c$

$$\frac{\chi_{ij}}{\chi_N} = \delta_{ij} - \tfrac{7}{8}\frac{\zeta(3)}{\pi^2 T^2} d_{\alpha i} d_{\alpha j} \cdots \tag{IV.85}$$

There are three important cases. For BW $\chi$ is isotropic and equal to

$$\frac{\chi_{\mathrm{BW}}}{\chi_N} = \tfrac{2}{3} + \tfrac{1}{3} Y\left(\frac{\Delta}{T}\right). \tag{IV.86}$$

For ABM all of the principal axes are equal and along the axis we have chosen as $w$, so that

$$\left(\frac{\chi_{\mathrm{ABM}}}{\chi_N}\right)_u = \left(\frac{\chi_{\mathrm{ABM}}}{\chi_N}\right)_v = 1$$

$$\left(\frac{\chi_{\mathrm{ABM}}}{\chi_N}\right)_w = \langle Y(\beta d)\rangle \tag{IV.87}$$

while, finally, $\chi(\mathrm{A}')$ has one "ESP" axis

$$\left(\frac{\chi_{\mathrm{A}}}{\chi_N}\right)_w = 1$$

$$\left(\frac{\chi_{\mathrm{A}'}}{\chi_N}\right)_u = \left(\frac{\chi_{\mathrm{A}'}}{\chi_N}\right)_v = \tfrac{1}{2} + \tfrac{1}{2}\langle Y(\beta d)\rangle.$$

As will be seen, A′ and ABM will be very strongly oriented in a magnetic field because of this anisotropy, the orientation energy being of the order $\chi_N H^2 = \mu^2 H^2/E_F$ per atom at low temperatures; this in $He^3$ is of order 1 erg cm$^{-3}$ for $10^3$ gauss. The A′ phase will be uniquely oriented as far as its spin axes are concerned, but $w$ for the ABM or "A" phase can lie anywhere in a plane perpendicular to **H**.

For orientation of **L**, and for all orientation effects in the BW or "B" phase, we must go on to study the nuclear dipole–dipole interaction. In every possible phase, the spatial correlation for a given spin configuration is anisotropic, which means that (as opposed to the ordinary atomic forces, which depend on spin only through statistics and thus are completely invariant to spin orientation) the nuclear dipole–dipole interaction

$$V_{dd}(\mathbf{r}) = \frac{(\boldsymbol{\mu}_1 \cdot \mathbf{r})(\boldsymbol{\mu}_2 \cdot \mathbf{r})}{r^5} - \frac{3\boldsymbol{\mu}_1 \cdot \boldsymbol{\mu}_2}{r^3} \tag{IV.88}$$

depends on relative orientation.

As shown by Leggett[22, 23, 55] the correlations derived in Section IV.2a imply a nuclear dipole–dipole interaction which may be written

$$V_{dd} = \frac{2}{3}\frac{\mu^2}{4}\sum_{\mathbf{kk}'}[\mathbf{x_k}\cdot\mathbf{x_{k'}} - 3\mathbf{x_k}\cdot(\mathbf{k}-\mathbf{k}')\mathbf{x_{k'}}\cdot(\mathbf{k}-\mathbf{k}')]. \qquad \text{(IV.89)}$$

Here we have defined an **x**-vector in analogy with the **d**-vector

$$\hat{x} = \begin{pmatrix} x_v - ix_u & ix_w \\ ix_w & x_v + ix_u \end{pmatrix}.$$

Note however that in (IV.89) for the first time vectors in $k$ and in spin space are related to each other and we are no longer free to use a separate $u$, $v$, $w$ coordinate system for spins. At this point we equate the axis systems $(u, v, w)$ and $(x, y, z)$.

In any unitary state $\mathbf{x_k}$ is proportional to $d_{\hat{\mathbf{k}}}$:

$$\mathbf{d}(\mathbf{k}) = -\tfrac{1}{2}\sum_{\mathbf{k}'} V_{\mathbf{kk}'}\mathbf{x}_{\mathbf{k}'} \qquad \text{(IV.90)}$$

We now express $d_{\hat{\mathbf{k}}}$ in terms of the bivector $d_{\alpha i}$, and do the angular averages in (IV.89) (Engelsberg, Brinkman, and Anderson[96]):

$$V_{dd} \simeq \frac{2\pi\mu^2}{12}(d_{ii}d_{jj}^* + d_{\alpha i}^*d_{i\alpha})/(2\lambda\omega_c)^2. \qquad \text{(IV.91)}$$

Here $\lambda$ is the "coupling constant" $N(0)V_l[(2l+1)/2]$. We have neglected the dependence of $F(E)$ on angles which then allows us to use the gap equation to evaluate the $k$-sum approximately. There is no approximation in the angle-dependence. The magnitude of the coefficient is of order $10^{-4}$ erg cm$^{-3}$, or a little larger if evaluated directly from resonance data ($5 \times 10^{-27}$ erg per atom).

The orientation effects caused by this dipolar interaction are straightforward for the ABM phase. This phase has two unique axes, which we have called $w$ and $z$, and rotating, say, $w$ away from $z$ into the $y$ direction we obtain

$$d_{i\alpha} = \begin{pmatrix} 0 & 0 & 0 \\ d\sin\theta & id\sin\theta & 0 \\ d\cos\theta & id\cos\theta & 0 \end{pmatrix}$$

and

$$d_{ii}^*d_{jj} + d_{i\alpha}^*d_{\alpha i} = 2|d|^2\sin^2\theta$$

which is minimized with $\sin\theta = 0$: $w$ and $z$ must be parallel, and in a magnetic field both perpendicular to **H**.

In the A′ phase, again, there are two unique axes, but the parallel state written down by Balian and Werthamer is the least favourable. The most favourable appears to be a 90° rotation about the $z$ axis to obtain

$$d_{i\alpha} = \begin{pmatrix} 0 & -d & 0 \\ d & 0 & 0 \\ 0 & 0 & 0 \end{pmatrix}$$

with $d^*_{ii}d_{jj} + d^*_{i\alpha}d_{\alpha i} = -d^2$. As we shall see, this resembles the BW case and will naturally arise from it when one component is depaired, as near a wall.

By definition, there is always a pair of sets of coordinate axes $x$, $y$, $z$ and $u$, $v$, $w$ in which the $d_{\alpha i}$ of the BW phase is

$$d_{\alpha i} = \begin{pmatrix} d & 0 & 0 \\ 0 & d & 0 \\ 0 & 0 & d \end{pmatrix}.$$

Two such sets of axes define an axis of rotation $z$ and a rotation angle $\theta$ about $z$ which carries one into the other. If we set our coordinate system along this axis, the two coordinates are related by the simple rotation through $\theta$. Since in the original system $d$ was a unit matrix, total rotations leave it so and only when we make the relative rotation $u$, $v$ into $x$, $y$ do we modify $d$. Thus in this system its matrix is

$$d_{i\alpha} = d \begin{pmatrix} \cos\theta & -\sin\theta & 0 \\ +\sin\theta & \cos\theta & 0 \\ 0 & 0 & 1 \end{pmatrix}$$

and

$$d^*_{ii}d_{jj} + d^*_{i\alpha}d_{\alpha i} = 4d^2(\cos\theta + 2\cos^2\theta),$$

which is minimized when $\theta = 104{\cdot}5° = \cos^{-1}(-\frac{1}{4})$. This condition (since other orientation effects in BW are weak, it is likely to be well satisfied in most regions of the liquid) leaves the BW "B" phase with the symmetry of an axial vector, the rotation $\hat{\mathbf{n}}\theta$. The vector $\hat{\mathbf{n}}$ defines the local state in the analysis of "textures".

The magnetic field orients the BW phase weakly through the following effect.[22, 23, 55, 96] Since it is the component of $d_{\alpha i}$, $d_w$, along the field $\mathbf{H}$ (taken in the $w$ direction) which reduces the susceptibility, clearly there is a term in the free energy $-\frac{1}{2}\Delta\chi H^2$ with $\Delta\chi$ a function of $d_w$. For temperatures near $T_c$ this term was written down by Ambegaokar and Mermin[28] and by the above authors:

$$\Delta F = c_1|\mathbf{M}\cdot\mathbf{d}_\alpha|^2 \qquad \text{(IV.92)}$$

with $c_1$ an approximate constant, at least near $T_c$. Thus in the presence of a magnetic field, the free energy will be minimized by a $d$-matrix of the form (IV.91) but in which the $z$–$z$ matrix element is increased and the $x$ and $y$

portions decreased by a factor. Then the dipolar energy (IV.92) will in turn depend on the field. The appropriate expression is given by Engelsberg, Brinkman and Anderson,[96]

$$F_A \simeq \frac{0{\cdot}5(\hbar\omega_0)^2}{E_F}\left(\frac{\mu\mathbf{H}\cdot\hat{\mathbf{n}}}{kT_c}\right)^2, \tag{IV.93}$$

where the observed longitudinal frequency shift is $\omega_0[(T_c - T)/T_c]^{\frac{1}{2}}$, and is a measure of the dipolar anisotropy. The resulting orientation is with $\hat{\mathbf{n}}$ parallel to $\mathbf{H}$.

Note that $F_A$ is independent of $T$, approximately, hence, though small, finite near $T_c$. In general, as we shall see later, textures in the BW "B" phase are controlled by wall free energies, or in some cases by interactions with heat or fluid flow, etc., all of which are probably proportional to $\Delta^2$ and hence to $T - T_c$. Thus as Webb *et al.*[97, 98] have observed, only extremely near to $T_c$ will one have good orientation in weak magnetic fields.

(c) *Specific heats*

In subsection (1) we derived expressions for the condensation free energy $F$. It is in principle not a separate question to calculate the specific heat

$$\Delta C = -T\frac{\mathrm{d}^2\Delta F}{\mathrm{d}T^2}. \tag{IV.94}$$

What we shall do here is briefly to mention the salient diagnostic features in which the anisotropic case differs from BCS and brings out the relevance of measuring the specific heat.

First we consider the discontinuity near $T_c$. Here we use the expression (IV.51) to obtain

$$\frac{\mathrm{d}F}{\mathrm{d}T} = \frac{\mathrm{d}F}{\partial T} + \frac{\partial F}{\partial \Delta}\frac{\mathrm{d}\Delta}{\mathrm{d}T} = \frac{\partial F}{\partial T}$$
$$= \frac{N(0)}{T}\frac{\partial}{\partial T}\int\frac{\mathrm{d}\Omega_{\mathbf{k}}}{4\pi}\mathrm{Tr}\left[\hat{\Delta}\hat{\Delta}^\dagger - \frac{7\zeta(3)}{8\pi^2}\frac{(\hat{\Delta}\hat{\Delta}^\dagger)^2}{2T^2} + \ldots\right]$$

and the second term is of order $(T - T_c)^2$ so may be neglected. Picking out the discontinuity, then, we have

$$\Delta(C)|_{T_c} = -N(0)\int\frac{\mathrm{d}\Omega_{\mathbf{k}}}{4\pi}\frac{\mathrm{d}}{\mathrm{d}T}\mathrm{Tr}\,\hat{\Delta}\hat{\Delta}^\dagger. \tag{IV.95}$$

We see already from (IV.95) an obvious fact: that the discontinuity in $C$ is precisely proportional to the relative condensation free energy of the different states.

For any given solution in this limit, we can (as we did towards the end of subsection 1) parameterize the free-energy expression (IV.50) or (IV.51) and find an effective $|\Delta|^4$ term. For example, for the ABM or $A_1$ solutions[91]

$$\frac{\langle(\Delta\Delta^\dagger)^2\rangle_{\mathrm{ABM}}}{(\langle\Delta\Delta^\dagger\rangle)^2} = \frac{6}{5}\frac{\langle(\Delta\Delta^\dagger)^2\rangle_{\mathrm{BW}}}{(\langle\Delta\Delta^\dagger\rangle)^2}. \tag{IV.96}$$

Using $\partial F/\partial\langle\Delta\Delta^\dagger\rangle = 0$, this gives that $(\Delta C_{sp})_{\mathrm{ABM}}$ is only $\frac{5}{6}$ as great as for BW, which is in fact the same as for BCS, since the two are thermally totally equivalent.

When, in the early measurements, it became clear that $C_{sp}$ had a greater jump experimentally than for BW–BCS, it became certain that the weak-coupling structure of the theory had to be abandoned. Any exotic phase was as certain to have a lower $C_{sp}$ as to have a higher free energy. We will see that this is the case in Section V. Ordinary BCS theory, in fact, for most metals gives too low a $C_{sp}$ (Bardeen and Schrieffer[99]). This is the result of the effects of strongly coupled phonons, and similar effects may be expected here, in addition to those we discuss in Section V.

A brief footnote to the specific heat discontinuity question: in a magnetic field the transition splits, both experimentally and theoretically, for reasons given by Ambegaokar and Mermin[28] and Varma and Werthamer.[29] The reason is the presence of two extra free-energy terms in the magnetic field; the one equivalent to the susceptibility decrease due to opposite-spin pairing already mentioned a few pages ago:

$$\begin{aligned} F_{\Delta\chi} &= C_1 H^2 |\Delta_{\uparrow\downarrow}|^2 \\ &= C_1 |\mathbf{H}\cdot\mathbf{d}_\alpha|^2 \end{aligned} \tag{IV.97}$$

and a second effect due to (presumably and primarily) the change in density of states at the Fermi level caused by the shift $\pm\mu H$ in $E_F$:

$$F_{\Delta E_F} = \eta H(|\Delta\uparrow|^2 - |\Delta\downarrow|^2) = i\eta\mathbf{H}\cdot(\mathbf{d}_\alpha \times \mathbf{d}_\alpha^*). \tag{IV.98}$$

(The change in density of states will of course modify $T_c$ and hence $\ln T_c/T$.)

As observed, two transitions appeared, with unequal specific heat jumps adding up to the total in the absence of a field. Low-field theory gives two equal jumps, at higher fields [since (IV.97) is bigger than (IV.98) at most fields] there are three: $A_1$ ($\uparrow\uparrow$ pairing only), A (ESP), and finally B = BW. This phenomenon will be analysed in detail for the real ABM case in Section V.

At low temperatures Anderson and Morel pointed out (note however that their formal specific heat expression is wrong) that all phases expect $l = 1$ BW will have algebraic rather than exponential specific heats, because of the existence of nodes of the gap. For instance, in the ABM state the effective single-particle density of states as $T \to 0$ is

$$N(E) = N(0) \int \frac{d\Omega}{4\pi} \left[ E^2 - \Delta_0^2(\hat{k}_x^2 + \hat{k}_y^2) \right] \simeq N(0) \frac{E^2}{\Delta_0^2}. \tag{IV.99}$$

This leads to a specific heat as $T \to 0$

$$(C_{sp})_{ABM}|_{T\to 0} \simeq 3{\cdot}5(\gamma T_c)(T/T_c)^3. \tag{IV.100}$$

The entropy also varies as $T^3$. The entropy of states with gap nodes, then, will of course tend to be higher, but not, of course, sufficiently so to make them stable in weak coupling.

### (d) *Superfluid and normal fluid densities*

Anderson and Morel gave a correct expression for the superfluid density as normally defined in the weak-coupling, perturbation limit. In fact the supercurrent structure is considerably more complicated than that of an ordinary superfluid, as we shall see in Section VI, but for many practical purposes we expect that the straightforward approach followed by Anderson and Morel,[9, 10] Balian and Werthamer,[12] and later Saslow[45] will suffice.

The idea is to exploit the BCS perturbation-theory calculation of the electromagnetic response of a superconductor to a vector potential **A**. This is formally simple, so long as one neglects the Fermi-liquid corrections which we have so far omitted and will replace in the next subsection, but conceptually somewhat tricky, and in fact strictly speaking incorrect. In essence, as Saslow makes clear, this is a calculation of the normal fluid density $\rho_n$: we impose a velocity $\mathbf{v}_n$ on the quasiparticles, calculate $E(\mathbf{v}_n)$ and hence

$$\mathbf{P}_n = \frac{\partial E}{\partial \mathbf{v}_n} = \overleftrightarrow{\rho}_n \mathbf{v}_n. \tag{IV.101}$$

Then we deduce $\rho_s$ by the standard Galilean invariance arguments which give $\rho_s = \rho - \rho_n$, i.e. the condensate has velocity

$$\mathbf{v}_s = \frac{\hbar}{m} \nabla \phi \tag{IV.102}$$

and density $\rho_s$, while $\phi$ is the phase of the condensate wavefunction and obeys the Landau acceleration–Josephson frequency equation

$$h \nabla \dot{\phi} = \nabla \mu. \tag{IV.103}$$

All of this is indeed true (although existing proofs are incomplete, Cross[100] has completed the logical argument) if the condensate has a unique velocity $\mathbf{v}_s$. For $l = 1$, unfortunately, there are nine components $d_{\alpha i}$ of the condensate and there is no guarantee that there is a single $\phi$ or a single $v_s$. Thus the condition for validity of the AM–Saslow theory is that $d_{\alpha i} \propto (d_{\alpha i})^0 e^{i\varphi}$, i.e. $(v_s)_{\alpha i} \neq f(\alpha, i)$. In any case the calculation of $\rho_n$ is strictly correct, and is, in analogy to the corresponding BCS expression

$$\vec{\rho}_n = \sum_{\mathbf{k}} \mathbf{kk} \frac{\partial f(E_{\mathbf{k}})}{\partial k} \tag{IV.104}$$

where $f$ is the Fermi factor $(e^{\beta E} + 1)^{-1}$. In the ABM state $\rho_n$ is anisotropic, with deviations from $\rho$ differing by a factor 2 near $T_c$:

$$\begin{aligned} \rho_n^{\parallel}/\rho &= \left(1 - \frac{7}{10\pi^2}\zeta(3)\frac{\langle\Delta^2\rangle}{T_c^2}\right) \\ \frac{\rho_n^{\perp}}{\rho} &= \left(1 - \frac{7}{5\pi^2}\zeta(3)\frac{\langle\Delta^2\rangle}{T_c^2}\right) \\ T &\to T_c \end{aligned} \tag{IV.105}$$

where $\parallel$ is aparallel and $\perp$ perpendicular to the orbital axis $z$. Thus $\rho_s$ as defined within this theory is anisotropic by a factor 2:

$$\rho_s^{\parallel}/\rho = \tfrac{1}{2}\rho_s^{\perp}/\rho = \frac{7}{10\pi^2}\zeta(3)\frac{\Delta^2}{T_c^2} + \ldots \tag{IV.106}$$

As $T \to 0$, however, $\rho_n \to 0$, although differently for $\parallel$ and $\perp$; so $\rho_s = \rho$, as required for Galilean invariance.

These quantities are then, one imagines, (after Fermi-liquid corrections if necessary) to be inserted into a two-fluid model on the HeII pattern, in which the kinetic free energy is

$$F_{\text{kin}} = \tfrac{1}{2}(\rho_s v_s^2 + \rho_n v_n^2) \tag{IV.107}$$

with $v_s$ obeying (IV.102, IV.103), and $v_n$ pinned to the walls in narrow channels. $v_n$ carries the entropy of the fluid, since it represents the quasiparticle gas, etc. From this one may, as Saslow[45] and Wölfle[53] have suggested, derive an anisotropic fourth sound ($\rho_s$ alone moving), second sound (a hydrodynamic counterflow oscillation), etc.

Except for fourth sound, it appears to the present authors that direct use of these two-fluid concepts will be at best unwise and at most misguided or incorrect. (See Wölfle[53] for a very sophisticated discussion.) Aside from the difficulty of the unique $v_s$, one also has different relative orders of magnitude of relaxation times, coherence lengths, diffusion lengths, etc. The many more degrees of freedom will allow many more collective modes, both hydrodynamic and collisionless, about which a massive literature is already growing. Since there is available a basically almost exact microscopic theory it seems usually possible, and far wiser, to trust for quantitative work to direct microscopic calculation rather than two-fluid heuristics until these can be sorted out.

Meanwhile, however, there are a few measurements which do have their natural discussion in the two-fluid picture: fourth sound[48, 101] and, particularly, early heat transport experiments[41] which demonstrated clearly a two-fluid counterflow. In these very interesting experiments non-linear instabilities were observed in the A phase highly suggestive of the expected interaction between anisotropic superflow and gap orientation.

## 3. FERMI–LIQUID CORRECTIONS TO STATIC AND LOW-FREQUENCY EFFECTS

Section 2 describes the very strong interaction effects between quasiparticles in normal $He^3$, which severely modify most of that liquid's properties from those of a free Fermi gas, which is the model we have used so far for the pairing phenomena. How can we treat the system in weak coupling when these strong interactions exist?

Leggett[13, 14] was the first to study the interaction of pairing and Fermi-liquid effects, and to develop a scheme for handling them together. The quantitative effects on many properties—in particular $\chi$ and $\rho_n$—are fairly large, but in a very specific sense the qualitative picture of handling the pairing independently is unchanged. We will follow Leggett's remarkably clear and complete discussion here. (It is striking that the susceptibility of the B phase is still compared with a figure given by Leggett[13] at that time)—see Fig. 6.

The philosophy of Landau is that states of the interacting Fermi system at temperatures $\ll T_F$ can be referred back and enumerated as adiabatic continuations of weakly-excited states of a non-interacting Fermi sea: i.e. labelled in terms of occupation numbers $n(\mathbf{k}, \sigma)$ of "quasiparticles" near the Fermi surface. The residual interactions of these quasiparticles may not be neglected but their total effects are weak: the large parts of the interactions are taken into account in renormalization effects.

Landau uses an effective hamiltonian containing the renormalized single-particle energy

$$H_k = \sum_{\mathbf{k},\sigma} \varepsilon(\mathbf{k}) n(\mathbf{k}, \sigma) \tag{IV.108}$$

plus a residual quasiparticle interaction between changes $\delta n$ in quasiparticle occupation at the Fermi surface:

$$H_L = \tfrac{1}{2} \sum_{\mathbf{k}\mathbf{k}'\sigma\sigma'} f(\mathbf{k}, \mathbf{k}'; \sigma, \sigma') \delta n(\mathbf{k}, \sigma) \delta n(\mathbf{k}', \sigma'). \tag{IV.109}$$

This was expanded in Section II (II.2) in terms of direct and exchange-like

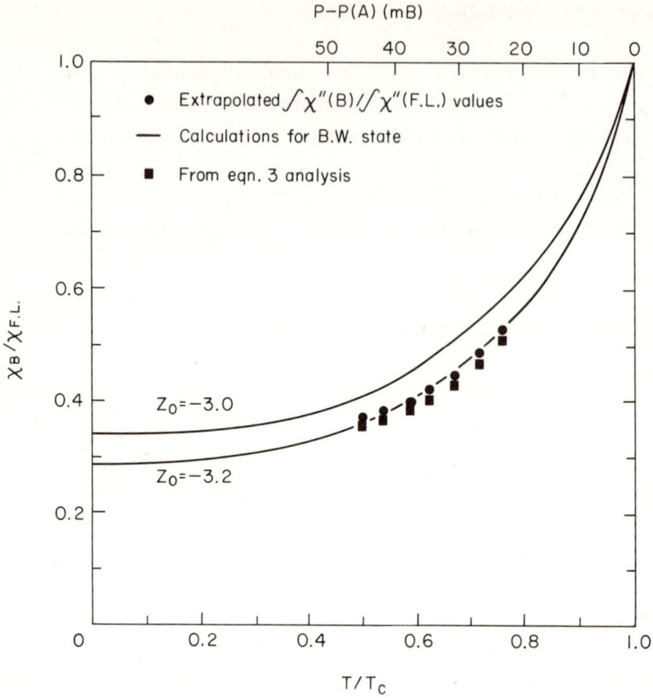


FIG. 6. A plot of the susceptibility versus reduced temperature for the B phase. Two values of $z_0 = 4F_0^a$ were used. The two sets of data are simply two slightly different ways of analyzing the saturation data discussed in Section VIII. See also Fig. 17.

parts $F^s$ and $F^a$ and in terms of spherical harmonics of $(\hat{\mathbf{k}}.\hat{\mathbf{k}}')$. This interaction represents the forward scattering amplitude of the quasiparticles and has the effect of modifying the effective single-particle energy $\varepsilon(\mathbf{k})$ proportionally to changes in the occupations of other states. As a result each quasiparticle moves in a "mean field" created by all the others.

A second type of residual interaction which cannot be neglected is the pairing interaction which causes the BCS transition (IV.1):

$$V_p = \sum_{\mathbf{kk}'\sigma\sigma'} V_{\mathbf{kk}',\sigma\sigma'} a_{\mathbf{k}\sigma}^{\dagger} a_{-\mathbf{k}\sigma'}^{\dagger} a_{-\mathbf{k}'\sigma'} a_{\mathbf{k}'\sigma}. \tag{IV.110}$$

Although Leggett's basic work (like the later work by Betbeder-Matibet and Nozières[102] and Wölfle[53] along the same lines) proceeds from a formal diagrammatic approach, the principle may be seen from a much less formal treatment given in his Section IV. In this section he points out that in a real sense the Fermi-liquid interactions do not interfere with the pairing phenomenon because pairing does not change the net occupation for a given direction on the Fermi surface:

$$\delta n(\mathbf{k}, \sigma) = N(0) \int d\varepsilon_{\mathbf{k}}\, \delta n(\mathbf{k}, \sigma) \simeq 0, \qquad \text{(IV.111)}$$

if, as is always the case in weak coupling, approximate hole–particle symmetry is maintained. Thus in fact the Fermi-liquid interactions remain unchanged upon pairing and the pairing proceeds according to the sum of $H_k$ and $V_p$. Thus, for instance, all the thermal phenomena are unaffected by $H_L$.

The pairing does, however, affect most of the characteristic Fermi-liquid transport, response, and collective-mode effects, because pairing changes precisely the long-wavelength response functions which are relevant to these. It is the essential result of the work of Leggett, Nozières, and Wölfle that one form or another of mean field theory can be used to calculate these effects. In rather general terms, we can suppose that, due to some applied field, the one-particle energies $\varepsilon$ undergo a small change:

$$\varepsilon \to \varepsilon + \delta\varepsilon(\mathbf{k}, \sigma). \qquad \text{(IV.112)}$$

We calculate the response of the non-interacting Fermi gas to this perturbation (as in subsection 2 of this section IV):

$$n(\mathbf{k}', \sigma) = \sum_{\mathbf{k}, \sigma} \chi^0(\mathbf{k}', \sigma'; \mathbf{k}, \sigma)\delta\varepsilon(\mathbf{k}, \sigma) \qquad \text{(IV.113)}$$

and, finally, we used the Fermi-liquid relation (II.1)

$$\delta\varepsilon(\mathbf{k}, \sigma) = (\delta\varepsilon)_{\text{ext}} + \sum_{\mathbf{k}', \sigma'} f(\mathbf{k}\mathbf{k}'; \sigma\sigma')\delta n(\mathbf{k}', \sigma') \qquad \text{(IV.114)}$$

to obtain a renormalized response which can be schematically written

$$\chi = \frac{\chi^0}{1 + f\chi^0}. \qquad \text{(IV.115)}$$

In general, the self-consistent solution of (IV.112–IV.114) will involve an integral equation and also must be carried out dynamically, but in most cases it is much simpler than that. If, with Leggett, we ignore higher spherical harmonics than $l = 1$ in the series of $F$'s (II.2), we obtain for the spin susceptibility exactly the standard Fermi-liquid "Stoner" result equivalent to (II.3)

$$\chi = \frac{\chi^0}{1 + [\chi^0/N(0)]F_0^a} \qquad \text{(IV.116)}$$

and for the normal fluid density

$$\rho_n = \frac{\rho_n^0}{1 + (\rho_n^0/Nm^*)\cdot\frac{1}{3}F_1}. \tag{IV.117}$$

Here $\chi^0$ is the susceptibility calculated in the last subsection including pairing only, and $\rho_n^0$ is the normal fluid density so calculated; $F_0^a$ is the Stoner-type exchange Landau parameter (often called $Z_0/4$ or $-\bar{I}$) and $F_1$ the effective mass parameter, for which

$$\frac{m^*}{m} = 1 + F_1/3. \tag{IV.118}$$

These expressions are very successful experimentally, as may be seen from Fig. 6 for the susceptibility (Osheroff;[27] one should note that prior susceptibility measurements[103] in the B phase were in a confused state) and Fig. 7 for the superfluid density $\rho_s = \rho - \rho_n$.

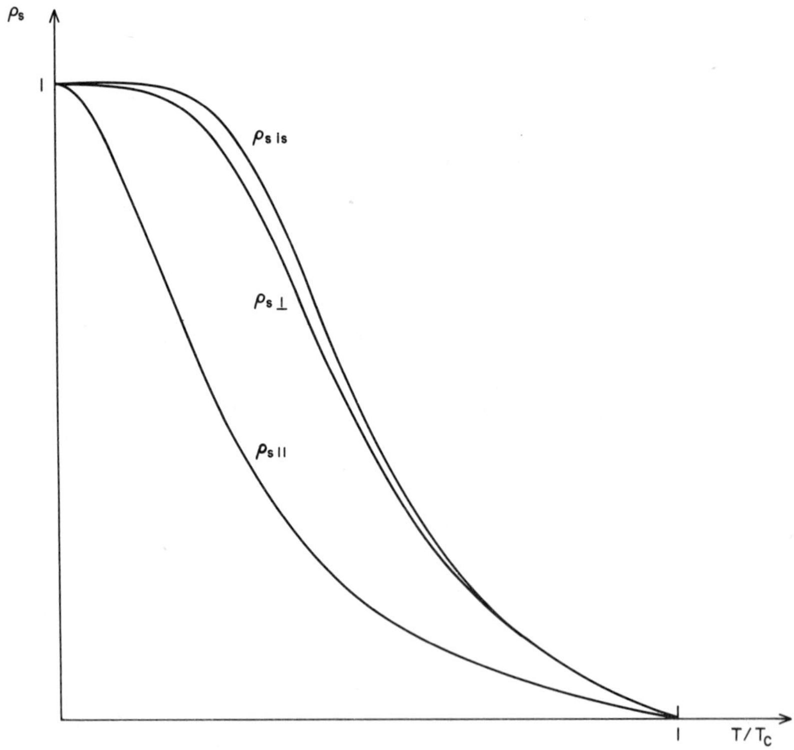


FIG. 7. A plot of $\rho_s$ including the Fermi-liquid corrections. Note that the actual $\rho_s$ is considerably smaller than the weak-coupling value and that it is quite anisotropic.[57]

The effect of these corrections may well be understood from the effective-field picture. When there is a large negative $Z^0$ (large enhancement) the reduction of the susceptibility by the pairing reduces the enhancement factor severely as well, and the susceptibility drops rapidly. The normal fluid density, on the other hand, is reduced rather than enhanced, so reaches its zero-temperature value more slowly at first.

Leggett himself already gave the type of expression which is correct if higher $Z$'s and $F$'s are not negligible.

$$\chi(T) = \int \frac{d\Omega_{\hat{k}}}{4\pi} \frac{\chi^0(T, \hat{k})}{1 + \int [\chi^0(T, \hat{k}')/N(0)] F^a(\hat{k}, \hat{k}')(d\Omega_{\hat{k}'}/4\pi)}. \qquad \text{(IV.119)}$$

As he pointed out, although (IV.119) reduces to (IV.117) (even as to slope) as $T \to T_c$, for the BW state there is no reason to be sure that $F^a(\hat{k}, \hat{k}')$ is not severely cut off in $|\hat{k} - \hat{k}'|$, which would have the effect of changing low-temperature susceptibilities rather severely—one would guess, incidentally, in quite the opposite direction to that suggested by Vuorio:[104] i.e. nearer for the BW state, not farther. The reason for this effect is that then the electrons on the different parts of the Fermi surface would act almost independently and of order $\frac{2}{3}$ of them would see no change in their effective fields at all.

For this reason we suggest that in fact the effective fields are, for whatever reason, of quite short range in space and of long range in $k$-space, and that higher Fermi-liquid corrections are empirically not very big. We await with interest more data on the temperature dependence of $\rho_s$.

## 4. SOUND IN ANISOTROPIC SUPERFLUID HE$^3$

Although sound, as a collective mode, properly might be considered to belong in a later section, both chronologically and from a theoretical point of view it seems appropriate to discuss it here—chronologically, because zero-sound attenuation was a very early measurement which gives a striking peak at the A transitions which was used as a convenient signal in early Cornell measurements[105] and theoretically, because it is closely related to Fermi-liquid theory and because the relevant theory is not highly dependent on the anisotropic effects, or on the spin-fluctuation theory.

Normal He$^3$, of course, at high temperatures and low frequencies has only ordinary "hydrodynamic" sound with velocity given by the classical formula

$$c_0^2 = \frac{p_F^2}{3mm^*}(1 + F_0^s), \qquad \text{(IV.120)}$$

where the quantity

$$\frac{p_F^2}{3m^*}(1 + F_0^s) = \kappa \tag{IV.121}$$

is the modulus of compressibility per atom. $F_0^s$ is quite large and $c_0 \gg v_F$. As we lower the temperature and raise the frequency we observe a peak in attenuation and a transition to "zero sound" which is a collisionless collective mode of the Fermi system in which the quasiparticles do not come into thermal equilibrium internally but respond only dynamically. This mode satisfies the integral equation for the zero-sound dispersion relation

$$(\omega - \mathbf{k}\cdot\mathbf{v}_0)\nu(\hat{\mathbf{k}}) = \mathbf{k}\cdot\mathbf{v}_0 \int F^s(\hat{\mathbf{k}},\hat{\mathbf{k}}')\hat{\nu}(\hat{\mathbf{k}}')\frac{d\Omega_{\hat{\mathbf{k}}'}}{4\pi}, \tag{IV.122}$$

where $\mathbf{v}_0$ is a vector of magnitude $v_F$ along the propagation direction, and $\nu(\hat{\mathbf{k}})$ is the modification of the distribution function in the $\hat{\mathbf{k}}$ direction. Because $\omega/k = c$ is large compared to $v_F$, because of the high compressibility modulus, $c$ for zero sound is only 4% larger at $T = 0$ than $c_0$.

Wölfle[32] has given a very clear discussion of sound propagation in the anisotropic case, restating in part some of the results of Leggett.[106] Earlier Wölfle[31] had also given a full treatment of the rather complicated problem of sound attenuation.

Briefly, the physical effect goes back to the simple observation of Anderson[107] and Bogoliubov[108] that at absolute zero the BCS superfluid should exhibit an ordinary sound mode with the normal velocity $c_0$ (IV.120) rather than zero sound. The physical reason is the fact that the Fermi continuum of hole–particle states $a^\dagger_{\mathbf{k}+\mathbf{q}}a_{\mathbf{k}}$ of energies up to $q \cdot v_F$ is removed by pairing, and thus the sound mode itself must saturate the sum rule: it must be a uniform compressional excitation because there are no quasiparticle modes at low frequency which can distort the Fermi surface, i.e. the mode is not repelled by a lower-frequency continuum (see Fig. 8).

Since the dispersion law changes as the mode passes through the region of the gap (well above the gap frequency, of course, zero sound will return) this implies a strong absorption, which in fact peaks very severely near $T_c$, and is entirely due to simple quasiparticle pair creation. In the normal state, momentum conservation prevents hole–particle pair excitation and only allows the scattering of excited particles, but the presence of a gap breaks this conservation law. As long as there is a region with $\Delta < \omega$ hole–particle creation can occur, although this is obviously a very anisotropic effect for the sound wave vector $\mathbf{q}$ with respect to the positions of the nodes of the gap in $k$-space. For the BW state, $\omega > \Delta$ only for ordinary ultrasonic frequencies and only very near $T_c$ and one expects a very narrow peak (see Fig. 9), but for ABM there are always nodes and one can understand the observed long tail of absorption for $T < T_c$.[109]

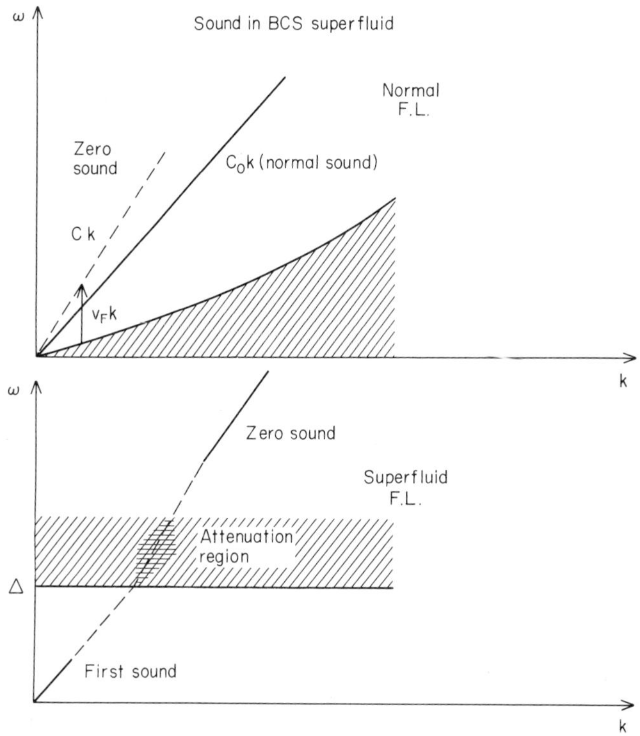


FIG. 8. A plot of the particle–hole continuum which repels the zero-sound wave in the normal state so that its velocity is not the same as the first-sound velocity.

We will only give a short discussion of the actual calculations. In the first place, at any reasonable frequency and for $T$ not quite near $T_c$ the conditions $\omega \ll \Delta$, $q\xi_0 \ll 1$ are satisfied and it is possible to use the general scheme of Betbeder-Matibet and Nozières.(102) In this case the gas of excited quasi-particles can be treated just precisely similarly to the Landau theory of normal Fermi liquids, which obey the kinetic equation (essentially (IV.122))

$$(\omega - \mathbf{q}\cdot\nabla_{\mathbf{k}}E_{\mathbf{k}})\nu_{\mathbf{k}} + \mathbf{q}\cdot\nabla_{\mathbf{k}}E_{\mathbf{k}}N(0)\delta E_{\mathbf{k}}(\mathbf{q},\omega) = I_{\mathbf{k}}(\nu)$$

where $I(\nu)$ is the collision integral and $\delta E_{\mathbf{k}}$ is the change in quasiparticle energy due to external forces and to the modified distribution $\nu$ itself:

$$\delta E_{\mathbf{k}} = \frac{\varepsilon}{E}\left(\frac{F_0^s}{N_F}\delta\rho - \delta\mu\right) + \frac{F_1^s}{3}\left(\frac{m}{m^*}\right)\mathbf{k}\cdot\frac{\overleftrightarrow{\rho}_n}{\rho}\cdot(\mathbf{v}_n - \mathbf{v}_s) + \mathbf{k}\cdot\mathbf{v}_s + \frac{\delta|\Delta^2|}{E_{\mathbf{k}}}. \qquad \text{(IV.123)}$$

The tensor $\overleftrightarrow{\rho}_n$ is the normal fluid density we already derived in the last section. $\mathbf{v}_s$ obeys the acceleration equation we also derived in the last section (and

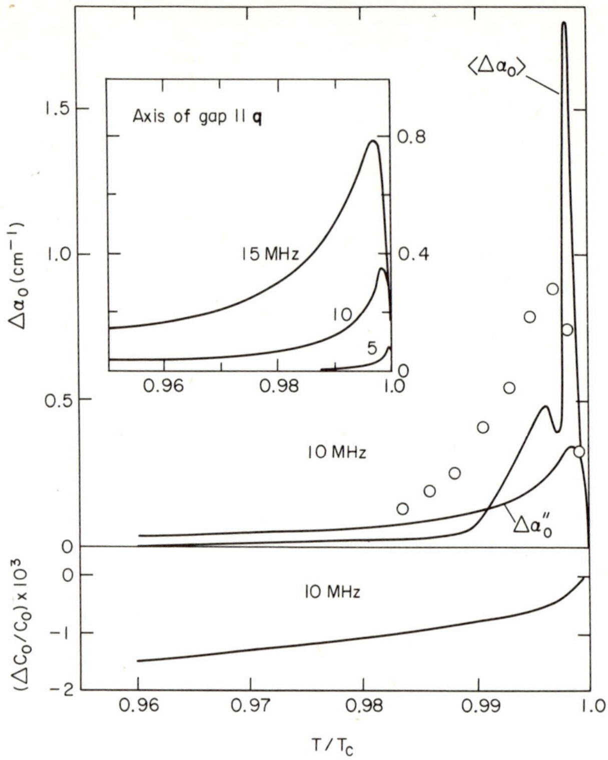


FIG. 9. The ultrasonic attenuation as calculated by Wölfle.[31] The points are from experiments of Paulson, Johnson and Wheatley.[109]

throughout we are of course assuming no $\alpha$, $i$-dependence of phase)

$$m\frac{\mathrm{d}\mathbf{v}_s}{\mathrm{d}t} = \hbar\nabla\delta\mu$$

and the continuity equation

$$\frac{\partial\delta\rho}{\partial t} = \nabla\cdot(\rho_n\mathbf{v}_n + \rho_s\mathbf{v}_s). \tag{IV.124}$$

$\mathbf{v}_n$ is defined in the obvious way as the mean quasiparticle velocity. Finally, the change in density is

$$\delta\rho = \frac{\Sigma(\varepsilon/E)\nu + N(0)\langle\lambda_k\rangle\delta\mu}{1 + F_0\langle\lambda\rangle}. \tag{IV.125}$$

$\langle\lambda_k\rangle$ is a kind of effective average BCS superfluid density for the direction $\hat{\mathbf{k}}$

$$\lambda(\hat{\mathbf{k}}) = \int_{-\infty}^{\infty} \mathrm{d}\varepsilon_{\mathbf{k}} n(E_{\mathbf{k}}) \frac{|\Delta_{\mathbf{k}}|^2}{E_{\mathbf{k}}^3} \tag{IV.126}$$

which varies from 0 at $T_c$ to 1 at $T = 0$. For the "collisionless" sound velocity, Wölfle obtains

$$c^2 = c^2(T = 0) + (3 + F_1)\left[-\langle \lambda(\hat{\mathbf{k}})(\hat{\mathbf{k}}\cdot\hat{\mathbf{q}})^4\rangle + \frac{\langle \lambda(\hat{\mathbf{k}})(\hat{\mathbf{k}}\cdot\hat{\mathbf{q}})^2\rangle^2}{\langle\lambda\rangle}\right], \tag{IV.127}$$

which has the appropriate behaviour but makes the zero-to-first-sound transition different for different orientations.

The hydrodynamic regime $\omega\tau < 1$ is very limited in extent, because of the presence of the gap and the resulting low quasiparticle density, to either extremely low frequencies or $T \simeq T_c$. In this limited regime first sound is normal, and a second sound exists with a highly anisotropic velocity

$$u_2^2 = v_F^2 \left(\frac{m^*}{m}\right) \frac{T}{2E_F} [\rho \mathbf{q}\cdot\overleftrightarrow{\sigma}\cdot\overleftrightarrow{\rho}_n^{-1}\cdot\overleftrightarrow{\sigma}\cdot\mathbf{q} - (\mathbf{q}\cdot\overleftrightarrow{\sigma}\cdot\mathbf{q})^2], \tag{IV.128}$$

where $\overleftrightarrow{\sigma}$ is a tensor entropy density of the normal fluid fraction, defined by the same sum (IV.104) as $\overleftrightarrow{\rho}_n$ multiplied by $\varepsilon_{\mathbf{k}}$. $u_2$ is very small, $<10^{-2}\, v_F$. For fourth sound, with the normal fluid pinned, Wölfle's results agree with those of Saslow[45] and de Gennes,[50, 51] which we have already discussed.

As for attenuation, Wölfle's earlier calculation is quite complex because it was necessary to go entirely outside the effective Landau model of Nozières and Betbeder-Matibet and use the general self-consistent dynamical theory which Leggett first gave in detail (it is amusing that in essence this is identically the "generalized RPA" scheme of Anderson[107]). This is no serious difficulty in principle, but one must take into account self-consistently modifications $\delta\hat{\mathscr{E}}_{\mathbf{k}}$ and $\delta\hat{W}_{\mathbf{k}}$ in the Balian–Werthamer matrices. Results for the attenuation peaks are shown in Fig. 9.

## V. Spin-fluctuation Stabilization of the "A" Phase

As discussed in the previous section on weak-coupling theory, Balian and Werthamer[12] have shown that their state is the lowest free-energy state at all temperatures. Clearly, there must be a strong coupling effect which tends to stabilize the A phase, which is now fairly conclusively established to be the Anderson–Morel state with the orbital angular momentum of both up and down spin pairs aligned. Since the estimates made in Section III indicate that the spin fluctuations are important in determining $T_c$ we examine a possible feedback mechanism involving these fluctuations which, as we shall see, stabilizes the ABM state.[34] The attractive interaction via spin fluctuations is similar to the attractive interaction via the virtual exchange of phonons in electronic superconductors. In the case of superconductors, because the phonons are mostly a property of the ions plus some high screening coming from the electrons, one does not expect large changes in the phonon spectrum unless one has a very strong coupling situation. On the other hand, in $He^3$ the spin fluctuations are part of the Fermi liquid itself and, as such, may change below the superfluid transition temperature. This is especially true at long wavelengths where the spin fluctuations are at very low frequencies. As was discussed in the previous section, the static susceptibilities $\chi^0$ become anisotropic and undergo large changes in the superfluid state. Since the effective interaction for zero momentum transfer depends on these susceptibilities, we expect the interaction to change on going into the superfluid state.

To see how such an effect occurs we examine the effective interaction more carefully. In the superfluid state the dynamic susceptibility is not isotropic, so that below $T_c$ each component of the triplet experiences a different effective interaction. This effect can be described by making $\Gamma_s(\mathbf{q})$ dependent on the components of the spin operators in (III.12), i.e. we write the exchange term as

$$\frac{1}{2}\sum_{\substack{\mathbf{k}\mathbf{k}'\\ \mathbf{q}}} \Gamma_s^i(\mathbf{q})\sigma^i_{\alpha\beta}\sigma^i_{\gamma\delta}a^\dagger_{\mathbf{k}+\mathbf{q}\alpha}a^\dagger_{\mathbf{k}'-\mathbf{q}\gamma}a_{\mathbf{k}'\delta}a_{\mathbf{k}\beta}. \tag{V.1}$$

We have not written down the most general form but rather one with which we can easily visualize the effect. In fact, we assume that $\Gamma_s^x = \Gamma_s^y \neq \Gamma_s^z$. Since the total spin is no longer conserved we must calculate the interaction for each component of the triplet. Using (III.2, III.3) we find

$$v_{m_l=\pm 1}(\mathbf{k}-\mathbf{k}') = -\Gamma_s^z(\mathbf{k}-\mathbf{k}') \tag{V.2}$$

$$v_{m_l=0}(\mathbf{k}-\mathbf{k}') = \Gamma_s^z(\mathbf{k}-\mathbf{k}') - 2\Gamma_s^x(\mathbf{k}-\mathbf{k}'). \tag{V.3}$$

Here $m_l$ denotes the three components of the triplet. If the weak-coupling susceptibilities have the same form as $\Gamma_s$, i.e. $\chi^0_{xx} = \chi^0_{yy} \neq \chi^0_{zz}$, then summing the diagrams in Figs. 3 and 4 leads to the results that

$$\Gamma_s^i(\mathbf{q}) = \frac{1}{2}\frac{I}{(1 - I\chi^0_{ii}(\mathbf{q},0))}$$

so that

$$v_{m_l=\pm 1}(\mathbf{q}') = -\frac{1}{2}\frac{I}{(1 - I\chi^0_{zz}(\mathbf{q},0))} \tag{V.4}$$

and

$$v_{m_l=0}(\mathbf{q}') = \frac{1}{2}\left(\frac{I}{(1 - I\chi^0_{zz}(\mathbf{q},0))} - \frac{2I}{(1 - I\chi^0_{xx}(\mathbf{q},0))}\right). \tag{V.5}$$

The feedback effect is very easily seen in these equations. At $q = 0$ we can use the results for the static values of $\chi_{ij}^0$ from the previous section. If we consider the BW state, the three susceptibilities are equal and reduced. Consequently, the effective interactions are also equal and reduced. If fed back into the gap equation, the value of the gap will be reduced and the free energy raised. However, in the ABM state, only $\chi_{zz}^0$ is reduced and only $m_l = 0$ pairs have a non-zero amplitude, so that as $\chi_{zz}^0$ decreases the relevant interaction *increases*. Thus the magnitude of the energy gap is increased and the free energy of the ABM state is lowered. Therefore, if the spin-fluctuation term is large enough, one can expect the ABM state to be energetically more favourable than the BW state. Of course, one cannot simply look at the values of the susceptibility at $q = 0$. In the first paper(34) on this effect, the values of $\chi_{ij}^0(\mathbf{q}, 0)$ were guessed from earlier work on the superconductors by Anderson and Suhl,(110) and it was shown that this feedback mechanism gives rise to fourth-order terms in the free energy which are such that if the spin fluctuations are sufficiently strong the ABM state can be stabilized.

More recently, Brinkman, Serene, and Anderson(37) and Kuroda(38, 39) have extensively investigated the feedback effect by calculating properly the susceptibility tensor for finite $q$ and $\omega$ and then calculating the free-energy changes. In order to do this they studied the change in the spin-fluctuation contribution to the free energy induced by the transition to the superfluid state. The spin-fluctuation contribution to the free energy can be written as

$$\Delta F = \frac{T}{2} \operatorname{Tr} \sum_m \int \frac{d^3 q}{(2\pi)^3} [\ln \{1 - I\chi_{ij}^0(\mathbf{q}, i\omega_m)\} + I\chi_{ij}^0(\mathbf{q}, i\omega_m)], \qquad \text{(V.6)}$$

where $\omega_m$ are the Matsubara frequencies $2m\pi T$ and the logarithm is defined in terms of its expansion. The trace is taken over $i$ and $j$. $\Delta F$ represents the sum of the diagrams in Fig. 10. The susceptibilities are defined as

$$\chi_{ij}^0(\mathbf{q}, i\omega_m) \equiv \int_0^\beta \exp(i\omega_m \tau) \langle T_\tau [M_i(\mathbf{q}, \tau) M_j(-\mathbf{q}, 0)] \rangle \, d\tau \qquad \text{(V.7)}$$

where

$$M_i(\mathbf{q}, \tau) = \frac{\sigma_{\alpha\beta}^i}{2} \sum_{\mathbf{p}} \exp[(H_0 - \mu N)\tau] a_{\mathbf{p}+\mathbf{q}/2,\alpha}^\dagger a_{\mathbf{p}-\mathbf{q}/2,\beta} \exp[-(H_0 - \mu N)\tau]. \text{(V.8)}$$

$T_\tau$ designates the usual time ordering and $H_0$ is the reduced hamiltonian about which we do perturbation theory. In order to calculate the change in $\Delta F$ we define $\delta\chi_{ij}^0$ to be the difference between the normal-state susceptibility functions and those of the superfluid, i.e.,

$$\delta\chi_{ij}^0 = \chi_{ij}^0|_N - \chi_{ij}^0|_S \qquad \text{(V.9)}$$

where $N$ and $S$ stand here for normal and superfluid respectively. We define

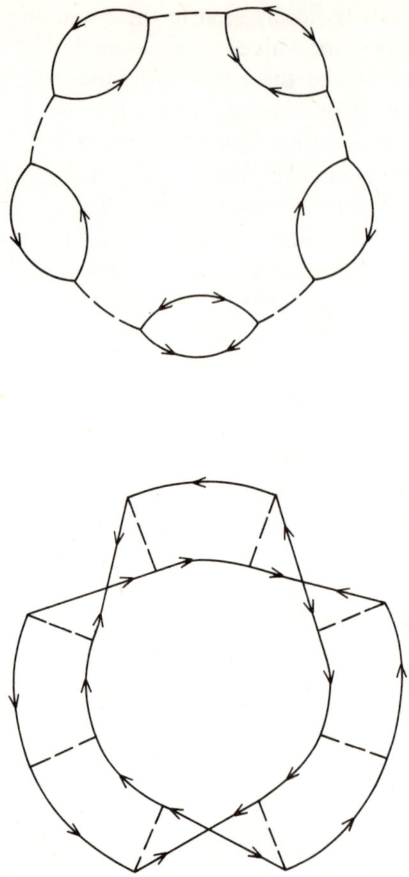

FIG. 10. The spin-fluctuation diagrams which contribute to the free energy. Note that both the normal and anomalous diagrams are included.

$\chi^0_{ij}|_N = \delta_{ij}\chi^0$. $\Delta F$ is then expanded in powers of $\delta\chi$. The first-order term is

$$\Delta F_1 = \frac{T}{2}\sum_i\sum_m \int \frac{\mathrm{d}^3q}{(2\pi)^3} \frac{I^2\chi^0(\mathbf{q}, i\omega_m)}{\{1 - I\chi^0(\mathbf{q}, i\omega_m)\}} \delta\chi^0_{ij}(\mathbf{q}, i\omega_m). \qquad \text{(V.10)}$$

The physical interpretation of $\Delta F^1$ is that it represents a contribution to the free-energy difference between the superfluid and normal states when the effective interaction is taken to be that obtained by virtual exchange of normal-state spin fluctuations. We assume that the effect of $\Delta F^1$ can be incorporated into the attractive interaction of the unperturbed weak-coupling calculation. This assumption has recently been investigated by Serene and Rainer[111] who find that although there are some non-negligible corrections to the results presented here they do not change the overall result that the ABM state becomes stable.

The term second order in $\delta\chi^0$ is

$$\Delta F^2 = -\frac{T}{4}\sum_{ij}\sum_{m}\int \frac{d^3q}{(2\pi)^3}\left[\frac{I\delta\chi^0_{ij}(\mathbf{q}, i\omega_m)}{1 - I\chi^0(\mathbf{q}, i\omega_m)}\right]^2. \tag{V.11}$$

One can convince oneself that the higher-order terms in the expansion of $\Delta F$ are negligible compared to $\Delta F^2$ because the important region of integration with respect to $q$ is for $2k_F\ q \gg \xi^{-1}$, where $\xi$ is the temperature-dependent coherence length. In this region it can be shown that $\delta\chi \propto N(0)/(q\xi)$, so one is expanding in powers of $\bar{I}/(k_F\xi)(1 - \bar{I})$ which is much less than one.

Examining equation (V.11) we see that the change in the spin-fluctuation free energy is always negative since $\delta\chi(\mathbf{q}, i\omega_m)$ is real. The principle we therefore use to study the feedback effect is the maximization of the change in the spin-fluctuation free energy.

The simple discussion in terms of the effective interaction is not entirely correct as it implies that the BW state will increase in energy via the feedback. This can, however, be rectified if one studies the gap equation in a strong-coupling formulation of the problem.

The calculation of $\delta\chi^0_{ij}$ is fairly complicated[37] so we shall not repeat it here. Rather we simply discuss the results of the calculations of the free energies in the various regimes. To do this we first rewrite (V.11) so that we can separate out the dependence of $\Delta F^2$ on various parameters. As already mentioned, the integral over $q$ is dominated by the $q > \xi^{-1}$ region and in this region we can write

$$\delta\chi^0_{ij} = \frac{N(0)}{qv_F}\,\delta\bar{\chi}^0_{ij}(\hat{\mathbf{q}}, i\omega_m), \tag{V.12}$$

i.e. we separate out the explicit dependence on the magnitude of $\mathbf{q}$. Then we can write

$$\Delta F^2 = -\frac{T}{4}\frac{1}{2\pi^2}\frac{1}{v_F^2}\int_0^{2k_F} dq\left(\frac{\bar{I}}{(1 - \bar{I} + \bar{I}q^2/12k_F^2)}\right)^2 \times \sum_{ij}\sum_{m}\int\frac{d\Omega_{\hat{q}}}{4\pi}[\delta\bar{\chi}_{ij}(\hat{\mathbf{q}}, i\omega_m)]^2. \tag{V.13}$$

Defining

$$\delta = \frac{150\pi^2}{7\zeta(3)}\left(\frac{T_c}{\varepsilon_F}\right)\int_0^{2k_F}\frac{dq}{2k_F}\left(\frac{\bar{I}}{(1 - \bar{I} + \bar{I}q^2/12k_F^2)}\right)^2, \tag{V.14}$$

we can write

$$\Delta F^2 = -N(0)\left(\frac{T}{T_c}\right)\delta\left(\frac{7\zeta(3)}{150\pi^2}\right)\sum_{ij}\sum_{m}\int\frac{d\Omega_{\hat{q}}}{4\pi}\,[\delta\bar{\chi}_{ij}(\hat{\mathbf{q}}, i\omega_m)]^2. \qquad (V.15)$$

Since $\delta\bar{\chi}$ depends only on $T$ and $\Delta$, we have separated off the spin-fluctuation parameters into a single variable $\delta$. The particular choice of normalization of $\delta$ made above turns out to be convenient in discussing the terms in $\Delta F^2$ that are fourth order in the free energy.

The stability of various phases just below $T_c$ is determined by an examination of the fourth-order terms in $\Delta$. As discussed in Section IV, from general invariance arguments[91, 35, 36] one can show that there are five fourth-order invariants. These can be written in terms of the $d_{\alpha i}$ previously introduced

$$F^4 = \frac{a_1}{2}\left|\sum_{\alpha i} d_{\alpha i}^2\right|^2 + \frac{a_2}{2} d^*_{\beta i} d^*_{\beta j} d_{\alpha i} d_{\alpha j} + \frac{a_3}{2} d^*_{\alpha i} d^*_{\beta i} d_{\alpha j} d_{\beta j}$$

$$+ \frac{a_4}{2}\{\sum_{\alpha i}|d_{\alpha i}|^2\}^2 + \frac{a_5}{2} d^*_{\alpha i} d^*_{\beta j} d_{\alpha j} d_{\beta i}. \qquad (V.16)$$

Using the weak-coupling theory presented in the previous section one can easily determine the weak-coupling values of the $a$'s. We find that $a_2 = a_4 = a_5 = -a_3 = -2a_1$, and assuming that the $d_{\alpha i}$ are normalized so that $\Sigma_{\alpha i}|d_{\alpha i}|^2 = 3|\Delta|^2$, i.e., $\Delta$ is defined as the average energy gap then

$$a_1^{wc} = -\frac{N(0)}{(\pi T)^2}\left(\frac{7\zeta(3)}{120}\right). \qquad (V.17)$$

The result for the spin-fluctuation contribution is

$$a_1^{SF} = a_1^{wc}\left(\frac{\delta}{10}\right)\left(\frac{T}{T_c}\right) \qquad (V.18)$$

and the spin-fluctuation $a$'s are such that $a_2 = 0{\cdot}5a_1$; $a_3 = 7{\cdot}0a_1$; $a_4 = -2{\cdot}0a_1$, and $a_5 = 5{\cdot}5a_1$. We can then write the total fourth-order free energy as (setting $T = T_c$)

$$F^4 = \frac{|a_1^{wc}|}{2}\{-(1{\cdot}0 + 0{\cdot}1\delta)|\sum_{\alpha i} d_{\alpha i}^2|^2 + (2{\cdot}0 - 0{\cdot}05\delta) d^*_{\alpha i} d^*_{\alpha j} d_{\beta i} d_{\beta j}$$

$$- (2{\cdot}0 + 0{\cdot}7\delta) d^*_{\alpha i} d^*_{\beta i} d_{\alpha j} d_{\beta j} + (2{\cdot}0 + 0{\cdot}2\delta)(\sum_{\alpha i}|d_{\alpha i}|^2)^2$$

$$+ (2{\cdot}0 - 0{\cdot}55\delta) d^*_{\alpha i} d^*_{\beta j} d_{\alpha j} d_{\beta i}\}. \qquad (V.19)$$

Using this expression along with the expression for the second-order free energy

$$F^2 = (1/3)N(0)\ln\left(\frac{T}{T_c}\right)|d_{\alpha i}|^2 \tag{V.20}$$

allows one to make a number of predictions regarding the thermodynamic properties near $T_c$. The most important prediction is that for $\delta$ greater than $\delta = 0{\cdot}47$ the Anderson–Morel state is the stable state just below $T_c$. This result is obtained by simply comparing the fourth-order terms, since the normalization of $\Delta$ is such that the second-order terms are equal:

$$F^4_{\mathrm{BW}} = \frac{|a_1^{wc}|}{2} 9\Delta^4(\tfrac{5}{3} - 0{\cdot}33\delta) \tag{V.21}$$

$$F^4_{\mathrm{AM}} = \frac{|a_1^{wc}|}{2} 9\Delta^4(2 - 1{\cdot}05\delta). \tag{V.22}$$

Equating these gives $\delta = 0{\cdot}47$.

One can calculate the discontinuity in the specific heat from the free energy. Using the fact that the weak-coupling value of the BW state is the same as the BCS value we find that

$$\frac{\Delta C_v}{C_v^>} = 1{\cdot}43(\tfrac{5}{3})/(2 - 1{\cdot}05\delta) \tag{V.23}$$

for the ABM state. Here $C_v^>$ is the value of the specific heat just above $T_c$. For the BW state we find

$$\frac{\Delta C_v}{C_v^>} = 1{\cdot}43/(1 - \tfrac{1}{5}\delta). \tag{V.24}$$

At the critical value of $\delta = 0{\cdot}47$ we find $\Delta C_v/C_v^> \simeq 1{\cdot}6$. This appears to be slightly smaller than the experimental value $\approx 1{\cdot}7$. If we assume that the change in $\Delta C_v/C_v^>$ with pressure is due to the change in $\delta$, then we find that $\delta = 0{\cdot}63$ at the melting curve. One can also predict the behaviour of the splitting of the transition in the presence of a magnetic field,[(28, 35, 36)] as discussed in Section IV. Writing

$$\hat{\Delta}(\mathbf{k}) = (\hat{k}_x + i\hat{k}_y)\begin{pmatrix}\Delta_\uparrow & 0\\ 0 & \Delta_\downarrow\end{pmatrix},$$

we note that the field-dependent term can be written as

$$F(H) = -N(0)\eta H(\Delta_\uparrow^2 - \Delta_\downarrow^2)$$

and determining the $d$'s in terms of $\Delta_\uparrow$ and $\Delta_\downarrow$ one can calculate the transition temperatures for ordering of the up and down spin pairs respectively. The up spins always order at $T_c + \eta H$ while the down spins order at

$$T_c - \eta H\left(\frac{1 - \delta/2}{1 + 0{\cdot}35\delta}\right).$$

Estimates of $\delta \approx 0{\cdot}6$ lead to a slope of $-0{\cdot}6\eta H$. This effect due entirely to strong coupling has recently been verified in NMR experiments by Osheroff and Anderson.[44] It is important to note that the reason the slope is not minus one is that the spin fluctuations introduce a coupling between the up spin pairs and the down spin pairs so that the down spin pairs order more easily after the up spins have ordered. The coupling induced by spin fluctuations between the up spin and down spin pairs also favours the parallel alignment of the orbital angular momentum of the up and down spins. Thus in weak-coupling theory the $\mathbf{l}$ of the up spins can be arbitrarily directed relative to the down spins. This led to confusion in the early papers and the label Anderson–Morel was used for states with $\mathbf{l}_\uparrow$ and $\mathbf{l}_\downarrow$ parallel or anti-parallel with one another. We now regard AM states as the class of states with $\mathbf{l}_\uparrow$ pointing arbitrarily with respect to $\mathbf{l}_\downarrow$ and the A phase is regarded as a particular state of this class with $\mathbf{l}_\uparrow = \mathbf{l}_\downarrow$. This state minimizes the spin-fluctuation energy.

In order to obtain the complete phase diagram of $He^3$ versus pressure we assume that $\delta$ is an increasing function of pressure and first determine the ratio, $T_{AB}/T_c$, of the transition temperature between the ABM and BW states $T_{AB}$ to $T_c$ versus $\delta$. We can then attempt to determine $\delta(P)$ from the effects discussed above and test the consistency of the theory. To do this, the expression for $\Delta F^2$ in (V.15) is expanded to order $(\Delta)^6$ so that the slope of the $T_{AB}/T_c$ versus $\delta$ curve can be calculated near $T_c$. It turns out that the largest contribution to the slope comes from the fact that the fourth-order spin-fluctuation terms are proportional to $(T\omega_s)^{-1}$ while the weak-coupling terms are proportional to $T^{-2}$. Therefore, as the temperature is lowered below $T_c$ at the critical value of $\delta$ the weak-coupling terms become more important and the BW state becomes more stable. Thus we expect the phase diagram to be roughly what is seen experimentally, i.e., the A phase is suppressed at lower temperatures. By calculating the crossover point at zero temperature $\delta_c(T = 0) = 0{\cdot}87$ and its slope as $T \to 0$, i.e.,

$$T_{AB}/T_c = \left(\frac{\delta - 0{\cdot}87}{8}\right)^{1/4},$$

we can plot the $T_{AB}/T_c$ curve shown in Fig. 11.

There are several ways of obtaining experimental estimates of $\delta(P)$. The first is from the specific heat discontinuity on the melting curve which gives $\delta(P = 34\,\text{atm}) = 0{\cdot}63$. This value is probably somewhat large since state-independent strong-coupling corrections may affect the specific heat. Alternatively, one can determine the $\delta$ on the melting curve from the slope of the

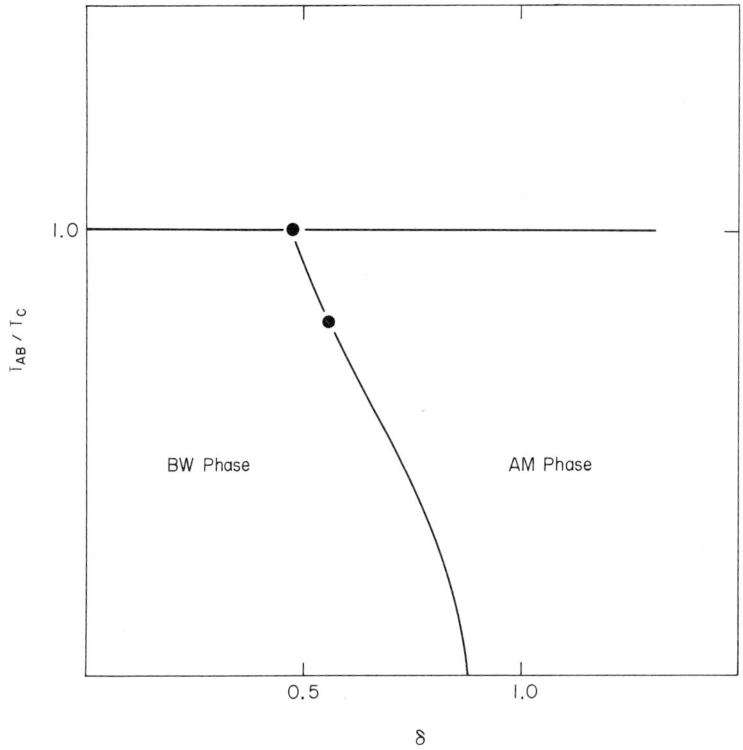


FIG. 11. A plot of $T_{AB}/T_c$ versus $\delta$ as described in the text. The experimental value of $\delta$ at $T_{AB}/T_c$ = 0·8 as determined from the NMR shifts[44] is marked on the theoretical curve.

NMR shift in the region of $A_1$ and $A_2$ when a large external field is present.[44] This gives $\delta(34\,\text{atm}) = 0{\cdot}54$. This value is used to insert an experimental point for $T_{AB}/T_c$ versus $\delta$ in Fig. 11. It is seen that the agreement between theory and experiment is quite good.

It should be noted that given the zero-field phase diagram the various aspects of the phase diagram in the presence of an external field, i.e. the splitting of $T_c$ into two transitions when going into the A phase and the suppression of the B phase at all pressures by an external field, are automatic consequences of the free-energy expansions when the field-dependent terms are included. Therefore, it appears that the effect on the spin fluctuations due to the change in the excitation spectrum in superfluid states does indeed give a complete description of the thermodynamics of the superfluid phases.

## VI. NMR IN THE SUPERFLUID PHASES

Of all the experimental work on the new phases of $He^3$, the study of the nuclear magnetic resonance has without question give us the most valuable clues as to the nature of these phases. The striking observation of a large shift in the transverse resonance frequency by Osheroff *et al.*[20] led to an enormous effort on the part of theorists throughout the world to understand the origin of this shift. To see why this observation was so important one must recognize the size of the effect. It was found that the transverse NMR frequency $\omega$ was given by the formula

$$\omega^2 = (\gamma H)^2 + \Omega_T^2 ,$$

where $\Omega_T$ is independent of field and depends on temperatures as $(T_c - T)^{\frac{1}{2}}$. Its largest value was found to be about 30 kHz and corresponds to an equivalent field of 10 gauss. Such a field is as though each $He^3$ nucleus is surrounded by 8–10 other $He^3$ nuclei whose magnetic dipoles are arranged so that their dipolar fields add.[23] Since such a configuration is unlikely to exist in $He^3$ one must find some more plausible explanation of the source of the shift. Such an explanation was given in a series of papers by Leggett.[21, 22, 23, 55, 112] We will not repeat all of his arguments and analysis here as these papers are quite extensive; rather we shall try to give a physical description of how such shifts arise and then give a simplified version of the derivation of the equation of motion applicable to the ABM and BW states which are now fairly conclusively established to represent the A and B phases.

There are several ingredients in the correct explanation of the shifts in the NMR frequencies. First, it is well known that without spin–orbit coupling there can be no shift in the resonant frequency, i.e., $\omega = \gamma H$. Consequently, the shift must involve the magnetic dipole interaction between the $He^3$ nuclei. It is the only appreciable spin–orbit interaction present. Second, in the anisotropic superfluid phases, the rotational invariance of the normal state in *both* spin and orbital space is broken, i.e., there spontaneously appear anisotropy axes such as the $\mathbf{l}$ and $\mathbf{w}$ vectors of the A phase. The importance of such broken spin–orbit symmetry was emphasized by Leggett.[23] In such states it is no longer true that the expectation value of the dipolar interaction

$$H_D = \tfrac{1}{2}\gamma^2\hbar^2 \sum_{i \neq j} \left\{ \frac{\boldsymbol{\sigma}_i \cdot \boldsymbol{\sigma}_j}{|r_{ij}|^3} - 3\frac{\boldsymbol{\sigma}_i \cdot \mathbf{r}_{ij}\, \boldsymbol{\sigma}_j \cdot \mathbf{r}_{ij}}{|r_{ij}|^5} \right\}. \tag{VI.1}$$

is zero. In fact, as we saw in Section IV, the expectation value of the dipolar interaction depends on the directions of the spin anisotropy axes relative to the orbital axes.

The system therefore establishes itself so that it minimizes the dipolar interaction in a fashion identical to the way a spin system orients along an easy axis. Finally, in spin resonance, the spin axes rotate relative to this minimum energy configuration and the dipolar interaction, therefore, acts as a restoring force and gives rise to a torque in the spin system. The spin system then acts like an oscillator whose "inertia" is the spin susceptibility and one obtains an additional contribution to the resonance frequency

$$\Omega_T^2 \propto \frac{1}{\chi}\langle H_D\rangle/\hbar^2. \qquad \text{(VI.2)}$$

The expectation value of $H_D$ is proportional to $(\gamma\hbar)/a^3$ times the number of superconducting pairs $(T_c/\varepsilon_F)^2$, so that

$$\Omega_T^2 \approx \frac{\gamma^2}{a^3}\frac{1}{\chi}\left(\frac{kT_c}{\varepsilon_F}\right)^2 \qquad \text{(VI.3)}$$

which is the order of magnitude of the observed shift. The large value of the shift comes about because of the combination of the first-order dipolar energy being nonzero and the response of the spin system being given by the susceptibility. Since $\chi^{-1} \propto \varepsilon_F$ we have $\Omega_T^2 \propto \varepsilon_F\langle H_D\rangle$, so that the shift is large compared to the dipole energy itself. Obtaining a shift which is the product of a large "exchange" energy ($\varepsilon_F$) and a small anisotropy energy is well known in antiferromagnetic resonance. Indeed, the resonance in the A phase is very analogous to antiferromagnetic resonance[24, 25] as we shall see.

In order to understand the NMR in the various phases we must derive the equations of motion for the spin angular momentum and the anisotropy axes. To do this we first write the dipolar energy in a form which makes the resonance calculation more tractable. Fourier transforming (VI.1) and ignoring the surface-dependent terms we find that

$$H_D = \frac{2\pi}{3}(\gamma\hbar)^2 \sum_{\mathbf{q}} (\boldsymbol{\sigma}_{\alpha\beta}\cdot\boldsymbol{\sigma}_{\gamma\delta} - 3(\boldsymbol{\sigma}_{\alpha\beta}\cdot\hat{\mathbf{q}})(\boldsymbol{\sigma}_{\gamma\delta}\cdot\hat{\mathbf{q}}))a^\dagger_{\mathbf{k}+\mathbf{q}\alpha}a^\dagger_{\mathbf{k}'-\mathbf{q}\gamma}a_{\mathbf{k}'\delta}a_{\mathbf{k}\beta}, \qquad \text{(VI.4)}$$

where $\hat{\mathbf{q}}$ is a unit vector in the direction of $\mathbf{q}$.

If we write this expression in terms of the anomalous expectation values

$$x_i(\mathbf{k}) \equiv 2\langle a_{-\mathbf{k}\alpha}a_{\mathbf{k}\beta}\rangle(\sigma_y\sigma_i)_{\alpha\beta} \qquad \text{(VI.5)}$$

and

$$x_i^*(\mathbf{k}) \equiv 2\langle a^\dagger_{\mathbf{k}\alpha}a^\dagger_{-\mathbf{k}\beta}\rangle(\sigma_i\sigma_y)_{\alpha\beta} \qquad \text{(VI.6)}$$

where the $\sigma$ are the Pauli matrices we find that

$$H_D = +\frac{24\pi}{3}(\gamma\hbar)^2 \sum_{\mathbf{k}\mathbf{k}'} (\mathbf{x}(\mathbf{k})\cdot\mathbf{x}^*(\mathbf{k}') - 3\mathbf{x}(\mathbf{k})\cdot\hat{\mathbf{q}}\mathbf{x}^*(\mathbf{k}')\cdot\hat{\mathbf{q}}) \qquad \text{(VI.7)}$$

where $\hat{\mathbf{q}} = (\mathbf{k} - \mathbf{k}')/|\mathbf{k} - \mathbf{k}'|$. Now we can go one step further to write for

$p$-wave pairing

$$x_i(\mathbf{k}) = \Sigma x_{\alpha i}\hat{k}_\alpha, \qquad \text{(VI.8)}$$

where $x_{\alpha i}$ are assumed to be roughly independent of the $|\mathbf{k}|$ in a region $v_F(k - k_F) = \pm\omega_c$ about the Fermi surface and zero elsewhere.

Doing the integrals over $\hat{\mathbf{k}}$ and $\hat{\mathbf{k}}'$ we find that the terms leading to spin–orbit coupling can be written as

$$\langle H_D \rangle = \Gamma_1(x_{ii}x_{jj}^* + x_{ij}x_{ji}^*) \qquad \text{(VI.9)}$$

where $\Gamma_1$ is a constant of order $(\gamma\hbar)^2\rho^2(\omega_c/\varepsilon_F)^2$. So far we have written the dipolar energy in terms of $x_{\alpha i}$ and not in terms of the $d_{\alpha i}$ related to the gap function $\Delta_{\alpha\beta}(\mathbf{k})$. In weak coupling these two functions are related by a simple factor of $(2\lambda\omega_c)$. Since we are accustomed to thinking in terms of the $d_{\alpha i}$ we will work with these variables and write

$$\langle H_D \rangle = \Gamma(d_{ii}d_{jj}^* + d_{ij}d_{ji}^*) \qquad \text{(VI.9}')$$

where $\Gamma = \Gamma_1/(2\lambda\omega_c)^2$.

If we add the energy required to induce a net spin angular momentum

$$H_S = (\gamma^2/2)\mathbf{S}\cdot\overleftrightarrow{\chi}^{-1}\cdot\mathbf{S} - \gamma\,\mathbf{S}\cdot\mathbf{H}, \qquad \text{(VI.10)}$$

where $\overleftrightarrow{\chi}^{-1}$ is the susceptibility tensor discussed in section IV, we have an energy function which when minimized with respect to the free variables in $d_{\alpha i}$ and the $\mathbf{S}$ gives us the equilibrium configuration for a given static field $\mathbf{H}$. As discussed in Section IV, for the A phase the equilibrium configuration is such that $\mathbf{w}$ is perpendicular to the external field. The dipole energy for this state reduces to

$$-3\Gamma\Delta^2((\mathbf{w}\cdot\mathbf{l})^2 - 1) \qquad \text{(VI.11)}$$

where $\Delta^2$ is the average of the square of the energy gap. Therefore $\mathbf{w}$ and $\mathbf{l}$ are parallel in equilibrium. They can lie anywhere in the plane perpendicular to $\mathbf{H}$ as shown in Fig. 12. In the BW state the situation is more complicated.

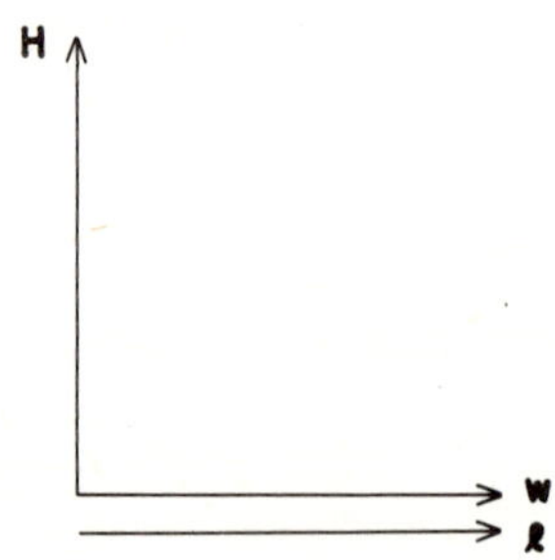


FIG. 12. The orientation of $\mathbf{l}$ and $\mathbf{w}$ in the presence of an external field $\mathbf{H}$ in the A phase.

The $d_{\alpha i}$ can be written as

$$d_{\alpha i} = \sqrt{3}\Delta \hat{O}(\mathbf{n}, \theta) \tag{VI.12}$$

where $\hat{O}(\mathbf{n}, \theta)$ is a rotation matrix specified by the axis of rotation $\mathbf{n}$ and the angle of rotation $\theta$. The dipolar energy is then

$$(H_D)_{\text{BW}} = 4\Gamma\Delta^2(\cos\theta + 2\cos^2\theta). \tag{VI.13}$$

It does not depend on the direction $\mathbf{n}$, but requires $\theta = \cos^{-1}(-\frac{1}{4})$.

Since the susceptibilities are isotropic in this phase there is nothing in (VI.10) to indicate how $\mathbf{n}$ should orient relative to the field. However, as Leggett[112] discussed (see also Engelsberg *et al.*[96]), there is a small term which is due to a combination of the depairing effects of the field acting back into the dipole energy which tends to make $\mathbf{n}$ parallel to $\mathbf{H}$. The equilibrium configuration for the B phase is therefore that shown in Fig. 13. The orbital

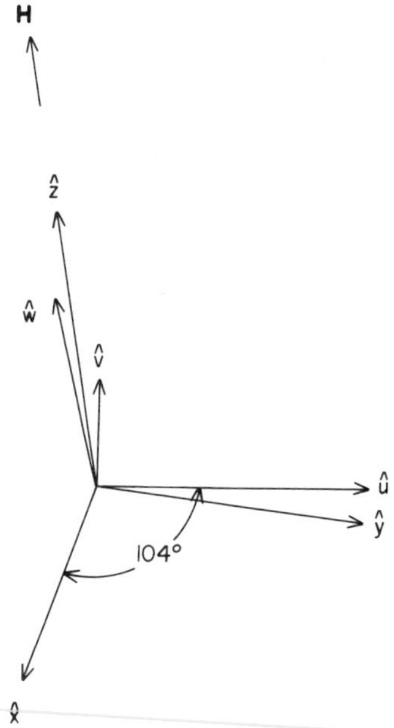


FIG. 13. The equilibrium orientation of the B phase. The spin axes $(u, v, w)$ are rotated relative to the orbital axis $(x, y, z)$ by 104°. The field is along the axis of rotation.

components $(x, y, z)$ are to be paired with the spin components $(u, v, w)$. The $(u, v, w)$ coordinates are rotated by $\cos^{-1}(-\frac{1}{4})$ about the external field.

Now given these equilibrium configurations we desire to obtain the equations of motion determining NMR. Leggett in his work studied the commutation relation between $\mathbf{S}$ and the $x_i(\mathbf{k})$ defined without the expectation values as in (VI.5, VI.6). He wrote the $x_i(\mathbf{k})$ operator averaged over the magnitude of $\mathbf{k}$ as $T_i(\hat{k})$ and showed that

$$[S_i, T_j(\hat{k})] = i\varepsilon_{ijk}T_k(\hat{k}). \tag{VI.14}$$

This equation is simply the statement that the $S_i$ act as generators of rotations in spin space. Therefore we derive the equations of motion using the $S_i$ as conjugate variables to the angles describing the rotation of $d_{\alpha i}$ in spin space. For small oscillations, we can consider rotations of the $d_{\alpha i}$ about the three coordinate axes to be independent. Such rotations change $d_{\alpha i}$ to

$$d_{\alpha i}(\theta_x, \theta_y, \theta_z) = d_{\alpha j}R_{ji}(\theta_x, \theta_y, \theta_z),$$

where $\hat{R}(\theta_x, \theta_y, \theta_z)$ is the sum of the small rotations about the three axes

$$\hat{R}(\theta_x,\theta_y, \theta_z) = \begin{pmatrix} 1 - \frac{1}{2}(\theta_y^2 + \theta_z^2) & \theta_z & -\theta_y \\ -\theta_z & 1 - \frac{1}{2}(\theta_x^2 + \theta_z^2) & \theta_x \\ \theta_y & -\theta_x & 1 - \frac{1}{2}(\theta_x^2 + \theta_y^2) \end{pmatrix}. \tag{VI.15}$$

Inserting this result into the expression for the dipolar energy one finds that it is quadratic in the $\theta$'s if the equilibrium configuration is correct. In the A phase

$$(H_D)_{\mathrm{ABM}} = +3\Gamma\Delta^2(\theta_x^2 + \theta_z^2), \tag{VI.16}$$

where we have chosen the $z$ axis to be along the field and the $y$ axis to be along $\mathbf{l}$. Since only the dipole energy changes under such a rotation, adding this expression to equation (VI.10) we obtain a total free energy

$$E = \frac{\gamma^2 S^2}{2\chi} - \gamma\mathbf{S}\cdot\mathbf{H} + 3\,\Gamma\Delta^2(\theta_x^2 + \theta_z^2). \tag{VI.17}$$

Using this as a hamiltonian we derive the equation of motion of $\mathbf{S}$ and $\theta_i$ by using the commutation relation $[S_i, \theta_j] = i\hbar\delta_{ij}$ and the usual commutator for spin angular momenta

$$[S_i, S_j] = i\hbar\varepsilon_{ijk}S_k. \tag{VI.18}$$

The resulting equations are

$$\frac{\mathrm{d}\mathbf{S}}{\mathrm{d}t} = \gamma\mathbf{S} \times \mathbf{H} - \nu^{-2}\chi\hat{\Omega}^2\cdot\boldsymbol{\theta} \tag{VI.19}$$

and

$$\frac{\mathrm{d}\boldsymbol{\theta}}{\mathrm{d}t} = \gamma(\gamma\mathbf{S}/\chi - \mathbf{H}). \tag{VI.20}$$

Here $\hat{\Omega}^2$ is a dyadic. For the ABM state, it is

$$\left(\frac{6\gamma^2\Gamma\Delta^2}{\chi}\right)\begin{pmatrix}1 & 0 & 0\\ 0 & 0 & 0\\ 0 & 0 & 1\end{pmatrix} = \hat{\Omega}^2. \tag{VI.21}$$

Assuming that $\mathbf{S} = \mathbf{S}_0 + \mathbf{S}_1(t)$ and $\mathbf{H} = \mathbf{H}_0 + \mathbf{H}_{rf}(t)$ and $S_1$ and $H_{rf}$ oscillate as $e^{i\omega t}$, we obtain the equation for $\mathbf{S}_1$

$$(\omega^2 - \hat{\Omega}^2\cdot)\mathbf{S}_1 - i\omega(\boldsymbol{\omega}_L \times \mathbf{S}_1) = -\gamma^{-1}\chi(\hat{\Omega}^2\cdot\mathbf{H}_{rf} + i\omega\boldsymbol{\omega}_L \times \mathbf{H}_{rf}) \tag{VI.22}$$

where $\boldsymbol{\omega}_L = \gamma\mathbf{H}_0$. This equation is identical to Eqn (5.7) of Leggett.[112] The relation between the matrix $\hat{\Omega}^2$ and the expansion of the dipolar energy in terms of the rotation angles is made transparent by our derivation. We have

$$\hat{\Omega}_{ij} = \frac{\gamma^2}{\chi}\frac{\partial^2 H_D}{\partial\theta_i\partial\theta_j}. \tag{VI.23}$$

Equation (VI.22) can now be used to calculate the dynamic susceptibilities defined as the solution to this equation

$$(S_1)_i = \gamma^{-1}\chi_{ij}H_{rfj}.$$

These susceptibilities have been discussed in detail by Leggett[112] and we will not discuss them here. Rather we concentrate on the frequencies of the normal modes and simply state their polarization. The modes are determined by the vanishing of the determinant of the coefficients of the $S_1$ on the left hand side of (VI.22). Returning to the particular form for the ABM state we find that the transverse resonance frequency is

$$\omega^2 = \omega_L^2 + \Omega_T^2 \tag{VI.24}$$

where $\Omega_T^2 = 6\gamma^2\Gamma\Delta^2/\chi$. Thus the shift in the transverse resonance $\Omega_T^2$ varies as $\Delta^2$. As discussed at the beginning of this section it is also of the right order of magnitude. The interesting feature of (VI.22) is that the equation for $S_{1z}$ is

$$(\omega^2 - \Omega_T^2)S_{1z} = -\gamma^{-1}\chi\Omega_T^2 H_{rfz} \tag{VI.25}$$

so that there is a longitudinal resonance in this state whose frequency is $\Omega_T^2$, i.e., exactly equal to the transverse resonance shift. The existence of this longitudinal resonance has recently been confirmed experimentally (Osheroff and Brinkman,[26] Bozler *et al.*,[113] and Webb *et al.*[97]).

Turning now to the B phase we need only calculate the $\hat{\Omega}^2$ tensor. With the equilibrium configuration discussed above we find that only the $\hat{\Omega}^2_{zz}$ component is nonzero and it is equal to

$$(6\gamma^2\Gamma\Delta^2/\chi_B)(5/2) = \hat{\Omega}^2_{zz} = \Omega_L^2. \tag{VI.26}$$

Substituting into (VI.22) we find that, in equilibrium, the B phase shows no shift in the transverse resonance, i.e., $\omega = \gamma H$, but that it should show a longitudinal resonance at $\Omega_L$. The frequency of this resonance written in terms of the A phase resonance is

$$\hat{\Omega}^2_{zz} = (5/2)\frac{\chi_A}{\chi_B}\left(\frac{\Delta_B^2}{\Delta_A^2}\right)\Omega_T^2, \tag{VI.27}$$

where $\chi_A/\chi_B$ is the ratio of the susceptibilities in the A and B phases and $(\Delta_B^2/\Delta_A^2)$ the ratio of the average of the square of their respective gaps. This relationship has been verified experimentally by Osheroff[27] to within 5%.

Work by Webb *et al.*[97,98] does not appear to give this agreement. There are, however, difficulties in observing the resonance in the B phase because of texture effects which we will discuss later.

The texture effects in the B phase will involve the direction of the axis of rotation **n** relative to the external field. We can calculate the resonance frequencies by calculating the $\hat{\Omega}^2$ matrix under circumstances where **n** makes an angle $\phi$ with respect to the external field. The result of this calculation is that one obtains two resonance frequencies. These are plotted versus the magnitude of the external field, in Fig. 14. At small fields there are two modes, one at $\Omega_L$ and the other at $\gamma H \cos\phi$. These modes are longitudinal and transverse (with respect to **n**) respectively. Therefore in this situation there is a mode

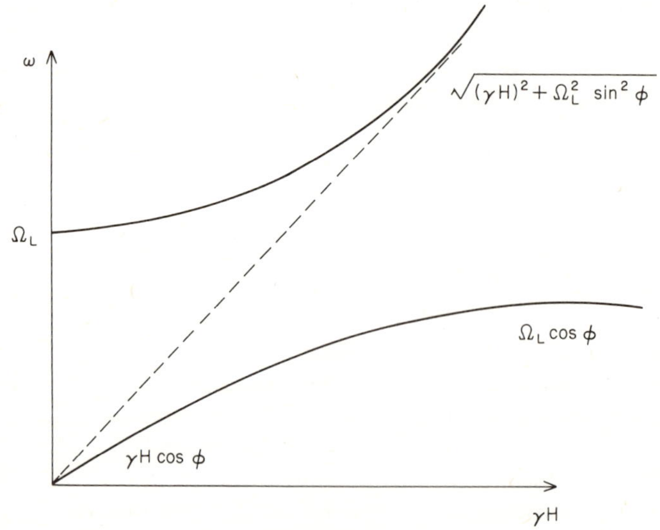


FIG. 14. A plot of the resonance frequencies as a function of field when the axis of rotation makes an angle $\phi$ with the external field.

with a component along the external field whose frequency is linear in $H$. At large fields $\gamma H \gg \Omega_L$ we have modes at

$$\omega = ((\gamma H)^2 + \Omega_L^2 \sin^2 \chi)^{\frac{1}{2}} \tag{VI.28}$$

and

$$\omega = \Omega_L \cos \chi, \tag{VI.29}$$

and these modes are polarized transversely and longitudinally with respect to the external field. The transverse-mode expression will be important in describing resonance line shape in situations where the texture of **n** is important.

Recently, Osheroff and Anderson[44] have investigated carefully the transverse shift in large fields where $T_c$ splits into two transitions $A_1$ and $A_2$. They find that the shift is proportional to the square of the sum of the up-spin order parameter and the down-spin order parameter, $(\Delta_\uparrow + \Delta_\downarrow)^2$. As already mentioned, they were able to accurately predict the resonant frequency using the spin-fluctuation model to obtain the fourth-order free energy. Their results are shown in Fig. 15.

Another interesting result of the Leggett equations is the prediction of nonlinear ringing associated with the longitudinal resonance (Leggett[112], Maki and Tsuneto[60] and Webb *et al.*[98]). Consider the A phase and assume only $\theta_z$ and $S_z$ to be nonzero. Then the total energy is

$$E = \frac{S_z^2}{2\chi} + \frac{\chi\Omega_T^2}{2} \sin^2 \theta_z. \tag{VI.30}$$

Since $\theta_z$ and $S_z$ are conjugate variables, the energy expression is that of a pendulum with $\theta_z$ the angle the pendulum makes with respect to the vertical axis and $S_z$ its angular momentum. The susceptibility acts like a moment of inertia. In the experiments to probe this effective hamiltonian one starts with an external field on, so that $S_z = \chi H$ and $\theta_z = 0$. One then suddenly turns off the field leaving the system with a "kinetic energy" $S_z^2/2\chi = \frac{1}{2}\chi H^2$. The system will then ring at a frequency determined by the amount of energy left in the system. Clearly, this frequency will become zero when the "kinetic energy" is such that the pendulum just swings into the upside down position. This occurs when $S_z^2/2\chi = \chi\Omega_T^2/2$ or $\gamma H = \Omega_T$. This was confirmed in the recent experiments by Webb *et al.*[97, 98]

The B phase is more complicated. When **n** is along **H** the $\theta$ in (VI.13) is the same as $\theta_z$, so that the energy is

$$\frac{S_z^2}{2\chi} + \chi\Omega_L^2(\tfrac{4}{15})(\cos\theta + 2\cos^2\theta). \tag{VI.31}$$

The equilibrium value is at $\theta = \cos^{-1}(-\frac{1}{4})$.

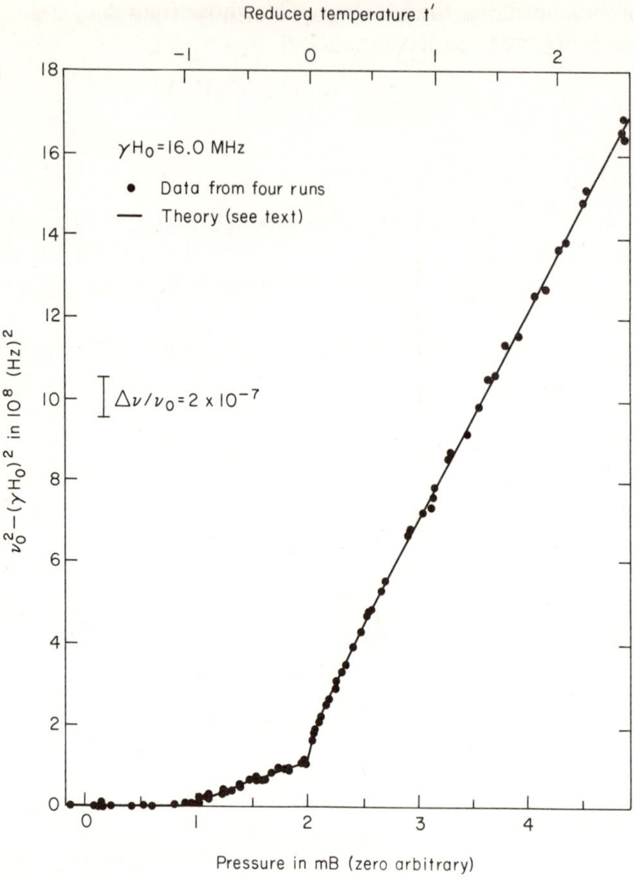


FIG. 15. The transverse resonance in the region of $A_1$ and $A_2$ as measured by Osheroff.[44]

The potential energy has two maxima in this case at $\theta = 0$ and $\pi$. These maxima should give rise to zeros in the ringing frequency at $(\gamma H/\Omega_L)^2 = \frac{3}{5}$ and $\frac{5}{3}$. It is not yet clear whether this prediction is correct or not.

The confirmation of all the above results leaves little question as to the validity of the theory of Leggett or that the two phases are any other than the ABM and BW states as first suggested by Anderson.[24] Several authors have tried to derive the expressions for the susceptibilities from microscopic theory (Maki and Ebisawa,[56, 114, 115] Takagi[116]) but have apparently enjoyed only partial success. We will consequently not attempt to review these approaches to the problem.

One aspect of the problem we have ignored in our discussion is the question of the role of the anisotropic susceptibilities in the calculation of the resonant frequencies. This question was discussed extensively by both Leggett[112] and Engelsberg *et al.*[96] The essential point in the discussion is that the $S_i$, as generators of rotation in spin space, rotate the susceptibility tensor as well as the magnetization and, consequently, the anisotropic susceptibilities do not appear in the equation of motion for **S**. They do appear in the equations for $\theta_i$ but do not affect the resonant frequencies.

It should be noted before we close this section that Combescot and Ebisawa[117] have given a rather successful theory of NMR line widths which gives a semiquantitative fit to Osheroff's longitudinal resonance data. Unfortunately we have no space to describe it here.

In concluding this section we should point out the relationship between the angles $\theta_i$ used in the present discussion and the Leggett discussion of the longitudinal resonance in terms of the Josephson effect. If one does a rotation of $\theta_z$ about the external field in the A phase the up-spin order parameter changes its phase by $\theta_z$ and the down-spin order parameter by $-\theta_z$. Therefore, the phase difference between the up-spin and down-spin superfluids is $2\theta_z$. In the longitudinal resonance the dipole interaction acts as a weak coupling between the two Fermi systems. The equation for $S_z$ is then the equation for the current between the two systems. The equation for $\theta_z$ is the equation for the phase difference. In it the $\gamma H$ acts as a potential difference between the two spin systems and, in general, the right-hand side of (VI.20) is the difference in chemical potential. This description of the equations of motion is undoubtedly closer to what is actually occurring than the classical approach presented here. The classical approach offers the advantage of exhibiting the full symmetry of the problem and a rather simple visualization of the modes.

## VII. Superflow and Spatial Variation: Generalized Ginzburg–Landau Approach

### 1. Generalized Ginzburg–Landau Theory

#### (a) *Ginzburg–Landau equation*

In Section IV we summarized the results of the BCS approach to superflow properties as carried through by Leggett[13, 14, 106] and Saslow.[45] The net result was expressions for $\rho_n$ near equilibrium, which we expected to be exact and which turned out to be anisotropic in the ABM state. Assuming the existence of a uniquely defined $v_s$, Galilean invariance then defines $\rho_s = \rho - \rho_n$ for us, again as an anisotropic tensor quantity. All of this was subject to straightforward Fermi-liquid corrections. In a final section we carried the problem of superflow a bit farther, setting up a hydrodynamics of the normal fluid in sufficient detail to derive second sound according to Wölfle.[32] All of this was limited by the $v_s$ assumption, as well as leaving out certain important relations which are not obvious in the perturbation approach.

In the Ginzburg–Landau theory of superconductivity the superflow properties are intimately related to the system's response to spatial variations, i.e. to an expansion of the free energy in terms of the gradient of the order parameter $\Psi$. One writes the free energy

$$F = \tfrac{1}{2}F_0(|\Psi|^2, T) + \tfrac{1}{2}F_{\mathrm{II}}(|\Psi^2, T)(\nabla\Psi)^2 + \dots. \tag{VII.1}$$

In the original Ginzburg–Landau theory,[118] as justified from BCS by Gor'kov,[119] $F_0$ was in turn expanded in a power series in $|\Psi|^2$, the first term being $a(T - T_c)|\Psi|^2$, while the coefficient $F_{\mathrm{II}}$ was a constant. This is useful only near $T_c$, and, ignoring $F_{\mathrm{II}}$, is the expansion in $|\Delta|^2$ we repeatedly used in Section IV. Werthamer[120, 90] and Eilenberger[88] have shown that expansions in $\nabla\Psi$ such as (VII.1) have a general validity when gradients are small compared to $(\xi_0)^{-1}$; this is the "Generalized Ginzburg–Landau" approach.

The term $F_{\mathrm{II}}$ not only controls the spatial fluctuations of the order parameter, it is within a factor the superfluid density $\rho_s$ as well. This is clear if we write

$$\mathbf{v}_s = \frac{\hbar}{2m}\nabla\varphi = \frac{\hbar}{2im}\frac{\nabla\Psi}{\Psi} \tag{VII.2}$$

and then clearly if

$$F_s = \frac{\rho_s v_s^2}{2}, \tag{VII.3}$$

$$\rho_s = F_{\mathrm{II}}\frac{4m^2}{\hbar^2}|\Psi|^2. \tag{VII.4}$$

The basic point here is one of gauge invariance. If we introduce a particle charge $q$ and a vector potential, the true one if we were speaking of electrons or a fictitious "baryon field" for neutral He atoms, gauge invariance tells us that $F_{\mathrm{II}}$ must multiply $|\nabla - (2q/i\hbar_c)\mathbf{A}|^2$, not $|\nabla|^2$, while the current must be defined as

$$\mathbf{j} = \frac{c}{q}\frac{\delta F}{\delta \mathbf{A}}. \tag{VII.5}$$

Hence the identity (VII.4).

De Gennes[50, 51] first pointed out that in the case of the complicated multidimensional order parameter of anisotropic superfluidity, a great deal of new physics could be obtained from the Ginzburg–Landau gradient expansion. A slightly more formal theory spelling out more aspects of the physics followed from Ambegaokar, de Gennes and Rainer.[52] Certain formal aspects of de Gennes' approach seem to us to be questionable and to have been more convincingly handled near $T_c$ by Wölfle,[53] while the most complete treatment is that of Blount, Anderson, and Brinkman[54, 25] which will be presented here. It is important to realize that in this case not only the supercurrent controls macroscopic (as opposed to fluctuation) phenomena; the angular elastic stiffness which follows from $F_{\text{II}}$-type terms is also of importance in many phenomena, especially NMR.[61]

We will specialize from the start to the $l = 1$ case, although most of the computation could be carried out in general. Here we have the gap bivector $d_{\alpha i}$ defined by

$$\Delta_{\sigma\sigma'}(\mathbf{k}) = \sum_{\alpha i} \hat{k}_\alpha d_{\alpha i}(\sigma_2\sigma_i)_{\sigma\sigma'}. \tag{VII.6}$$

As we have already explained, the major terms in the uniform free energy $F_0$ are invariant to rotations of both spin and orbit variables, so we define "states" in which $d_{\alpha i}$ can be reduced by rotation to certain canonical forms: the important ones being the BW

$$\text{BW:}\quad d_{\alpha i} = \begin{pmatrix} d & 0 & 0 \\ 0 & d & 0 \\ 0 & 0 & d \end{pmatrix} \tag{VII.7}$$

and the ABM

$$\text{ABM:}\quad d_{\alpha i} = \begin{pmatrix} 0 & 0 & d \\ 0 & 0 & id \\ 0 & 0 & 0 \end{pmatrix}. \tag{VII.8}$$

In order to minimize dipolar energy, the relative spin and orbit orientation must be fixed in such a way that (VII.8) remains as it is while (VII.7) becomes

$$d\begin{pmatrix} \frac{1}{4} & -\sqrt{\frac{15}{4}} & 0 \\ \sqrt{\frac{15}{4}} & \frac{1}{4} & 0 \\ 0 & 0 & 1 \end{pmatrix}. \tag{VII.9}$$

Let us first, following de Gennes[50, 51] and Wölfle,[53] make the standard Ginzburg–Landau expansion in both $d$ and $\nabla d$. In this case the $F_{\text{II}}$ term is quadratic in $d$ and its form is severely restricted by symmetry. The relevant energy terms are, like those responsible for $F_0$, separately spin and orbit invariant, and thus there are only three possible terms:

$$F_1 = \sum_{\alpha\beta}\sum_i |\nabla_\alpha d_{\beta i}|^2 \tag{VII.10}$$

$$F_2 = \sum_i |\nabla_\alpha d_{\alpha i}|^2 = |\nabla \cdot \mathbf{d}|^2 \tag{VII.11}$$

$$F_3 = \sum_{\alpha\beta} \sum_i (\nabla_\alpha d^*_{\beta i})(\nabla_\beta d_{\alpha i}), \tag{VII.12}$$

and one could write $(F)^{\text{II}} = K_1F_1 + K_2F_2 + K_3F_3$. Note that in each case the form is a simple dot product on the spin index; as in (VII.11) we will often omit the spin index which is always simply summed over, and treat $\hat{d}$ as a vector in its first orbital index $\alpha$.

De Gennes[51] and Ambegaokar, de Gennes and Rainer[52] follow liquid-crystal practice in using instead of (VII.12) the combination

$$(F_3)_{\text{DG}} = |\nabla \times \mathbf{d}|^2 = F_1 - F_3 \tag{VII.13}$$

as the third invariant. In principle this is equivalent but in practice leads to confusion.

From (VII.10) to (VII.12) one can construct one total divergence

$$F_3 - F_2 = \nabla \cdot [\mathbf{d}^* \cdot \nabla \mathbf{d} - \mathbf{d}^* \nabla \cdot \mathbf{d}]. \tag{VII.14}$$

In the total free energy such a term can lead only to a surface free energy, and since there are other, rather large sources of surface energies (VII.14) might as well be omitted from the free energy expression.

In particular, from the free energy there follows the Ginzburg–Landau differential equation for the spatial variation of the gap parameter, which is all-important in problems of textures, singularities, etc., as we shall see, and in this equation the form (VII.14) drops out upon partial integration. Thus for these purposes there are only two independent coefficients and one may use the form given by de Gennes

$$(F_{\text{DG}})^{\text{II}} = \tfrac{1}{2}K_L F_2 + \tfrac{1}{2}K_T(F_1 - F_3) \tag{VII.15}$$

instead of the form which is natural upon eliminating (VII.14) from (VII.10) to (VII.12)

$$(F)^{\text{II}} = \tfrac{1}{2}K_0 F_1 + \tfrac{1}{2}K_{\text{A}}(F_2 + F_3). \tag{VII.16}$$

Since $K_0$ and $K_{\text{A}}$ are more directly related to the supercurrents (as we shall see) we will use (VII.16). From (VII.16) we may derive the equilibrium "Ginzburg–Landau" equation for the gap bivector, $(\delta F/\delta d_{\alpha i})\delta d_{\alpha i} = 0$. We introduce $F_0(|d|^2)$ which is the free energy (IV.50) as a function of a uniform gap, and obtain for the coefficient of $\delta d_{\alpha i}$

$$\frac{\delta F_0}{\delta d_{\alpha i}} - K_0 \nabla^2 d^*_{\alpha i} - 2K_{\text{A}} \nabla_\alpha (\nabla \cdot \mathbf{d}^*_i) = 0 \tag{VII.17}$$

[since the form $\nabla(\nabla \cdot \mathbf{d})$ results from either $F_2$ or $F_3$, de Gennes' form (VII.15) would lead to the same equation with $K_A = K_L - K_T$, $K_0 = K_T$, but we repeat that $K_L$ and $K_T$ are not correct coefficients for the currents].

(b) "*Ordinary*" *supercurrents*

From (VII.10) to (VII.12) one can derive three formal expressions for supercurrents by introducing the gauge field $\mathbf{A}_B$ and setting

$$\nabla \rightarrow \nabla - \frac{2q}{i\hbar c}\mathbf{A}_B$$

and finally using (VII.4). (Equally, one could do this by doing a uniform Galilean transformation, setting

$$d \rightarrow d \exp\left(i\frac{2m\mathbf{v}_s}{\hbar}\cdot\mathbf{r}\right)$$

and $\mathbf{j}_s = \delta F/\delta \mathbf{v}_s$, i.e. the terms multiplying $\nabla\varphi$ where $\varphi$ is the phase of $d$.) We obtain thence

$$\mathbf{J}_1 = \frac{2i}{\hbar}[(d^*_{\alpha i}\nabla d_{\alpha i}) - \text{C.C.}] \tag{VII.18}$$

$$\mathbf{J}_2 = \frac{2i}{\hbar}[\mathbf{d}^*(\nabla\cdot\mathbf{d}) - \text{C.C.}] \tag{VII.19}$$

$$\mathbf{J}_3 = \frac{2i}{\hbar}[(\mathbf{d}^*\cdot\nabla)\mathbf{d} - \text{C.C.}] \tag{VII.20}$$

As Blount pointed out, the fact that $F_2 - F_3$ is a divergence does not prevent us from using the general *formal* expression for $F$, nor can we see any unique way to choose a free energy "orthogonal" to $F_2 - F_3$. Thus *a priori* any three coefficients for (VII.18) to (VII.20) are compatible with symmetry, and we could write

$$\mathbf{J} = j_1\mathbf{J}_1 + j_2\mathbf{J}_2 + j_3\mathbf{J}_3$$

where (VII.16) merely enforces the conditions

$$j_1 = \tfrac{1}{2}K_0$$
$$j_2 + j_3 = \tfrac{1}{2}K_A. \tag{VII.21}$$

Nonetheless we (as well as Wölfle[53]) find from microscopic calculations, in accord with the intuitive (VII.16), that the major portion of the supercurrent is given by

$$\mathbf{J} = \tfrac{1}{2}[K_0\mathbf{J}_1 + K_A(\mathbf{J}_2 + \mathbf{J}_3)]. \tag{VII.22}$$

Given that the form of the supercurrent is (VII.22), the coefficients $K_0$ and $K_A$ could be determined simply from Saslow's calculation[45] of the anisotropy of $\rho_s$ in the ABM state, since that determines two independent coefficients. Wölfle[53] in this limit, and Blount for general $T$, have calculated the free energy and current coefficients (without Fermi-liquid corrections, which we will discuss shortly). Blount's method was a generalization of the Eilenberger[88] scheme which we discussed in Section IV and which was used by him to derive the generalized Ginzburg–Landau theory. His result is

$$J_\alpha = \frac{\hbar\rho}{2m} T_{\alpha\beta\gamma\delta} \frac{1}{2i}(d^*_{\beta i}\nabla_\gamma d_{\delta i} - \text{C.C.}) \tag{VII.23}$$

where the coefficient $T_{\alpha\beta\gamma\delta}$ is given, to lowest order in $T_c/E_F$, by

$$T_{\alpha\beta\gamma\delta} = \frac{3}{4\pi}\int \mathrm{d}\Omega \frac{f_s(\hat{\mathbf{k}})}{|\Delta|^2} \hat{k}_\alpha \hat{k}_\beta \hat{k}_\gamma \hat{k}_\delta. \tag{VII.24}$$

where

$$f_s(\hat{\mathbf{k}}) = \Delta^2(\hat{\mathbf{k}}) \int \frac{\mathrm{d}\varepsilon_{\mathbf{k}}}{E(\mathbf{k})} \frac{\mathrm{d}}{\mathrm{d}E_{\mathbf{k}}} \left( \frac{\tanh(\beta E_{\mathbf{k}}/2)}{E_{\mathbf{k}}} \right). \tag{VII.25}$$

In the Ginzburg–Landau region proper, $T \simeq T_c$,

$$T_{\alpha\beta\gamma\delta} \simeq \frac{7\zeta(3)}{20\pi^2 T_c^2} (\delta_{\alpha\beta}\delta_{\gamma\delta} + \delta_{\alpha\gamma}\delta_{\beta\delta} + \delta_{\alpha\delta}\delta_{\gamma\beta}). \tag{VII.26}$$

Comparing (VII.26) with (VII.23) and (VII.21), we find that

$$K_0 = K_A = \frac{\hbar^2\rho}{m^2} \frac{7\zeta(3)}{80\pi^2 T_c^2}. \tag{VII.27}$$

To this order Blount finds that (VII.23) and (VII.24) hold at general temperatures, but of course in general (VII.24) is anisotropic. So long as we maintain axial symmetry, however, the general form of $\delta$-function products in (VII.24) is still maintained if one calculates in the axis system of $\hat{d}$, and very similar current expressions are retained. For BW, where isotropy of $\Delta$ continues to hold, the expressions (VII.22), (VII.16) and $K_0 = K_A$ are correct for all $T$.

### (c) *Balian–Werthamer: Complete expressions*

The BW case, because of its isotropy and because its gap function is essentially real, is a simple case. The general gap function is

$$d_{\alpha j} = d e^{i\phi} O_{ij}(\mathbf{n}, \theta)\delta_{\alpha i} \tag{VII.28}$$

where $O_{ij}$ is the orthogonal matrix for rotation by $\theta$ about the axis $\mathbf{n}$.

The current terms $\mathbf{J}_1 - \mathbf{J}_3$ all lead to the same current:

$$\mathbf{J}_1 = 3\mathbf{J}_2 = 3\mathbf{J}_3 = \frac{4}{\hbar}\,\mathrm{Tr}\,(\hat{d}^2)\nabla\varphi \tag{VII.29}$$

($d$ and $O_{ij}$ are real, hence gradients of them cancel in $\mathbf{J}$; we also use the orthogonality of $\hat{O}$). Note that $\mathbf{J}_2 - \mathbf{J}_3 \equiv \mathbf{0}$ so that there the ambiguity of $K$'s does not exist. We can then identify $K_0 = K_A = \frac{1}{5}\rho_s \times$ consts., where $\rho_s$ is the superfluid density already calculated in Section IV, since the current consists entirely of ordinary supercurrent with a well-defined $\mathbf{v}_s$ as assumed there.

Here the treatment of Fermi-liquid corrections requires some discussion. So far as we know, no rigorous treatment of Fermi-liquid corrections to the Ginzburg–Landau equation, as opposed to those to normal supercurrents, exists. From two rather qualitative but suggestive arguments we would suggest that $K_0$ and $K_A$ for the free energy (VII.16) and for the current (VII.22) must remain identical even after Fermi-liquid corrections. First, the expressions $F_1$, $F_2$, $F_3$ were derived as rotational invariants and the Fermi-liquid corrections bring in no rotational asymmetries so should maintain their rotationally invariant structure; gauge invariance of course connects some of the $F_i$ structure to the $J_i$. Second, the form of the Fermi-liquid correction is

$$\rho_s = \frac{\rho_s^0}{1 + (F_1/3)(\rho_n^0/Nm^*)} = \rho_s^0\left(1 - \frac{(F_1/3)N_n}{1 + (F_1/3)N_n}\right), \tag{VII.30}$$

where $\rho_s^0$ is the superfluid density calculated in weak coupling and $N_n$ is an effective relative density of normal quasiparticles. This suggests that we think heuristically of superfluid kinetic energies being reduced by a fraction for each normal particle: i.e. of a kind of "obstruction" effect by the normal fraction. Such observations should work equally well to reduce kinetic energies depending on $\nabla O_{ij}$ as those depending on $\nabla\varphi$. Thus in (VII.16) we will use the Fermi-liquid corrected $K$'s, proportional to $\rho_s$ as given in Section IV.3, recognizing however that this may be incorrect.

Brinkman *et al.*,[61] have given the form of $F^{\mathrm{II}}$ terms which results upon substitution of (VII.29) into (VII.16) (omitting even more terms which come from $\nabla\theta$ and $\nabla d$)

$$\begin{aligned} F^{\mathrm{II}}_{\mathrm{BW}} = \frac{\hbar^2\rho_s}{8m^2}(\nabla\varphi)^2 + \frac{\hbar^2\rho_s}{20\,m^2}(1 - \cos\theta)[4(\mathbf{n}\times\nabla\times\mathbf{n})^2 \\ + (3 + \cos\theta)(\mathbf{n}\cdot\nabla\times\mathbf{n})^2 + (3 - \cos\theta)(\nabla\cdot\mathbf{n})^2 \\ - 2\sin\theta\,(\nabla\cdot\mathbf{n})(\mathbf{n}\cdot\nabla\times\mathbf{n})]. \end{aligned} \tag{VII.31}$$

If we specialize as Brinkman *et al.* did to $\cos\theta = -\frac{1}{4}$, which will usually be fairly accurately fixed, we obtain the still complicated expression which will be used in BW texture problems:

$$F = \frac{\hbar^2 \rho_s}{8\,m^2}\left\{2(\nabla \times \mathbf{n})^2 + \tfrac{13}{4}(\nabla \cdot \mathbf{n})^2 - \frac{\sqrt{15}}{4}(\mathbf{n} \cdot \nabla \times \mathbf{n})(\nabla \cdot \mathbf{n}) - \tfrac{5}{8}(\mathbf{n} \cdot \nabla \times \mathbf{n})^2 + 16\nabla \cdot ((\mathbf{n} \cdot \nabla)\mathbf{n} - \mathbf{n}(\nabla \cdot \mathbf{n}))\right\}. \qquad \text{(VII.32)}$$

(d) *ABM: Orbital currents and orbital supercurrents*

Complicated as (VII.31) may appear, it pales beside the real complications of the ABM case. In that case, entirely aside from the difficulties of anisotropy we have those which result from the finite angular momentum of the pairs. The first difficulty is one of principle: the ambiguity of the $F_2 - F_3$ term. The fact that $F_2 - F_3$ is a divergence leads automatically to the conclusion that $\mathbf{J}_2 - \mathbf{J}_3$ is divergenceless,

$$\nabla \cdot (\mathbf{J}_2 - \mathbf{J}_3) = 0, \qquad \text{(VII.33)}$$

and in fact by actual expansion one can see that $J_2 - J_3$ is a curl:

$$\begin{aligned}\mathbf{J}_2 - \mathbf{J}_3 &= \frac{i}{\hbar}\nabla \times (\mathbf{d}^* \times \mathbf{d}) \\ &= \frac{i}{\hbar}\nabla \times \mathbf{l} \qquad \text{(VII.34)}\end{aligned}$$

where have introduced the vector $\mathbf{l}$ of "angular momentum of the pairs", which is of course *not* equal to $\langle L \rangle$.

Since $(F_2 - F_3)$ does not enter the Landau–Ginzburg equation nor, presumably, any hydrodynamic equations of motion, the corresponding current must be conserved automatically rather than as a consequence of the configuration; this is the physical meaning of (VII.33). This type of current is precisely analogous to the magnetization current $\mathbf{J}_M = \nabla \times \mathbf{M}$ of a ferromagnet: it transports nothing and follows simply as a consequence of the configuration. In particular, it can have no coupling with the conventional superflow transport processes, since transport implies a coupling with a source or sink, i.e. an energy term proportional to $\nabla \cdot \mathbf{J}$. In the ferromagnetic case there is an appropriate field with which it interacts; here, there is only its inertia, which is, however, all-important to the frequencies of orbit waves, as pointed out by Anderson[24, 25] and Wölfle,[53] just as the angular momentum of a ferromagnet is crucial to its spin waves.

Thus we must identify the coefficient of the current $(\mathbf{J}_2 - \mathbf{J}_3)$ with the "intrinsic" orbital angular momentum. We emphasize that this identification is neither rigorous nor unambiguous, since it is easy to invent configurations in which **d** varies in such a way as to give the sample as a whole a total angular momentum from the $(\mathbf{J}_2 + \mathbf{J}_3)$ term or even the $\mathbf{J}_1$ one. (One of these may in fact be a common texture in a tube.) Nonetheless, if the concept of an $\langle L \rangle$ has a meaning, it must be this. As we have already seen ((VII.23–VII.25), to lowest order in $\Delta/E_F$ this coefficient vanishes; Anderson and Morel estimated that in fact the coefficient was $\propto \Delta/E_F$ (see Section IV) but more recent estimates by Blount and by Cross place it one order lower. These estimates were direct calculations of the coefficient $K_l$ of $\mathbf{J}_2 - \mathbf{J}_3$ in

$$\mathbf{J} = \tfrac{1}{2}K_0\mathbf{J}_1 + \tfrac{1}{2}K_A(\mathbf{J}_2 + \mathbf{J}_3) + \tfrac{1}{2}K_l(\mathbf{J}_2 - \mathbf{J}_3)$$

$$K_l \simeq \frac{T_c^2}{E_F^2} K_0. \tag{VII.35}$$

But even omitting the effects of $\mathbf{J}_2 - \mathbf{J}_3$ as small, the $K_A$ term has rather interesting properties in that unlike (VII.34) the current depends on gradients of direction angles as well as of $\varphi$. When $F$ is expanded in terms of gradients of angles and $\nabla\varphi$, there are cross-terms and as a result there are what de Gennes calls "orbital supercurrents." As opposed to the orbital current $(\mathbf{J}_2 - \mathbf{J}_3)$, these currents are just as big as normal supercurrents. Near $T_c$ these have been correctly calculated by Wölfle and by Blount, and the structure of the current may be simply summarized as follows:

$$\mathbf{J} = \overleftrightarrow{\rho}_s \mathbf{v}_s + \overleftrightarrow{C}\frac{\hbar\nabla}{2m} \times \hat{\mathbf{l}} \tag{VII.36}$$

where

$$\frac{\overleftrightarrow{\rho}_s}{\rho} = \begin{pmatrix} 2 & 0 & 0 \\ 0 & 2 & 0 \\ 0 & 0 & 1 \end{pmatrix} \cdot \frac{7\zeta(3)}{10\,\pi^2}\frac{\langle\Delta^2\rangle}{T_c^2}\frac{m}{m^*}$$

and

$$\overleftrightarrow{C} = \tfrac{1}{2}\begin{pmatrix} 1 & 0 & 0 \\ 0 & 1 & 0 \\ 0 & 0 & -1 \end{pmatrix} \cdot \frac{7\zeta(3)}{10\,\pi^2}\frac{\langle\Delta^2\rangle}{T_c^2}\frac{m}{m^*}.$$

In each case the $z$-axis is taken parallel to **l**. The $\overleftrightarrow{C}$ term is the orbital supercurrent term; it never vanishes if $\nabla \times \mathbf{l}$ is finite but points in a different direction in general.

Blount has carried the current calculations to absolute zero within the weak-coupling approximation. Although, as we remarked in Section IV.2, at

$T = 0$ $\rho_s = \rho$, which is isotropic, the orbital supercurrent does not vanish but still retains exactly the same form:

$$\mathbf{J}_{T=0} = \frac{\rho\hbar}{2m}\left(\nabla\phi + \frac{1}{2}\begin{pmatrix}1 & 0 & 0\\ 0 & 1 & 0\\ 0 & 0 & -1\end{pmatrix}\nabla \times \mathbf{l}\right). \qquad \text{(VII.37)}$$

The superfluid density isotropy is achieved not by vanishing of the terms corresponding to $K_A$ but by virtue of the anisotropy of the $K_0$ coefficient [see (VII.38)].

The value of the second term in (VII.37) is again not absolutely independent of assumptions about Fermi-liquid corrections, but this is uncertain rather at intermediate temperatures—it is evident that at $T = 0$ the whole fluid must follow the equation of motion of the pairs. We shall not hazard a guess as to the correct temperature dependence of the Fermi-liquid coefficients, but will quote here Blount's general result for weak-coupling theory only:

$$\mathbf{J} = \frac{\hbar\rho}{2m}[4\alpha\nabla\varphi + (2\beta - 4\alpha)\hat{\mathbf{l}}(\hat{\mathbf{l}}\cdot\nabla\varphi) + 2\beta\nabla \times \mathbf{l} - 2\hat{\mathbf{l}}[\hat{\mathbf{l}}\cdot(\nabla \times \mathbf{l})]]. \qquad \text{(VII.38)}$$

Here

$$\alpha = \frac{3}{32\pi}\int d\Omega\, f_s \sin^2\theta$$

$$\beta = \frac{3}{8\pi}\int d\Omega\, f_s \cos^2\theta; \qquad \text{(VII.39)}$$

$f_s(\hat{\mathbf{k}})$ is as defined in (VII.25).

Blount has also worked out the general form of the Ginzburg–Landau free energy, again without Fermi-liquid corrections. We quote the rather complicated result:

$$F = \frac{\hbar^2\rho}{8m^2}\Big\{\alpha[4(\mathbf{l} \times \nabla\varphi)^2 + (\nabla\cdot\mathbf{l})^2]$$
$$+ \beta[2(\mathbf{l}\cdot\nabla\varphi)^2 + 2\nabla\varphi\cdot(\nabla \times \mathbf{l} - 2\mathbf{l}(\mathbf{l}\cdot\nabla \times \mathbf{l}))]$$
$$+ \left(\beta + \frac{2\gamma}{3}\right)(\mathbf{l} \times \nabla \times \mathbf{l})^2 + \left(\frac{\alpha + 2\beta}{3}\right)(\mathbf{l}\cdot\nabla \times \mathbf{l})^2\Big\}. \qquad \text{(VII.40)}$$

$\alpha$ and $\beta$ are as in (VII.39) while

$$\gamma = \frac{3}{4\pi}\int d\Omega\, f_s(\hat{\mathbf{k}})\frac{\cos^4\theta}{\sin^2\theta}. \qquad \text{(VII.41)}$$

Oddly, this coefficient diverges logarithmically as $T \to 0$, as a consequence of the zeros in the gap function. There appears to be an anomalous stiffness against the quantity $(\mathbf{l} \times \nabla \times \mathbf{l})$. Perhaps if the configuration were driven to have finite $(\mathbf{l} \times \nabla \times \mathbf{l})$, $\uparrow\downarrow$ pairing would be enforced in order to fill up the zeros.

Again, as for the current, we hesitate to speculate on Fermi-liquid effects. At $T = 0$ (VII.40) must be taken literally (here $\alpha = \frac{1}{4}$, $\beta = \frac{1}{2}$, $\gamma \to \ln T$). As $T \to T_c$,

$$\alpha = \beta = \frac{\gamma}{3} = \frac{d^2}{5} \frac{\rho_s}{\langle \Delta^2 \rangle} = \frac{7}{20\,\pi^2} \zeta(3) \frac{d^2}{T_c^2} \qquad \text{(VII.42)}$$

($d^2$ is the coefficient of $\sin^2\theta$ in $\Delta^2 = d^2 \sin^2\theta$)

and, as for the current, in this limit Fermi-liquid effects amount merely to multiplying all $\rho$'s by $m/m^*$ after calculating in weak coupling. Given the two limits, not much leeway is left for estimating the variation between them.

It is too early to decide which of the many striking properties of the orbital supercurrents will be of most interest. We note that they are by no means either conserved or irrotational by nature. The former means that they will play a role in transport processes, and indeed we can expect that externally applied currents will cause twists in the orientation over and above the free-energy minimizing effects, especially for heat currents, discussed by de Gennes and Rainer(121) [who seem to have ignored the cross terms of (VII.40)]; equally, any means of twisting the orientation will drive currents. The latter means (as pointed out by Ambegaokar, de Gennes and Rainer(52)) that the superflow is by no means a potential irrotational flow, which may lead to interesting differences. The whole problem of Josephson-type interference effects raised by those authors becomes far more complicated because of the non-Abelian character of the group of transformations of the order parameter which can affect its phase. Since it is not out of the question that a reasonable sized orifice ($\approx 1$ micron) could have Josephson properties, this is an experimentally interesting question.

## 2. COLLECTIVE MODES IN ANISOTROPIC SUPERFLUID $HE^3$

In Section IV.4 we discussed fairly completely the subject of soundlike modes in $He^3$. Those modes could be done reasonably completely using only conventionally defined $\rho_s$ and $\rho_n$ values. Here we wish to give a brief survey of those few calculations which have used more complete and rigorous theories, such as the Ginzburg–Landau techniques above, to discuss the various collective excitations involving the other (non-phase) degrees of freedom of $d$. The reader will recall that zero and first sound were at very high frequencies ($c \gg v_F$) and that second sound is confined to a very small frequency and temperature regime, and has a very slow velocity ($<1$ m sec$^{-1}$). Therefore the other modes we discuss will be separated from these easily.

Brinkman and Smith,[59] Maki,[122] Maki and Ebisawa,[114, 115] Combescot,[57, 58] and Wölfle[53] have all given treatments of one aspect or another of the spin-wave problem, which on the whole are in good agreement. Maki and Tsuneto[60] have discussed very simply some aspects of the ABM phase which agrees with Combescot's more complete treatment of the hydrodynamic regime $\omega\tau_D \ll 1$ (where $\tau_D$ is the collision time between quasiparticles, which near $T_c$ is of order $10^{-6}$–$10^{-7}$ sec). Brinkman and Smith, as well as Combescot, have discussed the BW case. The former treatment is particularly simple and we will reproduce it here to derive the spin waves for both phases.

For a rigorous theory including all Fermi-liquid effects, and especially for the collisionless regime at high frequencies, it is essential to go through one version or another of a full generalized RPA calculation, and this has been done by Maki and Ebisawa[56] and, particularly, by Combescot.[57] These theories are quite complicated and we will not discuss them in detail here, but note that they agree—with a suitable, but not unique, interpretation of Fermi-liquid corrections—with the heuristic theories at $T = 0$ and thus qualitatively in terms of number of modes, etc., they contribute nothing new.

In the theory of Brinkman and Smith the conjugate relationship between the spin angular momentum and the angles of rotation and spin space are further exploited to derive the spin waves. As already discussed in Section VI, the rotation angles are related to the phase difference between the up and down spin gap functions, so that this approach is identical to that of Maki and Tsuneto and may alternatively be thought of in terms of a Josephson effect between the up and down spin fermion reservoir. The advantage in thinking in terms of rotation angles is that the full symmetry of the problem is explicit.

Thus we introduce the same set of angles $(\theta_u, \theta_v, \theta_w)$ representing small rotations about the three axes. These angles now depend on position in space as well as time. The order parameter is

$$d_{\alpha i} = d^0_{\alpha j} O_{ji}(\theta_u, \theta_v, \theta_w), \tag{VII.43}$$

where $d^0_{\alpha j}$ is the equilibrium order parameter assumed to be uniform. The equations of motion are obtained from the local energy density $F_T$ which includes the same terms as the NMR calculations in Section VI (VI.17),

$$E = \frac{\gamma^2 S^2}{2\chi} - \gamma \mathbf{S} \cdot \mathbf{H} + H_0(\theta_u, \theta_v, \theta_w) \tag{VII.44}$$

along with the stiffness energy described by (VII.16)

$$F^{\mathrm{II}} \propto \nabla_\alpha d^*_{\beta i} \nabla_\alpha d_{\beta i} + \nabla_\alpha d^*_{\alpha i} \nabla_\beta d_{\beta i} + \nabla_\alpha d^*_{\beta i} \nabla_\beta d_{\alpha i}. \qquad \text{(VII.16)}$$

Then, using the commutation relations

$$[S_i, \theta_j] = i\hbar\delta_{ij} \qquad \text{(VII.45)}$$

and $F_T$ as a hamiltonian, we obtain the equations of motion

$$\frac{\partial}{\partial t} S_i = \gamma(\mathbf{S} \times \mathbf{H})_i - \frac{\partial}{\partial \theta_i}(H_0 + F^{\mathrm{II}}) \qquad \text{(VII.46)}$$

$$\frac{\partial}{\partial t} \theta_i = \frac{\partial F_T}{\partial S_i} = \gamma\left(\frac{\gamma S_i}{\chi} - H_i\right). \qquad \text{(VII.47)}$$

Let us now restrict ourselves to the ABM state. Then, as given by (VI.16)

$$H_0 = 3\Gamma\Delta^2(\theta_u^2 + \theta_v^2) \qquad \text{(VII.48)}$$

and after some algebra

$$F^{\mathrm{II}} = \frac{\hbar^2}{8m^2}(\nabla\theta_u \cdot \overleftrightarrow{\rho}_s \cdot \nabla\theta_u + \nabla\theta_v \cdot \overleftrightarrow{\rho}_s \cdot \nabla\theta_v), \qquad \text{(VII.49)}$$

where $\rho_s$ is the superfluid density discussed in Section IV and we have again taken $w$ to be the axis along which $m_s = 0$. Then

$$\frac{\partial}{\partial t} S_u = \gamma(\mathbf{S} \times \mathbf{H})_u - 6\Gamma\Delta^2(\theta_u) - \frac{\hbar^2}{4(m)^2}\mathbf{q} \cdot \overleftrightarrow{\rho}_s \cdot \mathbf{q}\theta_u. \qquad \text{(VII.50)}$$

Solving these equations we find two solutions just as in the uniform resonance case, a longitudinal mode in which **S** is parallel to **H** with a frequency

$$\omega_q^2 = \Omega_L^2 + \frac{\gamma^2}{\chi}\mathbf{q} \cdot \overleftrightarrow{\rho}_s \cdot \mathbf{q} \qquad \text{(VII.51)}$$

and the transverse mode with frequency

$$\omega_q^{T^2} = \omega_q^2 + (\gamma H)^2. \qquad \text{(VII.52)}$$

These results are identical to those of Maki and Tsuneto[60] for the wave associated with the longitudinal resonance. As we know, the expression used for $F^{\mathrm{II}}$ is only valid in the Landau–Ginzburg regime. However, the work of Combescot[57, 58] which starts from a less phenomenological approach gives an indication that the correct expression for the spin stiffness is (VII.49) with $\rho_s$ given by the weak-coupling form divided by $(1 + F_1^s/3)$, provided one ignores the higher antisymmetric Fermi-liquid corrections $F_1^a$, etc. To see this we note that (VII.50) is simply the continuity equation for the spin density and therefore the last term must be identified with $\nabla \cdot \mathbf{j}_s$ where $\mathbf{j}_s$ is the spin current. Combescot has shown that the superfluid density occurring in the spin current is not given by the same expression as the superfluid mass density because spin currents are not conserved quantities. He shows that, if you ignore higher-order antisymmetric Fermi-liquid factors other than $F_0^a$, the spin density occurring in the current is just $\rho_s^{wc}/(1 + F_1^s/3)$, where $\rho_s^{wc}$ is calculated without Fermi-liquid effects ($\rho_s^{wc}/\rho(1 - d)$ in his notation). Therefore, this superfluid "spin density" does not go to one as $T$ goes to zero but to $1/(1 + F_1^s/3)$.

Turning now to the collective spin waves in the $B$ phase we remember that the $d_{\alpha i}^*$ is a rotation matrix with the axis of rotation along the field and the angle of rotation $\theta = \cos^{-1}(-\frac{1}{4})$. The spin-wave calculation is simplified if we measure our spin variables in a coordinate system rotated by this same rotation. Thus we have **u**, **v** axes that are rotated by $\theta$ about the $z$ axis. Then the $d_{\alpha i}^*$ matrix is the unit matrix and the evaluation of $F^{\text{II}}$ is quite simple:

$$F^{\text{II}} = \rho_s \frac{\hbar^2}{20\, m^2}(4(\nabla_\alpha \theta_\beta)^2 - 2\nabla_\alpha \theta_\beta \nabla_\beta \theta_\alpha). \tag{VII.53}$$

Ignoring for the moment the dipolar and external field terms in $E_T$ we obtain the following equations:

$$\frac{\partial}{\partial t}\mathbf{S} = -\rho_s \frac{\hbar^2}{4m^2}(\tfrac{2}{5})(4q^2\boldsymbol{\theta} - 2\mathbf{q}(\mathbf{q}\boldsymbol{\theta})) \tag{VII.54}$$

$$\frac{\partial}{\partial t}\boldsymbol{\theta} = \mathbf{S}/\chi. \tag{VII.55}$$

These equations give three modes, two modes polarized transverse to **q** and one longitudinal, with velocities given by

$$v_L = \left(\frac{\hbar^2 \rho_s}{5\, m^2 \chi}\right)^{\frac{1}{2}} \tag{VII.56}$$

and

$$v_T = \sqrt{2}\, v_L. \tag{VII.57}$$

It must be remembered that the "polarization" is measured in the rotated coordinate system. As pointed out by Combescot, the polarization found for such waves would determine the form for $d_{\alpha i}^*$.

The analysis of inclusion of the dipolar energy leads to the conclusion that one transverse mode is unaffected, while the longitudinal and the other transverse mode mix in a $\hat{\mathbf{q}}$-dependent way and one of them becomes the longitudinal resonance as $q \to 0$. For details see Brinkman and Smith.[59] Finally, for the situation in the presence of a magnetic field the analysis becomes even more complicated.

As we have already remarked above: (a) the $\rho_s$ used in these equations is undoubtedly close to that in the absence of Fermi-liquid effects, so that the velocities are of order $v_F$ divided by (enhancement factor)$^{\frac{1}{2}}$; (b) the equations are valid in the hydrodynamic regime or near $T = 0$ or, with considerable quantitative change, in the collisionless regime $\omega \gg 10^7$. Where $\omega\tau_D \approx 1$ there will be severe damping by diffusion of the normal component. This assumes that $\omega \approx v_F q$; for modes of very long wavelength the behaviour depends strongly on magnetic field.) The best regime for observation should be $\omega \approx 10^5$–$10^6$, $\lambda \approx 10^{-1}$–$10^{-2}$ cm.

Orbit waves have been discussed only very briefly by Wölfle,[53] Anderson[24, 25] and de Gennes.[50, 51] The spin waves and the magnetic resonances we have discussed in the ABM state so far are all in essence oscillations of the spin portion of the gap bivector in the presence of an assumed stationary structure of the orbital portion. (This is not the case in the B phase, where only one rotational structure exists. As we have emphasized, *total* rotation in that case not only leaves the energy invariant, it leads to an identical state.) In actual fact, of course, the torque imparted by the spin system to the orbits must lead to (d/d$t$) $(\mathbf{L} + \mathbf{S}) \equiv \mathbf{0}$, in the absence of spin currents, but because there is a finite orbital angular momentum the net rotation of $d_\alpha$ is very small.

At some considerably lower frequency, one presumes, there must be a sequence of modes in which the orbit system alone precesses, or it carries the spin system with it, and the restoring forces are provided only by the intrinsic Landau–Ginzburg stiffness of the gap system and the pinning of the vector $\mathbf{l}$ to the walls of the container.

In the simplest case there is no magnetic field and thus literally there is no external orienting force. As Wölfle remarks, in this case the direction $\mathbf{l}$ of the orbital angular momentum $\mathbf{L}$, and $\mathbf{L}$ itself, are conjugate dynamical variables, so that

$$\frac{\partial}{\partial t}\,\delta L_x = -\frac{\delta F}{\delta l_y} = \frac{\rho_s}{8m}(-2\nabla_x\nabla_z\,\delta\varphi + (2\nabla_z^2 + \nabla^2)\delta l_y) \qquad \text{(VII.58)}$$

$$\frac{\partial}{\partial t}\,\delta L_y = -\frac{\rho_s}{8m}(2\nabla_y\nabla_z\delta\varphi + (2\nabla_z^2 + \nabla^2)\delta l_x) \qquad \text{(VII.59)}$$

where we have put $\mathbf{l}$ along $z$. Note the coupling of $\mathbf{L}$ to $\nabla\varphi = \mathbf{v}_s$, which means that orbital oscillations can be driven by superfluid flow.

Now the further progress depends on the ratio $\langle L\rangle = \mathbf{L}/\hat{\mathbf{l}}$. In AM this was estimated by the assumption

$$\langle L\rangle = \int \mathrm{d}^3\,r\; r\langle \mathbf{j}(\mathbf{r})\rho(0)\rangle_\varphi, \qquad \text{(VII.60)}$$

which Wölfle evaluates as

$$\langle L \rangle_W = \frac{5m}{16m^*} \frac{T_c}{E_F} \frac{\rho_s}{\rho}. \qquad \text{(VII.61)}$$

Cross and Blount, however, find no rigorous justification for (VII.60), and in particular suggest that if particle–particle and hole–hole contributions to $\langle n\mathbf{j} \rangle$ are nearly equal, the net coefficient $K_l$ which we find in the preceding section to represent $\langle L \rangle$ is an order smaller than (VII.61). Leaving this coefficient indeterminate, Wölfle's orbit-wave frequencies become

$$\omega(q) = \left(\frac{\langle L \rangle_W}{\langle L \rangle}\right) \frac{2\hbar}{5m^*} \frac{E_F}{kT_c} (3q_\parallel^2 + q_\perp^2)^{\frac{1}{2}} \left[3q_\parallel^2 + q_\perp^2 - 2\frac{q_\parallel^2 q_\perp^2}{2q_\perp^2 + q_\parallel^2}\right]^{1/2}.$$

For $q \simeq 1\,\text{cm}^{-1}$, this amounts to $\omega \approx 10^{-2}(\langle L \rangle_W / \langle L \rangle)$. As Anderson observed, this range of frequency matches, if the $L$ ratio were unity, rather nicely to the observed low-frequency fluctuations of NMR and zero-sound amplitudes in the A phase,[105, 123] if one assumes that the few lowest-frequency $L$-wave modes are involved. On the other hand, the Cross and Blount estimate leaves us with about 100 times higher frequencies, or $\omega \approx$ 1 Hz, which is altogether too high relative to the experimental results. Other aspects of these results, such as their magnetic-field dependence, are also rather questionable under this interpretation. Any simple interpretation suggests that **H** should raise the frequency by pinning **L** in a plane.[24, 25] That $L$-waves may be driven by the cross-coupling of $\hat{\mathbf{l}}$ and $\mathbf{v}_s$ is of course in good agreement both with these results (where random heat currents are likely to be present) and with Wheatley's heat transport instabilities.

One possible way out of this dilemma goes as follows. This is that often the A-phase texture may involve a curvature of $\hat{\mathbf{l}}$ near the vessel wall which would imply a rather large orbital supercurrent near the vessel wall. If $\hat{\mathbf{l}}$, for instance, must curve from perpendicular to the wall to axial along a tube (see Fig. 16c), the total surface current flowing is $\propto \int \nabla \times \hat{\mathbf{l}} \approx 1$ relative to $\Delta^2/E_F^2$, as for the weak orbital currents. The inertia of this circulation will be $10^2$–$10^4$ times that of the orbital moment, and thus macroscopic oscillations of $\hat{\mathbf{l}}$ will again revert to the $10^{-2}$ Hz frequency range. This, then, would suggest that not all, but a subset of orbit-wave modes will be found in the low-frequency range: a situation rather like the "Walker modes" of ferromagnetic insulators.[124]

Clearly this question is one of those most deserving of further study, both experimentally and theoretically, in the $He^3$ field. It seems possible that this is one of the few areas where the fundamentals of the theory are not yet established.

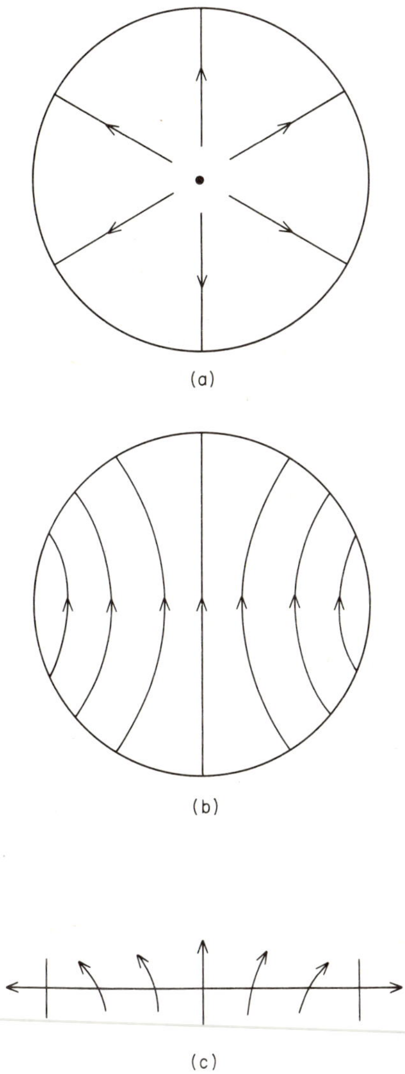


FIG. 16. The three textures discussed in the text as possible stable textures.

## VIII. Textures and Singularities of the Order Parameter

### 1. SINGULARITIES, VORTEX LINES, DISGYRATIONS AND POINT DEFECTS

The possible singularities in an order parameter as complicated as those of the A and B phases of $He^3$ will undoubtedly stretch our imagination. So far, however, only a limited amount of speculation by de Gennes[50, 51, 121] has been published. The simplest way to discuss these singularities is to consider the two phases separately.

(i) "*A*" *phase*

Without dipolar interactions the vectors **l** and **w** can be considered as independent vectors. Since the **l**-vector characterizes the orbital angular momentum it has a sense to it, i.e. $-\mathbf{l}$ is not equivalent to **l**. By contrast $-\mathbf{w}$ is equivalent to **w**, so that **w** is like the optical axis in the nematic phase of a liquid crystal. This means that if **w** varies in space along a closed path it is not necessary that **w** must always return to its original direction; it can return as $-\mathbf{w}$. The **l**-vector must always return to its beginning direction. Another difference that must be kept in mind in constructing singularities in the **l**-field is that **l** is the cross product of the real and imaginary vectors of the order parameter i.e.

$$d_{\alpha i} = (\mathbf{\Delta}' + i\mathbf{\Delta}'')_\alpha w_i$$

and

$$\mathbf{l} \propto \mathbf{\Delta}' \times \mathbf{\Delta}''.$$

Therefore, if we wish to contruct a singularity in the **l**-field we must be careful to ensure that the $\mathbf{\Delta}'$- and $\mathbf{\Delta}''$-fields do not have singularities. The simplest example is the case of a point singularity. We can have a singularity in the **w**-field in which **w** is directed radially about a single point but it is not possible to have such a singularity for the **l**-vector since one cannot make $\mathbf{\Delta}'$ and $\mathbf{\Delta}''$ unique functions on a sphere.

The more interesting singularities are line singularities. One can construct the usual vortex line which is a singularity in the phase $\phi$. However, it is only simple when **l** is along the line of the vortex. Then the current is simply tangential and proportional to the reciprocal of the distance from the line. Otherwise, because of the cross terms in (VII.40) between $(\nabla\phi)$ and $(\nabla \times \mathbf{l})$, **l** will be slightly distorted and the current will not be rotationally symmetric.

More interestingly, there can be line singularities in the **l**-field which de Gennes calls "disgyrations". These can be of two types; one called a tangential disgyration has **l** directed tangentially about the line axis ($\mathbf{\Delta}'$ along the line axis and $\mathbf{\Delta}''$ radial). It has a net current flowing along the line axis. Its magnitude can be obtained from (VII.36) to be

$$J = \frac{\hbar}{m^*}\left\{\frac{7\zeta(3)}{40\pi^2}\right\}\frac{\langle\Delta^2\rangle}{T_c^2}\frac{1}{r}.$$

The second, a radial disgyration, has $\mathbf{l}$ pointing radially outward from the line ($\mathbf{\Delta}'$ along the line axis and $\mathbf{\Delta}''$ tangential). There are no currents associated with a radial disgyration.

Just as it is possible for usual vortices to have more than one quantum of circulation, it is possible for $\mathbf{l}$ to rotate through multiples of $2\pi$ about the line axis. Such higher-order disgyrations would have a current along the line axis.

Similar behaviour can be obtained for the $\mathbf{w}$-vector. However, as discussed in section VI the dipolar energy tends to keep $\mathbf{w}$ and $\mathbf{l}$ parallel. We can estimate a length $\xi_D$ which determines the scale on which $\mathbf{w}$ will no longer follow $\mathbf{l}$ by equating the dipolar energy to the bending energy required to bend $\mathbf{w}$. This gives

$$\xi_D = \xi\left(\frac{\Delta F}{E_D}\right)^{\frac{1}{2}} \approx 10\mu, \tag{VIII.1}$$

where $\Delta F$ is the free energy gained in the superfluid state and $E_D$ is the dipole energy. In regions where $\mathbf{w}$ or $\mathbf{l}$ bends rapidly on the scale of $\xi_D$ they will not follow one another. For example, in small pores ($\leqslant 10\mu$) $\mathbf{l}$ will be fixed by the walls but $\mathbf{w}$ may actually be constant in space and be oriented by an external field.

### (ii) "$B$" phase

In this phase one has possible singularities in the phase, the axis of rotation $\mathbf{n}$, and possibly in the angle of rotation $\theta$. Since $\theta$ is fixed by the dipolar energy the last can only occur if the scale of the distortion is small compared to the dipolar coherence length $\xi_D$. Since the direction of the vector $\mathbf{n}$ specifies the sense of rotation, $\mathbf{n}$ is like $\mathbf{l}$ in that its variation along a closed path must be such that it always returns to its original direction. Unlike $\mathbf{l}$, however, it can exhibit a point singularity as described for the vector $\mathbf{k}$ since there are no difficulties with vectors perpendicular to it. In fact, in recent experiments reported by Osheroff a dip in the intensity of the NMR signal as a function of position along the axis of his cylindrical sample may be due to the presence of such a singularity. The point singularity could occur if two regions on the two ends of the cylinder grew together with $\mathbf{n}$ in opposite directions.

The vector $\mathbf{n}$ can also exhibit disgyration line-type singularities just as

**l** does. These may give rise to spin currents but they will not be discussed here.

An interesting feature of any singularities in the vector **n** is that within $\xi_D$ of the singularity the angle of rotation $\theta$ may begin to rotate and reduce the kinetic energy due to the **n** distortion. At the centre of the singularity $\theta$ may be zero and the bending energy reduced to something of the order of the dipolar energy. A somewhat related suggestion was made previously by de Gennes for the A phase. He suggested that the core of the disgyration in that phase may be the BW state.

### (iii) *Textures*

Having found that the order parameter in these new phases can exhibit singularities it is not too surprising that in the presence of boundaries the various vectors involved will exhibit textures similar to those found in liquid crystals. The possibility of such effects was first discussed by de Gennes[(50, 51, 121)] and in more detail by Ambegaokar, de Gennes and Rainer.[(52)] Recently, textures have been discussed for the B phase by Brinkman, Smith, Osheroff and Blount.[(61)]

The first question that must be answered is what are the boundary conditions for the order parameter. The answer to this question is a relatively simple one.[(52)] If one has spectral scattering of $He^3$ atoms then a wall only acts to depair that component of the triplet for which $m_l = 0$ normal to the wall. Said differently, when two particles forming a pair approach a wall they must circulate around one another in the plane of the wall. We will not go into the derivation of this result. It is obtained by studying the linearized gap equation and introducing the boundary conditions through the non-local kernel in the integral equation. A semiclassical approximation to the kernel allows one to obtain the solutions. Under circumstances when the scattering is diffuse it was found that all of the components of the order parameter are greatly reduced at the boundary but those with $m_l = 0$ still become zero.

Therefore, in terms of $d_{\alpha i}$ we must impose the condition that

$$s_\alpha d_{\alpha i} = 0 \qquad \text{(VIII.2)}$$

at the surface, where **s** is a unit vector. In addition to these boundary conditions we must also make sure that the currents into the wall vanish. This requires that

$$(\mathbf{s}\cdot\nabla)d_{\alpha i} = 0 \qquad \text{(VIII.3)}$$

for those components $d_{\alpha i}$ not going to zero by the first boundary condition. In the A phase the first condition is automatically satisfied if **l** is normal to the wall. Having **l** at any other direction than normal costs an energy of order $\Delta F\xi$; therefore, there is a strong force pinning **l** in this position.

Given this boundary condition we can examine various possible textures of **l** in restricted geometries. For a cylinder the following seems to be the situation. For small radii $r \leqslant 100$ the stable texture will be either one with **l** directed radially outward (or inward) from the axis of the cylinder with a disgyration down the centre or one with two tangential disgyrations running down opposite sides of the cylinder as illustrated in Fig. 16a, b. For larger radii, we expect to find a texture in which **l** simply rotates upward (or downward) as a function of radius and ends up pointing parallel to the cylindrical axis at the centre of the cylinder Fig. 16c. The form of the order parameter for such a texture can be obtained by starting with $\boldsymbol{\Delta}'$ and $\boldsymbol{\Delta}''$ in the plane perpendicular to the cylinder axis and then defining

$$\boldsymbol{\Delta}'(r) = \hat{O}(\hat{\boldsymbol{\phi}}, \theta(r))\boldsymbol{\Delta}' \qquad \text{(VIII.4)}$$

$$\boldsymbol{\Delta}(r) = \hat{O}(\hat{\boldsymbol{\phi}}, \theta(r))\boldsymbol{\Delta}'', \qquad \text{(VIII.5)}$$

where $\hat{O}$ is a rotation matrix and the axis of rotation is $\hat{\boldsymbol{\phi}}$, the tangential direction. The angle $\theta(r)$ is a function of position chosen so that $\theta(R) = \pi/2$ at the surface of the cylinder. This texture becomes more stable at large radius because it does not involve a disgyration whose energy becomes proportional to the $\ln(R/\xi)$ as the above textures do.

It should also be pointed out that such a texture has an angular momentum of order $\hbar\rho_s(\pi R^2)$ per unit length of the cylinder because it has a supercurrent flowing in the tangential direction.

When the radius of the cylinder is larger than $\approx 10\mu$, **w** and **l** will tend to bend together. Then for relatively small axial fields the last texture described (*c*) will be suppressed by the magnetic anisotropy associated with **w**. A crude estimate of when this will occur can be obtained by setting the magnetic anisotropy energy $\delta\chi H^2$ equal to the bending energy $\Delta F(\xi/R)^2$, giving

$$H_c \simeq \left(\frac{\xi}{R}\right)\left(\frac{\Delta F}{\delta\chi}\right)^{\frac{1}{2}} \approx \frac{\xi}{R}\left(\frac{\Delta}{\hbar\gamma}\right). \qquad \text{(VIII.6)}$$

This field is typically much less than one gauss.

Perhaps the best geometry for obtaining essentially uniform textures is that of two flat surfaces separated by a distance $\chi$. The **l** and **w** vectors will be normal to the surfaces if the external field is applied parallel to the surface. If the field is applied perpendicular to the surface **w** and **l** will tend to bend. This situation has been studied recently by Rainer.[121]

In the same geometry de Gennes and Rainer[121] studied the effect of a current and showed that, because of anisotropy in $\rho_s$ above a critical current given by

$$J_{\text{critical}} \propto \frac{\hbar}{m}\Delta^2\left(\frac{\pi}{\chi}\right),$$

the **l** vector will tend to bend across the sample and at high enough currents it will be pointing essentially along the direction of flow. The prediction of a critical field appears to be incorrect because of the cross coupling between **l** and the gradient of $\phi$ in (VII.40).

A similar situation is expected for heat flow since the quasiparticles carry the entropy and their distribution is anisotropic. This can be seen in the definition of the entropy tensor discussed earlier (IV.128) and obtained from the work of Wölfle[31, 32]

$$\sigma_{ij} = -\frac{1}{T}\sum_{\mathbf{k}}\frac{k_i}{m}\frac{k_j}{m^*}\frac{\mathrm{d}n(E_{\mathbf{k}}/kT)}{\mathrm{d}E}\,\varepsilon_{\mathbf{k}}. \tag{VIII.7}$$

Here $\mathrm{d}n/\mathrm{d}E$ is the derivative of the Fermi factor. Clearly this expression is large along **l** where the gap is small. Indeed, the effect of a heat flow appears to have been seen in the recent data of Osheroff and Brinkman,[26] where both **l** and **w** appear to rotate parallel to the external field when the field is sufficiently weak.

Turning now to the B phase, since no matter what the values of **n** and $\theta$ are one has a component of the order parameter (such that $m_l = 0$ normal to the surface) which must go to zero, there are no large orientational effects due to a surface. However, because the component of $d_{\alpha i}$ normal to the surface is suppressed, i.e. $s_\alpha d_{\alpha i} = 0$, the susceptibility becomes anisotropic in this region and the dipolar energy changes. By constructing a trial function which properly satisfies this boundary condition Brinkman, Smith, Osheroff and Blount[61] calculated these terms. The term arising from the dipolar energy is of the form

$$F_D^s = -b\int_S \mathrm{d}^2r[(\mathbf{s}\cdot\mathbf{n})^2 - \tfrac{5}{18}(\mathbf{s}\cdot\mathbf{n})^4], \tag{VIII.8}$$

where $b$ is of order $E_D\xi$. This term favours **n** being perpendicular to the surface. From the susceptibility anisotropy they obtained the term

$$F_H^s = -d\int_S \mathrm{d}^2r[\mathbf{s}\cdot\hat{O}(\mathbf{n},\theta)\cdot\mathbf{H}]^2 \tag{VIII.9}$$

where $\hat{O}(\mathbf{n},\theta)$ is the rotation matrix in (VII.28). The $d$ is of order $\xi(\chi_N - \chi_S)$. This term favours a direction of **n** such that the rotation matrix takes **H** into $\pm\mathbf{s}$. With $\theta = \cos^{-1}(-\frac{1}{4}) = 104°$ one can always find a direction of **n** such that this condition is satisfied. Since this term does not tend to change $\theta$ and since a surface dipole energy, which in fact would like to have $\theta = 90°$ at the surface, cannot compete against the volume dipole energy, the textures of the B phase can be described completely in terms of **n** variation. In the absence of a magnetic field the characteristic length in the problem is

$$R_c \simeq \xi\left(\frac{\Delta F}{E_D}\right) \simeq 0{\cdot}1(1 - T/T_c)^{\frac{1}{2}}\,\text{cm}.$$

Again considering a cylindrical sample we find that the two textures $b$ and $c$ in Fig. 16 compete, the $b$ texture being most stable for $R < 74R_c$.

In the presence of magnetic field along the cylinder we must include the anisotropy energy discussed in Section IV, eq. (IV.93). From it we obtain a field-dependent characteristic length $R_H = R_cH_B/H$, where $R_cH_B$ is estimated to be of order 10–20 cm gauss. As the field is turned on, the surface effects become less important and the texture $c$ becomes more stable and the **n** vector becomes more and more parallel to the cylinder axis.

As we discussed in Section VI, the transverse resonance frequency shifts when **n** is not parallel to the external field, so the texture of **n** being on the scale of cm has a profound effect on the NMR resonance. The simplest effect is that $R_H \propto (1 - T/T_c)^{\frac{1}{2}}$, so that the characteristic length becomes small near $T_c$. If one is looking at either transverse or longitudinal resonance at fields that are small compared to those necessary to align **n** at low temperature one should see the signal recover as the temperature is raised near to $T_c$. Indeed, such behaviour of both the transverse and longitudinal resonances has been seen by Ahonen *et al.*[(125)] and by Webb *et al.*[(97)]

Operating at a given temperature one expects the amount of unshifted resonance signal to increase as a function of increasing field. This has been found in the recent experiments of Osheroff and Brinkman.[(26)]

We can actually make a quantitative theory of the transverse line shape in the high-field limit if we assume that each region of the sample resonates at its local frequency which is given by

$$\omega^2 = (\gamma H)^2 + \Omega_L^2 \sin^2[\mathbf{n}(\mathbf{r})\cdot\mathbf{H}]. \tag{VIII.10}$$

In view of the fact that one expects collective modes as discussed in the previous section, the local approximation is not entirely justified. However, since we will be looking at small shifts of order a few hundred hertz which must be driven by the local shift, we assume it is a reasonable approximation.

A very good approximation to the variation of $[\mathbf{n}(\mathbf{r})\cdot\mathbf{H}]$ can be obtained by using the simple formula

$$\mathbf{n}(\mathbf{r})\cdot\mathbf{H} = A\sinh\left(\frac{rH}{R_cH_B}\right). \tag{VIII.11}$$

The constant $A$ is given by $A_0/\sinh(RH/R_cH_B)$, where $A_0$ is the angle of **n** relative to **H** at the surface and $R$ is the radius of the sample. The calculated lineshape is

$$P(\omega) = 2\left(\frac{H_B R_c}{HR}\right)^2 \frac{1}{\Gamma} f\left[2\left(\frac{\omega - \gamma H}{\Gamma}\right)\right]; \quad \omega > \gamma H \qquad \text{(VIII.12)}$$

here $\Gamma$ is the effective width,

$$\Gamma = (\Omega_L^2/\gamma H)A^2. \qquad \text{(VIII.13)}$$

Because of the sinh $(RH/R_c H_B)$ in this expression $\Gamma$ can indeed be quite small. The function $f(x)$ is found to be

$$f(x) = \ln(x^{\frac{1}{2}} + \sqrt{x+1})/\sqrt{x+x^2} \qquad \text{(VIII.14)}$$

The integrated intensity between $\gamma H$ and $\gamma H + \Delta\omega$ can be calculated as a function of field:

$$N(\Delta\omega) = \int_{\gamma H}^{\gamma H + \Delta\omega} P(\omega)\,\mathrm{d}\omega = \left\{1 - \frac{H_B R_c}{2HR} \ln\left(\frac{\Omega^2}{2\Delta\omega\gamma H} A_0^2\right)\right\}^2. \qquad \text{(VIII.15)}$$

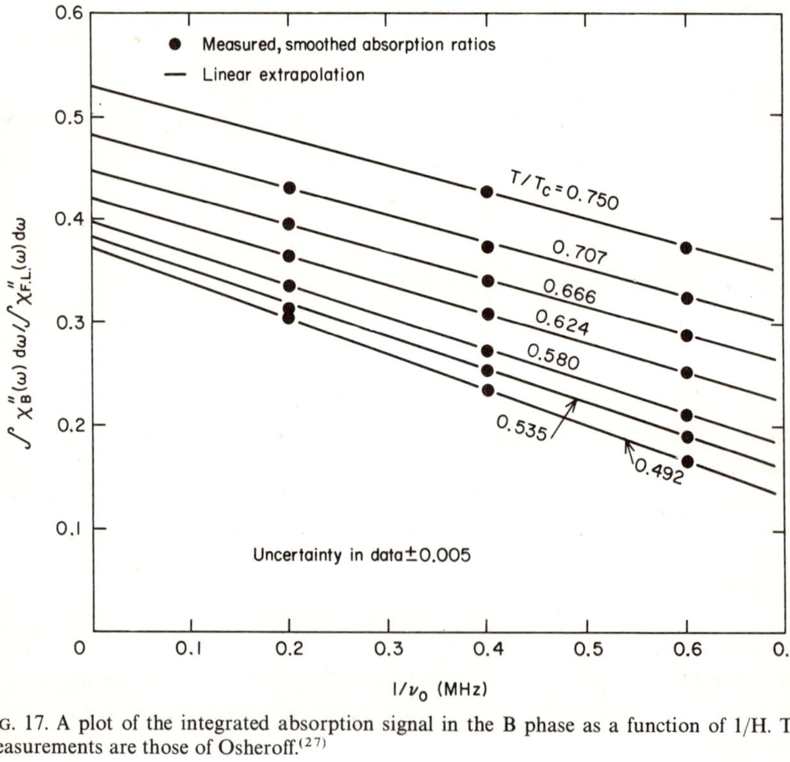


FIG. 17. A plot of the integrated absorption signal in the B phase as a function of 1/H. The measurements are those of Osheroff.[27]

This gives rise to an essentially $1/H$ saturation of the integrated intensity. Recent data of Osheroff[27] is shown in Fig. 17. The slopes of the various curves should be proportional to the coefficient of the bending energy in (VII.32) i.e. $\rho_s^{\frac{1}{2}}$. This expression does not, however, include Fermi-liquid effects. Recent calculations by Combescot[57, 58] and by Brinkman and Smith[59] suggest that this coefficient should be $\rho_s^0$, the superfluid density calculated in weak-coupling theory. The experimental curves vary on a scale quite close to $\rho_s$, which in the region of observation is varying more rapidly than $\rho_s^0$. This result could possibly be due to the neglect of dispersion already mentioned, since dispersion would presumably cause the signal to shift further than this calculation predicts. This effect may be stronger at lower temperatures where the dispersion is more important.

## Acknowledgements

The authors would like to acknowledge contributions and discussions from E. I. Blount, D. D. Osheroff, H. Smith, M. C. Cross, S. Engelsberg and C. M. Varma. We also would like to thank Mrs N. J. Hines for her patience in preparing the manuscript.

## References

1. J. Bardeen, L. N. Cooper and J. R. Schrieffer. *Phys. Rev.* **108**, 1175 (1957).
2. D. J. Thouless. *Ann. Phys. (New York)* **10**, 553 (1960).
3. J. C. Fisher. Private communication (1958).
4. L. P. Pitaevskii. *Sov. Phys. JETP* **10**, 1267 (1960).
5. K. A. Brueckner, T. Soda, P. W. Anderson and P. Morel. *Phys. Rev.* **118**, 1442 (1960).
6. V. J. Emery and A. M. Sessler. *Phys. Rev.* **119**, 43 (1960).
7. L. N. Cooper, R. L. Mills and A. M. Sessler. *Phys. Rev.* **114**, 1377 (1959).
8. W. R. Abel, A. C. Anderson, W. C. Black and J. C. Wheatley. *Physics* **1**, 337 (1965).
9. P. W. Anderson and P. Morel. *Physica* **26**, 671 (1960).
10. P. W. Anderson and P. Morel. *Phys. Rev.* **123**, 1911 (1961).
11. K. Yosida. *Phys. Rev.* **110**, 769 (1958).
12. R. Balian and N. R. Werthamer. *Phys. Rev.* **131**, 1553 (1963).
13. A. J. Leggett. *Phys. Rev. Letts* **14**, 536 (1965).
14. A. J. Leggett. *Phys. Rev.* **140**, A1869 (1965).
15. V. J. Emery. *Ann. Phys. (New York)* **28**, 1 (1964).
16. N. F. Berk and J. R. Schrieffer. *Phys. Rev. Letts* **17**, 433 (1966).
17. W. F. Brinkman and S. Engelsberg. *Phys. Rev.* **169**, 417 (1968).
18. A. J. Layzer and D. Fay. Proc. *Int. Low Temp. Conf. St. Andrews* (LT-11) (1968).
19. D. D. Osheroff, R. C. Richardson and D. M. Lee. *Phys. Rev. Letts* **28**, 885 (1972).
20. D. D. Osheroff, W. J. Gully, R. C. Richardson and D. M. Lee. *Phys. Rev. Letts* **29**, 920 (1972).

21. A. J. Leggett. *Phys. Rev. Letts* **29**, 1227 (1972).
22. A. J. Leggett. *Phys. Rev. Letts* **31**, 352 (1973).
23. A. J. Leggett. *J. Phys. C* **6**, 3187 (1973).
24. P. W. Anderson. *Phys. Rev. Letts* **30**, 368 (1973).
25. P. W. Anderson. *Proc. Nobel Symp.* **24**, 103 (1973) (Uppsala, Nobel Foundation).
26. D. D. Osheroff and W. F. Brinkman. *Phys. Rev. Letts* **32**, 584 (1974).
27. D. D. Osheroff. *Phys. Rev. Letts* **33**, 1009 (1974).
28. V. Ambegaokar and N. D. Mermin. *Phys. Rev. Letts* **30**, 81 (1973).
29. C. M. Varma and N. R. Werthamer. *Bull. Am. Phys. Soc.* **18**, 8 (1973).
30. B. R. Patton. Unpublished preprint (1973).
31. P. Wölfle. *Phys. Rev. Letts* **30**, 1169 (1973).
32. P. Wölfle. *Phys. Rev. Letts* **31**, 1437 (1973).
33. R. C. Richardson and D. M. Lee. *Proc. Nobel Symp.* **24**, 84 (1973) (Uppsala, Nobel Foundation).
34. P. W. Anderson and W. F. Brinkman. *Phys. Rev. Letts* **30**, 1108 (1973).
35. W. F. Brinkman and P. W. Anderson. *Proc. Nobel Symp.* **24**, 116 (1973) (Uppsala, Nobel Foundation).
36. W. F. Brinkman and P. W. Anderson. *Phys. Rev.* **A8**, 2732 (1973).
37. W. F. Brinkman, J. Serene and P. W. Anderson. *Phys. Rev.* **A10**, 2386 (1974).
38. Y. Kuroda. *Prog. Theor. Phys.* **51**, 1269 (1974).
39. Y. Kuroda. *Prog. Theor. Phys.* **53**, 349 (1975).
40. R. A. Webb, T. J. Greytak, R. T. Johnson and J. C. Wheatley. *Phys. Rev. Letts* **30**, 210 (1973).
41. T. J. Greytak, R. T. Johnson, D. N. Paulson and J. C. Wheatley. *Phys. Rev. Letts* **31**, 452 (1973).
42. J. C. Wheatley. *Physica* **69**, 218 (1973).
43. D. N. Paulson, H. Kojima and J. C. Wheatley. *Phys. Rev. Letts* **32**, 1098 (1974).
44. D. D. Osheroff and P. W. Anderson. *Phys. Rev. Letts* **33**, 686 (1974).
45. W. M. Saslow. *Phys. Rev. Letts* **31**, 870 (1973).
46. T. A. Alvesalo, Yu. D. Anufriyev, H. K. Collan, O. V. Lounasmaa and P. Wennerström. *Phys. Rev. Letts* **30**, 962 (1973).
47. T. A. Alvesalo, Yu. D. Anufriyev, H. K. Collan, O. V. Lounasmaa and P. Wennerström. *Proc. Nobel Symp.* **24**, 91 (1973) (Uppsala, Nobel Foundation).
48. H. Kojima, D. N. Paulson and J. C. Wheatley. *Phys. Rev. Letts* **32**, 141 (1974).
49. A. W. Yanof and J. D. Reppy. *Phys. Rev. Letts* **33**, 631 (1974).
50. P. G. de Gennes. *Phys. Letts* **44A**, 271 (1973).
51. P. G. de Gennes. *Proc. Nobel Symp.* **24**, 112 (1973) (Uppsala, Nobel Foundation).
52. V. Ambegaokar, P. G. de Gennes and D. Rainier. *Phys. Rev.* **A9**, 2676 (1974).
53. P. Wölfle. *Phys. Letts* **47A**, 224 (1974).
54. E. I. Blount, P. W. Anderson and W. F. Brinkman. Unpublished notes incorporated into Section VII of this article (1974).
55. A. J. Leggett. *Proc. Nobel Symp.* **24**, 109 (1973) (Uppsala, Nobel Foundation).
56. K. Maki and H. Ebisawa. *Prog. Theor. Phys.* **50**, 1452 (1973).
57. R. Combescot. *Phys. Rev.* **A10**, 1700 (1974).
58. R. Combescot. *Phys. Rev. Letts* **33**, 946 (1974).
59. W. F. Brinkman and H. Smith. *Phys. Rev.* **A10**, 2325 (1974).
60. K. Maki and T. Tsuneto. *Prog. Theor. Phys.* **52**, 773 (1974).
61. W. F. Brinkman, H. Smith, D. D. Osheroff and E. I. Blount. *Phys. Rev. Letts* **33**, 624 (1974).
62. G. Baym and C. J. Pethick. To be published (1975).

63. J. C. Wheatley. *In* "Quantum Fluids" (D. F. Brewer, Ed.), Wiley, New York (1966).
64. S. Doniach and S. Engelsberg. *Phys. Rev. Letts* **17**, 750 (1966).
65. M. J. Rice. *Phys. Rev.* **159**, 153 (1967).
66. M. J. Rice. *Phys. Rev.* **162**, 189 (1967).
67. S. K. Ma, M. T. Beal-Monod and D. R. Fredkin. *Phys. Rev.* **174**, 227 (1968).
68. C. J. Pethick. *In* "Lectures in Theoretical Physics" (K. T. Mahunthappa, Ed.), Gordon and Breach, New York (1969).
69. V. J. Emery. *Phys. Rev.* **170**, 205 (1968).
70. J. Amit, J. W. Kane and H. Wagner. *Phys. Rev.* **175**, 313 (1968).
71. N. F. Mott. *Proc. Phys. Soc.* **62**, 416 (1949).
72. M. C. Gutzwiller. *Phys. Rev.* **137**, A1726 (1965).
73. W. F. Brinkman and T. M. Rice. *Phys. Rev.* **B2**, 4302 (1970).
74. W. F. Brinkman and T. M. Rice. *In* "Battelle Colloquium on Critical Phenomena" (1971), p. 593.
75. S. Babu and G. E. Brown. *Ann. Phys. (New York)* **78**, 1 (1973).
76. E. Feenberg. "Theory of Quantum Fluids", Academic Press, New York and London (1969).
77. K. A. Brueckner and J. L. Gammel. *Phys. Rev.* **109**, 1040 (1958).
78. E. Østgaard. *Phys. Rev.* **187**, 371 (1969).
79. L. P. Gor'kov and L. P. Pitaevskii. *Sov. Phys. Dokl.* **8**, 788 (1964).
80. H. Bookbinder. *Phys. Rev.* **A3**, 372 (1971).
81. A. J. Layzer and D. Fay. *Int. J. Mag.* **1**, 135 (1971).
82. S. Nakajima. *Prog. Theor. Phys.* **50**, 1101 (1973).
83. A. J. Layzer and D. Fay. *Solid State Commun.* **15**, 599 (1974).
84. J. R. Schrieffer. "Superconductivity," Benjamin, New York (1963), Ch. 5.
85. L. Tewordt. *J. Law Temp. Phys.* **15**, 349 (1974); *Zeit. f. Phys.* **268**, 207 (1974).
86. G. M. Eliashberg. *Sov. Phys. JETP* **11**, 696 (1960).
87. E. P. Wigner. "Theory of Groups in Quantum Mechanics", Springer, Berlin (1973).
88. G. Eilenberger. *Zeit. f. Phys.* **182**, 427 (1965).
89. J. M. Luttinger and J. C. Ward. *Phys. Rev.* **118**, 1417 (1960).
90. N. R. Werthamer. *In* "Superconductivity" (R. D. Parks, Ed.), Dekker, New York (1969), p. 321.
91. N. D. Mermin and G. Stare. *Phys. Rev. Letts* **30**, 1135 (1973).
92. I. S. Gradshteyn and I. M. Ryzhik. "Table of Integrals, Series and Products", Academic Press, New York and London (1965), 4.116.
93. J. M. Delrieu. Private communication (1973).
94. E. I. Blount and M. C. Cross. Private communication.
95. P. W. Anderson. *Rev. Mod. Phys.* **38**, 298 (1966).
96. S. Engelsberg, W. F. Brinkman and P. W. Anderson. *Phys. Rev.* **A9**, 2592 (1974).
97. R. A. Webb, R. L. Kleinberg and J. C. Wheatley. *Phys. Letts* **48A**, 421 (1974).
98. R. A. Webb, R. L. Kleinberg and J. C. Wheatley. *Phys. Rev. Letts* **33**, 145 (1974).
99. J. Bardeen and J. R. Schrieffer. *Prog. Low Temp. Phys.* **3**, 212 (1961).
100. M. C. Cross. Private communication.
101. A. W. Yanof, E. Smith, D. M. Lee, R. C. Richardson and J. D. Reppy. *Bull. Am. Phys. Soc.* **19**, 435 (1974).
102. O. Betbeder-Matibet and P. Nozières. *Ann. Phys. (New York)* **51**, 392 (1969).
103. D. N. Paulson, R. T. Johnson and J. C. Wheatley. *Phys. Rev. Letts* **31**, 746 (1973).
104. M. Vuorio. *J. Phys. C* **7**, L5 (1974).

105. D. T. Lawson, W. J. Gully, S. Goldstein, R. C. Richards and D. M. Lee. *Phys. Rev. Letts* **30**, 541 (1973).
106. A. J. Leggett. *Phys. Rev.* **147**, 146 (1966).
107. P. W. Anderson. *Phys. Rev.* **112**, 1900 (1958).
108. N. N. Bogoliubov, V. Tolmachev and D. V. Shirkov. "New Method in the Theory of Superconductors", Consultants Bureau, New York, (1959).
109. D. N. Paulson, R. T. Johnson and J. C. Wheatley. *Phys. Rev. Letts* **30**, 829 (1973).
110. P. W. Anderson and H. Suhl. *Phys. Rev.* **116**, 898 (1959).
111. J. Serene and D. Rainer. To be published (1975).
112. A. J. Leggett. *Ann. Phys.* (*New York*) **85**, 11 (1974).
113. H. M. Bozler, M. E. R. Bernier, W. J. Gully, R. C. Richardson and D. M. Lee. *Phys. Rev. Letts* **32**, 875 (1974).
114. K. Maki and H. Ebisawa. *J. Low Temp. Phys.* **15**, 212 (1974).
115. K. Maki and H. Ebisawa. *Phys. Rev. Letts* **32**, 520 (1974).
116. S. Takagi. *Prog. Theor. Phys.* **51**, 674 (1975).
117. R. Combescot and H. Ebisawa. *Phys. Rev. Letts* **33**, 810 (1974).
118. V. L. Ginzburg and L. D. Landau. *Zh. Eksp. Teor. Fiz.* **20**, 1064 (1950).
119. L. P. Gor'kov. *Sov. Phys. JETP* **9**, 1364 (1959).
120. N. R. Werthamer. *Phys. Rev.* **132**, 663 (1963).
121. P. G. de Gennes and D. Rainer. *Phys. Letts* **46A**, 429 (1974).
122. K. Maki. *Prog. Theor. Phys.* **52**, 745 (1974).
123. D. D. Osheroff, Thesis, Cornell University (1972).
124. L. R. Walker. *Phys. Rev.* **105**, 390 (1957).
125. A. I. Ahonen, M. T. Haikala and M. Kousins. *Phys. Letts* **47A**, 215 (1974).

# 9. Dilute Solutions of $He^3$ in Superfluid $He^4$ at Low Temperatures

GORDON BAYM

*Department of Physics, University of Illinois, Urbana, Illinois 61801, USA*

## I. INTRODUCTION

$He^3$ is the only substance that can be dissolved in superfluid $He^4$; at concentrations less than 6·5% it remains in solution down to absolute zero.[1] At very low temperatures the $He^3$ forms a degenerate Fermi liquid, with the unique property that one can vary both its density and temperature over a wide range, enabling it to be studied from the fully degenerate quantum regime to the "high temperature" classical regime. Figure 1 shows the phase

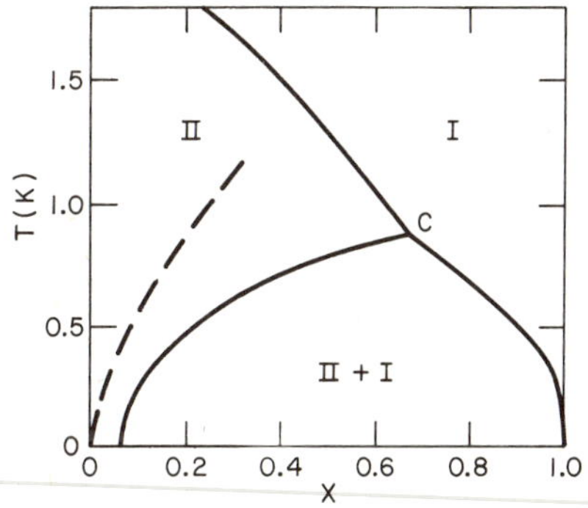


FIG. 1. Phase diagram of solutions of $He^3$ in $He^4$ in the fractional $He^3$ concentration, temperature plane. Region I is normal, II superfluid, while I + II is the region of phase separation. Along the dashed line the $He^3$ Fermi temperature equals the temperature. C is the tricritical point.

diagram[2] in the $He^3$ concentration $[x = N_3/(N_3 + N_4)]$, temperature plane. The $He^4$-rich region at the left, marked II, is superfluid, while region I is normal. The $\lambda$ line which separates these two regions intersects the $T$ axis at the pure $He^4$ $\lambda$ point. In the regime labelled I + II the system physically separates into two phases, a $He^4$-rich phase (in region II) below and a lighter $He^3$-rich phase (in I) which floats above. The point labelled $C$, the meeting of the $\lambda$ line and the phase separation curve, is the tricritical point. The dashed line in Fig. 1 is the $He^3$ degeneracy line, along which the temperature equals the $He^3$ degeneracy temperature $T_f$. Below this line the $He^3$ is degenerate.

In these lectures I shall focus on dilute solutions at low temperatures, describing their elementary excitations and how their thermal and transport properties can be derived from the elementary excitations and their interactions. The theory of dilute solutions at low temperatures, below 1·5 K generally, is greatly simplified by the fact that one can, in the spirit of the Landau Fermi-liquid theory, describe the low-lying excited states of the system in terms of weakly interacting elementary excitations above the ground state. These excitations are of two types: phonon and roton excitations of the $He^4$, whose parameters are nearly equal to those of pure $He^4$, and $He^3$ quasiparticle excitations, characterized at low momentum by the Landau–Pomeranchuk spectrum[(3)]

$$\varepsilon_p = \varepsilon_0 + \mathbf{p}^2/2m^*, \tag{I.1}$$

where $\mathbf{p}$ is the momentum of the excitation, and $m^*$ is the quasiparticle effective mass. At low momentum the $He^3$ quasiparticles are kinematically stable against emission of $He^4$ excitations.

The system behaves at very low temperatures as a Fermi liquid of $He^3$ quasiparticles moving through a background $He^4$ ether; in this régime the full machinery of the Landau Fermi-liquid theory is applicable. The freedom to vary the $He^3$ density experimentally enables one, as we shall see, to deduce a great deal of information about the effective interactions of $He^3$ quasiparticles. The $He^3$-phonon system, which determines the properties of the system at temperatures below $\approx$ 0·6 K, is rather analogous to non-relativistic quantum electrodynamics in the sense that phonon velocities are large compared with characteristic $He^3$ velocities; as we shall see, the theory of the $He^3$-phonon interaction can be developed by very general thermodynamic and Galilean invariance arguments, analogous to those used in the Landau Fermi-liquid theory.

To define the energies of the elementary excitations, let us consider a low-lying macroscopic state, in a box of volume $V$, in which the density is uniform and the superfluid velocity $\mathbf{v}_s$ is everywhere zero. Such a state is characterized by the (smoothed) distribution functions $n_{\mathbf{p}\sigma}$, the mean number of $He^3$ quasiparticles of momentum $\mathbf{p}$ and spin $\sigma$, and $f_{\mathbf{q}}$ the mean number of $He^4$ excitations of momentum $\mathbf{q}$, and in addition, the number density $n_4$ of $He^4$ atoms. The energy $E$ (per unit volume) of the state is a functional only of the $n_{\mathbf{p}\sigma}$, $f_{\mathbf{q}}$ and $n_4$:

$$E = E(\{n_{\mathbf{p}\sigma}\}, \{f_{\mathbf{q}}\}, n_4). \tag{I.2}$$

Since the number of He$^3$ quasiparticles equals the number of He$^3$ atoms, we have

$$N_3 = n_3 V = \sum_{\mathbf{p}\sigma} n_{\mathbf{p}\sigma}, \tag{I.3}$$

where $n_3$ is the He$^3$ number density.

The He$^3$ quasiparticle energy $\varepsilon_{\mathbf{p}\sigma}$ is then given by the variation of the energy $E$ with respect to $n_{\mathbf{p}\sigma}$, holding all other $n_{\mathbf{p}'\sigma'}$, all $f_{\mathbf{q}}$ and $n_4$ fixed:

$$\varepsilon_{\mathbf{p}\sigma} = V(\delta E/\delta n_{\mathbf{p}\sigma})_{n_{\mathbf{p}'\sigma'}, f_{\mathbf{q}}, n_4} \tag{I.4}$$

Similarly the He$^4$ excitation energies are given by

$$\omega_{\mathbf{q}} = V(\delta E/\delta f_{\mathbf{q}})_{n_{\mathbf{p}\sigma}, f_{\mathbf{q}'}, n_4}, \tag{I.5}$$

while the derivative of $E$ with respect to $n_4$ defines $\mu_4$, the He$^4$ chemical potential:

$$\mu_4 = (\delta E/\delta n_4)_{n_{\mathbf{p}\sigma}, f_{\mathbf{q}}}. \tag{I.6}$$

Thus the total variation of the energy, at fixed volume and $\mathbf{v}_s$, is

$$\delta E = \mu_4 \delta n_4 + \frac{1}{V} \sum_{\mathbf{p}\sigma} \varepsilon_{\mathbf{p}\sigma} \delta n_{\mathbf{p}\sigma} + \frac{1}{V} \sum_{\mathbf{q}} \omega_{\mathbf{q}} \delta f_{\mathbf{q}}. \tag{I.7}$$

Note that $\mu_3$, the He$^3$ chemical potential, does not occur in the variation of $E$, since in the quasiparticle picture $n_3$ does not occur as a variable independent of the $n_{\mathbf{p}\sigma}$. Note also that $\varepsilon_{\mathbf{p}\sigma}$ and $\omega_{\mathbf{q}}$ depend on the number of He$^3$ quasiparticles and He$^4$ excitations; this leads to interactions among the elementary excitations, and hence to the temperature dependence of $\varepsilon_{\mathbf{p}\sigma}$ and $\omega_{\mathbf{q}}$ in states of thermodynamic equilibrium.

It is often quite useful to regard the He$^4$ chemical potential $\mu_4$ as the independent variable instead of $n_4$. Then if we write

$$E'(\{n_{\mathbf{p}\sigma}\}, \{f_{\mathbf{q}}\}, \mu_4) = E(\{n_{\mathbf{p}\sigma}\}, \{f_{\mathbf{q}}\}, n_4) - \mu_4 n_4 \tag{I.8}$$

we have

$$\delta E' = -n_4 \delta \mu_4 + \frac{1}{V} \sum_{\mathbf{p}\sigma} \varepsilon_{\mathbf{p}\sigma} \delta n_{\mathbf{p}\sigma} + \frac{1}{V} \sum_{\mathbf{q}} \omega_{\mathbf{q}} \delta f_{\mathbf{q}}. \tag{I.9}$$

The quasiparticle energies are thus equivalently defined by

$$\begin{aligned} \varepsilon_{\mathbf{p}\sigma} &= V(\delta E'/\delta n_{\mathbf{p}\sigma})_{n_{\mathbf{p}'\sigma'}, f_{\mathbf{q}}, \mu_4} \\ \omega_{\mathbf{q}} &= V(\delta E'/\delta f_{\mathbf{q}})_{n_{\mathbf{p}\sigma}, f_{\mathbf{q}'}, \mu_4}. \end{aligned} \tag{I.10}$$

For momenta $p$ small compared with $m_4 s$, where $m_4$ is the He$^4$ atomic mass, and $s = 238\,\text{m sec}^{-1}$ is the He$^4$ first-sound velocity, the He$^3$ excitation spectrum has the form (I.1). At low $T$ and $q$ the He$^4$ excitations are phonons with energies $\omega_{\mathbf{q}}$ given by

$$\omega_{\mathbf{q}} = sq, \qquad q \ll q_D; \tag{I.11}$$

here $q_D = 1{\cdot}09\text{Å}^{-1}$ is the He$^4$ Debye wavenumber, defined by $n_4^0 = q_D^3/6\pi^2$, where $n_4^0$ is the pure He$^4$ number density at $T = P = 0$. For $q \simeq q_0 = 1{\cdot}91\text{Å}^{-1}$ the He$^4$ excitations are of the form

$$\omega_{\mathbf{q}} = \Delta + \frac{(q - q_0)^2}{2\mu_r}, \qquad q \sim q_0 \tag{I.12}$$

which is the roton spectrum. In pure He$^4$ at $T = 0$ and zero pressure, $\Delta = 8{\cdot}65\,\text{K}$, $\mu_r = 0{\cdot}16 m_4$.

The entropy $S$ of the macroscopic state depends only on the number of ways of distributing a given number of elementary excitations among the various possible momentum states. Thus $S$ is a sum of contributions from each individual mode of excitation, given by the same formulae as for non-interacting particles; we have

$$S = -k\sum_{\mathbf{p}\sigma}[n_{\mathbf{p}\sigma}\ln n_{\mathbf{p}\sigma} + (1 - n_{\mathbf{p}\sigma})\ln(1 - n_{\mathbf{p}\sigma})] - k\sum_{\mathbf{q}}[f_{\mathbf{q}}\ln f_{\mathbf{q}} - (1 + f_{\mathbf{q}})\ln(1 + f_{\mathbf{q}})], \tag{I.13}$$

where $k$ is Boltzmann's constant. The entropy has no explicit dependence on $N_4$ or $\mathbf{v}_s$; its total variation is given by

$$\delta S = -k\sum_{\mathbf{p}\sigma}(\delta n_{\mathbf{p}\sigma})\ln[n_{\mathbf{p}\sigma}/(1 - n_{\mathbf{p}\sigma})] - k\sum_{\mathbf{q}}(\delta f_{\mathbf{q}})\ln[f_{\mathbf{q}}/(1 + f_{\mathbf{q}})]. \tag{I.14}$$

It should be stressed here that the "free particle" form of the entropy is completely consistent, as we shall see, with temperature-dependent excitation energies, including a temperature-dependent effective mass, and with weak interactions between all the elementary excitations.

In thermal equilibrium the smoothed distribution functions $n_{\mathbf{p}\sigma}$ and $f_{\mathbf{q}}$ are determined from the relation

$$\delta E = T\delta(S/V) + \mu_3\delta n_3 + \mu_4\delta n_4, \tag{I.15}$$

valid for any variation about thermodynamic equilibrium at temperature $T$. Using (I.7) and (I.14) we find, comparing coefficients of $\delta n_{\mathbf{p}\sigma}$, that

$$n_{\mathbf{p}\sigma} = \frac{1}{\exp[(\varepsilon_{\mathbf{p}\sigma} - \mu_3)/kT] + 1}, \tag{I.16}$$

and from comparing coefficients of $\delta f_{\mathbf{q}}$ that

$$f_{\mathbf{q}} = \frac{1}{\exp(\omega_{\mathbf{q}}/kT) - 1}. \tag{I.17}$$

The He$^3$ chemical potential $\mu_3$, which enters the theory only as an equilibrium quantity, is determined by (I.16) and (I.3). The energy $E$ of a state of thermodynamic equilibirum depends on the temperature *only* through the He$^3$ and He$^4$ excitation distribution functions (I.16) and (I.17).

The total variation of the free energy density $F = E - TS/V$ is, from (I.7), (I.15)–(I.17),

$$\delta F = \mu_3 \delta\left[\frac{1}{V}\sum_{\mathbf{p}\sigma} n_{\mathbf{p}\sigma}\right] + \mu_4 \delta n_4 - (S/V)\delta T. \tag{I.18}$$

Also, in the quasiparticle picture the pressure $P$ is given by

$$P = -(\partial EV/\partial V)_{n_{\mathbf{p}\sigma}, f_{\mathbf{q}}, n_4 V} \tag{I.19}$$

(this is equivalent to differentiating at constant entropy). Equation (I.18) agrees with the usual thermodynamic result, and leads in the limit of large volume to the Gibbs–Duhem relation

$$\delta P = n_3 \delta\mu_3 + n_4 \delta\mu_4 + (S/V)\delta T. \tag{I.20}$$

The relations (I.13)–(I.20) tell one how to compute the thermodynamics in terms of the elementary excitations. Let us turn now to studying the details of the excitation spectrum, beginning with the properties of a single He$^3$ quasiparticle.†.

## II. Properties of One He$^3$ Atom in He$^4$ at T = 0

The energy of a single He$^3$ quasiparticle in superfluid He$^4$ at $T = 0$ is given, for $p \ll m_4 s$, by $\varepsilon_p = \varepsilon_0 + p^2/2m$, where at zero pressure $\varepsilon_0 \simeq -2{\cdot}785$ K, the He$^3$ effective mass $m$ is $\simeq 2{\cdot}34 m_3$, and $m_3$ is the He$^3$ atomic mass. These values are determined experimentally from the heat-of-mixing[11] of He$^3$ in He$^4$, and from the specific heat[12] of dilute solutions. In the limit of zero He$^3$ concentration $\varepsilon_0$ is just the He$^3$ chemical potential.

### 1. VOLUME OCCUPIED BY HE$^3$

The volume occupied by a single He$^3$ atom in pure He$^4$ is $n_4^{-1} \simeq 45{\cdot}8$ Å$^3$.

† Valuable further discussions of the properties of dilute solutions of He$^3$ in He$^4$ can be found in Khalatnikov's book,[4] and in the review article by Ebner and Edwards.[2] The latter contains an extensive review of experiments on the thermal properties of dilute solutions, as well as a detailed bibliography. Additional reviews are found in Refs. 5–10.

Let us write the volume occupied by a $He^3$ added to $He^4$ at constant *pressure*, as $v_3 = (1 + \alpha)n_4^{-1}$. Experimentally[(2)] $\alpha \simeq 0{\cdot}285$ at $T = 0$ and $P = 0$. If $N_4$ $He^4$ atoms occupy a volume $V$, then at the same pressure $N_4$ $He^4$ atoms plus $N_3$ $He^3$ atoms will occupy a volume $V + N_3 v_3$. Thus the change in the mean $He^4$ density on adding $He^3$ atoms at constant pressure is

$$\left(\frac{\partial n_4}{\partial n_3}\right)_P = -(1 + \alpha). \tag{II.1}$$

If we note from the Gibbs–Duhem relation (I.20) that, in the limit of zero $He^3$ concentration and constant $T$, keeping $P$ constant is equivalent to keeping $\mu_4$ constant, we see that the left side of (II.1) is equivalently

$$\left(\frac{\partial n_4}{\partial n_3}\right)_{\mu_4} = -\left(\frac{\partial \mu_4}{\partial n_3}\right)_{n_4}\left(\frac{\partial n_4}{\partial \mu_4}\right)_{n_3}. \tag{II.2}$$

Since

$$\left(\frac{\partial \mu_4}{\partial n_3}\right)_{n_4} = \frac{\partial^2 E}{\partial n_3 \partial n_4} = \left(\frac{\partial \mu_3}{\partial n_4}\right)_{n_3}, \tag{II.3}$$

and as $n_3 \to 0$

$$\frac{\partial \mu_4}{\partial n_4} = \frac{m_4 s^2}{n_4} \tag{II.4}$$

(the relation between the compressibility and the sound velocity), we find the simple and useful relation between $\partial \varepsilon_0/\partial n_4$ and $\alpha$:

$$\frac{\partial \varepsilon_0}{\partial n_4} = \frac{m_4 s^2}{n_4}(1 + \alpha). \tag{II.5}$$

Physically the reason a $He^3$ occupies a 30% larger volume than a $He^4$ is due to the smaller mass and hence the greater zero point motion of the $He^3$ atom. We can do a simple microscopic calculation that makes this idea quantitative. The exact Hamiltonian of the mixture is

$$H = \sum_{He^4} \frac{p_i^2}{2m_4} + \tfrac{1}{2}\sum_{ij} u(\mathbf{r}_i - \mathbf{r}_j) + \sum_{He^3} \frac{p_j^2}{2m_3}, \tag{II.6}$$

where $u$ is the bare interatomic potential, which is the same for all pairs of helium atoms, whether 4–4, 4–3 or 3–3; the sum in the middle term is over all pairs. Since $m_3 = 3m_4/4$ we can equivalently write $H$ as

$$H = \sum_i \frac{p_i^2}{2m_4} + \tfrac{1}{2}\sum_{ij} u(\mathbf{r}_i - \mathbf{r}_j) + \xi \sum_{He^3} \frac{p_i^2}{2m_4} \tag{II.7}$$

where the first sum is now over both the $He^3$ and the $He^4$ atoms, and $\xi \equiv$

$(m_4 - m_3)/m_3 = \frac{1}{3}$. The first two terms in (II.7) are formally the Hamiltonian for $N = N_3 + N_4$ $He^4$ atoms. The final term $\propto \xi$ represents the additional kinetic energy of the $He^3$ atoms due to their lighter mass; it is essentially a small perturbation.

We may estimate the ground-state energy of the system by choosing as a trial wave function the true ground-state wave function of pure $He^4$ at the same particle density

$$n = n_3 + n_4 \tag{II.8}$$

as in the mixture. This trial wave function does not take into account the Pauli principle between $He^3$ of parallel spin, and thus our result will be valid for one $He^3$ atom, or for two $He^3$ atoms of opposite spin. With this trial function we find

$$E = E_0(n) + n_3 E_1(n), \tag{II.9}$$

where $E_0$ is the ground-state energy density of pure $He^4$ at density $n$, and

$$E_1(n) = \xi\langle p^2/2m_4\rangle = \langle p^2/6m_4\rangle, \tag{II.10}$$

i.e., $\frac{1}{3}$ the mean kinetic energy per particle in pure $He^4$ at density $n$ at $T = 0$. For a single $He^3$ present we identify $\varepsilon_0$ as

$$\varepsilon_0 = \frac{\partial E}{\partial n_3} = \frac{\partial E_0}{\partial n_4} + E_1(n_4). \tag{II.11}$$

The value of $\partial E_0/\partial n_4 = \mu_4$ is $-7{\cdot}17$ K, and McMillan[13] gives the theoretical value $\langle p^2/2m_4\rangle \simeq 14{\cdot}16\,K$. We thus estimate in this simple model $\varepsilon_0 \simeq -2{\cdot}45$ K. The result (I.27) is more generally an upper bound to $\varepsilon_0$. McMillan also estimates $n_4\partial\langle p^2/2m_4\rangle/\partial n_4 \simeq 25{\cdot}1$ K, and thus we find from (II.11) and (II.5) the estimate

$$\alpha \simeq \tfrac{1}{3}\frac{n_4}{m_4 s^2}\frac{\partial}{\partial n_4}\left\langle\frac{p^2}{2m_4}\right\rangle \simeq 0{\cdot}31. \tag{II.12}$$

It is interesting to note that were $m_3 = m_4$, Eq. (II.9) with $E_1 \equiv 0$ would be exact, since then the ground-state wave function of the one $He^3$, $N_4$ $He^4$ atom state would be the $N_4 + 1$ $He^4$ atom ground-state wave function, i.e., completely symmetric.[14] Similarly the ground state of two $He^3$ of opposite spin and $N_4$ $He^4$ is the same as that of $N_4 + 2$ $He^4$ atoms.

Massey and Woo[15] have done a more complete calculation of $\alpha$, as a function of density, which agrees remarkably well with experiment.[16] They have also derived, using the work of Davison and Feenberg,[17] the theoretical result $\varepsilon_0 = -2{\cdot}70$ K.

2. $HE^3$ EFFECTIVE MASS

The effective mass of a $He^3$ in $He^4$ is larger than $m_3$ because as the $He^3$ atom moves through the fluid it creates a flow in the fluid which also carries momentum. The simplest estimate of the effective mass is from the classical result that a sphere of mass $m_3$ in a fluid of mass density $m_4 n_4$ behaves as if it has a mass[18]

$$m_{cl} = m_3 + \tfrac{1}{2} m_4 n_4 v_3, \tag{II.13}$$

where $v_3$ is the volume of the fluid displaced by the sphere, i.e., the mass associated with the fluid flow is half the mass displaced by the sphere. From $n_4 v_3 = 1 + \alpha$ we find $m_{cl} = 1{\cdot}85\, m_3$.

A microscopic calculation by Woo, Tan and Massey,[19] using the method of correlated basis functions, yields the result $m = 2{\cdot}37\, m_3$ at $P = 0$. Recently Pandharipande and Itoh[20] have done a microscopic calculation that gives a good explanation of the pressure dependence of $m$. They write the energy of a $He^3$ quasiparticle of momentum $p$ as

$$\varepsilon_p = \varepsilon_0(n_4) + \frac{p^2}{2m_3} + \Sigma(n_4)p^2, \tag{II.14}$$

and find that the coefficient of the momentum-dependent interaction term, $\Sigma(n_4)$, is to good accuracy a sum of attractive pairwise interactions of the $He^3$ with the individual $He^4$ and of the form

$$\Sigma(n_4) = -an_4/(1 + m_3/m_4)^2, \tag{II.15}$$

where $a$ is a positive constant. Their result for the $He^3$ effective mass is thus

$$m/m_3 = [1 - 2a\, m_3 n_4/(m_3 + m_4)^2]^{-1}. \tag{II.16}$$

If one fits (II.16) to the experimental value $m = 2{\cdot}34\, m_3$ at $P = 0$, one finds a form

$$m/m_3 = [1 - 0{\cdot}57\, n_4/n_4^0]^{-1}, \tag{II.17}$$

for the dependence of $m$ on the $He^4$ density, where $n_4^0$ is the $He^4$ density at $T = P = 0$. Equation (II.17) is in excellent agreement with the experiments of Brubaker *et al.*[21] on second sound in dilute solutions, from which they deduce the dependence of $m$ on pressure; their results at $P = 0$, 10 and 20 atm can be fitted to (II.17) with a coefficient 0·56 instead of 0·57.

## III. INTERACTIONS OF $HE^3$ AT VERY LOW TEMPERATURE

At temperatures below $\approx 0{\cdot}6$ K, the $He^4$ is essentially in its ground state; the excitations of phonons and rotons can be neglected. The He$^3$ in solution then behave as a pure interacting Fermi liquid in the presence of a He$^4$ background "ether", which we can describe by the Landau Fermi-liquid theory.

## 1. HE$^3$ LANDAU PARAMETERS

The He$^3$ Landau parameters, which describe the He$^3$–He$^3$ Fermi-liquid interactions, are defined by

$$f_{\mathbf{p}\sigma,\,\mathbf{p}'\sigma'} \equiv V^2\left(\frac{\delta^2 E'}{\delta n_{\mathbf{p}\sigma}\delta n_{\mathbf{p}'\sigma'}}\right)_{\mu_4} = V\left(\frac{\delta \varepsilon_{\mathbf{p}\sigma}}{\delta n_{\mathbf{p}'\sigma'}}\right)_{\mu_4}. \tag{III.1}$$

We can equivalently define another set of "direct" Landau parameters in terms of variations of $E$ by

$$f^{\text{dir}}_{\mathbf{p}\sigma,\,\mathbf{p}'\sigma'} \equiv V^2\left(\frac{\delta^2 E}{\delta n_{\mathbf{p}\sigma}\delta n_{\mathbf{p}'\sigma'}}\right)_{n_4} = V\left(\frac{\delta \varepsilon_{\mathbf{p}\sigma}}{\delta n_{\mathbf{p}'\sigma'}}\right)_{n_4}. \tag{III.2}$$

According to the rules for partial derivatives the two sets of Landau parameters are simply related by

$$f_{\mathbf{p}\sigma,\,\mathbf{p}'\sigma'} = f^{\text{dir}}_{\mathbf{p}\sigma,\,\mathbf{p}'\sigma'} + \left(\frac{\partial \varepsilon_{\mathbf{p}\sigma}}{\partial n_4}\right)_{n_{\mathbf{p}'\sigma'}}\left(\frac{\delta n_4}{\delta n_{\mathbf{p}'\sigma'}}\right)_{\mu_4}. \tag{III.3}$$

Now

$$\left(\frac{\delta n_4}{\delta n_{\mathbf{p}'\sigma'}}\right)_{\mu_4} = -\frac{\delta^2 E'}{\delta\mu_4\delta n_{\mathbf{p}'\sigma'}} = -\left(\frac{\partial \varepsilon_{\mathbf{p}'\sigma'}}{\partial \mu_4}\right)_{n_{\mathbf{p}\sigma}} = -\left(\frac{\partial n_4}{\partial \mu_4}\right)_{n_{\mathbf{p}\sigma}}\left(\frac{\partial \varepsilon_{\mathbf{p}'\sigma'}}{\partial n_4}\right)_{n_{\mathbf{p}\sigma}}, \tag{III.4}$$

so that the relation becomes

$$f_{\mathbf{p}\sigma,\,\mathbf{p}'\sigma'} = f^{\text{drr}}_{\mathbf{p}\sigma,\,\mathbf{p}'\sigma'} - \left(\frac{\partial \varepsilon_{\mathbf{p}\sigma}}{\partial n_4}\right)_{n_{\mathbf{p}''\sigma''}}\left(\frac{\partial n_4}{\partial \mu_4}\right)_{n_{\mathbf{p}''\sigma''}}\left(\frac{\partial \varepsilon_{\mathbf{p}'\sigma'}}{\partial n_4}\right)_{n_{\mathbf{p}''\sigma''}}. \tag{III.5}$$

In the derivatives the He$^3$ quasiparticle distribution is kept fixed. This equation has a simple physical interpretation. It says that the interaction between two He$^3$ quasiparticles, described by $f_{\mathbf{p}\sigma,\,\mathbf{p}'\sigma'}$, can take place either by the He$^3$ quasiparticles interacting at constant $n_4$, the "direct" term, or via processes in which the quasiparticle $\mathbf{p}'\sigma'$ creates a He$^4$ density fluctuation, described by (III.4), which then interacts with the quasiparticle $\mathbf{p}\sigma$. We shall refer to this latter process as the "phonon-exchange" or "phonon-induced" interaction, since it corresponds to the exchange of a phonon between the two He$^3$ quasiparticles; the full form of the phonon-induced interaction is given in Section IV.2(a). See Fig. 2. Then for small $\mathbf{p}$ and $\mathbf{p}'$ the contribution to $f_{\mathbf{p}\sigma,\mathbf{p}'\sigma'}$ from the phonon-induced process is

$$-\frac{m_4 s^2}{n_4}(1+\alpha)^2, \tag{III.6}$$

where we have used (II.4) and (II.5).

Unlike in a pure translationally invariant Fermi liquid, such as liquid $He^3$, $f_{\mathbf{p}\sigma,\mathbf{p}'\sigma'}$ is not in general a function only of $\mathbf{p}-\mathbf{p}'$. This is because the rest frame

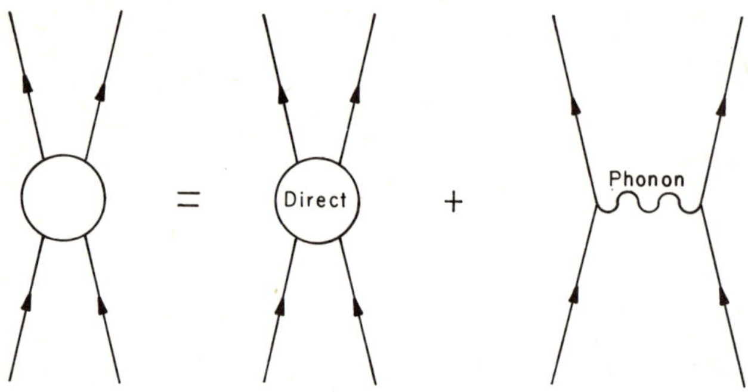


FIG. 2. $He^3$–$He^3$ effective interaction as a sum of a direct scattering plus a phonon-exchange scattering.

of the background $He^4$, the frame in which the superfluid velocity vanishes, establishes a preferred frame, and destroys Galilean invariance for the $He^3$ system alone.

In dilute solutions of $He^3$ it is sufficient to expand the total energy (I.2) to second order in $n_{\mathbf{p}\sigma}$. Thus we write

$$E'(\{n_{\mathbf{p}\sigma}\},\{f_{\mathbf{q}}\},\mu_4) = E'(\{f_{\mathbf{q}}\},\mu_4) + \frac{1}{V}\sum_{\mathbf{p}\sigma} \varepsilon^0_{\mathbf{p}\sigma} n_{\mathbf{p}\sigma} + \tfrac{1}{2}\frac{1}{V^2}\sum_{\mathbf{p}\sigma,\mathbf{p}'\sigma'} f_{\mathbf{p}\sigma,\mathbf{p}'\sigma'} n_{\mathbf{p}\sigma} n_{\mathbf{p}'\sigma'} + \cdots, \tag{III.7}$$

where $\varepsilon^0_{\mathbf{p}}$ and $f_{\mathbf{p}\sigma,\mathbf{p}'\sigma'}$ depend only on $\{f_{\mathbf{q}}\}$ and $\mu_4$, and $E'(\{f_{\mathbf{q}}\},\mu_4)$ is the energy $E-\mu_4 n_4$ of pure $He^4$. With (III.7) the $He^3$ quasiparticle energy is given by

$$\varepsilon_{\mathbf{p}\sigma} = \varepsilon^0_{\mathbf{p}} + \frac{1}{V}\sum_{\mathbf{p}'\sigma'} f_{\mathbf{p}\sigma,\mathbf{p}'\sigma'} n_{\mathbf{p}'\sigma'}. \tag{III.8}$$

When dealing with quasiparticles whose spins are all eigenstates of the spin $\sigma$ along a given direction, say $z$, the quasiparticle interaction assumes the simple form:

$$
\begin{aligned}
f_{\mathbf{p}\uparrow,\mathbf{p}'\uparrow} &= f_{\mathbf{p}\downarrow,\mathbf{p}'\downarrow} = f^s_{\mathbf{pp}'} + f^a_{\mathbf{pp}'} \\
f_{\mathbf{p}\uparrow,\mathbf{p}'\downarrow} &= f_{\mathbf{p}\downarrow,\mathbf{p}'\uparrow} = f^s_{\mathbf{pp}'} - f^a_{\mathbf{pp}'},
\end{aligned}
\tag{III.9}
$$

where the arrows denote the eigenstates of $\sigma_z$, and the superscripts $s$ and $a$ stand for *symmetric* and *antisymmetric*.

At very low temperatures the $He^3$ form a completely degenerate Fermi liquid, filling a Fermi sea with radius the Fermi momentum

$$p_f = (3\pi^2 n_3)^{1/3}\hbar, \tag{III.10}$$

with density of states $N(0) = m^* p_f/\pi^2\hbar^3$, where $m^*$ is the $He^3$ effective mass, on the Fermi surface, defined by writing

$$v_f = (\partial\varepsilon_{p\sigma}/\partial p)_{p=p_f} = p_f/m^* \tag{III.11}$$

where $v_f$ is the Fermi velocity. At $T = 0$ the distribution function $n_{p\sigma}$ takes the familiar form

$$n^0_{p\sigma} = \begin{cases} 1, & \varepsilon_{p\sigma} < \mu \\ 0, & \varepsilon_{p\sigma} > \mu. \end{cases} \tag{III.12}$$

The $He^3$ degeneracy or Fermi temperature [the dashed line in Fig. 1] is

$$T_f = \frac{p_f^2}{2m^*\kappa} \simeq 0{\cdot}119\left(\frac{x}{0{\cdot}01}\right)^{2/3}. \tag{III.13}$$

The quantity

$$x = \frac{n_3}{n_3 + n_4} = \frac{n_3}{n} \tag{III.14}$$

is the fractional concentration, by number, of $He^3$ in solution. The latter formula in (III.13), in kelvin, is valid to lowest order in $x$, and the numerical coefficient is that for $P = 0$.

When one deals with the interactions of quasiparticles both basically on the Fermi surface it is useful to expand $f^s_{\mathbf{pp}'}$ and $f^a_{\mathbf{pp}'}$ in terms of the angle $\theta$ between $\mathbf{p}$ and $\mathbf{p}'$:

$$
\begin{aligned}
f^s_{\mathbf{pp}'} &= \sum_{l=0}^{\infty} f_l^s P_l(\cos\theta) \\
f^a_{\mathbf{pp}'} &= \sum_{l=0}^{\infty} f_l^a P_l(\cos\theta),
\end{aligned}
\tag{III.15}
$$

where the $P_l$ are Legendre polynomials. The $f_l$ are given in terms of the $f_{\mathbf{pp}'}$ by the inverse relations

$$f_l^s = \frac{2l+1}{2}\int_{-1}^{1} d\mu\, P_l(\mu)\tfrac{1}{2}(f_{\mathbf{p}\uparrow,\mathbf{p}'\uparrow} + f_{\mathbf{p}\uparrow,\mathbf{p}'\downarrow}) \tag{III.16}$$

$$f_l^a = \frac{2l+1}{2}\int_{-1}^{1} d\mu\, P_l(\mu)\tfrac{1}{2}(f_{\mathbf{p}\uparrow,\mathbf{p}'\uparrow} - f_{\mathbf{p}\uparrow,\mathbf{p},\downarrow}), \tag{III.17}$$

where $\mu$ is the cosine of the angle between $\mathbf{p}$ and $\mathbf{p}'$ (assumed to be on the Fermi surface).

The conventional Landau parameters, defined by

$$F_l^s \equiv N(0)f_l^s, \qquad F_l^a \equiv N(0)f_l^a, \tag{III.18}$$

provide useful dimensionless measures of the strengths of the interactions between quasiparticles on the Fermi surface.

From (III.8) and (III.11) we may relate the effective mass to the Landau parameters by

$$\frac{\mathbf{p}}{m^*} = \frac{\mathbf{p}}{m} + \frac{1}{V}\sum_{\mathbf{p}'\sigma'} (\nabla_p f_{\mathbf{p}\sigma,\,\mathbf{p}'\sigma'})n_{\mathbf{p}'\sigma'}. \tag{III.19}$$

Here we have taken $\varepsilon_p^0 = \varepsilon_0 + p^2/2m$, where $m$ is the effective mass of a single $He^3$ in pure $He^4$. Multiplying both sides by $\nabla_p n_{\mathbf{p}\sigma}$, summing over all $\mathbf{p}\sigma$ and integrating this $\nabla_p$ by parts in the right hand terms we find

$$\frac{1}{m^*} = \frac{1}{m} + \frac{1}{3n_3V^2}\sum_{\substack{\mathbf{p}\sigma\\ \mathbf{p}'\sigma'}} (\nabla_p^2 f_{\mathbf{p}\sigma,\mathbf{p}'\sigma'})n_{\mathbf{p}\sigma}n_{\mathbf{p}'\sigma'}. \tag{III.20}$$

Were $f_{\mathbf{p}\sigma,\mathbf{p}'\sigma'}$ to depend only on $\mathbf{p} - \mathbf{p}'$, as would be the case were the $He^3$ themselves to form a Galilean invariant system, we could carry the $\nabla_p$ through to the $n_{\mathbf{p}'\sigma'}$ in (III.19), and use (III.12) to reduce (III.20) to the usual Galilean-invariant Fermi-liquid relation,

$$\frac{m^*}{m} = 1 + \frac{F_1^s}{3}, \tag{III.21}$$

where $F_1^s$ is defined by (III.18).

## 2. LOW TEMPERATURE PROPERTIES OF DILUTE SOLUTIONS

We now describe briefly the low temperature properties of the dilute solutions and their dependence on the interaction between the $He^3$ quasiparticles.

After this we will discuss phenomenological fits to and theories of the interaction.

### (a) *Specific heat*

For He$^3$ concentrations $x \gtrsim 1\%$ the specific heat at $T \lesssim 0{\cdot}5$ K is dominated entirely by the He$^3$ excitations. For $T \ll T_f$, the specific heat is linear in $T$ and given at constant volume by the usual Fermi-liquid result

$$c_V = T\left(\frac{\partial s}{\partial T}\right)_V = s = \frac{m^* p_f}{3\hbar^3} k^2 T. \tag{III.22}$$

It is also convenient to express $c_V$ in terms of the Fermi temperature, $T_f \equiv p_f^2/2m^*k$, and the total particle density $n$, as

$$c_V = \frac{\pi^2}{2} nk \frac{T}{T_f}, \tag{III.23}$$

which is simply a factor $\pi^2 T/(3T_f)$ times the classical specific heat. Measurements[12] in this regime determine $m^*(x) \simeq 2{\cdot}38\, m_4$ for $x = 1{\cdot}3\%$ and $\simeq 2{\cdot}46$ for $x = 5{\cdot}0\%$. As the temperature is increased one observes that the He$^3$ contribution to the specific heat is, as expected, essentially that of an ideal Fermi gas, approaching $\approx \frac{3}{2} n_3 k$ for $T \gg T_f$. The effects of He$^3$–He$^3$ interactions on the high temperature specific heat have not been well studied, since it is difficult to sort out these corrections from the contribution to the specific heat of the He$^4$ excitations.

### (b) *Chemical potentials and molar volumes*

The He$^3$ chemical potential at $T = 0$ equals the energy of a quasiparticle at the top of the Fermi sea:

$$\mu_3(n_3, \mu_4) = \varepsilon_0(\mu_4) + \frac{p_f^2}{2m(\mu_4)} + \frac{2}{V}\sum_{\mathbf{p}'} (f^s_{\mathbf{p},\mathbf{p}'}(\mu_4))_{p=p_f} n_{\mathbf{p}'}. \tag{III.24}$$

The change in $\mu_3$ with He$^3$ density, at constant $\mu_4$, is computed exactly as in a pure Fermi liquid, and the result is

$$\left(\frac{\partial \mu_3}{\partial n_3}\right)_{\mu_4} = \frac{1 + F_0^s}{N(0)}, \tag{III.25}$$

where $F_0^s$ is defined by (III.18). More generally at finite temperature $T$, the He$^3$ chemical potential is of the form

$$\mu_3(n_3, \mu_4, T) = \varepsilon_0(\mu_4, T) + \mu_f(n_3, \mu_4, T) + \varepsilon_{\text{int}}(n_3, \mu_4, T), \tag{III.26}$$

where $\mu_f$ is the chemical potential of a free Fermi gas of mass $m$ at temperature $T$ and density $n_3$, and $\varepsilon_{\text{int}}$ is a correction depending on the $He^3$–$He^3$ interaction. Note that $\mu_f$ has the structure

$$\mu_f = kT_f^0 g(T/T_f^0), \tag{III.27}$$

where $T_f^0 = p_f^2/2mk$, $n_3 = p_f^3/3\pi^2\hbar^3$, and $g$ is a known dimensionless function obeying $g(0) = 1$.

It is often more convenient experimentally to use the variables $x, P, T$ in place of $n_3, \mu_4, T$. As $He^3$ are added to $He^4$ at constant pressure the $He^4$ chemical potential changes; this change is most easily computed from the Gibbs–Duhem relation (I.20), which allows us to write

$$0 = n_3\left(\frac{\partial\mu_3}{\partial x}\right)_{T,P} + n_4\left(\frac{\partial\mu_4}{\partial x}\right)_{T,P}. \tag{III.28}$$

Thus dividing by $n_4$ and integrating with respect to $x$ we have

$$\mu_4(x, T, P) = \mu_4(0, T, P) - \int_0^x dx' \frac{x'}{1 - x'} \frac{\partial\mu_3(x', T, P)}{\partial x'}. \tag{III.29}$$

This relation is the starting point for the calculation of the osmotic pressure $\Pi$ defined by

$$\mu_4(x, T, P + \Pi) = \mu_4(0, T, P). \tag{III.30}$$

Detailed calculations of the osmotic pressure are given by Ebner and Edwards.[(2)]

In a dilute solution at $T = 0$, $\mu_3(x, P)$ is less than the chemical potential $\mu_3(1, P)$ of pure $He^3$ at the same pressure. As $He^3$ are added to the system they enter the solution up to a limiting concentration $x_s$ where

$$\mu_3(x_s, P) = \mu_3(1, P). \tag{III.31}$$

Beyond this point the system separates into two phases: one a dilute solution at concentration $x_s$ and the other pure $He^3$, which, because it is lighter, floats on top of the dissolved phase. The limiting concentration $x_s$ is sensitive to the interaction between the $He^3$ quasiparticles. To see this, we first note that at $T = 0$ the difference between $\mu_4(x, P)$ and $\mu_4(x = 0, P)$ is, from (III.29), at least of order $x^{5/3}$, and thus to terms of relative order $x$ we can replace $\mu_4 = \mu_4(x, P)$ in (III.24) by $\mu_4(0, P) \equiv \mu_4(P)$. Now at $P = 0$, $\mu_3(x = 1, P = 0) = -2{\cdot}47$ K while $\mu_3(x = 0, P = 0) = -2{\cdot}79$ K. Neglecting the interaction term in (III.26) we would predict that phase separation occurs when $T_f(x) = \mu_3(1, 0) - \mu_3(0, 0)$, or from (III.13) at $x = 4{\cdot}3\%$. The fact that the experimental $x_s$ is $\simeq 6{\cdot}5\%$ tells one immediately that a $He^3$ atom in solution at finite $x$ is more tightly bound than it is as $x \to 0$; in other words, there must be a net attractive interaction between the $He^3$. Experimental study[(11)] of the heat-of-mixing of dilute solutions shows that $\varepsilon_{\text{int}}(n_3, \mu_4(P = 0), T = 0)$ is $\approx -2x$ (in kelvin).

The density $n = n_3 + n_4$, or molar volume $v = n^{-1}$, of the solution at given $x$, $T$ and $P$ can be found from the thermodynamic relation

$$n^{-1} = (\partial\mu/\partial P)_{x,\,T} \tag{III.32}$$

where

$$\begin{aligned}\mu(x, T, P) &= (1 - x)\mu_4(x, T, P) + x\mu_3(x, T, P)\\ &= (1 - x)\mu_4(0, T, P) + (1 - x)\int_0^x dx' \frac{\mu_3(x', T, P)}{(1 - x')^2}\end{aligned} \tag{III.33}$$

is the Gibbs free energy per particle. To derive the latter form we use (III.29). Then evaluating $\mu_3(x, T, P)$ at zero temperature from (III.26), replacing $\mu_4(x, P)$ by $\mu_4(P)$, we find that to order $x^2$,

$$\begin{aligned}\mu(x, P) = (1 - x)\mu_4(P) + x\left\{\varepsilon_0[\mu_4(P)] + \tfrac{3}{10}\frac{p_f^2}{m}\right\}\\ + (1 - x)\int_0^x dx' \frac{\varepsilon_{\text{int}}[x'n_4(P), \mu_4(P)]}{(1 - x')^2},\end{aligned} \tag{III.34}$$

where $n_4(P)$ is the density of pure $He^4$ at pressure $P$.

Let us write the molar volume in the form

$$n^{-1}(x, T, P) = n_4^{-1}(0, T, P)[1 + x\alpha(x, T, P)], \tag{III.35}$$

defining $\alpha(x, T, P)$. Using (III.34) and (III.32) we then derive the result for $\alpha(x, T = 0, P)$:

$$\begin{aligned}\alpha(x, P) = \alpha(x = 0, P) + \frac{p_f^2}{5m}\frac{1}{m_4 s^2}\left(1 - \frac{3}{2}\frac{n_4}{m}\frac{\partial m}{\partial n_4}\right)\\ + (1 - x)n_3\int_0^x dx' \frac{\partial\varepsilon_{\text{int}}[x'n_4(P), \mu_4(P)]/\partial P}{(1 - x')^2},\end{aligned} \tag{III.36}$$

where

$$\alpha(x = 0, P) = \frac{\partial\varepsilon_0(\mu_4)}{\partial\mu_4} - 1 \tag{III.37}$$

is the fractional excess volume occupied by a single $He^3$ atom in pure $He^4$ at $T = 0$. We see from (III.36) that a calculation of the contribution of $He^3$–$He^3$ interactions to the molar volume requires a knowledge of the pressure dependence of the Landau parameters; these corrections to $\alpha$ at $T = 0$ are $\lesssim 1\%$ for $x \lesssim 5\%$. Measurements of the molar volume of dilute solutions are not delicate enough to give one information about the interactions of $He^3$ quasiparticles. More generally, at finite $T$

$$\alpha(x, T, P) = \alpha(0, T, P) + \frac{P_f}{n_3}\frac{\partial n_4}{\partial P}\left(1 - \frac{3}{2}\frac{n_4}{m}\frac{\partial m}{\partial n_4}\right) + \text{interaction terms}, \quad \text{(III.38)}$$

where $P_f(x, T, P)$ is the pressure of a free Fermi gas of mass $m$ at temperature $T$ and density $n_3$ equal to that of the $He^3$ in solution. Note that since $\partial \ln m / \partial \ln n_4 \approx 1{\cdot}2$ the coefficient of the "kinetic" term $\propto P_f$ in (III.38) is negative; this term tends to make dilute solutions have a negative coefficient of thermal expansion at low $T$.

(c) *Spin susceptibility*

The $He^3$ spin susceptibility in dilute solutions of $He^3$ in $He^4$ at low temperatures is given by the Fermi-liquid result

$$\chi = \frac{\hbar^2}{4}\frac{\gamma^2 N(0)}{1 + F_0^a}, \quad \text{(III.39)}$$

where $\gamma$ is the gyromagnetic ratio, and $F_0^a$ is defined by (III.18). $F_0^a$ appears experimentally[(2)] to be positive and $\approx 0{\cdot}05$–$0{\cdot}10$.

(d) *Low temperature $He^3$ transport properties*

The specific heat, molar volume, and spin susceptibility are only weakly sensitive to the interactions among the $He^3$ quasiparticles. On the other hand the very low temperature transport properties are determined entirely by the $He^3$–$He^3$ interactions, and these, together with heat-of-mixing data, provide the best handle by which to deduce the effective interactions from experiment.

At temperatures $T \ll T_f$ the spin diffusion coefficient of dilute solutions of $He^3$ is $\propto T^{-2}$ and is given explicitly in terms of the $He^3$–$He^3$ scattering rates in the neighbourhood of the Fermi surface by the exact Fermi-liquid result [see (III.55) below]. Thermal conduction in dilute solutions is by both the $He^4$ excitations and the $He^3$ quasiparticles; at low $T < 0{\cdot}01$ K $\ll T_f$ conduction by $He^3$, $\propto T^{-1}$, is the dominant mechanism. The $He^3$ thermal conductivity in this régime is given by the exact Fermi-liquid result (III.57) below. Similarly the dominant viscosity in this temperature régime is that of the $He^3$, which is $\propto T^{-2}$ and is given by (III.59). The reason that the viscosity and thermal conductivity due to $He^4$ thermal excitations can be neglected at low $T$ is that the number of $He^4$ excitations at very low $T$ is $\propto T^3$, while phonon mean free paths are determined primarily by scattering from the walls.

We turn now to the description of the effective interactions between He$^3$ quasiparticles and how they can be deduced from experiment.

## 3. PHENOMENOLOGICAL EFFECTIVE INTERACTION

Imagine that we have just two He$^3$ quasiparticles in pure He$^4$ at $T = 0$, one of momentum $\mathbf{p}$ and spin $\uparrow$ and the other of momentum $\mathbf{p}'$ and spin $\downarrow$; and that they scatter together to a final state in which the two quasiparticles have momenta and spin $\mathbf{p} + \mathbf{q}, \uparrow$ and $\mathbf{p}' - \mathbf{q}, \downarrow$. Due to the weakness of microscopic He$^3$ interactions involving the spin, such as spin–orbit interactions or direct spin–spin interactions,† we can neglect the possibility of a spin flip in this process and regard the two quasiparticles as distinguishable. Let us denote the scattering amplitude for this process by $t_{\mathbf{p},\mathbf{p}'}(\mathbf{q})$, where we assume states to be normalized in unit volume. (Due to the lack of Galilean invariance this amplitude does not depend only on the difference $\mathbf{p} - \mathbf{p}'$.) Now if the two quasiparticles have parallel spin initially, then the scattering amplitude is the sum of a term $t_{\mathbf{p},\mathbf{p}'}(\mathbf{q})$ minus an exchange term $t_{\mathbf{p},\mathbf{p}'}(\mathbf{p}' - \mathbf{p} - \mathbf{q})$, in which $\mathbf{p} + \mathbf{q} \leftrightarrow \mathbf{p}' - \mathbf{q}$ in the final state; for the exchange process the momentum transfer is $\mathbf{p}' - \mathbf{p} - \mathbf{q}$. Thus we may write the scattering amplitude for arbitrary initial spins (up or down) as

$$\langle \bar{\mathbf{p}}\sigma, \bar{\mathbf{p}}'\sigma' | t | \mathbf{p}\sigma, \mathbf{p}'\sigma' \rangle = t_{\mathbf{p},\mathbf{p}'}(\bar{\mathbf{p}} - \mathbf{p}) - \delta_{\sigma,\sigma'} t_{\mathbf{p},\mathbf{p}'}(\bar{\mathbf{p}}' - \mathbf{p}) \tag{III.40}$$

(from momentum conservation $\bar{\mathbf{p}} + \bar{\mathbf{p}}' = \mathbf{p} + \mathbf{p}'$ and from energy conservation $\varepsilon_{\bar{\mathbf{p}}} - \varepsilon_{\mathbf{p}} = \varepsilon_{\mathbf{p}'} - \varepsilon_{\bar{\mathbf{p}}'}$).

The scattering amplitude (III.40) is related to the He$^3$ Landau parameters as follows. In the high frequency limit ("$\omega$-limit") in which the energy transfer $\omega = \varepsilon_{\bar{\mathbf{p}}} - \varepsilon_{\mathbf{p}}$ is much greater than the momentum transfer $q = |\bar{\mathbf{p}} - \mathbf{p}|$ times the He$^4$ sound velocity, $\omega \gg qs > qv_f$, the He$^4$ density $n_4$ as well as the He$^3$ distribution function $n_{\mathbf{p}''}$, for $\mathbf{p}'' \neq \mathbf{p}, \mathbf{p}'$, remains fixed during the scattering; then the scattering amplitude reduces to the bare interaction $f^{\text{dir}}_{\mathbf{p}\sigma,\mathbf{p}'\sigma'}$ between the quasiparticles:

$$\lim_{\omega\to 0} \lim_{q\to 0} \langle \bar{\mathbf{p}}\sigma, \bar{\mathbf{p}}'\sigma' | t | \mathbf{p}\sigma, \mathbf{p}'\sigma' \rangle = f^{\text{dir}}_{\mathbf{p}\sigma,\mathbf{p}'\sigma'}. \tag{III.41}$$

On the other hand, in the low frequency "$k$-limit", in which $\omega \ll qv_f < qs$, the He$^4$ and other He$^3$ will have time to readjust so as to keep the He$^4$ and He$^3$ chemical potentials fixed in the scattering; then

$$\lim_{q\to 0} \lim_{\omega\to 0} \langle \bar{\mathbf{p}}\sigma, \bar{\mathbf{p}}'\sigma' | t | \mathbf{p}\sigma, \mathbf{p}'\sigma' \rangle = V\left(\frac{\delta\varepsilon_{\mathbf{p}\sigma}}{\delta n_{\mathbf{p}'\sigma'}}\right)_{\mu_3,\mu_4}. \tag{III.42}$$

† As Leggett[22] has pointed out, such interactions do play an important role in determining the pairing state in pure superfluid He$^3$.

If we evaluate (III.42) in the limit of vanishing concentration, then from (III.7) we find,

$$\lim_{q\to 0}\lim_{\omega\to 0}\langle\bar{\mathbf{p}}\sigma,\bar{\mathbf{p}}'\sigma'|t|\mathbf{p}\sigma,\mathbf{p}'\sigma'\rangle = f_{\mathbf{p}\sigma,\mathbf{p}'\sigma'}. \tag{III.43}$$

If one looks at the analogous relation in pure $He^3$ one finds additional terms on the right side which represent the repeated scattering of particle–hole pairs; such terms are absent here since the $He^3$ density of states $N(0)$ vanishes as $x \to 0$.

From the form of the right side of (III.31) we see that the Landau parameters obey

$$f_{\mathbf{p}\sigma,\mathbf{p}'\sigma'} = t_{\mathbf{p},\mathbf{p}'}(0) - \delta_{\sigma,\sigma'}t_{\mathbf{p},\mathbf{p}'}(\mathbf{p}'-\mathbf{p}), \tag{III.44}$$

where in $t_{\mathbf{p},\mathbf{p}'}$ the energy transfer is set equal to zero, and then, in $t_{\mathbf{p},\mathbf{p}'}(0)$, the momentum transfer is taken to be zero. Note that

$$f_{\mathbf{p}\uparrow,\mathbf{p}\uparrow} = 0. \tag{III.45}$$

Two basic approaches have been used in constructing a form for the scattering amplitude. The first is that of Emery,[(23)] who assumes an effective potential $V(r)$ between $He^3$ atoms, and then determines the scattering phase shifts by solving the two-body Schrödinger equation. Emery pointed out that the parameters of the effective potential could be fitted directly to the spin diffusion data since there is no contribution to spin diffusion from the $He^4$ excitations. He specifically chose the potential to be of the form of a hard core of radius $r_c$ inside an attractive square well of depth $V_0 = s_0(\pi\hbar/2b)^2/m$ between $r = r_c$ and $r = r_c + b$; outside $r_c + b$ the potential was taken to vanish. The parameters that gave the best fit to the measured spin diffusion at $x = 1{\cdot}3\%$ and $5\%$, from the degenerate to the classical régime, are $r_c = 1{\cdot}8$ Å, $b = 0{\cdot}8$ Å, $s_0 = 0{\cdot}81$, $m = 2{\cdot}34\, m_3$; then $V_0 = 22$ K.

The second approach is that of Bardeen, Baym and Pines,[(24)] (BBP) who assume that $t_{\mathbf{p},\mathbf{p}'}(\mathbf{q})$ depends only on the momentum transfer $\mathbf{q}$, and write

$$t_{\mathbf{p},\mathbf{p}'}(\mathbf{q}) = v(q); \tag{III.46}$$

$v(q)$ is called the "effective interaction" between the $He^3$. (One can imagine $v(q)$, of dimension energy × volume, to be the Fourier transform of an effective pseudopotential $v(r)$ acting between $He^3$ quasiparticles of opposite spin; because $v(q)$ is already the scattering amplitude it would be double counting to attempt to solve the Schrödinger equation with $v(r)$.)

The effective interaction can be determined from experiment as follows. From (III.44) and (III.46) we have

$$f_{\mathbf{p}\sigma,\mathbf{p}'\sigma'} = v(0) - \delta_{\sigma,\sigma'}v(|\mathbf{p}-\mathbf{p}'|), \tag{III.47}$$

and

$$f^s_{\mathbf{p},\mathbf{p}'} = v(0) - \tfrac{1}{2}v(|\mathbf{p} - \mathbf{p}'|)$$
$$f^a_{\mathbf{p},\mathbf{p}'} = -\tfrac{1}{2}v(|\mathbf{p} - \mathbf{p}'|). \tag{III.48}$$

Then from (III.26) and (III.24) we see that

$$\varepsilon_{\text{int}}(n_3, \mu_4, T = 0) = 2\int_0^{p_f} \frac{d^3p'}{(2\pi\hbar)^3}(f^s_{\mathbf{p},\mathbf{p}'})_{p=p_f}$$
$$= 6n_3\int_0^1 dy(1 - y)y^2[2v(0) - v(2p_f y)]. \tag{III.49}$$

In the limit of small $p_f$, $\varepsilon_{\text{int}} \simeq \frac{1}{2}v(0)n_3$; comparing with the heat-of-mixing result that at $P = 0$, $\varepsilon_{\text{int}} \approx -2x(K)$, we see that $v(0)$ is *negative* and $\approx -\frac{1}{10}m_4s^2/n_4$ for zero pressure. Note that $m_4s^2 = 27{\cdot}3\,\text{K}$ sets a natural scale of energies in He$^4$ at $T = 0$. The effective interaction is thus relatively quite weak; we shall see later how this weakness results from the fact that He$^3$ and He$^4$ are isotopes.

The $q$ dependence of $v$ can be determined by fitting the theoretical transport coefficients to the measured He$^3$ spin diffusion coefficient $D_\sigma$ and to the He$^3$ contribution to the thermal conductivity $K$ in solution. To compute the transport coefficients in terms of $v(q)$ we note that in a quasiparticle scattering $\mathbf{p} + \mathbf{p}' \to \bar{\mathbf{p}} + \bar{\mathbf{p}}'$, the scattering amplitude (III.40) is given by $v(|\bar{\mathbf{p}} - \mathbf{p}|) - \delta_{\sigma,\sigma'}v(|\bar{\mathbf{p}}' - \mathbf{p}|)$. It is most convenient, in constructing the exact solutions of the He$^3$ quasiparticle kinetic equation for the transport coefficients, to express the scattering amplitude on the Fermi surface in terms of the scattering angles $\theta$ and $\phi$, where the "polar" angle $\theta$ is the angle between the incident momenta $\mathbf{p}$ and $\mathbf{p}'$, and the "azimuthal" angle $\phi$ is the angle between the plane containing the incident momenta $\mathbf{p}, \mathbf{p}'$ and the plane containing the final momenta $\bar{\mathbf{p}}, \bar{\mathbf{p}}'$. In terms of these angles

$$|\bar{\mathbf{p}} - \mathbf{p}| = 2p_f \sin\frac{\theta}{2}\sin\frac{\phi}{2}$$
$$|\bar{\mathbf{p}}' - \mathbf{p}| = 2p_f \sin\frac{\theta}{2}\cos\frac{\phi}{2}. \tag{III.50}$$

Then the scattering rates $W_{\sigma\sigma'}$ on the Fermi surface are given by

$$W_{\uparrow\uparrow}(\theta, \phi) = \frac{2\pi}{\hbar}\left|v\left(2p_f \sin\frac{\theta}{2}\sin\frac{\phi}{2}\right) - v\left(2p_f \sin\frac{\theta}{2}\cos\frac{\phi}{2}\right)\right|^2$$
$$W_{\uparrow\downarrow}(\theta, \phi) = \frac{2\pi}{\hbar}\left|v\left(2p_f \sin\frac{\theta}{2}\sin\frac{\phi}{2}\right)\right|^2. \tag{III.51}$$

The exact solutions for the transport coefficients are constructed in terms of $W_{\sigma\sigma'}$ as follows.[25–27] The average scattering probability for $\sigma'$ unpolarized with respect to $\sigma$ is given by

$$W(\theta, \phi) = \tfrac{1}{4}W_{\uparrow\uparrow} + W_{\uparrow\downarrow}. \tag{III.52}$$

Let us define an average $W$ by

$$\langle W \rangle = \int \frac{d\Omega}{4\pi} \frac{W(\theta, \phi)}{\cos(\theta/2)}, \tag{III.53}$$

where $d\Omega = \sin\theta \, d\theta \, d\phi$ and the integration is over $4\pi$ solid angle, and define a characteristic relaxation time for $He^3$–$He^3$ scattering in the immediate neighbourhood of the Fermi surface by

$$\tau \equiv \frac{8\pi^4\hbar^6}{m^{*3}\langle W \rangle (KT)^2}. \tag{III.54}$$

Then the exact solutions for the low temperature $He^3$ transport coefficients are, for the spin diffusion coefficient:

$$D_\sigma = \frac{2}{3\pi^2} v_f^2 \tau (1 + F_0^a) \sum_{v=1,3,5,\ldots} \frac{(2v + 1)}{v(v + 1)[v(v + 1) - 2\lambda_D]} \tag{III.55}$$

where

$$\lambda_D = 1 - \frac{1}{\langle W \rangle} \int \frac{d\Omega}{4\pi} \frac{W_{\uparrow\downarrow}(1 - \cos\theta)(1 - \cos\phi)}{2\cos(\theta/2)}; \tag{III.56}$$

for the thermal conductivity:

$$K = \frac{2}{\pi^2} c_V v_f^2 \tau \sum_{v=2,4,6,\ldots} \frac{(2v + 1)}{v(v + 1)[v(v + 1) - 2\lambda_K]} \tag{III.57}$$

where

$$\lambda_K = \frac{1}{\langle W \rangle} \int \frac{d\Omega}{4\pi} \frac{W(\theta, \phi)}{\cos(\theta/2)} (1 + 2\cos\theta); \tag{III.58}$$

and for the viscosity:

$$\eta_3 = \frac{2}{5\pi^2} n_3 \tau p_f v_f \sum_{v=1,3,5,\ldots} \frac{(2v + 1)}{v(v + 1)[v(v + 1) - 2\lambda_\eta]}, \tag{III.59}$$

where

$$\lambda_\eta = \frac{1}{\langle W \rangle} \int \frac{d\Omega}{4\pi} \frac{W(\theta, \phi)}{\cos(\theta/2)} \left(1 - 3\sin^4\frac{\theta}{2}\sin^2\phi\right). \tag{III.60}$$

The first fits of $v(q)$ were made from the measured spin diffusion coefficient at saturated vapour pressure using Hone's result for the spin diffusion coefficient, based on the original Abrikosov–Khalatnikov approximate solution of the transport equation. (This result was given by the $\nu = 1$ term alone of (III.55) times 4/3.) A fit to transport coefficients can at best determine $v(q)$ to within an overall sign; the fact that $\varepsilon_{int} \approx -2x$ determines that $v(q)$ must be attractive for small $q$. The BBP fit was

$$v_{BBP}(q) = -v_0 \cos(\beta q/\hbar). \tag{III.61}$$

where $v_0 = -0{\cdot}0754 m_4 s^2/n_4$ and $\beta = 3{\cdot}16$ Å$^{-1}$.

The interactions derived by the Emery and BBP procedures give similar results for calculated properties of dilute solutions. It was found though that neither of these potentials yielded a consistent fit to the subsequently measured thermal conductivity[28] at $x = 1{\cdot}3\%$ and 5%. This led Baym and Ebner[29] to re-examine the Abrikosov–Khalatnikov approximate solution,

TABLE 1. Comparison of experimental and theoretical values of the transport coefficients and Landau parameters of dilute solutions of He$^3$ in He$^4$.

| | 1·3% | | 5·0% | |
|---|---|---|---|---|
| | Theory | Experiment | Theory | Experiment |
| $DT^2$ (cm$^2$ K$^2$/sec) | $18{\cdot}6 \times 10^{-6}$ | $17{\cdot}2 \times 10^{-6}$ | $80 \times 10^{-6}$ | $90 \times 10^{-6}$ |
| $KT$ (erg/sec cm) | 9·6 | 11 | 27 | 24 |
| $F_1^s/3$ | 0·0174 | | 0·055 | |
| $m^*/m_3$ | 2·38 | $2{\cdot}38 \pm 0{\cdot}04$ | 2·47 | $2{\cdot}46 \pm 0{\cdot}04$ |
| $m/m_3 = 2{\cdot}34$ | | | | |
| $F_0^a$ | 0·09 | $0{\cdot}09 \pm 0{\cdot}03$ | 0·04 | $0{\cdot}08 \pm 0{\cdot}03$ |
| $F_0^s$ | −0·20 | | −0·42 | |

and to construct a fit to $v(q)$ using the variational solution of the kinetic equation (corresponding to the leading ($\nu = 1$ or 2) terms in the exact solutions), which provided a much better description of the experiments. Ebner[30] has also constructed a $v(q)$, using the exact solution for $K$ and $D_\sigma$, in the form of a power series

$$v(q) = -\frac{m_4 s^2}{n_4} \sum_{n=0}^{4} a_n (q/2p_5)^{2n}, \tag{III.62}$$

where $p_5$ is the Fermi momentum of an $x = 5{\cdot}0\%$ solution; the resulting best choice for the coefficients $a_n$ is

$$a_0 = 0{\cdot}0780, \quad a_1 = -0{\cdot}2627, \quad a_2 = 0{\cdot}4893, \quad a_3 = -0{\cdot}7329, \quad a_4 = 0{\cdot}4109. \tag{III.63}$$

The comparison with experiment of the transport coefficients as well as other Fermi-liquid properties calculated from (III.62) are shown in Table 1. The best fit, (III.63), produces only 10% agreement between theory and experiment for $K$ and $D_\sigma$; any adjustment of the parameters from (III.63) only worsens the overall agreement. This lack of agreement must arise from the approximation of the true scattering amplitude $t_{\mathbf{p},\mathbf{p}'}(\mathbf{q})$ by a velocity (and concentration) independent effective interaction $v(\mathbf{q})$; the velocity and concentration dependence should produce terms in $t_{\mathbf{pp}'}(\mathbf{q})$ of the order of the concentration, and hence are capable of producing corrections of the order of 10% in the transport coefficients. Formulas and graphs of thermodynamic quantities calculated with (III.62) and (III.63), together with a complete discussion of the fitting problem, are given by Ebner and Edwards.[(2)]

The Landau parameters are given in terms of the effective interaction, from (III.48), by

$$F_0^a = -N(0)\int_0^1 v(2p_f y)y\,\mathrm{d}y$$

$$F_0^s = N(0)v(0) + F_0^a$$

$$F_1^s = -N(0)\int_0^1 v(2p_f y)y(1-2y^2)\,\mathrm{d}y, \qquad \text{(III.64)}$$

where $N(0) = m^* p_f/\pi^2\hbar^3$. Since the effective interaction has been assumed to depend only on the momentum transferred in the interaction, the effective mass $m^*$ is given in terms of $F_1^s$ by the result (III.21) for a Galilean invariant system. Note that $F_0^a$ is slightly positive, implying a small reduction in the spin susceptibility from its value for a free Fermi gas. Since $F_0^s$ is negative the $He^3$ is not expected to have a well defined zero sound mode at zero pressure.

Part of the original theoretical interest in dilute solutions centred on the possibility that the $He^3$ may themselves undergo a superfluid pairing transition. So far there has been no experimental evidence that the $He^3$ become superfluid. The effective interaction between $He^3$ quasiparticles is only weakly attractive. One may estimate, from BCS theory, that the superfluid transition temperature is given by

$$T_c \simeq T_f \exp\left[1/N(0)v_l\right],$$

where $v_l$ is the most attractive spherical harmonic of the effective interaction. For $s$-state pairing one has $N(0)v_{l=0} = -F_0^a$, while for $p$-state pairing $N(0)v_{l=1} = -F_1^s/3$. The highest $T_c$ at zero pressure occurs for $x \approx 1{\cdot}6\%$, where one expects an $s$-state pairing transition at $T_c \approx 2 \times 10^{-6}\,\mathrm{K}$. The crucial point in estimating $T_c$ for opposite spin pairing is that the spin diffusion experiments place so small an upper bound on the magnitude of the effective interaction that the transition temperature at $P = 0$ is unlikely to be above a few microdegrees. There has been some suggestion that $T_c$ may increase to $\approx 10^{-4}$ K under pressure.[31, 32] Transport coefficient measurements at elevated pressure would be very useful in assessing the possibility of an experimentally accessible $He^3$ superfluid transition in dilute solutions.

## 4. MICROSCOPIC APPROACHES TO THE EFFECTIVE INTERACTION

Measured on the scale of characteristic interaction strengths, $m_4 s^2/n_4$, in liquid helium, the effective $He^3$–$He^3$ interaction (III.62, III.63) is very weak. The coefficient $a_0 = 0{\cdot}0780$ is surprisingly close to $\alpha^2 = 0{\cdot}0812$ (for $\alpha = 0{\cdot}285$ at $P = 0$), so that

$$v(q = 0) \simeq -\alpha^2 m_4 s^2/n_4. \tag{III.65}$$

One can understand physically the weakness of the effective interaction by asking how the $He^3$ atoms in solution become dynamically aware of each other. Since the force fields produced by $He^3$ atoms and $He^4$ atoms are identical, a $He^3$ atom cannot tell whether the potential it feels is produced by a $He^3$ or $He^4$ atom. It is only through the ways that the $He^3$ differ from their $He^4$ environment that an effective interaction is produced; only the differences in $He^3$–$He^4$ mass and statistics enable the $He^3$ atoms to identify each other.

Even though $He^3$ are fermions and $He^4$ are bosons, statistics plays a minor role in determining the effective interaction. This is because the effective interaction is determined by the behaviour of two opposite spin $He^3$ in the solution. The scattering of two $He^3$ particles of the same spin includes a further exchange term. But $He^3$ atoms of opposite spin do not obey the exclusion principle, and thus statistics alone offer a $He^3$ atom of one spin orientation no way of distinguishing a $He^3$ atom of opposite spin from a $He^4$ atom in the solution.

The important source of the interaction is thus the $He^3$–$He^4$ mass difference, which has the consequence that a $He^3$ atom occupies a volume in the solution $(1 + \alpha)$ times that occupied by a $He^4$ atom. Thus, in a sense, the effective interaction between two $He^3$ is equivalent to that between two "holes" of relative volume $\alpha$ in the liquid. Their interaction is expected to be of relative order $\alpha^2$. (That the interaction is attractive does not, however, follow from so general a consideration.)

To discuss the theoretical basis for the result (III.65), we note that from (III.46) and (III.44)

$$v(q = 0) = t_{\mathbf{p},\mathbf{p}'}(0) = f_{\mathbf{p}\uparrow,\mathbf{p}'\downarrow} = \left(\frac{\partial \varepsilon_{\mathbf{p}\uparrow}}{\partial n_{\mathbf{p}'\downarrow}}\right)_{\mu_4} \tag{III.66}$$

in the limit of zero concentration. In this limit (3.66) reduces to

$$v(q=0) = \lim_{x\to 0}\left(\frac{\partial \varepsilon_\uparrow}{\partial n_\downarrow}\right)_{\mu_4} = \lim_{x\to 0}\left(\frac{\partial^2 E'}{\partial n_\uparrow \partial n_\downarrow}\right)_{\mu_4}, \tag{III.67}$$

where $E' = E - \mu_4 n_4$, $n_\uparrow$ and $n_\downarrow$ denote the density of $He^3$ with up and down spins, and $\varepsilon_\uparrow$ is the energy of an up-spin $He^3$ quasiparticle of zero momentum. If we write the ground-state energy in the form

$$E(n_4, n_\uparrow, n_\downarrow) = E_0(n) + n_3 E_1(n) + n_\uparrow n_\downarrow E_2(n) + \tfrac{1}{2}(n_\uparrow^2 + n_\downarrow^2)E_3(n) \tag{III.68}$$

(where $n = n_3 + n_4$, $n_3 = n_\uparrow + n_\downarrow$), which (aside from $He^3$ quasiparticle kinetic energy terms of order $x^{5/3}$, irrelevant here) is perfectly general up to terms of order $x^2$, then to order $x$

$$\mu_4 = E_0' + n_3 E_1' \tag{III.69}$$

while

$$\varepsilon_\uparrow = \mu_4 + E_1 + n_\downarrow E_2. \tag{III.70}$$

$E_0(n)$ is the ground-state energy of pure $He^4$ at density $n$, and in (III.69) the prime denotes differentiation. From (III.70) with (II.4) and (II.5) we have the exact results at $x = 0$,

$$\alpha = E_1'/E_0'' \tag{III.71}$$

and

$$v(0) = \left(\frac{\partial \varepsilon_\uparrow}{\partial n_\downarrow}\right)_{\mu_4} = -(E_1')^2/E_0'' + E_2 = -\frac{m_4 s^2}{n_4}\alpha^2 + E_2. \tag{III.72}$$

We see then that the empirical result (III.65) is equivalent to the term $E_2$ in (III.68) being negligible.

The calculation below (II.7) is the simplest realization of the idea that $He^3$ (of opposite spin) become aware of each other only because their mass is different from $m_4$. The energy in that calculation, (II.9), is essentially the first order perturbation due to the mass difference. In (II.9) the term $E_2$ vanishes, and the relation (III.65) holds exactly. One should also note that, since in the hypothetical case $m_3 = m_4$ the exact ground state of $N_4$ $He^4$ plus two $He^3$ of opposite spin is the $N_4 + 2$ particle $He^4$ ground state, $E_1$ and $E_2$ both vanish, and thus $v(0) \equiv 0$. (We shall see below that in this case $E_3 \equiv 0$ as well.)

The first correction to (III.65) is second order in the mass difference $\xi = (m_4 - m_3)/m_3$. This correction is simply related to dependence of the pure $He^4$ kinetic energy on the $He^4$ mass, and in fact is an order of magnitude smaller than $\alpha^2 m_4 s^2/n_4$. We see this in the following. From second-order perturbation theory we see that up to terms second order in $\xi$,

$$n_\uparrow n_\downarrow E_2(n) = \frac{\xi^2}{V^2} \sum_b \frac{[\langle 0|\Sigma_\uparrow p_i^2/2m_4|b\rangle\langle b|\Sigma_\downarrow p_j^2/2m_4|0\rangle + \text{c.c.}]}{E_0 - E_b} \tag{III.73}$$

or

$$E_2(n) = \frac{2\xi^2}{(2m_4)^2} \operatorname{Re} \sum_b \frac{\langle 0|p_1^2|b\rangle\langle b|p_2^2|0\rangle}{E_0 - E_b}, \tag{III.74}$$

where the sum is over all excited states $|b\rangle$ of the Hamiltonian $H_0(N, m_4)$ of $N$ pure He$^4$ atoms (not just the symmetric states), and $H_0|b\rangle = VE_b|b\rangle$; also $|0\rangle$ denotes the symmetric ground state of $H_0$. In (III.74) we may think of the index 1 as referring to an up-spin He$^3$ and 2 to a down-spin He$^3$. The important point is that particles 1 and 2 are different. (Terms where 1 and 2 are the same are contributions to $E_1$.)

Since $H_0$ is a completely symmetric Hamiltonian, the excited states $|b\rangle$ in (III.74) can be classified according to the representations of the permutation group of $N$ objects. Because the ground state $|0\rangle$ is completely symmetric, the states $|b\rangle$ either can be completely symmetric, belonging to the Young tableau □□□□□, or else must belong to the Young tableau $\begin{smallmatrix}\square\square\square\square\\ \square\phantom{\square\square\square}\end{smallmatrix}$. These are the only two representations one can reach by acting on the symmetric state $|0\rangle$ with a single particle operator $p_i^2$. The symmetric states, which we denote by $|s\rangle$, are allowable excited states of pure He$^4$, while the latter states correspond to "quasiparticle" states of a single He$^3$ of mass $m_4$ in a background of $N - 1$ He$^4$ of mass $m_4$. These are not allowable states of pure He$^4$.

In the limit of the total particle number $N \to \infty$, the quasiparticle states reached by acting with $p_1^2$ on the ground state are orthogonal to those reached by acting with $p_2^2$ on the ground state, i.e. spin-up and spin-down quasiparticle states are orthogonal. Thus only the symmetric states contribute to $E_2$. To see this, we write (III.74) as

$$E_2 = \frac{2\xi^2}{(2m_4)^2} \frac{1}{N(N-1)} \sum_{i,j=1}^{N} \sum_b \frac{\langle 0|p_i^2|b\rangle\langle b|p_j^2|0\rangle(1 - \delta_{ij})}{E_0 - E_b}. \tag{III.75}$$

In the term proportional to 1 the intermediate states are those reached by acting with the symmetric operator $\Sigma_i p_i^2$ on $|0\rangle$ and hence they must be themselves completely symmetric. The second term, $\propto \delta_{ij}$, which includes contributions from the quasiparticle states, is clearly of order $1/N$ relative to the first term and it may therefore be dropped. Hence as $N \to \infty$:

$$E_2(n) = 2\xi^2 \sum_s \frac{|\langle 0|p^2/2m_4|s\rangle|^2}{E_0 - E_s} < 0, \tag{III.76}$$

where the sum is over symmetric states only.

We now observe that the expression (III.76) is the coefficient of the expansion to second order in $\xi$ of the energy of a system of $N$ $He^4$ atoms of mass $m_4(1+\xi)^{-1}$. For such a system

$$H_0(N, m_4(1+\xi)^{-1}) = H_0(N, m_4) + \xi \sum_{i=1}^{N} p_i^2/2m_4, \tag{III.77}$$

and doing perturbation theory to second order in $\xi$ we find that the ground state energy density in fixed volume is

$$E_0(n, m_4(1+\xi)^{-1}) = E_0(N, m_4) + \frac{n\xi}{2m_4}\langle p^2\rangle_{n,\,m_4} + \frac{n^2}{2}E_2(n, m_4), \tag{III.78}$$

where $E_2$ is given by (III.76). Thus

$$E_2(n) = (\xi/n)^2 \partial^2 E_0(n, m_4(1+\xi)^{-1})/\partial\xi^2, \tag{III.79}$$

evaluated at $\xi = 0$. Since quite generally

$$\begin{aligned}\partial E_0(n, m_4(1+\xi)^{-1})/\partial\xi &= \langle 0|\partial H_0/\partial\xi|0\rangle/V \\ &= (n/2m_4)\langle p^2\rangle_{n,\,m=m_4(1+\xi)^{-1}},\end{aligned}$$

we arrive at the result for $E_2$ to second order in $\xi$:

$$\begin{aligned}E_2(n) &= \frac{\xi^2}{2m_4 n}\frac{\partial\langle p^2\rangle_{n,\,m_4(1+\xi)^{-1}}}{\partial\xi} \\ &= -\frac{\xi^2}{2n}\frac{\partial}{\partial m_4}\frac{2m_4}{\xi}E_1^{(1)}\end{aligned} \tag{III.80}$$

where $E_1^{(1)} = \xi\langle p^2\rangle/2m_4$ is the first order result for $E_1$. Thus from (III.72) we see that to terms of order $\xi^2$

$$v(0) + \alpha^2 m_4 s^2/n_4 = -(\xi^2/2n_4)\partial\langle p^2\rangle/\partial m_4 \equiv -\beta m_4 s^2/n_4, \tag{III.81}$$

where $\langle p^2\rangle$ is the mean value of $p^2$ in the ground state of pure $He^4$ at density $n_4$, and the derivative is at fixed $n_4$. To terms of order $\xi^2$ the results (III.80) and (III.81) are exact.

The expectation value $\langle p^2\rangle$ at given density should be only weakly dependent on the $He^4$ mass, since the mean momentum is due essentially to the relative localization of a $He^4$ atom by the hard cores of its neighbouring atoms. (The zero point kinetic energy $\langle p^2/2m_4\rangle$ *is* strongly dependent on the mass.) Thus to a first approximation one expects the correction term (III.81) to the result (III.65) to be very small. There is no direct way to measure $\partial\langle p^2\rangle/\partial m_4$ in pure $He^4$, and so to have a more accurate estimate one must turn to calculations of the $He^4$ kinetic energy as a function of mass. From calculations by Massey[33] and Smith and Pandharipande[34] one finds $\partial \ln \langle p^2 \rangle / \partial \ln m_4 \approx 0{\cdot}2$, increasing slowly with pressure. This implies that the second order correction term $\beta$ is $\approx 0{\cdot}01$, decreasing slightly with pressure. Detailed results are given in Ref. 35. The second order correction term $\beta$ is always positive, so that it increases the magnitude of $V(0)$ from $\alpha^2 m_4 s^2/n_4$. Terms of order $\xi^3$ and higher could be of either sign and of order 5% of $m_4 s^2/n^4$.

The only experimental information on the relation (III.65) at higher pressure is from measurements of Landau *et al.*[36] on the osmotic pressure. Landau *et al.* fit a "speculative" He$^3$–He$^3$ effective interaction, and find $v(q = 0,\ P = 10\ \text{atm}) = 0{\cdot}76\ v(q = 0,\ P = 0)$, $v(q = 0,\ P = 20\ \text{atm}) = 0{\cdot}70\ v(q = 0, P = 0)$. They also find that $m_4 s^2 \alpha^2/n_4$ is 3% larger in magnitude than $|v(0)|$ at $P = 0$, 18% larger at $P = 10$ atm, and 21% larger at $P = 20$ atm. These differences are in the opposite direction to the second order correction (III.81).

The study of the dilute gas model of dilute solutions yields strong support for the relation (III.65). Saam[37] has computed the ground state energy to order $x^2$ in the dilute gas model and finds, in the notation of (III.68), the following results, for $n_3 \ll n_4$ and $m_3/m_4 = \frac{3}{4}$:

$$E_0(n) = \left[\frac{v_0}{2} + \tfrac{4}{15}\lambda\right] n^2$$

$$E_1(n) = \left[\frac{v_0}{6} + \tfrac{2}{3}(0{\cdot}1326)\lambda\right] n$$

$$E_2(n) = (0{\cdot}0009)\lambda$$

$$E_3(n) = \frac{v_0}{36} + (0{\cdot}0164)\lambda, \tag{III.82}$$

where

$$\lambda = 16 v_0 (na^3/\pi)^{1/2}$$

$$v_0 = 4\pi a/m_4, \tag{III.83}$$

and $a$ is the He$^4$–He$^4$ scattering length. The results (III.82) are valid to first order in the dilute gas parameter $(na^3)^{1/2}$. The numerical coefficients in (III.82) are functions only of the mass ratio $m_3/m_4$, given explicitly in Ref. 37. Then from (III.82) we have

$$\alpha = \tfrac{1}{6} - (0{\cdot}0340)\lambda/v_0, \tag{III.84}$$

$$v(0) = -\frac{v_0}{36} + (0{\cdot}0173)\lambda, \tag{III.85}$$

while

$$\frac{m_4 s^2}{n_4}\alpha^2 = \frac{v_0}{36} + (0{\cdot}0164)\lambda \simeq -v(0) + \lambda\xi^2/105. \tag{III.86}$$

The latter is the leading term in an expansion in $\xi^2$. Thus in the dilute gas model, Eqn (III.65) is exact to order $(\lambda/v_0)^0$, and it gives the term of order $\lambda$ correctly to 5%.

The case $m_3 = m_4$ of the dilute gas model has the interesting result that $E_1 = E_2 = E_3 \equiv 0$ so that to order $x^2$

$$E(n_4, n_\uparrow, n_\downarrow) = E_0(n), \qquad m_3 = m_4, \tag{III.87}$$

where $E_0(n)$ is the same as in (III.82).† Then $\alpha = 0$ and $v(0) = 0$. The result (III.87) is completely general, as we can see by the following argument. Imagine that we add just one $He^3$ of mass $m_4$ and spin up to pure $He^4$. The ground-state wave function of the $N_4 + 1$ particle state is the completely symmetric $He^4$ ground state, and (III.87) is the ground-state energy. Now imagine that we add a second $He^3$; if its spin is down, then the ground-state wave function is again the completely symmetric $He^4$ ground-state wave function for $N_4 + 2$ particles, the $He^3$ behave as marked $He^4$, and (III.87) is valid. On the other hand if the second $He^3$ has spin up, while the Hamiltonian is completely symmetric, the ground-state wave function must be anti-symmetric with respect to interchange of the two $He^3$, i.e., it must be the lowest energy state belonging to the representation of the $N_4 + 2$ particle permutation group with Young tableau $\yng(4,1)$ (the lowest energy state belonging to the representation $\yng(4,1,1)$ is expected to be of higher energy). The point now is that in a system with $N_4 + 1$ $He^4$ atoms and one $He^3$ of mass $m_3 = m_4$, the $He^3$ quasiparticle states, which belong to the Young tableau $\yng(4,1)$, have energy equal to the ground-state energy plus $p^2/2m$, that is, there is no energy gap between the lowest state belonging to $\yng(4,1)$ and the symmetric ground state. Since this lowest state is the ground state of the $N_4$ $He^4$ plus two parallel-spin $He^3$ particle system, one expects it to have the same energy as the $N_4 + 2$ particle $He^4$ ground state, and thus (III.87) to be valid in the parallel-spin case as well.

Several microscopic calculations of the effective interaction have been based on the idea of a $He^3$ acting as a $He^4$ plus a hole of relative size $\alpha$. Campbell[(38)] estimated from this point of view the modification of the bare

† It should be pointed out that the derivation of (III.58) in the dilute gas model simply to order $(\lambda/v_0)^0$ is nontrivial, requiring examination of terms with explicit $v_0^2$ coefficients.[(37)]

$He^3$–$He^3$ interaction due to polarization of the $He^4$ background, using the experimental $\alpha$ as well as the average nearest neighbour distances in dilute solutions as input parameters. He found in his calculation that for small $He^3$–$He^3$ separation $r \lesssim 2{\cdot}5$ Å the effective potential (as used by Emery) is the same as the bare potential, while for $r \gtrsim 4{\cdot}8$ Å it is $\alpha^2$ times the bare potential. The modification of the potential by the medium is qualitatively similar to that proposed by Emery.(23) The $q$-dependent scattering amplitude he derived in this way is in reasonably good agreement up to $q \gtrsim p_5$ with the effective interaction (III.62), (III.63), except at small $q$ where it is too attractive. Woo, Tan and Massey,(19) and McMillan(39) have made considerable headway in deriving the $He^3$–$He^3$ scattering amplitude from first principles using the method of correlated basis functions. This work has so far yielded a qualitative effective interaction consistent with (III.62), (III.63). In McMillan's calculation the effective interaction is a sum of two terms, $-m_4 s^2\alpha^2/n_4$, plus a dipolar interaction [cf. (IV.36)].

## IV. Interaction Between the $He^3$ Quasiparticles and the $He^4$ Fluid

So far we have considered the purely Fermi-liquid aspects of dilute solutions of $He^3$ in superfluid $He^4$, assuming that the background $He^4$ plays the role of an inert ether. We have always assumed that the $He^4$ superfluid velocity $\mathbf{v}_s$ is fixed at zero, and neglected, at low temperatures, real phonon and roton excitations of the $He^4$ fluid. We consider now the role played by the $He^3$ in modifying phenomena involving excitation of the $He^4$ fluid. We shall determine the effects of superfluid flow on the $He^3$ quasiparticles, from which we can determine the $He^3$ contribution to the normal mass density of the system and the interaction of $He^3$ with long wavelength phonons, and then describe, as a detailed illustrative example, the effects of the $He^3$ on first sound propagation in the system, and discuss their effects on other transport properties.

### 1. EFFECTS OF SUPERFLUID FLOW ON $HE^3$ QUASIPARTICLES

In determining the $He^3$ quasiparticle properties so far we have only discussed situations in which the $He^4$ superfluid velocity $\mathbf{v}_s$ remains fixed at zero. For example, the quasiparticle energies (I.1) and the effective $He^3$ interactions are all computed at $\mathbf{v}_s = 0$. The superfluid velocity defines a preferred frame and destroys Galilean invariance of the $He^3$ subsystem, even when there are no $He^4$ excitations present. To compute the effect of superfluid flow on the

$He^3$ quasiparticle properties, we begin by determining, through applying Galilean invariance arguments to the entire $He^3$–$He^4$ system, the dependence of the quasiparticle energy on $\mathbf{v}_s$.

If we observe the system from a frame (denoted by a prime) moving with velocity $\mathbf{u}$ then the Hamiltonian in the primed frame is related to that in the lab frame by

$$H' = H - \mathbf{P}\cdot\mathbf{u} + \tfrac{1}{2}Mu^2, \tag{IV.1}$$

where $\mathbf{P}$ is the momentum operator in the lab frame and $M = N_4 m_4 + N_3 m_3$ is the total mass of the system. Thus the total energy and momentum of a state in the primed frame are given by

$$\mathscr{E}' = \mathscr{E} - \mathbf{P}\cdot\mathbf{u} + \tfrac{1}{2}Mu^2 \tag{IV.2}$$

$$\mathbf{P}' = \mathbf{P} - M\mathbf{u}. \tag{IV.3}$$

where $\mathbf{P}$ is the momentum of the state in the lab frame.

Consider now the change in energy due to adding a $He^3$ quasiparticle of momentum $\mathbf{p}$ in the lab frame. This increases the total mass of the system by $m_3$, the *bare* $He^3$ mass, since the addition of a quasiparticle involves the addition of one bare particle. In the lab frame the momentum of the system increases by $\mathbf{p}$, and the energy increases by $\varepsilon_\mathbf{p}$. From (IV.3) we see that in the primed frame the momentum increases by $\mathbf{p} - m_3\mathbf{u}$ while the energy increases by $\varepsilon_\mathbf{p} - \mathbf{p}\cdot\mathbf{u} + \tfrac{1}{2}m_3u^2$. Thus the quasiparticle energy, in the primed frame, is given by

$$\varepsilon'_{\mathbf{p}-m_3\mathbf{u}} = \varepsilon_\mathbf{p} - \mathbf{p}\cdot\mathbf{u} + \tfrac{1}{2}m_3u^2, \tag{IV.4}$$

or

$$\varepsilon'_\mathbf{p} = \varepsilon_{\mathbf{p}+m_3\mathbf{u}} - \mathbf{p}\cdot\mathbf{u} - \tfrac{1}{2}m_3u^2. \tag{IV.5}$$

If the superfluid velocity vanishes in the initial (unprimed) frame, then in the primed frame $\mathbf{v}_s = -\mathbf{u}$. Hence from (IV.5) we may write

$$\varepsilon'_\mathbf{p} = \varepsilon_{\mathbf{p}-m_3\mathbf{v}_s} + \mathbf{p}\cdot\mathbf{v}_s - \tfrac{1}{2}m_3v_s^2. \tag{IV.6}$$

The quantity $\varepsilon'_\mathbf{p} = \varepsilon_\mathbf{p}\{n'_{\mathbf{p}'}, \mathbf{v}_s\}$ is the quasiparticle energy in the presence of a superfluid velocity $\mathbf{v}_s$, and a $He^3$ quasiparticle distribution

$$n'_\mathbf{p} = n^0_{\mathbf{p}-m_3\mathbf{v}_s}. \tag{IV.7}$$

Thus (with spin indices suppressed)

$$\varepsilon_\mathbf{p}\{n^0_{\mathbf{p}''-m_3\mathbf{v}_s}, \mathbf{v}_s\} = \varepsilon_{\mathbf{p}-m_3\mathbf{v}_s}\{n^0_{\mathbf{p}''}, \mathbf{v}_s = 0\} + \mathbf{p}\cdot\mathbf{v}_s - \tfrac{1}{2}m_3\mathbf{v}_s^2. \tag{IV.8}$$

This equation is the fundamental consequence of Galilean invariance for the dependence of the quasiparticle energies on $\mathbf{v}_s$.

The generalization of (IV.8) to include a background of He$^4$ excitations is straightforward; the point to bear in mind is that under a Galilean transformation the momentum $\mathbf{k}$ of an excitation is the *same* in the primed and unprimed frames, since creating a He$^4$ excitation doesn't change the number of He$^4$ atoms. Thus

$$f'_{\mathbf{q}} = f_{\mathbf{q}}. \tag{IV.9}$$

The excitation energy in the primed frame is given by

$$\omega'_{\mathbf{q}} = \omega_{\mathbf{q}} - \mathbf{q}\cdot\mathbf{u}. \tag{IV.10}$$

[This is equivalent to (IV.5) with $m_3 = 0$.] We can generally however neglect effects of He$^4$ excitations on the dependence of the He$^3$ energies on $\mathbf{v}_s$.

For a single He$^3$ quasiparticle in He$^4$ we find from (IV.8) and (I.1) that to order $v_s^2$

$$\varepsilon_{\mathbf{p}}\{\mathbf{v}_s\} = \varepsilon_0 + \frac{p^2}{2m} + \frac{\delta m}{m}\mathbf{p}\cdot\mathbf{v}_s - \frac{1}{2}\frac{\delta m}{m} m_3 v_s^2 \tag{IV.11}$$

where

$$\delta m = m - m_3. \tag{IV.12}$$

As we shall see in the next section, this result is very useful in determining the interactions between long wavelength phonons and He$^3$ quasiparticles.

Expanding both sides of (IV.8) to first order in $\mathbf{v}_s$ we derive the first variation of $\varepsilon_{\mathbf{p}}$ with $\mathbf{v}_s$:

$$\varepsilon_{\mathbf{p}}\{\mathbf{v}_s\} = \varepsilon_{\mathbf{p}}\{0\} + \left(1 - \frac{m_3}{m^*}\right)\mathbf{p}\cdot\mathbf{v}_s + m_3 \sum_{\mathbf{p}'} f_{\mathbf{p}\mathbf{p}'}\mathbf{v}_s \cdot \nabla_{\mathbf{p}} n^0_{\mathbf{p}'}, \tag{IV.13}$$

where we have used $\nabla_{\mathbf{p}}\varepsilon_{\mathbf{p}} = \mathbf{p}/m^*$. This result enables us to compute the normal mass density carried by the He$^3$ quasiparticles in a hydrodynamic process. Consider a flow situation in which the superfluid velocity is non-zero and the He$^3$ Fermi sea is one of local equilibrium displaced by a momentum $\mathbf{p}_0$, i.e.

$$n_{\mathbf{p}} = n^0_{\mathbf{p}-\mathbf{p}_0} \tag{IV.14}$$

Then the He$^3$ normal velocity $\mathbf{v}_3$ is defined in terms of the He$^3$ quasiparticle current (in unit volume) by

$$\mathbf{v}_3 n_3 \equiv \sum_{\mathbf{p}} (\nabla_{\mathbf{p}}\varepsilon_{\mathbf{p}}) n_{\mathbf{p}}, \tag{IV.15}$$

and the He$^3$ normal mass density $\rho_{n3}$ is defined in terms of the He$^3$ momentum density (measured with respect to that in the frame in which the superfluid is at rest) by

$$\sum_{\mathbf{p}} (\mathbf{p} - m_3\mathbf{v}_3)n_{\mathbf{p}} \equiv \rho_{n3}(\mathbf{v}_3 - \mathbf{v}_s). \tag{IV.16}$$

This normal mass density is measured in second sound experiments as well as in rotating bucket experiments. Using (IV.14) in the left side of (IV.16) we find

$$n_3(\mathbf{p}_0 - m_3\mathbf{v}_s) = \rho_{n3}(\mathbf{v}_3 - \mathbf{v}_s). \tag{IV.17}$$

To compute $\rho_{n3}$ we determine (IV.15) to first order in $\mathbf{p}_0$ and $\mathbf{v}_s$; after a short calculation using (IV.13) and (IV.14) one finds

$$\begin{aligned}\sum_{\mathbf{p}} (\nabla_{\mathbf{p}}\varepsilon_{\mathbf{p}})n_{\mathbf{p}} &= \sum_{\mathbf{p}} \frac{\mathbf{p}}{m^*}\,\delta n_{\mathbf{p}} + \left(\nabla_{\mathbf{p}} \sum_{p'} f_{\mathbf{pp}'}\right)\delta n_{\mathbf{p}'}n^0_{\mathbf{p}} + \nabla_{\mathbf{p}}(\varepsilon_{\mathbf{p}}\{\mathbf{v}_s\} - \varepsilon_{\mathbf{p}}\{0\})n^0_{\mathbf{p}} \\ &= \frac{n_3}{m^*}\mathbf{p}_0 + n_3\left(1 - \frac{m_3}{m^*}\right)\mathbf{v}_s + \sum_{\mathbf{pp}'} f_{\mathbf{pp}'}(\{\mathbf{p}_0 - m_3\mathbf{v}_s\}\cdot\nabla n^0_{\mathbf{p}})\nabla n^0_{\mathbf{p}'}.\end{aligned} \tag{IV.18}$$

The final term is simply $(\mathbf{p}_0 - m_3\mathbf{v}_s)n_3F_1^s/3m^*$, so that

$$\mathbf{v}_3 - \mathbf{v}_s = \left(1 + \frac{F_1^s}{3}\right)\frac{\mathbf{p}_0 - m_3\mathbf{v}_s}{m^*}. \tag{IV.19}$$

Substituting in (IV.17) we see that the normal mass density is given by

$$\rho_{n3} = \frac{m^*n_3}{1 + F_1^s/3}. \tag{IV.20}$$

We may use (III.20) to write

$$\frac{n_3}{\rho_{n3}} \equiv \frac{1}{m_n^*} = \frac{1}{m} + \frac{1}{6n_3}\sum_{\mathbf{pp}'} n^0_{\mathbf{p}}n^0_{\mathbf{p}'}(\nabla_{\mathbf{p}} + \nabla_{\mathbf{p}'})^2 f_{\mathbf{pp}'}. \tag{IV.21}$$

As in pure $He^3$, the effective mass $m^*$ entering the specific heat differs from the mass $m_n^*$ [Eqn (IV.17)] entering flow experiments. The mass $m_n^*$, sometimes called the inertial mass, differs from the "bare" effective mass $m$ by terms which reflect the lack of Galilean invariance of the $He^3$–$He^3$ interaction. It is clear from this form that if we assume Galilean invariance of the $He^3$ subsystem, so that $f_{\mathbf{pp}'}$ depends only on $\mathbf{p} - \mathbf{p}'$, then $m^*/m = 1 + F_1^s/3$ and

$$\rho_{n3} = mn_3. \tag{IV.22}$$

After we have developed the full interaction between $He^3$ and phonons, we shall give [Eqn (IV.40)] a first estimate of $m_n^*$ assuming that the breaking of

Galilean invariance arises purely through the phonon-exchange interaction [(Fig. 4) on p. 453].

## 2. INTERACTION OF HE$^3$ QUASIPARTICLES WITH LONG WAVELENGTH PHONONS

### (a) *Effective interaction Hamiltonian*

The form of the coupling between He$^3$ quasiparticles and long wavelength phonons can be determined by a combination of thermodynamic and Galilean-invariance arguments. In this derivation we shall neglect the He$^3$ concentration dependence of the He$^3$–phonon interaction.

The long wavelength low-frequency deviations of the He$^4$ from equilibrium may be described by a He$^4$ density *variation* $\rho_4(\mathbf{r}, t)$, and a local superfluid velocity $\mathbf{v}_s(\mathbf{r}, t)$ (given microscopically by the gradient of the phase of the He$^4$ condensate wave function). The interaction between a long wavelength phonon and a He$^3$ quasiparticle is determined by the dependence of the quasiparticle energy on the local He$^4$ density and velocity.

The dependence of $\varepsilon_{\mathbf{p}}$ on $\mathbf{v}_s$ is given by (IV.11). The quasiparticle energy depends on $\rho_4$ through the dependence of $\varepsilon_0$ and $m$ on the He$^4$ density; the dependence of $\varepsilon_0$ and $\rho_4$ can be written to second order as

$$\varepsilon_0\{\rho_4\} = \varepsilon_0 + \frac{\partial \varepsilon_0}{\partial n_4}\rho_4 + \tfrac{1}{2}\frac{\partial^2 \varepsilon_0}{\partial n_4^2}\rho_4^2, \tag{IV.23}$$

with a similar result for $m\{\rho\}$. The partial derivatives are at constant He$^3$ density. Then from (IV.11) and (IV.23) the variation of $\varepsilon_{\mathbf{p}}$ with $\rho_4$ and $\mathbf{v}_s$ becomes, to second order

$$\begin{aligned}\delta\varepsilon_{\mathbf{p}}(\mathbf{r}, t) &= \left(\frac{\partial \varepsilon_0}{\partial n_4} + \frac{p^2}{2}\frac{\partial m^{-1}}{\partial n_4}\right)\rho_4(\mathbf{r}, t) + \frac{\delta m}{m}\mathbf{p}\cdot\mathbf{v}_s(\mathbf{r}, t) \\ &+ \tfrac{1}{2}\left[\left(\frac{\partial^2 \varepsilon_0}{\partial n_4^2} + \frac{p^2}{2}\frac{\partial^2 m^{-1}}{\partial n_4^2}\right)\rho_4(\mathbf{r}, t)^2 - \frac{\delta m}{m} m_3 \mathbf{v}_s(\mathbf{r}, t)^2\right] \\ &- m_3 \frac{\partial m^{-1}}{\partial n_4}\mathbf{p}\cdot\mathbf{v}_s(\mathbf{r}, t)\rho_4(\mathbf{r}, t).\end{aligned} \tag{IV.24}$$

The final term is the variation with $\rho_4$ of the $(\delta m/m)\mathbf{p}\cdot\mathbf{v}_s$ term in (IV.11).

The point now is that, to a quasiparticle, a long wavelength phonon looks like a uniform motion of the He$^4$ with a local velocity $\mathbf{v}_s$ and a local density

$n_4 + \rho_4$. Thus (IV.24) is the effective interaction Hamiltonian between a quasiparticle and a long wavelength phonon, where $\rho_4(\mathbf{r}, t)$ and $\mathbf{v}_s(\mathbf{r}, t)$ are the density variations and flow velocity associated with the phonon.

In order to calculate the amplitude for phonon–quasiparticle scattering we "quantize" the interaction (IV.24) by treating $\mathbf{r}$ and $\mathbf{p}$ as the quasiparticle position and momentum *operators*, and by expanding $\rho_4(\mathbf{r})$ and $\mathbf{v}_s(\mathbf{r})$ in terms of the operators $b_{\mathbf{q}}$ and $b_{\mathbf{q}}^\dagger$ that annihilate and create phonons of momentum $\mathbf{q}$. In a box of volume $V$ these expansions are[(4)]

$$\rho_4(\mathbf{r}) = \sum_{\mathbf{q}} \left( \frac{q^2 n_4}{2m_4 \omega_{\mathbf{q}} V} \right)^{1/2} [b_{\mathbf{q}} \exp(i\mathbf{q}\cdot\mathbf{r}) + b_{\mathbf{q}}^\dagger \exp(-i\mathbf{q}\cdot\mathbf{r})] \qquad \text{(IV.25)}$$

and

$$\mathbf{v}_s(\mathbf{r}) = \sum_{\mathbf{q}} \left( \frac{\omega_{\mathbf{q}}}{2m_4 n_4 V} \right)^{1/2} \hat{\mathbf{q}}[b_{\mathbf{q}} \exp(i\mathbf{q}\cdot\mathbf{r}) + b_{\mathbf{q}}^\dagger \exp(-i\mathbf{q}\cdot\mathbf{r})]. \qquad \text{(IV.26)}$$

The operators $b_{\mathbf{q}}$ and $b_{\mathbf{q}}^\dagger$ obey the commutation relations

$$[b_{\mathbf{q}}, b_{\mathbf{q}'}^\dagger] = \delta_{\mathbf{q}\mathbf{q}'}$$
$$[b_{\mathbf{q}}, b_{\mathbf{q}'}] = [b_{\mathbf{q}}^\dagger, b_{\mathbf{q}'}^\dagger] = 0. \qquad \text{(IV.27)}$$

The coefficients in the expansions are determined by (i) the particle conservation law for linearized free phonons

$$\left[\frac{\partial \rho_4(r)}{\partial t} + n_4 \nabla \cdot \mathbf{v}_s(\mathbf{r}) = 0\right],$$

and (ii) the commutation relation between the $He^4$ density and current operators.

Substituting (IV.25) and (IV.26) into (IV.24) we find that the first terms in (IV.24), linear in $\rho_4$ or $\mathbf{v}_s$, correspond to the absorption or emission of a phonon by a quasiparticle, as in Fig. 3a. We denote these terms in (IV.24) by $V_1$. In quantizing (IV.24) the $\mathbf{p}\cdot\mathbf{v}_s(r)$ term must be symmetrized as $\frac{1}{2}[\mathbf{p}\cdot\mathbf{v}_s(\mathbf{r}) + \mathbf{v}_s(\mathbf{r})\cdot\mathbf{p}]$, since the position and momentum of a quasiparticle fail to commute. This symmetrization derives from the fact that the momentum density operator for a collection of particles has the symmetrized form

$$\mathbf{g}(\mathbf{r}) = \tfrac{1}{2}\sum_i [\mathbf{p}_i \delta(\mathbf{r} - \mathbf{r}_i) + \delta(\mathbf{r} - \mathbf{r}_i)\mathbf{p}_i],$$

where $\mathbf{p}_i$ and $\mathbf{r}_i$ are the momentum and position operator of the $i$th particle. The terms $\propto \partial m^{-1}/\partial n_4$ in (IV.24) must be similarly symmetrized; this symmetrization operation is not unique. However we shall not need their explicitly symmetrized forms here. The terms quadratic in $\rho_4$ and $\mathbf{v}_s$, which we denote collectively by $V_2$, correspond to the scattering of a phonon by a quasiparticle, as in Fig. 3b; they lead in addition to two-phonon emission or absorption processes.

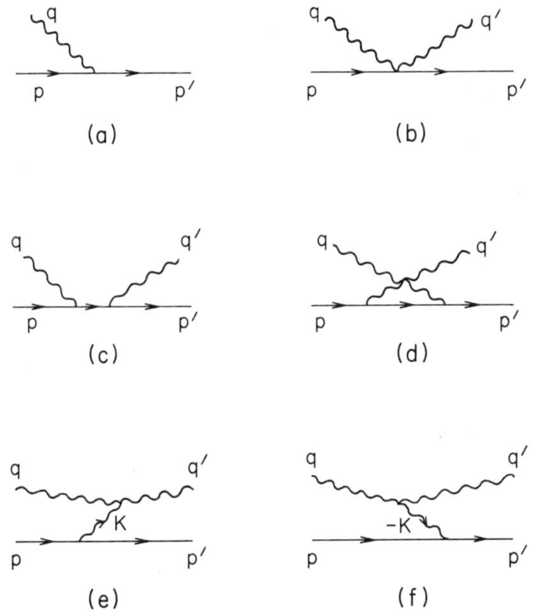


FIG. 3. Interactions of phonons (wavy lines) with quasiparticles (solid lines). (a) Phonon absorption; (b), (c), and (d), three processes leading to scattering of phonons by quasiparticles; (e) and (f), "phonon-induced" scattering of phonons by quasiparticles.

The matrix element for a quasiparticle of momentum $\mathbf{p}$ (with $p \ll q_D$) to absorb a phonon of momentum $\mathbf{q}$, ending up with momentum $\mathbf{p}'$, as in Fig. 3a, is then to leading order:

$$\langle \mathbf{p}' | V_1 | \mathbf{p}\,\mathbf{q} \rangle = \delta_{\mathbf{p}',\,\mathbf{p}+\mathbf{q}} \left( \frac{q n_4}{2 m_4 s V} \right)^{1/2} \left[ \frac{\partial \varepsilon_0}{\partial n_4} + \frac{\delta m s}{n_4 q} (\varepsilon_{\mathbf{p}'} - \varepsilon_{\mathbf{p}}) \right]. \tag{IV.28}$$

We have omitted the contribution $\propto p^2 \partial m^{-1}/\partial n_4$, which is of order $(p/ms)^2$ compared with the $\partial \varepsilon_0/\partial n_4$ term. No Fermi-liquid corrections are included in (IV.28); thus $\varepsilon_{\mathbf{p}'} - \varepsilon_{\mathbf{p}} = \mathbf{q} \cdot (\mathbf{p} + \mathbf{p}')/2m$.

The matrix element for the direct scattering process illustrated in Fig. 3b is

$$\langle \mathbf{p}'\mathbf{q}' | V_2 | \mathbf{p}\mathbf{q} \rangle = \delta_{\mathbf{p}+\mathbf{q},\,\mathbf{p}'+\mathbf{q}'} \frac{\sqrt{q q'}\, n_4}{2 m_4 s V} \left( \frac{\partial^2 \varepsilon_0}{\partial n_4^2} - \frac{\delta m}{m} \frac{m_3 s^2}{n_4^2} \hat{\mathbf{q}} \cdot \hat{\mathbf{q}}' \right). \tag{IV.29}$$

for $p, p' \ll q_D$; again we have neglected the terms $\propto \partial m^{-1}/\partial n_4$, since they vanish as $p$ and $p' \to 0$.

The coefficients in (IV.28) and (IV.29) may be deduced from thermodynamic measurements. Equation (II.5) states

$$\frac{\partial \varepsilon_0}{\partial n_4} = \frac{m_4 s^2}{n_4}(1 + \alpha), \tag{II.5}$$

and thus

$$\frac{\partial^2 \varepsilon_0}{\partial n_4^2} = \frac{m_4 s^2}{n_4^2}\left[(1 + \alpha)(2u - 1) + n_4 \frac{\partial \alpha}{\partial n_4}\right] \tag{IV.30}$$

where

$$u = \frac{n_4}{s}\frac{\partial s}{\partial n_4} \tag{IV.31}$$

is the $He^4$ Grüneisen constant, $\simeq 2{\cdot}84$ at vapour pressure at $T = 0{\cdot}1$ K.[40] Since $n_4 \partial\alpha/\partial n_4 \simeq -1{\cdot}03$ at zero pressure,[41] the square bracket in (IV.30) is $\simeq 4{\cdot}98$.

This general procedure for relating matrix elements of processes involving long wavelength phonons to thermodynamic and hydrodynamic quantities is known as the Landau *quantum hydrodynamic* method. The spirit of the method is the same as that of the Landau Fermi-liquid theory. Its validity depends on the assumption that the scattering amplitudes have no singular structure as $\omega$, the phonon frequency, and $\mathbf{q}$, the phonon wavenumber, both tend to zero. While this is an eminently reasonable property, the possibility of singular structure in the $He^3$–phonon scattering vertices has not, to my knowledge, been definitively ruled out.

### (b) *Phonon–exchange interaction between $He^3$ quasiparticles*

We have now developed sufficient machinery to write down the complete form of the phonon-exchange interaction between two $He^3$ quasiparticles. The complete process is the sum of the two terms shown in Fig. 4. The first process (4a), in which quasiparticle $\mathbf{p}$ emits a phonon of momentum $\mathbf{q}$ which is absorbed by $\mathbf{p}'$, has a matrix element given by second order perturbation theory as

$$\frac{\langle \bar{\mathbf{p}}' | V_1 | \mathbf{p}'\mathbf{q}\rangle\langle \bar{\mathbf{p}}\mathbf{q} | V_1 | \mathbf{p}\rangle}{\omega - sq} = \delta_{\mathbf{p}+\mathbf{p}', \bar{\mathbf{p}}+\bar{\mathbf{p}}'}\delta_{\mathbf{p}, \bar{\mathbf{p}}+\mathbf{q}} \frac{n_4 q}{2m_4 sV} \times \frac{[(\partial\varepsilon_0/\partial n_4) - (\delta m s\omega/n_4 q)]^2}{\omega - sq}, \tag{IV.32}$$

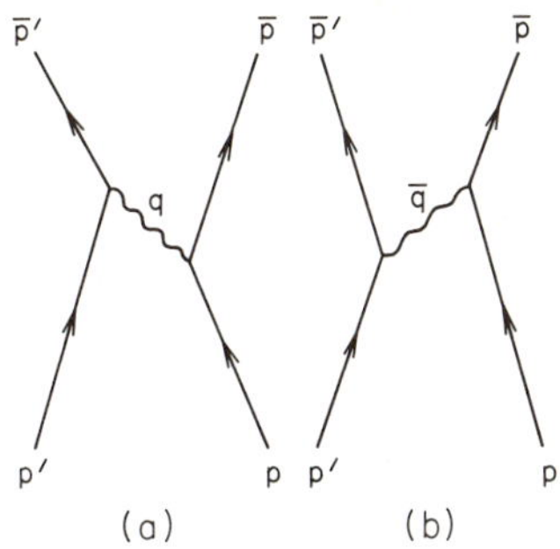


FIG. 4. The two processes producing the phonon (wavy line)-exchange scattering of two He$^3$. Time increases upwards.

where we have assumed energy conservation between the initial and final states, and have written

$$\omega = \varepsilon_{\mathbf{p}} - \varepsilon_{\bar{\mathbf{p}}} = \varepsilon_{\bar{\mathbf{p}}'} - \varepsilon_{\mathbf{p}'}. \tag{IV.33}$$

The second process has a matrix element

$$\frac{\langle\bar{\mathbf{p}}|V_1|\mathbf{p}, -\mathbf{q}\rangle\langle\bar{\mathbf{p}}', -\mathbf{q}|V_1|\mathbf{p}'\rangle}{-\omega - sq} = -\delta_{\mathbf{p}+\mathbf{p}',\bar{\mathbf{p}}+\bar{\mathbf{p}}'}\delta_{\mathbf{p},\bar{\mathbf{p}}+\mathbf{q}}\frac{n_4 V}{2m_4 sV} \times \frac{[(\partial\varepsilon_0/\partial n_4) + (\delta m s\omega/n_4 q)]^2}{\omega + sq}. \tag{IV.34}$$

We write the total matrix element for the phonon-exchange scattering, the sum of (IV.32) and (IV.34), as $\delta_{\mathbf{p}+\mathbf{p}',\bar{\mathbf{p}}+\bar{\mathbf{p}}'}\delta_{\mathbf{p},\bar{\mathbf{p}}+\mathbf{q}}\langle\bar{\mathbf{p}}\bar{\mathbf{p}}'|V_{\text{ph-ex}}|\mathbf{p}\mathbf{p}'\rangle/V$, where

$$\langle\bar{\mathbf{p}}\bar{\mathbf{p}}'|V_{\text{ph-ex}}|\mathbf{p}\mathbf{p}'\rangle = -\frac{m_4 s^2}{n_4}\left[(1+\alpha)^2 - \frac{\omega^2}{\omega^2 - s^2q^2}(1 + \alpha + \delta m/m_4)^2\right]. \tag{IV.35}$$

For scattering of quasiparticles with $\omega = 0$, (IV.35) reduces to the final term of (III.5).

The second term of (IV.31) is effectively a dipolar interaction between the quasiparticles, which depends on their momenta. For $q \ll p, p' \ll ms$ the second term is

$$\langle\bar{\mathbf{p}}\bar{\mathbf{p}}'|V_{\text{dip}}|\mathbf{p}\mathbf{p}'\rangle = -(\mathbf{p}\cdot\hat{\mathbf{q}}/m)(\mathbf{p}'\cdot\hat{\mathbf{q}}/m)(1 + \alpha + \delta m/m_4)^2 m_4/n_4. \tag{IV.36}$$

This interaction is the exact quantum analogue of the classical dipolar interaction of two hard spheres at large separation in a classical fluid. It is of order $(v_f/s)^2$ smaller than the first term in (IV.35).

[For a classical sphere of volume $v_3$, $\delta m = m_4 n_4 v_3/2$, while $n_4 v_3 =$

$1 + \alpha$; for this case (IV.6) reduces to

$$V_{\text{dip}} = -(3v_3/2)^2 m_4 n_4 (\mathbf{p}\cdot\hat{\mathbf{q}}/m)(\mathbf{p}'\cdot\hat{\mathbf{q}}/m). \tag{IV.37}$$

Because of the momentum dependence of this interaction the velocity of the particle with momentum $\mathbf{p}$ is

$$\mathbf{v}_{\mathbf{p}} = \frac{\mathbf{p}}{m} - \nabla_p V_{\text{dip}},$$

with a similar expression for $\mathbf{v}_{\mathbf{p}'}$. The total classical energy $p^2/2m + p'^2/2m + V_{\text{dip}}$ is then

$$\frac{m}{2}(\mathbf{v}_{\mathbf{p}}^2 + \mathbf{v}_{\mathbf{p}'}^2) + \left(\frac{3v_3}{2}\right)^2 m_4 n_4 (\mathbf{v}_{\mathbf{p}}\cdot\hat{\mathbf{q}})(\mathbf{v}_{\mathbf{p}'}\cdot\hat{\mathbf{q}}). \tag{IV.38}$$

This is the form in which the classical dipolar interaction is usually given.[18]]

The phonon–exchange interaction contributes directly to the scattering amplitude $t_{\mathbf{pp}'}(\bar{\mathbf{p}} - \mathbf{p})$ (in (III.40)). The total scattering amplitude must include the exchange term $\langle \bar{\mathbf{p}}'\bar{\mathbf{p}} | V_{\text{ph-ex}} | \mathbf{pp}' \rangle$ as well. When one computes the Landau parameters from (III.43), the direct term $\langle \bar{\mathbf{p}}\bar{\mathbf{p}}' | V_{\text{ph-ex}} | \mathbf{pp}' \rangle$ leads to the final term in (III.5); one should note though that the exchange term is included in the first part, $f^{\text{dir}}_{\mathbf{p},\sigma,\mathbf{p}'\sigma'}$, the Landau parameter defined at constant $n_4$. The total contribution of the phonon–exchange interaction to $f_{\mathbf{p}\sigma,\mathbf{p}'\sigma'}$ is

$$-\frac{m_4 s^2}{n_4}\left[(1+\alpha)^2(1-\delta_{\sigma\sigma'}) - \delta_{\sigma\sigma'}\frac{\omega^2}{s^2 q^2}(1+\alpha+\delta m/m_4)^2\right] \tag{IV.39}$$

where $\omega = \varepsilon_{\mathbf{p}} - \varepsilon_{\mathbf{p}'}$, $\mathbf{q} = \mathbf{p} - \mathbf{p}'$, and we have neglected the $\omega^2$ in the denominator in (IV.35).

We may make a first estimate of the effect of $He^3$–$He^3$ interactions on the $He^3$ normal mass density by assuming that the lack of the Galilean invariance of the quasiparticle interactions is due entirely to the phonon exchange term (IV.39). Substituting (IV.39) into (IV.21) we find

$$\begin{aligned}\frac{1}{m_n^*} &= \frac{1}{m}\left[1 + \tfrac{1}{6}\frac{n_3}{n_4}\frac{m_4}{m}(1+\alpha+\delta m/m_4)^2\right]\\ &\simeq \frac{1 + 0{\cdot}5x}{m},\end{aligned} \tag{IV.40}$$

an estimate in qualitatively good agreement with the concentration dependence of $m_n^*$ as determined by Brubaker *et al.*[21] A detailed comparison of theory with experiment is given in Ref. 2. Disatnik and Brucker[42] have also given a computation of the normal mass density taking into account the lack of Galilean invariance in the $He^3$ effective interaction.

## 3. SCATTERING OF PHONONS BY $HE^3$ QUASIPARTICLES

At temperatures below $\approx 0{\cdot}6$ K, where thermally excited phonons and rotons are few in number, transport processes in the $He^4$, such as thermal conduction and momentum transport (viscosity) by phonons, are limited by scattering and absorption of phonons by $He^3$ quasiparticles, as well as by boundary scattering of phonons for temperatures well below 100 mK. In this section we shall evaluate the amplitude for the scattering of a long wavelength phonon against a $He^3$ quasiparticle.

The complete scattering amplitude is a sum of the amplitudes for the five processes shown in Figs. 3b, c, d, e, f. The first three processes are analogous to the non-relativistic scattering of photons by electrons. The latter two processes involve phonon–phonon scattering. The amplitude for the process b is given by (IV.29). The amplitude for processes c and d is easily computed using (IV.28) in the standard second order perturbation-theory expressions, with the result

$$\delta_{\mathbf{p}+\mathbf{q},\,\mathbf{p}'+\mathbf{q}'}\frac{m_4 s\sqrt{qq'}}{2n_4 V}\left[-\left(\frac{\delta m}{m_4}\right)^2\frac{\hat{\mathbf{q}}\cdot\hat{\mathbf{q}}'}{m}\right.$$
$$\left.+\left(1+\alpha+\frac{\delta m}{m^4}\right)^2 s^2\left(\frac{1}{\varepsilon_{\mathbf{p}}+sq-\varepsilon_{\mathbf{p}+\mathbf{q}}}+\frac{1}{\varepsilon_{p'}-sq-\varepsilon_{\mathbf{p}'-\mathbf{q}}}\right)\right]. \tag{IV.41}$$

In deriving (IV.41) we have neglected the exclusion principle in the intermediate states; in dilute solutions exclusion of intermediate states reduces the scattering rate by terms of relative order $x$. Since the $He^3$ Fermi velocity $v_f \ll s$ it is sufficient to expand the energy denominators in (IV.41) to terms proportional to $(sq)^{-2}$. Furthermore in the temperature regime where $He^3$-phonon scattering is important the relevant phonon energies are much smaller than $m_4 s^2$. This implies that quasiparticle–phonon scattering on the energy shell is essentially elastic, i.e., $q = q'$ to terms of relative order $kT/ms^2$ or $v_f/s$. Thus the total amplitude for processes b, c and d, which we call the direct quasiparticle–phonon scattering, reduces to

$$\langle\mathbf{p}'\mathbf{q}'|T_{\text{dir}}|\mathbf{p}\mathbf{q}\rangle \simeq \delta_{\mathbf{p}+\mathbf{q},\,\mathbf{p}'+\mathbf{q}'}\frac{sq}{2n_4 V}\left\{\left[\left(1+\alpha+\frac{\delta m}{m_4}\right)^2\frac{m^4}{m}-\frac{\delta m}{m_4}\right]\cos\theta\right.$$
$$\left.+\frac{n_4^2}{m_4 s^2}\frac{\partial^3\varepsilon_0}{\partial n_4^2}\right\}, \tag{IV.42}$$

where $\theta$ is the angle between $\mathbf{q}$ and $\mathbf{q}'$.

The two processes in Figs. 3e and f are the scattering of a phonon by a quasiparticle through the exchange of a virtual phonon, analogous to the phonon–exchange interaction between two $He^3$ considered in the previous section. In order to calculate their contribution, denoted by $\langle\mathbf{p}'\mathbf{q}'|$

$T_{ph}|\mathbf{pq}\rangle$, to the total scattering amplitude, we need to find the three-phonon vertex as it occurs in e and f. This may be found for long wavelength phonons by the quantum hydrodynamic method by expanding the total energy of the system to third order in the density variation $\rho_4(\mathbf{r})$ and flow velocity $\mathbf{v}_s(\mathbf{r})$ associated with the phonons; the third order term in the energy is

$$V_3 = \int d^3r \left[ \tfrac{1}{2} m_4 \mathbf{v}_s(\mathbf{r}) \rho_4(\mathbf{r}) \cdot \mathbf{v}_s(\mathbf{r}) + \tfrac{1}{6} \frac{\partial}{\partial n_4}\left(\frac{m_4 s^2}{n_4}\right) \rho_4(\mathbf{r})^3 \right]. \tag{IV.43}$$

The first term is from the kinetic energy of flow, while the coefficient in the latter term is $(\partial^3 E_0/\partial n_4^3)/3!$, where $E_0(n_4)$ is the ground state energy density of pure $He^4$. With the expansions (IV.25) and (IV.26) we see that the three-phonon matrix element in Fig. 3e is

$$\langle \mathbf{q}'|V_3|\mathbf{q}\ \mathbf{K}\rangle = \delta_{\mathbf{q}+\mathbf{K},\mathbf{q}'} \left(\frac{qq'ks}{8m_4 n_4 V}\right)^{1/2} [2u - 1 + \hat{\mathbf{q}}\cdot\hat{\mathbf{q}}' + \hat{\mathbf{q}}\cdot\mathbf{K} + \hat{\mathbf{q}}'\cdot\mathbf{K}], \tag{IV.44}$$

while that in Fig. 3f is $\langle \mathbf{q}', -\mathbf{K}|V_3|\mathbf{q}\rangle = \langle \mathbf{q}|V_3|\mathbf{q}', -\mathbf{K}\rangle$.

The scattering amplitude due to phonon exchange $\langle \mathbf{p}'\mathbf{q}'|T_{ph}|\mathbf{pq}\rangle$ is simply evaluated on the energy shell when the scattering is elastic; then the $\delta m$ term in (IV.28) vanishes, as does $(\hat{\mathbf{q}} + \hat{\mathbf{q}}')\cdot\mathbf{K}$. The result is

$$\langle \mathbf{p}'\mathbf{q}'|T_{ph}|\mathbf{pq}\rangle = -\delta_{\mathbf{p}+\mathbf{q},\mathbf{p}'+\mathbf{q}'} \frac{sq}{2n_4 V}(1 + \alpha)(2u - 1 + \cos\theta). \tag{IV.45}$$

Adding (IV.45) and (IV.42), and using (IV.30), we see that the total phonon-quasiparticle scattering amplitude is

$$\langle \mathbf{p}'\mathbf{q}'|T_{qp-ph}|\mathbf{pq}\rangle = \delta_{\mathbf{p}+\mathbf{q},\mathbf{p}'+\mathbf{q}'} \frac{sq}{2n_4 V}\left[ n_4 \frac{\partial\alpha}{\partial n_4} + \left(1 + \alpha + \frac{\delta m}{m_4}\right) \times \left(1 + \alpha - \frac{m_3}{m_4}\right)\frac{m_4}{m}\cos\theta \right]. \tag{IV.46}$$

This result is valid for $q, q', p, p' \ll ms$.

The rate at which phonons are scattered from $\mathbf{q}$ to $\mathbf{q}'$ by the $He^3$ is given by

$$2\sum_{\mathbf{pp}'} 2\pi\delta(\varepsilon_{\mathbf{p}} + sq - \varepsilon_{\mathbf{p}'} - sq')|\langle \mathbf{p}'\mathbf{q}'|T_{qp-ph}|\mathbf{pq}\rangle|^2[1+f_{\mathbf{q}'}]n_{\mathbf{p}}(1-n_{\mathbf{p}'}); \tag{IV.47}$$

the initial factor of 2 accounts for the $He^3$ spin. If we assume the $He^3$ to be in thermal equilibrium and neglect the Pauli-principle corrections we find from (IV.47), by multiplying by $\delta(\cos\theta - \hat{\mathbf{q}}\cdot\hat{\mathbf{q}}')$ and summing over all $\mathbf{q}'$, that the rate at which phonons of momentum $\mathbf{q}$ are scattered through the angle $\theta$, divided by $1 + f_{\mathbf{q}'}$, is

$$\Gamma(q, \theta) = \frac{n_3 s q^4}{8\pi n_4^2}\left[n_4 \frac{\partial \alpha}{\partial n_4} + \left(1 + \alpha + \frac{\delta m}{m_4}\right)\left(1 + \alpha - \frac{m_3}{m_4}\right)\frac{m_4}{m}\cos\theta\right]$$

$$\simeq \frac{x s q^4}{8\pi n_4}(-1{\cdot}03 + 0{\cdot}70\cos\theta)^2. \qquad \text{(IV.48)}$$

This scattering rate has the $q^4$-dependence characteristic of Rayleigh scattering. Consequently its importance in limiting phonon transport phenomena increases rapidly with temperature. For very long wavelengths, however, and consequently at very low temperatures, the absorption of phonons by He$^3$ via ultrasonic attenuation, i.e., He$^3$ viscosity, plays a more dominant role in limiting rates of phonon transport. In Section IV.6 we shall summarize calculations of transport phenomena involving phonon-quasiparticle scattering, after we develop the theory of first-sound propagation in dilute solutions.

We indicate the relation of the calculations of this section to the Landau parameters describing the phonon-quasiparticle interaction. In the processes shown in Figs 3 b, c, d one never encounters vanishing energy denominators. We may therefore identify the forward phonon–quasiparticle scattering amplitude (IV.42) with the Landau parameter:

$$\left(\frac{\delta \omega_{\mathbf{q}}}{\delta n_{\mathbf{p}}}\right)_{f_{\mathbf{q}'}, n_{\mathbf{p}'}, n_4} = V\langle \mathbf{pq} | T_{\text{dir}} | \mathbf{pq}\rangle.$$

$$= \frac{sq}{2n_4}\left[\left(1 + \alpha + \frac{\delta m}{m_4}\right)^2 \frac{m_4}{m} - \frac{\delta m}{m_4} + (1 + \alpha)(2u - 1) + n_4 \frac{\partial \alpha}{\partial n_4}\right]. \qquad \text{(IV.49)}$$

The addition of the phonon–exchange scattering $V\langle \mathbf{pq} | T_{\text{ph}} | \mathbf{pq}\rangle$ to this result converts it from the Landau parameter defined at fixed $n_4$ to that defined at fixed $\mu_4$ [cf. (III.3), (III.4)]:

$$\left(\frac{\delta \omega_{\mathbf{q}}}{\delta n_p}\right)_{f_{\mathbf{q}'}, n_{\mathbf{p}'}, n_4} = V\langle \mathbf{pq} | T_{\text{qp–ph}} | \mathbf{pq}\rangle$$

$$= \frac{sq}{2n_4}\left[n_4 \frac{\partial \alpha}{\partial n_4} + \left(1 + \alpha + \frac{\delta m}{m_4}\right)\left(1 + \alpha - \frac{m_3}{m_4}\right)\frac{m_4}{m}\right]. \qquad \text{(IV.50)}$$

The shift in the $T = 0$ phonon frequency produced by the addition of He$^3$ to pure He$^4$, at constant $\mu_4$, is then, to lowest order in the He$^3$ concentration $x$,

$$\delta \omega_{\mathbf{q}} = \sum_{\mathbf{p}} \langle \mathbf{pq} | T_{\text{qp–ph}} | \mathbf{pq}\rangle n_{\mathbf{p}}. \qquad \text{(IV.51)}$$

Since (IV.50) is proportional to $q$, we can write (IV.51) as a shift in the $T = 0$

sound velocity:

$$s(x, \mu_4) = s(0, \mu_4)\left\{1 + \frac{x}{2}\left[n_4 \frac{\partial \alpha}{\partial n_4} + \left(1 + \alpha + \frac{\delta m}{m_4}\right)\left(1 + \alpha - \frac{m_3}{m_4}\right)\frac{m_4}{m}\right]\right\}$$

$$\simeq s(0, \mu_4)(1 - 0{\cdot}17x). \qquad \text{(IV.52)}$$

The latter result is at $P = 0$.

We turn now to the full theory of first-sound propagation.

## 4. FIRST SOUND IN DILUTE SOLUTIONS

The addition of $He^3$ atoms to superfluid $He^4$ modifies the first-sound velocity, and by providing additional degrees of freedom for energy absorption increases the attenuation of first sound over its value in pure $He^4$. For temperatures below about 0·3 K for $He^3$ concentrations of a few percent, the primary mechanism for sound attenuation is absorption by the $He^3$ (via the $He^3$ viscosity), a process which is kinematically possible because of $He^3$–$He^3$ collisions. Rayleigh scattering of sound by $He^3$ quasiparticles gives negligible attenuation. In the temperature range from 0·3 K to about 0·6–0·7 K, attenuation due to thermal phonons becomes important. This latter process depends critically on the scattering of thermal phonons by the $He^3$. For temperatures above 0·7 K rotons and phonon–phonon scattering become important; however we shall here assume temperatures below this and neglect thermal rotons.

The propagation of first sound in dilute solutions is described by the kinetic equations for the $He^3$ quasiparticles and phonons, together with the superfluid acceleration equation. The $He^3$ kinetic equation is

$$\frac{\partial n_{\mathbf{p}}}{\partial t} + \nabla_p \varepsilon_{\mathbf{p}} \cdot \nabla_r n_{\mathbf{p}} - \nabla_r \varepsilon_{\mathbf{p}} \cdot \nabla_p n_{\mathbf{p}} = I_{3-3} + I_{3-\text{ph}}, \qquad \text{(IV.53)}$$

where $I_{3-3}$ is the $He^3$–$He^3$ collision integral and $I_{3-\text{ph}}$ describes collisions between $He^3$ quasiparticles and phonons. The phonon kinetic equation is similarly

$$\frac{\partial f_{\mathbf{q}}}{\partial t} + \nabla_q \omega_{\mathbf{q}} \cdot \nabla_r f_{\mathbf{q}} - \nabla_r \omega_{\mathbf{q}} \cdot \nabla_q f_{\mathbf{q}} = I_{\text{ph}-3} + I_{\text{ph}-\text{ph}}, \qquad \text{(IV.54)}$$

where $I_{\text{ph}-3}$ describes phonon-$He^3$ collisions and absorption of phonons by the $He^3$, and $I_{\text{ph}-\text{ph}}$ describes phonon–phonon scattering. We discuss the form of the phonon–$He^3$ collision term in Section IV.6(a).

The linearized equation of motion for superfluid flow is

$$m_4 \frac{\partial \mathbf{v}_s}{\partial t} = -\nabla \mu_4, \tag{IV.55}$$

where $m_4$ is the bare He$^4$ mass and $\mu_4$ the He$^4$ chemical potential. In computing the propagation of first sound we shall neglect Fermi-liquid effects due to interactions between excitations; at low He$^3$ concentrations and low temperatures these are small. Then in the presence of superfluid flow $\varepsilon_\mathbf{p}$ is given by (IV.6), while

$$\omega_\mathbf{q}\{\mathbf{v}_s\} = sq + \mathbf{q}\cdot\mathbf{v}_s. \tag{IV.56}$$

(a) *Conservation laws*

We now write the linearized conservation laws for He$^3$ number, overall mass, energy and momentum, Summing (IV.53) over $\mathbf{p}$, using (IV.6), and noting that collisions conserve He$^3$ atoms, we have

$$\frac{\partial n_3}{\partial t} + n_3 \nabla\cdot\mathbf{v}_3 = 0; \tag{IV.57}$$

$n_3$ is the He$^3$ number density and the He$^3$ velocity $\mathbf{v}_3$ is defined by (IV.15).
The equation expressing conservation of mass is

$$m_4 \frac{\partial n_4}{\partial t} + m_3 \frac{\partial n_3}{\partial t} + \nabla\cdot\mathbf{g} = 0, \tag{IV.58}$$

where the total momentum density is given (in unit volume) by

$$\mathbf{g} = \sum_\mathbf{p} \mathbf{p} n_\mathbf{p} + \sum_\mathbf{p} \mathbf{q} f_\mathbf{q} + m_4 n_4 \mathbf{v}_s. \tag{IV.59}$$

In linear order

$$\sum_\mathbf{p} \mathbf{p} n_\mathbf{p} = n_3(m\mathbf{v}_3 - \delta m \mathbf{v}_s) = n_3[m_3\mathbf{v}_3 + \delta m(\mathbf{v}_3 - \mathbf{v}_s)] \tag{IV.60}$$

[compare with (IV.16) and (IV.22)]. Similarly

$$\sum \mathbf{q} f_\mathbf{q} = \rho_{\text{ph}}(\mathbf{v}_{\text{ph}} - \mathbf{v}_s), \tag{IV.61}$$

thus defining the phonon normal velocity $\mathbf{v}_{\text{ph}}$; the phonon normal mass density is given by

$$\rho_{\text{ph}} = -\int \frac{d^3q}{(2\pi)^3} \frac{q^2}{3} \frac{\partial f_q^0}{\partial \omega_q} = \frac{2\pi^2}{45} \frac{(kT)^4}{s^5 \hbar^3}. \tag{IV.62}$$

The momentum density is then

$$\mathbf{g} = \rho_s \mathbf{v}_s + n_3 m \mathbf{v}_3 + \rho_{\text{ph}} \mathbf{v}_{\text{ph}}, \tag{IV.63}$$

and the resulting equation of mass conservation is

$$m_4 \frac{\partial n_4}{\partial t} + m_3 \frac{\partial n_3}{\partial t} + \rho_s \nabla \cdot \mathbf{v}_s + m n_3 \nabla \cdot \mathbf{v}_3 + \rho_{\text{ph}} \nabla \cdot \mathbf{v}_{\text{ph}} = 0 \qquad \text{(IV.64)}$$

where

$$\rho_s = m_4 n_4 - \delta m n_3 - \rho_{\text{ph}} \qquad \text{(IV.65)}$$

in the superfluid mass density. Note that the effective mass addition $\delta m n_3$ of the $He^3$ quasiparticles (which is actually $He^4$ mass) is carried with the $He^3$ quasiparticles in low-frequency phenomena, and hence is not counted as superfluid mass density.

The equation for conservation of momentum may be derived by multiplying (IV.53) by $\mathbf{p}$, summing over all $\mathbf{p}$, multiplying (IV.54) by $\mathbf{q}$, summing over all $\mathbf{q}$, using (IV.59), (IV.55), (IV.56) and (IV.6), and noting that total momentum is conserved in all collisions; one finds in linear order

$$\frac{\partial g_i}{\partial t} + \nabla_j T_{ij} = -n_4 \nabla_i \mu_4 - n_3 \nabla_i \varepsilon_0(n_4) + \frac{5}{2} \frac{P_f}{m} \nabla_i m(n_4) - \rho_{\text{ph}} s \nabla_i s(n_4), \qquad \text{(IV.66)}$$

where the "kinetic" part of the stress tensor is

$$T_{ij} = \sum_{\mathbf{p}} \frac{p_i p_j}{m} n_{\mathbf{p}} + \sum_{\mathbf{q}} \frac{s q_i q_j}{q} f_{\mathbf{q}}, \qquad \text{(IV.67)}$$

and $P_f$ is the pressure of a free Fermi gas of mass $m$ at temperature $T$ and density $n_3$. The right side of (IV.66) can, from (I.7), be written as

$$\nabla_i \Big(E - n_4 \mu_4 - \sum_{\mathbf{p}} \varepsilon_{\mathbf{p}} n_{\mathbf{p}} - \sum_{\mathbf{q}} \omega_{\mathbf{q}} n_{\mathbf{q}}\Big)$$

where $E$ is the ground-state energy density.

The equation for energy conservation can be similarly derived by taking moments of the kinetic equations, using conservation of total excitation energies in collisions; in linear order

$$\sum_{\mathbf{p}} \varepsilon_{\mathbf{p}} \left(\frac{\partial}{\partial t} + \frac{\mathbf{p}}{m} \cdot \nabla_r\right) n_{\mathbf{p}} + \sum_{\mathbf{q}} \omega_{\mathbf{q}} \left(\frac{\partial}{\partial t} + s \hat{\mathbf{q}} \cdot \nabla_r\right) f_{\mathbf{q}}$$
$$= \left[(\varepsilon_0 n_3 + \tfrac{5}{2} P_f) \frac{\delta m}{m} + s^2 \rho_{\text{ph}}\right] \nabla \cdot \mathbf{v}_s. \qquad \text{(IV.68)}$$

If we neglect phonon and $He^3$ thermal conductivity, which is valid in considering first-sound propagation above $\approx$ 20–30 mK for $He^3$ concentrations $\gtrsim 1\%$, $\nabla_r n_{\mathbf{p}}$ and $\nabla_r f_{\mathbf{q}}$ can be replaced by their local equilibrium values, and

(IV.68) becomes

$$\sum_{\mathbf{p}} \varepsilon_{\mathbf{p}} \frac{\partial n_{\mathbf{p}}}{\partial t} + \sum_{\mathbf{q}} \omega_{\mathbf{q}} \frac{\partial f_{\mathbf{q}}}{\partial t} + (\varepsilon_0 n_3 + \tfrac{5}{2} P_f) \nabla \cdot \mathbf{v}_3 + s^2 \rho_{\mathrm{ph}} \nabla \cdot \mathbf{v}_{\mathrm{ph}} = 0. \tag{IV.69}$$

(b) *He$^3$ kinetic equation in relaxation-time approximation*

In order to use the conservation laws to describe the propagation of sound it is necessary to solve the He$^3$ and phonon kinetic equations in the presence of a sound wave. We shall carry out the solution here for the He$^3$ kinetic equation assuming a relaxation-time approximation for He$^3$–He$^3$ collisions, and neglecting thermal phonons. This is adequate to describe sound propagation below 0·3 K.

The primary problem is to determine $T_{ij} = \Sigma (p_i p_j/m) \delta n_{\mathbf{p}}$ in the presence of a sound wave. This can be done by solving the He$^3$ kinetic equation in the presence of an external potential $U \exp(i\mathbf{q}\cdot\mathbf{r} - i\omega t)$. In the relaxation-time approximation this equation takes the form

$$(\omega - \mathbf{q}\cdot\mathbf{v}_{\mathbf{p}})\delta n_{\mathbf{p}} + U\mathbf{q}\cdot\mathbf{v}_{\mathbf{p}} \frac{\partial n^0_{\mathbf{p}}}{\partial \varepsilon_{\mathbf{p}}} = -\frac{i}{\tau}(n_{\mathbf{p}} - n^{\mathrm{l.e.}}_{\mathbf{p}}) \tag{IV.70}$$

where $n^{\mathrm{l.e.}}_{\mathbf{p}}$ is the distribution function of He$^3$ in local equilibrium, with the same quasiparticle number, momentum and energy as the distribution function $n_{\mathbf{p}}$. Then

$$n_{\mathbf{p}} - n^{\mathrm{l.e.}}_{\mathbf{p}} = \delta n_{\mathbf{p}} - \delta n^0_{\mathbf{p}} + \mathbf{v}_3 \cdot \nabla_{\mathbf{p}} n^0_{\mathbf{p}}; \tag{IV.71}$$

$n^0_{\mathbf{p}} + \delta n^0_{\mathbf{p}}$ is a local equilibrium distribution spherically symmetric about the origin, with the same quasiparticle number and energy density as $n_{\mathbf{p}}$, and $\mathbf{v}_3$ is the local He$^3$ velocity.

From (IV.70) and (IV.71) we readily find

$$\delta n_{\mathbf{p}} = \frac{\delta n^0_{\mathbf{p}} - (\mathbf{v}_3 - i\tau U\mathbf{q})\cdot \mathbf{v}_{\mathbf{p}} \partial n^0_p/\partial \varepsilon_p}{1 - i\tau\omega + i\tau \mathbf{q}\cdot\mathbf{v}_{\mathbf{p}}}, \tag{IV.72}$$

together with the particle and momentum conservation laws

$$\omega \delta n_3 = m n_3 \mathbf{q}\cdot\mathbf{v}_3 \tag{IV.73}$$

$$\omega m n_3 \mathbf{v}_s - \sum_{\mathbf{p}} (\mathbf{q}\cdot\mathbf{v}_{\mathbf{p}})\mathbf{p}\delta n_{\mathbf{p}} = n_3 \mathbf{q} U, \tag{IV.74}$$

where $\delta n_3$ is the variation in the He$^3$ density. The first-sound velocity is large compared with typical He$^3$ velocities, and thus we are interested in $\omega \gg q v_p$, where $v_p$ is a typical quasiparticle velocity. Taking moments of (IV.72) with respect to $\mathbf{v}^2_{\mathbf{p}}$ and $(\mathbf{v}_{\mathbf{p}}\cdot\hat{\mathbf{q}})^2$, expanding in $qv_p/\omega$, keeping terms of order $q^0 \delta n^0_{\mathbf{p}}$, $\mathbf{q}\cdot\mathbf{v}_3$ and $q^2 U$, and using the conservation laws (IV.73) and

(IV.74), we discover

$$(1 - i\tau\omega)\sum_{\mathbf{p}} \mathbf{v}_{\mathbf{p}}^2 \delta n_{\mathbf{p}} = \sum_{\mathbf{p}} \mathbf{v}_{\mathbf{p}}^2 \delta n_{\mathbf{p}}^0 - \frac{5i\tau\omega}{mn_3} P_f \delta n_3 \tag{IV.75}$$

and

$$(1 - i\tau\omega)\sum_{\mathbf{p}} (\mathbf{v}_{\mathbf{p}} \cdot \hat{\mathbf{q}})^2 \delta n_{\mathbf{p}} = \tfrac{1}{3}\sum_{\mathbf{p}} \mathbf{v}_{\mathbf{p}}^2 \delta n_{\mathbf{p}}^0 - \frac{3i\tau\omega}{mn_3} P_f \delta n_3 . \tag{IV.76}$$

Since

$$\sum \mathbf{v}_{\mathbf{p}}^2 (\delta n_{\mathbf{p}}^0 - \delta n_{\mathbf{p}}) = 0 \tag{IV.77}$$

we find

$$\sum \mathbf{v}_{\mathbf{p}}^2 \delta n_{\mathbf{p}} = \frac{5P_f}{mn_3} \delta n_3 \tag{IV.78}$$

$$\begin{aligned} \sum (\mathbf{v}_{\mathbf{p}} \cdot \hat{\mathbf{q}})^2 \delta n_{\mathbf{p}} &= \frac{P_f}{3m}\left(\frac{5\delta n_3}{n_3} - \frac{4i\tau}{1 - i\tau\omega}\mathbf{q}\cdot\mathbf{v}_3\right) \\ &= \left(1 - \frac{4}{5}\frac{i\tau\omega}{1 - i\tau\omega}\right)\frac{5}{3}\frac{P_f}{mn_3}\delta n_3 . \end{aligned} \tag{IV.79}$$

Equation (IV.78) implies that for $\omega \gg qv_p$ the variation of the quasiparticle energy $\Sigma\varepsilon_{\mathbf{p}}\delta n_{\mathbf{p}}$ is given by the adiabatic result

$$\sum \varepsilon_{\mathbf{p}} \delta n_{\mathbf{p}} = \left(\varepsilon_0 + \frac{5}{2}\frac{P_f}{n_3}\right)\delta n_3 ; \tag{IV.80}$$

the coefficient $\varepsilon_0 + 5P_f/2n_3$ is the derivative of the quasiparticle energy $\Sigma\varepsilon_{\mathbf{p}} n_{\mathbf{p}}^0$ with respect to $n_3$ at fixed quasiparticle entropy density $s_3$. From (IV.79) we find that the variation in the stress tensor is given by

$$\sum \frac{p_i p_j}{m} \delta n_{\mathbf{p}} = \left(\frac{\partial P_f}{\partial n_3}\right)_{s_3} \delta n_3 \delta_{ij} - \frac{\tau P_f}{1 - i\tau\omega}(\nabla_i v_{3j} + \nabla_j v_{3i} - \tfrac{2}{3}\delta_{ij}\nabla\cdot\mathbf{v}_3). \tag{IV.81}$$

Equation (IV.81) is essentially the result for $T_{ij}$ in the hydrodynamic limit, only with the first viscosity $\eta$ replaced by $\tau P_f/(1 - i\tau\omega)$. The relaxation time $\tau$ can thus be identified with that appearing in the first viscosity:

$$\tau = \eta / P_f . \tag{IV.82}$$

One may verify by application of the conservation laws that the $He^3$ density–density correlation function $\chi_{33}(q, \omega)$ [which gives the variation of the $He^3$ density, $\delta n_3(\mathbf{q}, \omega)$, induced by an external potential $U(\mathbf{q}, \omega) \exp[i(\mathbf{q}\cdot\mathbf{r} - \omega t)]$ coupled to the $He^3$ density via $\delta n_3 = \chi_{33} U$], is given in the limit $\omega \gg qv_p$ by†

$$\chi_{33}(q,\omega) = \frac{n_3 q^2}{m\omega^2 - (5P_f/3n_3)q^2\,[1 - \frac{4}{5}(i\tau\omega/1 - i\tau\omega)]}. \qquad \text{(IV.83)}$$

It is also straightforward to verify that in the presence of a sound wave, which couples to the $He^3$ via (IV.24), Eqns (IV.80) and (IV.81) still correctly describe the linear response of the $He^3$ when $\omega \gg qv_p$.

### (c) *Dispersion relation for first sound*

We may now use the conservation laws to find the relation of $\omega$ to $q$ in a first-sound wave. The first step is to Fourier transform the conservation laws (IV.57), (IV.58), (IV.64) and (IV.66), neglecting the terms due to thermal phonons. Substituting $\mathbf{g}$ in (IV.58) from (IV.63) and using (IV.55) and (IV.57) to eliminate $\mathbf{v}_s$ and $\mathbf{v}_3$ we find

$$m_4\omega^2\delta n_4 - \rho_s q^2 \delta\mu_4/m_4 = \delta m\omega^2\delta n_3, \qquad \text{(IV.84)}$$

where $\delta n_3$, $\delta n_4$ and $\delta\mu_4$ are the local variations in the $He^3$ density and $He^4$ density and chemical potential caused by the sound wave. The variation of $\mu_4$ can be written with the aid of (I.7) as

$$\delta\mu_4 = \delta\frac{\partial}{\partial n_4}E(\{n_\mathbf{p}\}, n_4) = \left(\frac{\partial^2 E}{\partial n_4^2}\right)_{n_\mathbf{p}}\delta n_4 + \sum_\mathbf{p}\frac{\partial^2 E}{\partial n_4 \partial n_\mathbf{p}}\delta n_\mathbf{p}$$

$$= \left(\frac{\partial\mu_4}{\partial n_4}\right)_{s_3, n_3}\delta n_4 + \sum_\mathbf{p}\frac{\partial\varepsilon_\mathbf{p}}{\partial n_4}\delta n_\mathbf{p}; \qquad \text{(IV.85)}$$

the $n_4$ derivatives are carried out with the $He^3$ distribution function, or equivalently the $He^3$ number density and entropy density $s_3$, held fixed. We evaluate the sum using (IV.78) for the term $\propto p^2\delta n_\mathbf{p}$, and find

$$\sum_\mathbf{p}\frac{\partial\varepsilon_\mathbf{p}}{\partial n_4}\delta n_\mathbf{p} = \left(\frac{\partial\varepsilon_0}{\partial n_4} - \frac{5}{2}\frac{P_f}{n_3}\frac{w}{n_4}\right)\delta n_3 = \left(\frac{\partial\mu_3}{\partial n_4}\right)_{s_3, n_3}\delta n_3, \qquad \text{(IV.86)}$$

where

$$w = \frac{n_4}{m}\frac{\partial m}{\partial n_4}. \qquad \text{(IV.87)}$$

The derivative $(\partial\mu_3/\partial n_4)_{s_3, n_3}$ acts as an effective temperature-dependent coupling of first sound to the $He^3$ density. Substituting (IV.85) and (IV.86)

† This result is valid only for frequencies sufficiently small that the $He^3$ carry their screening clouds $\delta m$ with them. At very high frequencies $\chi_{33}$ must approach $n_3 q^2/m_3\omega^2$.

into (IV.84) we have

$$\left[m_4\omega^2 - \frac{\rho_s}{m_4}\left(\frac{\partial\mu_4}{\partial n_4}\right)_{s_3,n_3} q^2\right]\delta n_4 = \left[\delta m\omega^2 + \frac{\rho_s}{m_4}\left(\frac{\partial\mu_3}{\partial n_4}\right)_{s_3,n_3} q^2\right]\delta n_3. \quad \text{(IV.88)}$$

To lowest order in $x$, $\rho_s$ on the right side may be replaced by $m_4 n_4$.

It remains to compute $\delta n_3$ in terms of $\delta n_4$. We substitute (IV.63) into the momentum conservation law (IV.66), use (IV.81) for $T_{ij}$, eliminate $\mathbf{v}_s$ and $\mathbf{v}_3$ in terms of $\delta\mu_4$ and $\delta n_3$, and find

$$\delta n_3 = \chi_{33}(q,\omega)\left[\frac{\delta m}{m_4}\delta\mu_4 + \left(\frac{\partial\mu_3}{\partial n_3}\right)_{s_3,n_3}\right]\delta n_4 \quad \text{(IV.89)}$$

with $\chi_{33}$ given by (IV.83). For small concentrations $\delta\mu_4$ here may be replaced by $(m_4 s^2/n_4)\delta n_4$. Then from (IV.88) and (IV.89) we find the dispersion relation

$$\omega^2 = \frac{q^2}{m_4}\left[\frac{\rho_s}{m_4}\left(\frac{\partial\mu_4}{\partial n_4}\right)_{s_3,n_3} + \chi_{33}(q,\omega)\left(\frac{\delta m s^2}{n_4} + \left(\frac{\partial\mu_3}{\partial n_4}\right)_{s_3,n_3}\right)\right.$$
$$\left.\times\left(\delta m\frac{\omega^2}{q^2} + n_4\left(\frac{\partial\mu_3}{\partial n_4}\right)_{s_3,n_3}\right)\right]. \quad \text{(IV.90)}$$

As $n_3 \to 0$, this reduces to the dispersion relation $\omega^2 = s^2q^2$.

(d) *Sound attenuation*

The amplitude attenuation $\alpha_1$ of first sound is defined by writing the complex solution of (IV.90) for real $\omega$ in the form

$$q(\omega) = \frac{\omega}{s} + i\alpha_1. \quad \text{(IV.91)}$$

Then to leading order in the $He^3$ concentration we find from (IV.90)

$$\alpha_1 = -\frac{\omega}{2m_4 s^3}\left(\delta m s^2 + \frac{\partial\mu_3}{\partial n_4}\right)^2 \operatorname{Im}\chi_{33}(q = \omega/s, \omega). \quad \text{(IV.92)}$$

We may replace $\partial\mu_3/\partial n_4$ to lowest order by $\partial\varepsilon_0/\partial n_4$, and expand $\operatorname{Im}\chi$ to first order in $P_f$, with the result

$$\alpha_1 = \frac{2}{3}\frac{P_f(n_3,T)}{n_4 m s^2}\frac{m_4}{ms}(1 + \alpha + \delta m/m_4)^2\frac{\omega^2\tau}{1+\omega^2\tau^2}. \quad \text{(IV.93)}$$

The scattering time $\tau$ is that for viscosity; it is expected to have the form at low temperatures

$$\tau = \frac{A}{T^2}\left[1 + B\left(\frac{T}{T_f}\right)^2\right], \quad \text{(IV.94)}$$

where $T_f$ is the He$^3$ Fermi temperature; $A \simeq 20 \times 10^{-12}$ sec K$^2$ and $B \simeq 2$. Comparison of the theoretical result (IV.93) with experiment is given in Ref. 43. The extension of the theory of sound attenuation in dilute solutions to include attenuation by thermal phonons, whose mean free path is limited by He$^3$ scattering, is given in Ref. 44.

### (e) *Temperature dependence of the sound velocity*

From the dispersion relation (IV.90), together with Re $\chi_{33}$ evaluated from (IV.81) to first order in $P_f$, we find that the sound velocity has the form

$$s(n_3, n_4, T) = s(n_4) + \delta s(n_3, n_4, T), \tag{IV.95}$$

where $s(n_4)$ denotes the sound velocity in pure He$^4$ at density $n_4$ and

$$\delta s = \frac{1}{2m_4 s}\left\{\delta\left[\frac{\rho_s}{m_4}\left(\frac{\partial \mu_4}{\partial n_4}\right)_{s_3, n_3}\right] + \frac{n_3 n_4}{ms^2}\left(\frac{\delta m s^2}{n_4} + \left(\frac{\partial \mu_3}{\partial n_4}\right)_{s_3, n_3}\right)^2 \times \left[1 + \frac{P_f}{n_3 m s^2}\left(3 - \frac{4/3}{1 + \omega^2\tau^2}\right)\right]\right\} \tag{IV.96}$$

is the small shift in the sound velocity. The first term, the variation of $(\rho_s/m_4)$ $(\partial\mu_4/\partial n_4)$ from its value in pure He$^4$ is evaluated by writing

$$\left(\frac{\partial \mu_4}{\partial n_4}\right)_{s_3, n_3} = \frac{\partial^2}{\partial n_4^2}\left[E_0(n_4) + \sum_{\mathbf{p}} \varepsilon_{\mathbf{p}} n_{\mathbf{p}}\right] = \frac{m_4 s^2}{n_4} + \sum_{\mathbf{p}} \frac{\partial^3 \varepsilon_{\mathbf{p}}}{\partial n_4^2} n_{\mathbf{p}}$$
$$= \frac{m_4 s^2}{n_4} + \frac{\partial^2 \varepsilon_0}{\partial n_4^2} n_3 + \frac{3}{2} P_f m \frac{\partial^2 m^{-1}}{\partial n_4^2}, \tag{IV.97}$$

and using $\rho_s = m_4 n_4 - \delta m n_3$. To compare (IV.96) with experiments carried out at fixed pressure, it is necessary to include the modification of the He$^4$ density in $s(n_4)$ due to the presence of the He$^3$. This can be done by using (III.35) and (III.38) to write

$$s(n_4) = s_0 + \frac{\partial s}{\partial n_4} \delta n_4$$
$$= s_0\left[1 - u(1 + \alpha_0)x + \frac{P_f}{m_4 s^2 n_4}\left(1 - \tfrac{3}{2} w\right)\right], \tag{IV.98}$$

where $\alpha_0$ and $s_0$ refer to zero He$^3$ concentration values in pure He$^4$ at the given pressure and $u = (n_4/s)\partial s/\partial n_4$. The temperature dependence of $P_f$ in (IV.98) includes effects of thermal expansion on $s$. The net result for the shift in the sound velocity, at fixed pressure, is to first order in $x$ and $P_f$:

$$\frac{\delta s(x, T, \omega)}{s_0} = \frac{x}{2}\left[n_4 \frac{\partial \alpha_0}{\partial n_4} + \lambda(\lambda - 1)\frac{m}{m_4}\right]$$

$$+ \frac{P_f}{2m_4 s^2 n_4}\left[3(w - \lambda)^2 + w\lambda + (3w - 2)u - \frac{3}{2}\frac{n_4^2}{m}\frac{\partial^2 m}{\partial n_4^2} \qquad \frac{4\lambda^2/3}{1 + \omega^2\tau^2}\right]; \tag{IV.99}$$

in deriving (IV.99) we have used (IV.86), and written

$$\lambda = 1 + \alpha_0 + \delta m/m_4.$$

Note that $s$ decreases with temperature and increases with frequency. The temperature-dependent part of this result is compared with experiments in Ref. 45.

The first term $\propto x$ in (IV.99) is that already found in (IV.52) from the point of view of quantum-mechanical perturbation theory. In fact the complete expressions (IV.93) for the attenuation and (IV.99) for the temperature dependence of $s$ can be equivalently derived in this manner.[43, 45] Measurements of the low temperature shift in $s$ with $He^3$ concentration provide a determination of $n_4 \partial\alpha/\partial n_4$ at zero concentration; Abraham *et al.*[41] find

$$n_4 \partial\alpha/\partial n_4 = -1{\cdot}03. \tag{IV.100}$$

## 5. SECOND SOUND

Second sound in dilute solutions is a hydrodynamic mode in which there is a counter-oscillation of the normal and superfluid mass densities. The normal mass density is the sum of that due to $He^4$ phonons [Eqn (IV.89)] and rotons, plus that due to the $He^3$, $\rho_{n3}$ [Eqn (IV.20)]. Unlike in pure $He^4$, the normal mass density remains finite at $T = 0$.

The detailed calculation of first-sound propagation given above illustrates the general methods by which one can derive the hydrodynamic properties of dilute solutions. In the case of first sound one is interested in situations where the frequency of the sound varies from values much less than the collision frequency (the hydrodynamic regime) to values much greater than the collision frequency (the collisionless regime). On the other hand second sound exists only in the hydrodynamic regime, and its properties, which may be determined quite simply from the hydrodynamic equations, involve only thermodynamic derivatives, which may be found without having to solve the transport equations for the excitations. The basic theory of second sound is derived from the hydrodynamic equations by Khalatnikov.[(4)] Brubaker *et al.*[(21)] have done extensive measurements of the second-sound velocity, as a function of He$^3$ concentration, temperature and pressure, and from these measurements have deduced the He$^3$ inertial mass $m_n^* = \rho_{n3}/n_3$.

## 6. TRANSPORT PROPERTIES

### (a) *Phonon kinetic equation*

The presence of He$^3$ impurities in superfluid He$^4$ changes the transport properties of the superfluid in two ways. First, the He$^3$ quasiparticles, carrying energy, momentum, and spin, provide a new mechanism for transport. At extremely low temperature, where the thermal excitations of the He$^4$ are negligible, the contributions of the He$^3$ dominate the transport coefficients. Secondly, transport by phonons and rotons is limited by their interactions with the He$^3$, either by scattering or by absorption or emission by the He$^3$. In the temperature range up to $\approx$0·6 K for a few percent He$^3$ concentration, where rotons can be neglected, phonon mean free paths are determined entirely by the He$^3$–phonon interaction (except at very low temperatures where boundary scattering becomes important).

The phonon transport properties are determined by the phonon kinetic equation (IV.54). The phonon–He$^3$ collision term is a sum of a scattering term, $I_{\text{scatt}}$, which can be determined from the phonon–Ge$^3$ scattering theory developed in Sec. IV.3, and an absorption term, $I_{\text{abs}}$, which can be determined from the theory of sound attenuation in Sec. IV.4(d). Because He$^3$–He$^3$ scattering rates are so relatively rapid, these collision terms may be evaluated under the assumption that the He$^3$ quasiparticles maintain local equilibrium among themselves. Then in the frame in which $\mathbf{v}_s = \mathbf{0}$, denoted by a bar,

$$\bar{I}_{\text{scatt}} = -2 \sum_{\mathbf{pp'q}} 2\pi\delta(\bar{\varepsilon}_{\mathbf{p}} + \bar{\omega}_{\mathbf{q}} - \bar{\varepsilon}_{\mathbf{p}'} - \bar{\omega}_{\mathbf{q}'})|\langle \mathbf{p'q'}|T_{qp-ph}|\mathbf{pq}\rangle|^2 \times [\bar{f}_{\mathbf{q}}(1 + \bar{f}_{\mathbf{q}'})\bar{n}_{\mathbf{p}}(1 - \bar{n}_{\mathbf{p}'}) - (1 + \bar{f}_{\mathbf{q}})\bar{f}_{\mathbf{q}'}(1 - \bar{n}_{\mathbf{p}})\bar{n}_{\mathbf{p}'}], \quad \text{(IV.101)}$$

where $\langle|T_{\text{qp-ph}}|\rangle$ is the He$^3$-phonon scattering amplitude; the factor of 2 is the spin sum.

To compute phonon transport properties as well as sound attenuation one is interested only in small deviations from global equilibrium with $\mathbf{v}_s = \mathbf{0}$, Linearizing (IV.101), and using the fact that to lowest order He$^3$–phonon scattering is elastic ($q = q'$, $p = p'$), we have

$$\bar{I}_{\text{scatt}} = -2 \sum_{\mathbf{pp'q}} 2\pi\delta(\varepsilon_{\mathbf{p}} + \omega_{\mathbf{q}} - \varepsilon_{\mathbf{p}'} - \omega_{\mathbf{q}'})|\langle \mathbf{p'q'}|T_{\text{qp-ph}}|\mathbf{pq}\rangle|^2 \times \{n_{\mathbf{p}}^0(1 - n_{\mathbf{p}}^0)[\delta\bar{f}_{\mathbf{q}} - \delta\bar{f}_{\mathbf{q}'}] + f_{\mathbf{q}}^0(1 + f_{\mathbf{q}}^0)[\delta\bar{n}_{\mathbf{p}} - \delta\bar{n}_{\mathbf{p}'}]\}. \quad \text{(IV.102)}$$

The transformation of (IV.102) back to the (unbarred) lab frame is trivial. Since the phonon momentum $\mathbf{q}$ is independent of the frame in which it is measured we have $\bar{f}_{\mathbf{q}} = f_{\mathbf{q}}$; furthermore the rate of change of this distribution function is the same in both frames so that

$$I_{\text{scatt}} = \bar{I}_{\text{scatt}}. \tag{IV.103}$$

In addition $\bar{n}_{\mathbf{p}} = n_{\mathbf{p}+m_3\mathbf{v}_3}$ [cf. (IV.7)] so that in linear order for the $He^3$ in local equilibrium

$$\delta\bar{n}_{\mathbf{p}} = \mathbf{p}\cdot(\mathbf{v}_s - \mathbf{v}_3)\frac{\partial n^0_{\mathbf{p}}}{\partial \varepsilon_{\mathbf{p}}}, \tag{IV.104}$$

where $\mathbf{v}_3$ is the $He^3$ flow velocity. Thus

$$\bar{I}_{\text{scatt}} = -2\pi \sum_{\mathbf{p}\mathbf{p}'\mathbf{q}'} 2\pi\delta(\varepsilon_{\mathbf{p}} + \omega_{\mathbf{q}} - \varepsilon_{\mathbf{p}'} - \omega_{\mathbf{q}'})|\langle\mathbf{p}'\mathbf{q}'|T_{\text{qp–ph}}|\mathbf{p}\mathbf{q}\rangle|^2$$
$$\times\, n^0_{\mathbf{p}}(1 - n^0_{\mathbf{p}})\left[\delta f_{\mathbf{q}} - \delta f_{\mathbf{q}'} - \frac{\partial f^0_{\mathbf{q}}}{\partial \omega_{\mathbf{q}}}(\mathbf{q} - \mathbf{q}')\cdot(\mathbf{v}_s - \mathbf{v}_3)\right]. \tag{IV.105}$$

To lowest order (IV.46) for $T_{\text{qp–ph}}$ implies that the scattering contains only *s*, *p* and *d* waves. The detailed further reduction of the collision term (IV.105) into spherical harmonic components is given in Ref. 46.

The rate at which phonons are absorbed by $He^3$ is given in terms of the amplitude attenuation $\alpha_1$ by

$$\frac{1}{\tau_{\text{abs}}(q)} = 2s\alpha_1. \tag{IV.106}$$

To evaluate $\alpha_1$ for thermal phonons we use (IV.93) with the $He^3$–$He^3$ scattering rate $\tau^{-1}$ replaced by the more general quantum result $\tau^{-1}[1 + (\hbar\omega/2\pi kT)^2]$; thus

$$\frac{1}{\tau_{\text{abs}}(q)} = \frac{4}{3}\frac{m_f P_f(T)}{m^2 n_4}\frac{(1 + \alpha + \delta m/m_4)^2 q^2 \tau[1 + (\hbar sq/2\pi kT)^2]}{[1 + (\hbar sq/2\pi kT)^2]^2 + (sq\tau)^2}. \tag{IV.107}$$

When the $He^3$ are in local equilibrium with temperature $T_3(\mathbf{r}, t)$ and velocity $\mathbf{v}_3(\mathbf{r}, t)$ the absorption process causes the phonon distribution to relax to local temperature $T_3$ and velocity $\mathbf{v}_3$. The net rate of absorption $I_{\text{abs}}(q)$, the difference of the absorption and emission rates, then becomes

$$I_{\text{abs}} = -\frac{1}{\tau_{\text{abs}}}\delta f_{\mathbf{q}} - \frac{\partial f^0_{\mathbf{q}}}{\partial \omega_{\mathbf{q}}}[qT\delta(s/T_3) + \mathbf{q}\cdot(\mathbf{v}_s - \mathbf{v}_3)], \tag{IV.108}$$

where $\delta s$ and $\delta T_3$ are the first-order deviations from global equilibrium. The phonon kinetic equation with (IV.105) and (IV.108) is the starting point for calculating the phonon contributions to transport in dilute solutions. The application to sound attenuation is described in Ref. 47, to viscosity in Ref. 46, and to thermal conductivity in Ref. 43. The main results for the viscosity and thermal conductivity are the following.

### (b) *First viscosity*

The first viscosity is the sum of a $He^3$ contribution $\eta_3$ and a phonon and roton contribution $\eta_4$. The latter contribution is very unlike that in pure $He^4$ because of scattering of $He^4$ excitations by the $He^3$ quasiparticles.

For temperatures below $\approx 0{\cdot}1$ K for $x \gtrsim 1\%$, $\eta_4$ is completely negligible (except for boundary scattering effects) and $\eta \simeq \eta_3$. For $T \ll T_f$, the $He^3$ Fermi temperature, $\eta_3 \propto T^{-2}$, and is given exactly in terms of the $He^3$–$He^3$ scattering amplitude on the Fermi surface by (III.59). At higher temperatures one can compute $\eta_3$ in the relaxation-time approximation [cf. (IV.81)] as

$$\eta_3 = P_f(T)\tau. \tag{IV.109}$$

The phonon viscosity $\eta_{\text{ph}}$ which begins to become important above $\approx 0{\cdot}1$ K may be computed by solving the phonon kinetic equation, and one finds

$$\eta_{\text{ph}} = -\sum_{\mathbf{q}} \frac{s^2q^2}{15} \frac{\partial f^0_{\mathbf{q}}}{\partial \omega_{\mathbf{q}}} \tau_v(q), \tag{IV.110}$$

where

$$\tau_v^{-1}(q) = \tau_{\text{abs}}^{-1}(q) + \tau_2^{-1}(q) \tag{IV.111}$$

and $\tau_2^{-1}$ is the phonon-$He^3$ scattering rate for $\delta f_{\mathbf{q}} \propto P_2(\cos\theta_{\mathbf{q}})$, computed from (IV.105). Note that $\eta_{\text{ph}}$ diverges if the absorption term is neglected.

### (c) *Thermal conductivity*

The structure of the thermal conductivity, a sum of the $He^3$ plus phonon and roton thermal conductivities, is quite similar to that of the first viscosity. For $T \lesssim 0{\cdot}1$ K, $He^3$ thermal conductivity ($\propto T^{-1}$ for $T \ll T_f$) is dominant, except at very low $T$ where boundary scattering is important. The phonon thermal conductivity is computed analogously to the phonon viscosity; one finds

$$K_{\text{ph}} = -\sum_{\mathbf{q}} \frac{s^2q}{3} \frac{\partial f^0_{\mathbf{q}}}{\partial T} \tau_K(q) \tag{IV.112}$$

where

$$\tau_K^{-1}(q) = \tau_{\text{abs}}^{-1}(q) + \tau_1^{-1}(q), \tag{IV.113}$$

and $\tau_1^{-1}$ is the phonon–$He^3$ scattering rate for $\delta f_{\mathbf{q}} \propto P_1(\cos\theta_{\mathbf{q}})$.

The theory is in qualitative agreement with thermal conductivity experiments;[48, 49] however, it appears that (IV.46), with parameters taken from experiment, underestimates the scattering amplitude by possibly as much as 40%. The reason for this discrepancy is not understood at present.

## REFERENCES


1. D. O. Edwards, D. F. Brewer, P. Seligmann, M. Skertic and M. Yaqub. *Phys. Rev. Letts* **15**, 773 (1965).
2. C. Ebner and D. O. Edwards. *Phys. Reports* **2C**, 77 (1970).
3. L. D. Landau and I. Pomeranchuk. *Dokl. Akad. Nauk. SSSR* **59**, 669 (1948); I. Pomeranchuk. *Zh. Eksper. i. Teor. Fiz.* **19**, 42 (1949).
4. I. M. Khalatnikov. "An Introduction to the Theory of Superfluidity", Benjamin, New York (1965).
5. J. Wilks. "The Properties of Liquid and Solid Helium", Oxford (1967).
6. W. E. Keller. "Helium-3 and Helium-4," Plenum Press, New York (1969).
7. K. W. Taconis and R. de Bruyn Ouboter. *In* "Progress in Low Temperature Physics" Vol. IV (C. J. Gorter, Ed.), North-Holland Publishing, Amsterdam (1964), p. 38.
8. J. C. Wheatley. *Am. J. Phys.* **36**, 181 (1968); *in* "Progress in Low Temperature Physics", Vol. VI, (C. J. Gorter, Ed.), North-Holland Publishing, Amsterdam, (1964), p. 77.
9. D. O. Edwards. *In* "Proceedings of the Eleventh International Conference on Low Temperature Physics", St. Andrews, Scotland, 1968 (J. F. Allen, D. M. Finlayson and D. M. McCall, Eds), St. Andrews University (1968), p. 352.
10. R. Radebaugh. National Bureau Standards Report No. 362 (U.S. Government Printing Office, Washington, 1967).
11. P. Seligmann, D. O. Edwards, R. E. Sarwinski and J. T. Tough. *Phys. Rev.* **181**, 415 (1969).
12. A. C. Anderson, W. R. Roach, R. E. Sarwinski and J. C. Wheatley. *Phys. Rev. Letts* **16**, 263 (1966).
13. W. L. McMillan. *Phys. Rev.* **138**, A442 (1965).
14. R. P. Feynman. *Phys. Rev.* **94**, 362 (1954).
15. W. E. Massey and C. W. Woo. *Phys. Rev. Letts* **19**, 301 (1967).
16. G. E. Watson, J. D. Reppy and R. C. Richardson. *Phys. Rev.* **188**, 384 (1969). B. M. Abraham, O. G. Brandt and Y. Eckstein. *In* "Proceedings of the Twelfth International Conference on Low Temperature Physics", Kyoto, Japan, 1970, (E. Kanda, Ed.), Keigaku Publishing Co., Ltd. (1971). C. Boghosian and H. Meyer. *Phys. Letts* **25A**, 352 (1967).
17. T. B. Davison and E. Feenberg. *Phys. Rev.* **178**, 306 (1969).
18. L. D. Landau and E. M. Lifshitz. "Fluid Mechanics", Addison-Wesley, Reading, Mass. (1959).
19. C. W. Woo, H. T. Tan and W. E. Massey. *Phys. Rev. Letts* **22**, 278 (1969); *Phys. Rev.* **185**, 287 (1969).
20. V. R. Pandharipande and N. Itoh. *Phys. Rev.* **A8**, 2564 (1973).
21. N. R. Brubaker, D. O. Edwards, R. E. Sarwinski, P. Seligmann and R. A. Sherlock. *Phys. Rev. Letts* **25**, 715 (1970); *J. Low Temp. Phys.* **3**, 619 (1970).
22. A. J. Leggett. *Phys. Rev. Letts* **31**, 352 (1973).

23. V. J. Emery. *Phys. Rev.* **148**, 138 (1966); **161**, 194 (1967).
24. J. Bardeen, G. Baym and D. Pines. *Phys. Rev. Letts* **17**, 372 (1966); *Phys. Rev.* **156**, 207 (1967).
25. H. Højgaard Jensen, H. Smith and J. W. Wilkins. *Phys. Letts* **27A**, 532 (1968).
26. G. A. Brooker and J. Sykes. *Phys. Rev. Letts* **21**, 279 (1968).
27. J. Sykes and G. A. Brooker. *Ann. Phys.* (*N.Y.*), **56**, 1 (1970).
28. W. R. Abel, R. T. Johnson, J. C. Wheatley and W. Zimmermann, Jr. *Phys. Rev. Letts* **18**, 737 (1967).
29. G. Baym and C. Ebner. *Phys. Rev.* **170**, 346 (1968).
30. C. Ebner. *Phys. Rev.* **185**, 392 (1969); see also K. A. Kuenhold and C. Ebner, *Phys. Rev.* **A9**, 2724 (1974).
31. J. Landau, J. T. Tough, N. R. Brubaker and D. O. Edwards. *Phys. Rev.* **A2**, 2472 (1970).
32. M. B. Hoffberg. *Phys. Rev.* **A5**, 1963 (1972).
33. W. E. Massey. *Phys. Rev.* **151**, 153 (1966).
34. R. Smith and V. R. Pandharipande. Private communication.
35. G. Baym. *J. Low Temp. Phys.* **18**, 335 (1975).
36. J. Landau, J.T. Tough, N. R. Brubaker and D. O. Edwards. *Phys. Rev.* **A2**, 2472 (1970).
37. W. F. Saam. *Ann. Phys.* (*N.Y.*), **53**, 239 (1969).
38. L. J. Campbell. *Phys. Rev. Letts* **19**, 156 (1967).
39. W. L. McMillan. *Phys. Rev.* **175**, 266 (1968); **182**, 299 (1969); also, C. S. Hsu, *Phys. Rev.* **A8**, 376 (1973).
40. B. M. Abraham, Y. Eckstein, J. B. Ketterson, M. Kuchnir and P. R. Roach. *Phys. Rev.* **A1**, 250 (1970).
41. B. M. Abraham, Y. Eckstein, J. B. Ketterson, M. Kuchnir and J. Vignos. *Phys. Rev.* **181**, 347 (1969).
42. Y. Disatnik and H. Brucker. *J. Low Temp. Phys.* **7**, 501 (1972); Y. Disatnik and A. Yaniv. *J. Low Temp. Phys.* **10**, 595 (1973).
43. G. Baym and C. Ebner. *Phys. Rev.* **164**, 235 (1967).
44. B. M. Abraham, Y. Eckstein, J. B. Ketterson and J. H. Vignos. *Phys. Rev. Letts* **17**, 1254 (1966).
45. G. Baym. *In* "Proceedings of the Eleventh International Conference on Low Temperature Physics", St. Andrews, Scotland, 1968 (J. F. Allen. D. M. Finlayson and D. M. McCall, Eds.), St. Andrews University (1968), p. 385.
46. G. Baym and W. F. Saam. *Phys. Rev.* **171**, 172 (1968).
47. G. Baym, W. F. Saam and C. Ebner. *Phys. Rev.* **173**, 306 (1968).
48. W. R. Abel and J. C. Wheatley. *Phys. Rev. Letts* **21**, 1231 (1968).
49. R. L. Rosenbaum, J. Landau and Y. Eckstein. *J. Low Temp. Phys.* **16**, 131 (1974).

# 10. KAPITZA CONDUCTANCE

L. J. CHALLIS

*Department of Physics, University of Nottingham, Nottingham, England*

## I. INTRODUCTION

When heat flows across a boundary between a solid and liquid helium, a discontinuity in temperature appears between the two materials. For small temperature differences, $\delta T \ll T$, this is proportional to the heat current density $\dot{Q}/A$ and we may define a resistance $R_K$ and a conductance $h_K$ by

$$R_K^{-1} = h_K = \dot{Q}/(A\delta T).$$

The discontinuity which was first observed by Kapitza[1] can be very large, particularly at low temperatures, in comparison with the temperature drop across a thickness of, say, 1 mm of the solid or helium next to the interface so it has important technical implications as well as providing an intriguing problem in physics that has been very actively studied in recent years.

Conductance measurements have now been made for more than 20 different solids including superconducting, normal and ferromagnetic metals, dielectrics and paramagnetic salts. They include measurements to liquid $He^3$, liquid $He^4$II, liquid $He^4$I and mixtures and there are related measurements on gaseous and solid helium. The effects of surface treatments, changes in helium density and electric and magnetic fields have all been examined and measurements have been made with steady, alternating and pulsed heat currents (heat pulses) as well as with "monochromatic" phonon sources.

From this wealth of data and from theoretical considerations it appears that we can identify two main types of Kapitza conductance: phonon conductance and magnetic conductance. Phonon conductance, in which energy is transferred from phonon modes in the solid to phonon modes or other excitations in the liquid, will always be present although it becomes very small at low temperatures because of the rather rapid fall in the density of phonon modes as the frequency decreases. Magnetic conductance is the result of magnetic coupling across the interface and so is restricted to $He^3$ in contact with solids containing unpaired magnetic spins. The magnetic conductance can actually increase as the temperature falls so it is of considerable interest in the technology of very low temperature physics.

## II. Phonon Conductance

It is convenient to discuss the process of phonon conductance in terms of an average transmission coefficient $\bar{\alpha}_1$ which is the probability that the energy of a phonon in the solid incident on the interface should be transmitted into the liquid. The total phonon energy $W$ incident on the interface from the solid is given by kinetic theory as

$$W/A = \tfrac{1}{4}U_D\bar{c} \simeq \tfrac{1}{4}U_Dc_D,$$

where $U_D$ is the energy density per unit volume and $\bar{c}$ is the average velocity of the phonons present and is equal to $c_D$, the Debye velocity, to within a few percent. $A$ is the area of the interface. The energy flux transmitted is therefore

$$(W/A)_{\text{trans}} = \tfrac{1}{4}U_Dc_D\bar{\alpha}_1$$

and when the helium is at the same temperature as the solid, this is balanced by an equal flux in the opposite direction. However if there is a temperature difference $\delta T$ between them, there is a net flow

$$\dot{Q}/A = \frac{\mathrm{d}}{\mathrm{d}T}\,(\tfrac{1}{4}U_Dc_D\bar{\alpha}_1)\delta T \simeq \tfrac{1}{4}C_Dc_D\bar{\alpha}_1\delta T,$$

if we assume $\mathrm{d}\bar{\alpha}_1/\mathrm{d}T$ is small. $C_D$ is the specific heat. Hence the transmission coefficient

$$\bar{\alpha}_1 = 4h_K/C_Dc_D \tag{II.1}$$

may readily be determined from a measurement of the phonon conductance. The phonons in the heat current will usually have a distribution with a maximum at a frequency of $v \sim 80T$ GHz so unless $\bar{\alpha}_1(v)$ varies rapidly with frequency, we may take the value of $\bar{\alpha}_1$ determined experimentally from (II.1) to refer to a frequency of $80T$ GHz. We stress that $\bar{\alpha}_1$ obtained this way using steady heat flows is an average over all directions and polarizations. Information about these dependences can be obtained using pulse techniques.[2]

Acoustic theory[3] which normally works well for systems at lower frequencies ($\gtrsim$100 MHz) provides a value for $\bar{\alpha}_1$ of

$$\bar{\alpha}_1 = \left(\frac{\rho_2 c_2}{\rho_1 c_D}\right)^4 \tag{II.2}$$

where $c_2$ is the sound velocity in helium and $\rho_1$ and $\rho_2$ are the densities of the solid and helium respectively. This has a value between about $10^{-3}$ and $10^{-2}$ for most solids which is independent of frequency. However values of $\bar{\alpha}_1$ determined experimentally from (II.1) or from heat pulse measurements are invariably larger than these theoretical values and Fig. 1 which shows the largest reported values at $\sim$120 GHz (1·5K) indicates that $\bar{\alpha}_1$ can be up to $\sim$200 times larger than the theoretical value. The disagreement is less severe at the lower frequencies as can be seen for copper in Fig. 2, although the values shown are still 5–10 times greater than the acoustic values even at $\sim$5 GHz. The transmission coefficient increases with frequency at least up to $\sim$100 GHz but at and above these frequencies the dependence can apparently change. Measurements at most interfaces still show an increase with frequency ($h_K$ varying more rapidly than $T^3$) but others show approximately frequency independent transmission or even a decrease ($h_K$ varying less rapidly than $T^3$). This behaviour is shown schematically by the broken lines in Fig. 2. To our knowledge no measurements showing values of $\bar{\alpha}_1$ decreasing with frequency at higher temperatures have been extended down to low enough temperatures to show a maximum such as that in curve C† but a maximum is of course a necessary feature if the curve is to return to the acoustic or Khalatnikov limit at very low frequencies. The existence of a maximum could possibly indicate that the enhanced transmission has resonant features and that curve *A* refers to surfaces where the resonant frequencies lie above 150 GHz.

It seems then at frequencies above perhaps $\sim$1 GHz energy is either going into the phonon modes of liquid helium with a much higher probability than that predicted by acoustic theory or it is going into other modes. The similarity in the behaviour of $He^4$ and $He^3$ seen in Fig. 2 suggests that the normal modes such as rotons,‡ quasiparticles, vortices etc. which differ in the two liquids cannot be very important but one cannot perhaps rule out the possibility of exciting short lived modes as an intermediate state.

Further insight into the enhanced transmission can be obtained by seeing how it depends on the elastic properties of the solid. Figure 1 showed that there was little or no correlation between $\bar{\alpha}_1$ and the acoustic impedance

† A maximum at $\sim$8K observed with heat pulses[9] was attributed to the formation of gas at the interface.

‡ Transmission into roton modes should vary as $\exp(-\Delta kT)$—a detailed theory[10] suggests $T^{-3/2}\exp(-\Delta/kT)$—and the lack of evidence for a rapidly varying term of this sort provides further support for the view that rotons are of little importance in the transmission at least for $\nu \gtrsim 160$ GHz($T < 2K$). There is also no evidence in pulse work of any features in $\bar{\alpha}_1$ at $h\nu = \Delta$.[11]

$Z_1 = \rho_1 c_D$ which is again at variance with acoustic theory[2] but it has been shown that a degree of correlation is obtained when $\bar{\alpha}_1$ is plotted against $c_D$ or $\rho_1$ separately[4] and the form of this suggested that $\bar{\alpha}_1 \tilde{\propto} c_D/\rho_1$ which is

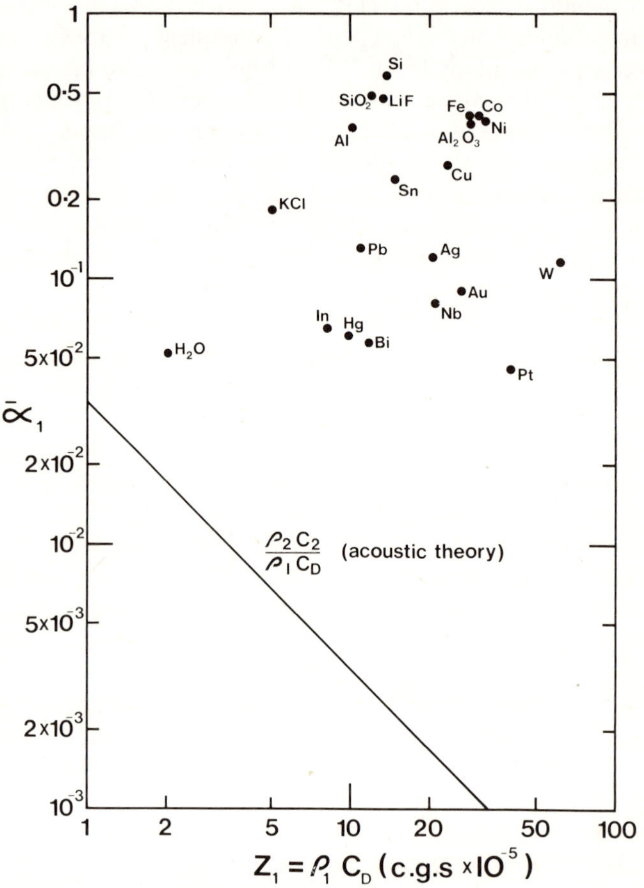


FIG. 1. The largest reported values of the mean transmission coefficient $\bar{\alpha}_1$ to $He^4$ at 120 GHz (1·5K) are shown plotted against the acoustic impedance of the solid ($\rho_1 c_D$). The line shows the predictions of acoustic theory. The plot is taken from reference 4 and the sources of data are given there though points for $H_2O$ (ice) and Bi have been added.[5] Recent data for Ga[6] were omitted by oversight.

plotted in Fig. 3. There again is clearly some fair degree of correlation. Dimensional arguments might suggest $\bar{\alpha}_1 \simeq (c_D\rho_2/c_2\rho_1)$ (although of course $\bar{\alpha}_1$ must always be $\gtrsim 1$) so this is also plotted and is seen to have the right order of magnitude. However $(\rho_2/c_2)$ is decreased by 23% by applying a pressure of 25 atm to liquid $He^4$ while experimentally a very small increase is observed.[12, 13, 7] One further feature of the experimental data is that quite wide variations in $\bar{\alpha}_1$ occur for different surfaces of the same material[14] although the extrinsic property causing this variation has not yet been identified

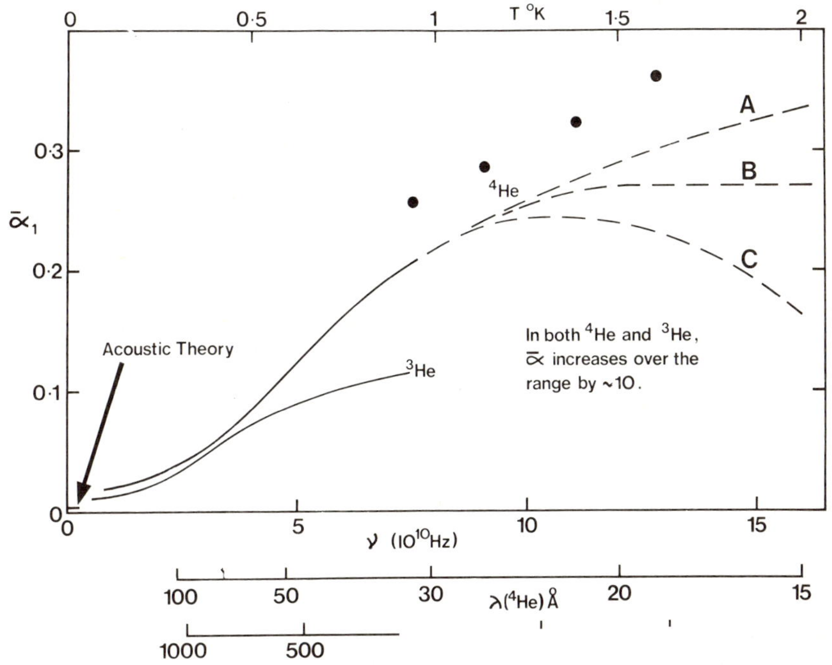


FIG. 2. The approximate frequency dependence of $\bar{\alpha}_1$ from copper to liquid $He^4$ and $He^3$. $\bar{\alpha}_1$ was obtained from the conductance measurements of Anderson *et al.*[7] The broken lines A, B and C represent possible extrapolations. The points are values of $\bar{\alpha}_1$ from Sherlock *et al*[8] for energy emitted *outside* the critical cone. The figure includes axes showing the phonon wavelengths in the copper and helium.

positively. A problem in doing this is that it is difficult to vary independently such properties as roughness, dislocation density, surface impurity level etc. which might be of importance. Nevertheless measurements at low temperatures[15] do seem to suggest that $\bar{\alpha}_1$ is increased by increasing the dislocation density although similar measurements at higher temperatures[16] show little such dependence.

Acoustic theory assumes translational symmetry at the interface. This implies conservation of wave vector parallel to the surface and this, since frequency is also conserved, leads directly to Snell's law ($c_1 \sin \theta_1 = c_2 \sin \theta_2$, where $\theta_1$ and $\theta_2$ are the angles of incidence and refraction). Now the velocity of sound in helium is only $\sim 1/10$ of that in a solid so that energy going into phonon modes in the helium must all be transmitted within a narrow cone of half angle $\sin^{-1}(c_1/c_D) \sim 5°$ which corresponds to only a small fraction, $(c_1/c_D)^2 \sim 1\%$ of the total solid angle of the half space. The restriction this provides on the heat flow becomes very apparent when we consider that this means that $\sim 99\%$ of the phonons in the helium which strike the interface

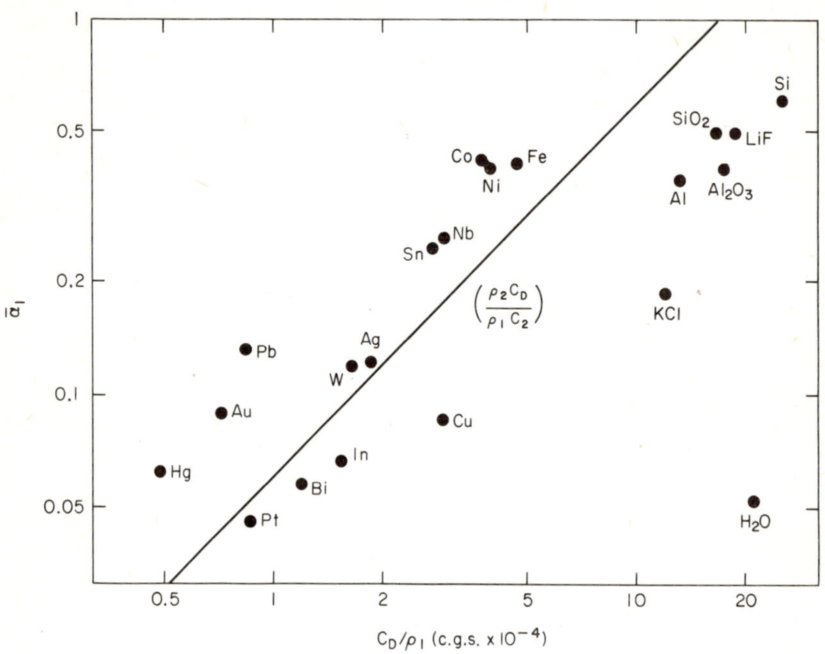


FIG. 3. The values of $\bar{\alpha}_1$ (Fig. 1) are shown plotted against $C_D/\rho_1$. The plot is from reference 4 but points have been added for $H_2O$ and Bi. The line shows the relationship $\bar{\alpha}_1 = \rho_2 c_D/\rho_1 c_2$ (see text).

have zero probability of transmission. This very important feature of the theory has recently been tested by Sherlock *et al*[(8)] using heat pulses. The energy emitted at a particular angle of "refraction" was measured by a bolometer in the helium and the bolometer signal was plotted as a function of angle. The surface was a cleaved face of NaF and the helium was sufficiently cold that the angular distribution measured was not perturbed by phonon scattering in the helium. The results are shown in Fig. 4. There is a pronounced peak in $\bar{\alpha}_1$ at near to normal emission whose width is in good agreement with the width of the critical cone. $\bar{\alpha}_1$ is much smaller outside the cone, but when the energy emitted is integrated over the whole of the solid angle the flux outside the cone turns out to be very much greater than that inside the cone. Absolute values of $\bar{\alpha}_1$ obtained by assuming the peak values to be those given

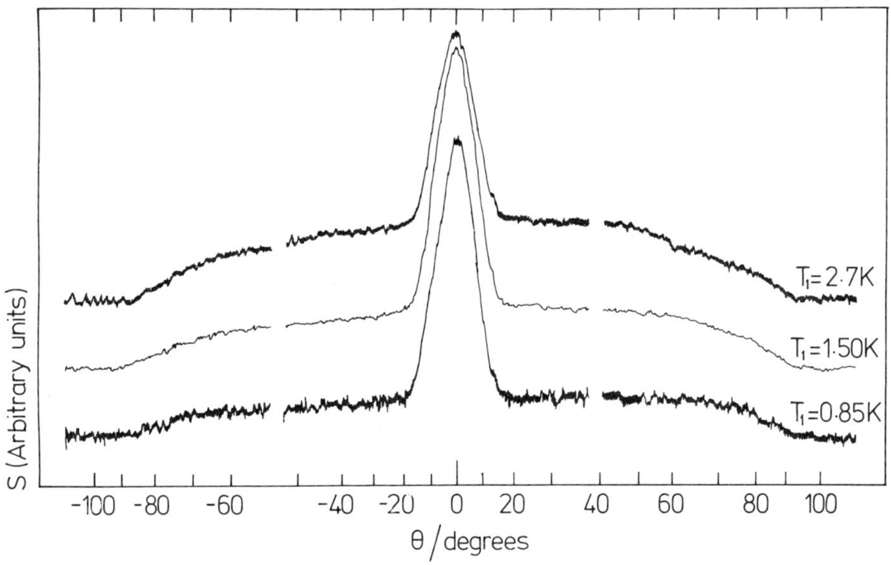


FIG. 4. The angular distribution of the intensity of phonons emitted from a NaF crystal[8] at different temperatures of the solid. The temperature of the helium was ~0·1K. The peak width is consistent with the critical cone calculated assuming $k_{\parallel}$ conservation. Since the solid angle inside the cone is less than 1% of the total, most of the energy is emitted outside the cone.

by acoustic theory are shown as points in Fig. 2 and are seen to be quite consistent with the conductance data. It seems clear then that at high frequencies at least, the enhanced transmission is very largely due to energy emitted outside the critical cone. This is an important advance and we need next to know whether this energy is emitted directly into phonon modes as a result of a breakdown of the $k_{\parallel}$ conservation rule or into other modes which must then presumably decay rapidly into phonon modes since the energy is known to travel to the bolometer at the speed of sound.

Some insight into this problem comes from studies of energy transfer across an interface between a solid and gaseous helium at low temperatures.†

† Many other examples of contact conductance have now been studied but in general the data are not very different from the predictions of acoustic theory.

It has been found that the transmission coefficient to both $He^3$ and $He^4$ gases is similar in both magnitude and frequency dependence to that to the liquids.[17] The three most recent experiments measured the reflection coefficient of phonon pulses at the solid side of the interface and can be interpreted as showing that as the pressure of gas is increased from zero, $\bar{\alpha}_1$ increases rapidly to very near the value for bulk liquid when the surface is covered with only a few atomic layers. Kinder and Dietsche have indicated difficulties with this interpretation in their experiments and suggested that the loss of energy from the phonon modes being detected might be due to conversion into other modes *in the solid*. However Guo and Maris, and Long, Sherlock and Wyatt interpret their data as an energy loss to the helium and this would certainly be suggested by the steady heat current experiment of Mate and Sawyer, and Anderson and Johnson, in which $h_K$ and hence $\bar{\alpha}_1$ was determined from the energy passing through the system. In all three pulse experiments, the loss rate was found to be quite different for the various modes and the nature of the differences led Long *et al.* to suggest that surface modes were being excited as intermediate states.

This very interesting result that $\bar{\alpha}_1$ is virtually the same for the gas above a certain pressure as for the liquid at S.V.P. (and so as we have noted as for the liquid under 25 at) can perhaps be interpreted in two main ways. It might, as Long *et al.* have suggested, indicate that in both cases the energy goes largely into single particle excitations and theoretical treatments of such processes will be discussed later. However it should be pointed out that it might also indicate that an intermediate state in the surface of the solid is very strongly coupled to the helium in both cases, though not necessarily by the same process, and that the transmission process is limited by the coupling between this state and the phonons in the solid. This of course is the same whether the state of the helium is gaseous or liquid.

We conclude this discussion of the experimental work by noting that we have not so far distinguished between metals and dielectrics. There is justification for this in that plots such as Fig. 1 and experiments in which a superconductor is driven normal do suggest that the presence of conduction electrons does not make a major difference to the conductance.[18] However the recent experiments by Wagner, Kollarits and Yaqub[6] on the effect of fields on single crystal specimens of gallium may focus further attention on this particular aspect of the problem. We mention too, but have not time to discuss, the recent work on the coating of the surface with "impurity" atoms[19, 17, 16] and also experiments[20] which show that $\dot{Q}$ can depend non-linearly on $\delta T$ and which are interpreted as evidence that turbulence can affect $h_K$.

The theoretical attack on the Kapitza problem has also been very active since the first application of acoustic theory in 1952 and particularly in the last three or four years. To discuss this work we look first at the assumptions implicit in acoustic theory. The two materials are assumed to be continuous, non-dissipative elastic media with properties which change discontinuously at the interface. In fact liquid helium is a particularly open compressible system so that the Van der Waals forces between solid and helium atoms will change considerably the properties of the helium within $\sim 15$Å of the surface, the phonons will have finite lifetimes in both media, there may be impurities at the surface and so on.

The first attempt to modify acoustic theory was concerned with the attractive potential and with the resulting compressed layer of helium near the interface.[12] This provides a matching section between the acoustic transmission lines of the solid and the helium and the resulting effect on the transmission is to replace the acoustic impedance $\rho_2 c_2$ of the helium by the input impedance $\mathrm{Re}[Z(v)]$ to the section as seen by solid. It is found that the enhancement this brings is only significant when the phonon wavelength $\lambda$ in the helium ($\sim 38/T$Å) is $\gtrsim l$, the length of the section ($\sim 15$Å). Thus when $kl = 0{\cdot}3$ (50 GHz, T = 0·6K) there is an enhancement of only 20% but this rises steadily to a factor of 4 at 150 GHz ($T = 1{\cdot}8$K). Recent calculations[21] broadly confirm this magnitude of the effect although in this work the transmission coefficient is found to reach a maximum at $T \sim 1$K and then fall. This later work also shows that a further enhancement is obtained if transverse wave propagation is allowed in the nearest layer to the interface which is possible if this is solid. So this enhancement of the phonon transmission is likely to be important at the higher frequencies generated at $T \gtrsim 0{\cdot}5$K although it is not enough even here to explain the whole of the enhancement observed experimentally nor does it account for the transmission observed outside of the critical cone. It also cannot of course explain the enhancement observed at frequencies of only a few GHz or so.

This approach achieved better matching by allowing for an increase in the acoustic impedance of the helium. The converse approach in which possible mechanisms are explored which might reduce the effective impedance of the solid has also been examined. Vuorio[22] has pointed out that the impedance is lowered at frequencies at which the solid surface vibrates resonantly. He noted that impurities loosely bound to the surface through a thin layer of helium could have resonant frequencies of $\sim 100$ GHz and showed that the transmission coefficient rose by a factor of $\sim(\rho_1/\rho_2)$ at resonance. This can be very substantial but as we have noted, the largest enhancements observed experimentally ($\sim 200$) are for solids of low density but high velocity of sound and this is contrary to what is expected on this theory.

Some insight into the rôle that might be played by single particle excitations comes from considering a gas model. The eigenstate of the gas atoms are determined by the attractive potential and, when this is modulated by a phonon in the solid, transitions can occur between the eigenstates of the helium atoms with the absorption or emission of a phonon in the solid. This type of approach was first tried by Challis and Toombs in 1970,[23] it is in fact essentially the accommodation problem studied in the thirties.[24] There are two main types of transition: those between two unbound states and those between an unbound and a bound state. (Transitions between bound states do not of course transfer energy into mobile excitations in the helium.) The calculations were carried out using Maxwell-Boltzmann statistics. The first type of transition leads essentially to the same sort of results as acoustic theory applied to a compressible continuum. Thus at low frequencies the Khalatnikov limit is obtained, with $c_2$ being now the velocity of sound in the gas, and at higher frequencies, a modest enhancement occurs. It is possible to imagine then that a phonon has been transmitted into the gas although in reality this will be very short lived at these frequencies. The second type of transition is the process of desorption which was first proposed by Johnson and Little.[25] (It may also be possible to imagine this as the emission of a "phonon" of reduced frequency). However the difficulty in calculating the conductance due to this process is that at low temperatures, gas atoms condense into the ground state i.e. into the first adsorbed layer so that although a very large conductance is obtained, the density in the adsorbed layer becomes unrealistically large. This is not too serious a problem if Fermi–Dirac statistics are used since the occupancy of the bound state is then limited and it was found by Toombs, Sheard and Rice[26] that transmission due to the desorption process was always less than that due to transitions between unbound states.[27]

To do the calculation from first principles for all statistics then we would need to include repulsion between the helium atoms but this complication is sensibly sidestepped by Long[28] who *starts* with the assumption that the adsorbed layer has a maximum density equal to that observed experimentally. Hence the probability of an atom being adsorbed is proportional to the number of vacant sites in the layer and he shows that this approach predicts a pressure dependence of $\bar{\alpha}_1$ at low pressures similar in form to that observed. The calculation of $\bar{\alpha}_1$ is then straightforward and he obtains values for longitudinal waves at normal incidence of $\sim 2\%$† which are similar to those observed by Long *et al.*[17]

However at other angles and for other modes, values of $\bar{\alpha}_1$ of up to $\sim 45\%$ were observed and although Long notes that if surface waves are generated through scattering from these bulk modes, $\bar{\alpha}_1$ could be increased by $\sim 3$ (see later) this is still far too little to account for the much larger values observed. We note too that it is difficult to see how the parametric dependence sug-

† We note that in an earlier discussion of the desorption process by Anderson and Johnson,[17] $\bar{\alpha}_1$ was taken to be 1 for phonons of energy greater than the binding energy.

gested by Fig. 3 might arise in the present form of the gas model.

The last approach to be discussed—the effect of the finite lifetime of a phonon in the solid—would seem the most promising. The very important analysis carried out independently by Haug and Weiss[29] and by Peterson and Anderson[30] showed that values of $\bar{\alpha}_1$ comparable to those observed could be obtained by numerical solution if the elastic constants and k-vectors in the solid were made complex to account for attenuation. They showed that energy could be transmitted outside the critical cone and indeed that when $k_{\|} l_s \sim 1$, ($l_s$ is the phonon mean free path) the transmission coefficient became very roughly independent of the angle of transmission and equal to its value inside the cone[29] ($\sim \rho_2 c_2/\rho_1 c_D$). The process was discussed further in Refs. 31 and 21 and also in 4 and the latter discussion will be summarised briefly below.

It was noted that for $k_{\|} l_s \sim 1$, the increase in $\bar{\alpha}_1$ outside the cone resulted in an increase in $\bar{\alpha}_1$ of $(\rho_2 c_2/\rho_1 c_D)(c_D/c_2)^2$ i.e. to the value shown by the line in Fig. 3. For $k_{\|} l_s < 1$ the full factor of $(c_D/c_2)^2$ is not obtained and this could be the reason why the observed pressure dependence of $\bar{\alpha}_1$ is intermediate between that of $\rho_2 c_2$ and $\rho_2/c_2$. So this approach does seem capable of explaining most of the major experimental features.

The physical explanation for the enhancement seems to be closely associated with Rayleigh modes of a surface. At a perfect surface, phonons from the helium incident outside the critical angle are reflected but part of their energy is stored in the solid surface. The amount stored is really quite negligible even when summed over all angles except for phonons incident within a very small solid angle between two cones where the angle of incidence is such that the helium phonon and the Rayleigh mode of the free surface can mix because they have the same $k_{\|}$. The helium energy stored at this angle can now be transferred to the solid if there is scattering in the solid surface. The energy transferred at first increases linearly with the rate of scattering and should be independent of the properties of the helium. Eventually however it saturates when the rate of scattering from the surface reaches the rate at which the energy can be brought up to the surface by the helium and this is essentially limited by the relatively few phonons that mix with the Rayleigh mode i.e. those that are incident in the very small solid angle. The value of $\bar{\alpha}_1$ at saturation is about three times bigger than that when there is no dissipation[3, 32]. However as the scattering increases so does the uncertainty $\Delta k_{\|}^s = l_s^{-1}$ which governs the precision to which $k_{\|}$ must be conserved. So more phonons can in fact mix with the Rayleigh modes, the small solid angle starts to widen out and $\bar{\alpha}_1$ rises again from its "saturation" value. The increase in solid angle will vary more rapidly than $\Delta k_{\|}^s$ i.e. $l_s^{-1}$ so that the bottleneck will again move to the solid and $\bar{\alpha}_1$ should again become roughly independent of the properties of the helium. But eventually when

$\Delta k^s_{\parallel} = k_{\parallel}$, all the helium phonons can store energy in the Rayleigh mode and $\bar{\alpha}_1$ must saturate or even fall since further increase in the scattering may impede the flow of phonons into the bulk solid.

So the transmission coefficient should be very sensitive to the attenuation of the surface waves. At high temperatures ~1K, these travel in the first few Å of the surface and it is easy to imagine that the most important scattering process might be roughness. At lower temperatures (~10 mK), the penetration is ~1000Å, microscopic roughness may be less important so that, as stressed in reference 31, it is not unexpected that $\bar{\alpha}_1$ should become sensitive to dislocation density.

We finish this discussion by taking this argument back to the gas model. We recall that there it was suggested[28] that scattering in the surface could

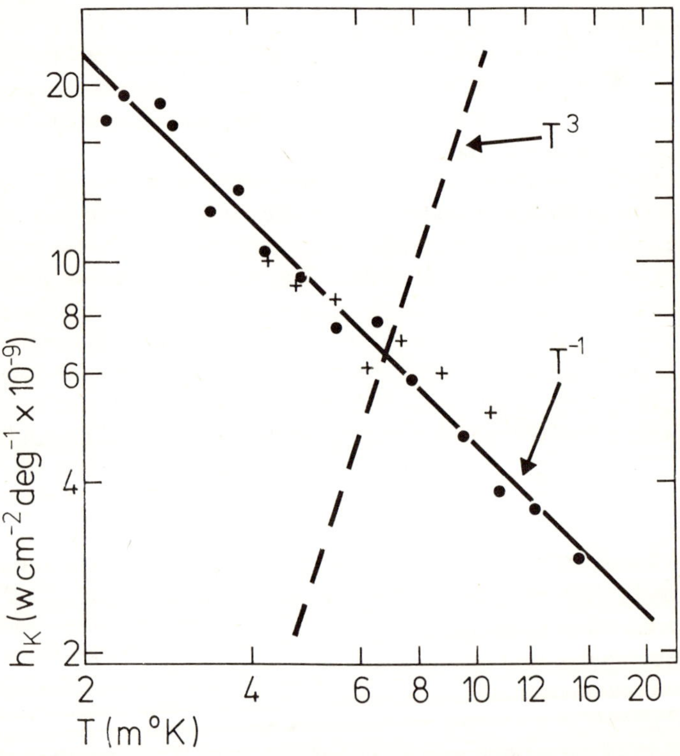


FIG. 5. The Kapitza conductance of CMN to liquid $He^3$. ● 0·28 atm. + 27 atm. The data are those of Abel *et al.* [38, 39] and are taken from reference 40. The solid line shows the dependence $h_K \propto T^{-1}$ and for comparison the broken line shows approximately the phonon conductance between Cu and $He^3$.[7] It appears however that the phonon conductance for CMN is smaller and does not dominate until $T \gtrsim 18$ mK.[41]

produce an enhancement of $\sim 3$ because essentially of the increase in the amplitude of vibration of the surface that results. However we can now see that a much more important effect there too is the non-conservation of $k_{\parallel}$ that results. Conservation of $k_{\parallel}$ provides a severe limit on the amount of phase space into which bound or unbound atoms may be excited and if this is removed we may expect a substantial enhancement of $\bar{\alpha}_1$. It would be very interesting to repeat the calculations of the gas model with some uncertainty $\Delta k_{\parallel}^{s}$ included† and also include specifically the surface mode aspects. We venture a prediction that when $\Delta k_{\parallel}^{s} \to k_{\parallel}$, it will be found that "phonons" in the gas incident at all angles can excite surface modes very easily and that the bottleneck may prove to be the scattering of these modes into bulk phonons in the solid.

In conclusion we note that a great deal of recent interesting theoretical work has been left out of this discussion. We mention particularly the development of a wholly quantum mechanical treatment of Kaptiza conductance. Until recently treatments were based on classical elastic theory or on the semiclassical Boltzmann equation approach but it has now been shown that many of the aspects discussed in this paper—the pure solid,(33) scattering in the surface,(34) the relative importance of zero sound modes and quasiparticles in $He^3$ and mixtures(35) etc. can be treated with advantage using a transfer Hamiltonian approach. However it is satisfying that this approach gives very much the same results in nearly all cases. We mention also further work on lifetime effects in both the solid(36) and the liquid.(37)

## III. MAGNETIC KAPITZA CONDUCTANCE

### 1. CMN–LIQUID $HE^3$

In 1966 Abel, Anderson, Black and Wheatley(38) made the interesting and technically very important discovery that at very low temperature, the conductance between cerium magnesium nitrate (CMN) and liquid $He^3$ is orders of magnitude smaller than that between $He^3$ and other solids. It also has a completely different temperature dependence varying approximately as $T^{-1}$ rather than $T^3$ (see Fig. 5). Clearly a second mechanism of conductance is present and, since the major additional feature is the presence of electronic spins in the CMN, it seems very likely as suggested by Wheatley(42) that this mechanism involves the magnetic coupling betweep the spins and the nuclear moments of the $He^3$ atoms. Strong support for this idea has come from measurements showing that the conductance appears to be field depen-

† Some uncertainty will arise due to collisions of gas atoms not included in the theories.

dent[41] and also from work on a CMN surface "plated" with a monolayer of $He^4$.[40]† The plated surface had a conductance which at the lowest temperatures was only $\sim 2\%$ of that for an unplated surface and with a temperature dependence quite similar to that for a non-magnetic solid. The additional mechanism is then strongly reduced when the $He^3$ is separated from the CMN by a short distance of non-magnetic material and this would be consistent with it being due to a relatively short range interaction such as a magnetic coupling.‡

A theoretical treatment of magnetic conductance given in 1970 by Leggett and Vuorio[44] was very successful in that it anticipated the field dependence and accounted for both the temperature dependence and approximate magnitude of the experimental results. Further treatments of the same process but using somewhat different approaches and improved representations of the interaction have been given by Guyer[45] and by Mills and Beal–Monod.[46] These theoretical treatments all use the Kubo formula but in discussing the mechanism here we use a simpler although much less rigorous approach.

The ground state of the $Ce^{3+}$ ion in CMN is a Kramers doublet and the separation of the ions, $a$, is sufficiently large that the interaction between them is dipolar. The interaction energy is $\sim \mu_e^2/a^3$ and ordering occurs at a temperature $T_c \sim 2$ mK given by $k_B T_c \sim \mu_e^2/a^3$. ($\mu_e$ is the $Ce^{3+}$ moment.) When $T \gg T_c$ and $H \sim 0$ the neighbouring moments are equally likely to be $\pm\mu_e$ and the average local field is zero.§ However at any instant in time, since the up spins are randomly placed on neighbouring sites, there is a distribution of fields and so of ground state splittings $\Delta$ over the ions in the sample. The distribution is independent of temperature in this limit and the splitting has an average value of $|\Delta| \sim kT_c$. The splitting for a particular ion will change in a time $T_2$ which is expected to be $\sim |\Delta|/\hbar \sim 10^{-8}\, s$ and this appears to be confirmed by experiment.[47] We shall see later that this is sufficiently fast that heat can reach the surface through the spin system with reasonable ease.

In liquid $He^3$, thermal conduction is almost entirely due to the independent motion of quasiparticles and the spin lattice relaxation time is long so to obtain energy transfer between $He^3$ and CMN there must be coupling between the kinetic energy of the quasiparticles and the spin energy of the CMN. This is provided by the fluctuating magnetic field at a CMN spin due to the rapid movements of the $He^3$ moments in the liquid near to the spin.

† Since in $He^3$–$He^4$ solutions the $He^4$ is adsorbed preferentially onto surfaces[43] the plating can be obtained by working with dilute solutions of $He^4$ in $He^3$.

‡ The $He^3$ is also less dense when $He^4$ is present since it then experiences a weaker attractive potential from the CMN.

§ The average local field is of course zero at lower temperatures provided $T > T_c$.

Thus a quasiparticle approaches the surface, flips the spin of a CMN ion and the energy $\Delta$ is given to or taken from the kinetic energy of the quasiparticle. Conservation of angular momentum can be obtained either by reversal of the nuclear moment (the energy required to do this is very much less than $\Delta$) or by a change in direction of the particle, and the change in linear momentum of the quasiparticle is taken up by the CMN sample as a whole.

Most of the energy transfer seems likely to take place from the $Ce^{3+}$ ions in the layer nearest the surface and we suppose that there are $n_a$ such ions per unit area ($n_a \sim a^{-2}$) each with a fixed splitting $\Delta$ equal to the mean value. This assumption amounts to taking averages earlier than we should and also to ignoring correlation effects but should not alter the result by more than a numerical factor of order unity. We take the $Ce^{3+}$ ions and the $He^3$ system to have temperatures of $T$ and $T - \delta T$ respectively. So there are $n_a Z^{-1} \exp(-\Delta/k_B T)\, Ce^{3+}$ ions per unit area in the upper state and $n_a Z^{-1}$ in the lower state where $Z^{-1} = 1 + \exp(-\Delta/k_B T)$. The net heat flow per unit area from the CMN to the $He^3$ is therefore

$$\dot{Q} = \sum_{kk'} \Delta n_a Z^{-1}\{\exp(-\Delta/k_B T) f_k(1 - f_{k'}) - f_{k'}(1 - f_k)\} P_{kk'} \qquad \text{(III.1)}$$

where $f_k$ and $f_{k'}$ are the Fermi functions for $He^3$ at $(T - \delta T)$ and $P_{kk'}$ is the transition rate of a $He^3$ atom moving from an occupied state $k$ to an empty state $k'$. Since $E_{k'} - E_k \simeq \Delta$ the expression in the curly bracket can readily be shown to equal

$$\Delta \exp(-\Delta/k_B T) f_k(1 - f_{k'})\delta T/k_B T^2,$$

and for $k_B T \gg \Delta$ as assumed here

$$Z^{-1} \exp(-\Delta/k_B T) \simeq \tfrac{1}{2} \qquad \text{and}$$

$$f_k(1 - f_{k'}) \simeq -k_B T \frac{df}{d\varepsilon} \simeq k_B T\, \delta(E_k - E_F)$$

so that from (III.1)

$$h_K = \dot{Q}/\delta T = \tfrac{1}{2} \sum_{kk'} n_a \Delta^2 T^{-1} P_{kk'} \delta(E_k - E_F). \qquad \text{(III.2)}$$

Now, by second order perturbation theory, the transition rate from the combined $|i\rangle$ state (e.g. $Ce^{3+} \uparrow, k$) to $|f\rangle$ $(\downarrow, k')$ is

$$P_{kk'} = \frac{2\pi}{\hbar} |\langle i|\mathscr{H}'|f\rangle|^2 \delta(E_{k'} - E_k - \Delta)$$

where $\mathscr{H}'$ is the Hamiltonian coupling the two systems. The interaction energy $\simeq g_n \mu_n \mu_e / r^3 \simeq g_0/r^3$ where $g_n \mu_n$ is the moment of the $He^3$ nucleus and $r$ is the separation of the $Ce^{3+}$ spin and the $He^3$ nucleus. To do the calculation properly we need to express $\mathscr{H}'$ as a product of raising or lowering operators for the spin and terms such as $a_{k'}^{\dagger}a_k$ which annihilate quasiparticles in state $k$ and create them in state $k'$. However, for our present purposes we jump this step and take

$$P_{kk'} \simeq \frac{2\pi}{\hbar}(g_0/b^3)^2\delta(E_{k'} - E_k - \Delta).$$

The interaction selects only those quasiparticles near to the surface and $b$ is a mean separation for these so $b^3$ is very approximately the volume of $He^3$ which interacts "effectively" with one spin. We now substitute this expression for $P_{kk'}$ into (III.2) to obtain

$$h_K = \pi\hbar^{-1}(g_0/b^3)^2 n_a \Delta^2 T^{-1} \sum_k \sum_{k'} \delta(E_{k'} - E_k - \Delta)\delta(E_k - E_F)$$

$$= \pi\hbar^{-1}(g_0/b^3)^2 n_a \Delta^2 T^{-1} \left(\frac{dn}{d\varepsilon}\right)^2 V^2$$

where $(dn/d\varepsilon)V$ is the density of states at the Fermi surface for a volume $V$ and, since the interaction from one spin is effectively restricted to a volume of $He^3$, $V = n^3$, we find

$$h_K \simeq \pi\hbar^{-1} g_0^2 n_a \Delta^2 \left(\frac{dn}{d\varepsilon}\right)^2 T^{-1}. \qquad \text{(III.3)}$$

This expression is very similar to that of Leggett and Vuorio in zero field

$$h_K = \pi\hbar^{-1} g_0^2 n_a \left(\frac{dn}{d\varepsilon}\right)^2 K_{\text{eff}}^2 k_B^2 T Q_s(T) \qquad \text{(III.4)}$$

since their $Q_s(T) \sim \Delta^2/k_B^2T^2$. They have also allowed for Fermi liquid effects which increase $h_K$ by $K_{\text{eff}}^2$($K_{\text{eff}}^2 \gtrsim 12$ for pure $He^3$ and $\sim 1$ for dilute solutions of $He^3$ in $He^4$).

As we noted earlier, the magnitude and temperature dependence predicted by this approach is in rather good agreement with experiment; the most recent measurements[(41)] suggest $h_K = 1{\cdot}1 \times 10^{-10}T^{-1} W\,\text{cm}^{-2}\,\text{deg}^{-1}$ while (III.4) gives $h_K = 0{\cdot}2 \times 10^{-10}T^{-1}$. We also see from (III.3) that if $\Delta$ is increased by applying a magnetic field, $h_K$ should increase and should vary as $H^2$ when $\Delta_H \gg \Delta$. Figure 6 shows that the experimental results of Bishop *et al.*[(41)] are not inconsistent with the predicted dependence at lower fields if one keeps in mind the experimental uncertainties referred to by the authors. Nevertheless it would clearly be valuable to have more data on the effects of magnetic fields.

The dependence of $h_K$ on $(dn/d\varepsilon)$ and $K_{\text{eff}}$ suggests that it should be increased by a factor of up to $\sim 5$[(44)] by an increase in the density of the liquid to near the solidification point but experimentally no significant increase has been observed (see Fig. 5). This disagreement may well be attributable[40] to the substantial increase in density of the $He^3$ in the first layer or so next to the CMN due to the presence of the attractive potential. The density of these layers which are of course those most strongly coupled to the spins will be rather insensitive to changes in pressure. Of course if the dense layer were

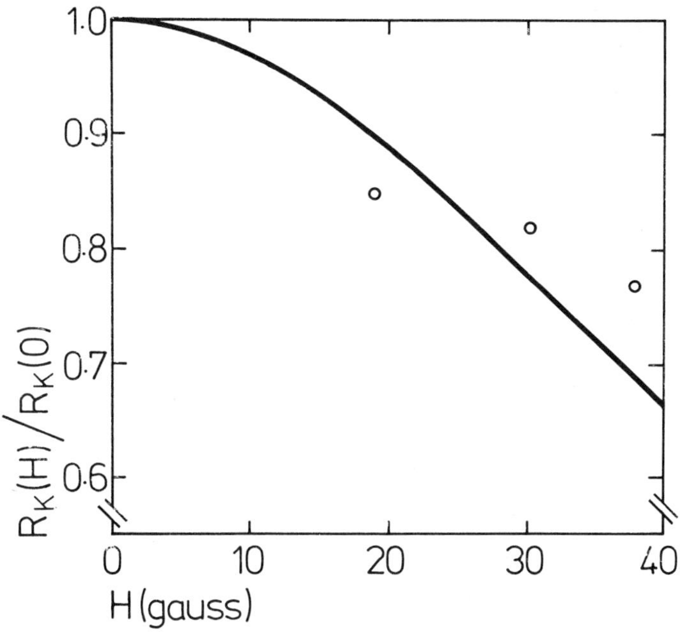


FIG. 6. The magnetic field dependence of $R_K$ for CMN to $He^3$ at 19 mK plotted as $R_K(H)/R_K(O)$.[41] The solid line represents $R_K(H)/R_K(O) = (1 + (g\beta H)^2/2\Delta^2)^{-1}$ (see reference 48).

solid as could well be the case, the treatment given here would hardly seem appropriate and indeed it might be more suitable to apply a treatment for CMN to solid $He^3$. However as we shall see later, this leads to a conductance $\propto T^{-2}$ which is in disagreement with experiment. This might then be evidence that the $He^3$ in the first adsorbed layer is still highly mobile.

We saw that in the phonon problem it was convenient to discuss the conductance in terms of a transmission coefficient for phonons and before leaving the liquid $He^3$ problem we derive an expression for the transmission coefficient of spin energy from CMN to $He^3$. We first consider how this energy is conducted through the spin system inside the CMN. Kinetic theory would suggest that

$$K \sim \tfrac{1}{3}C\bar{v}\lambda$$

where C is the specific heat of the spins per unit volume, $\bar{v}$ the mean velocity of a spin flip and $\lambda$ the mean free path. We can write $C = n_a C_s/a$ where $C_s$ is the specific heat of a spin, so that $C_s \sim \Delta^2/4k_B T^2$, $\bar{v} \sim a/T_2 \sim a\Delta h^{-1}$ and $\lambda \sim a$ and

$$K \sim \tfrac{1}{12} n_a a h^{-1} \Delta^3/k_B T^2.\dagger \tag{III.5}$$

We define an intrinsic conductance $h_i = K/a$ for CMN to describe the rate at which spin energy passes from one layer of spins to the next at a distance $a$ away when there is a temperature difference ($a$ grad $T$) between them and in terms of this, the transmission coefficient for spin energy can be written

$$\alpha_s = h_K/h_i.$$

From (III.3) and (III.5) this becomes

$$\alpha_s = 12\pi g_0^2 \left(\frac{\mathrm{d}n}{\mathrm{d}\varepsilon}\right)^2 k_B T/\Delta \sim 10^{-9}\, T \text{ where } T \text{ is in mK}$$

and we have used the parameters given in reference 44.

We see then that the magnetic transmission process is really very inefficient and that $h_K$ is only substantial because of the very large value of $h_i$. The reason for the small size of $\alpha_s$ can be seen if we write $\alpha_s$ in the form

$$\alpha_s = 12\pi \left[\left(\frac{g_0}{b^3}\right)^2 \Big/ \Delta^2\right] \left[\left(b^3 \left(\frac{\mathrm{d}n}{\mathrm{d}\varepsilon}\right) k_B T\right)\left(b^3 \left(\frac{\mathrm{d}n}{\mathrm{d}\varepsilon}\right)\Delta\right)\right].$$

The first square bracket gives the ratio of the squares of the interaction energies $Ce^{3+} - He^{3+}$ and $Ce^{3+} - Ce^{3+}$ and has a value $\sim 10^{-7}$. The second square bracket is the ratio of the number of $He^3$ nuclei and of $Ce^{3+}$ spins to which a $Ce^{3+}$ spin can transfer energy and has a value $\sim 10^{-2} T$ ($T$ in mK). The number of $Ce^{3+}$ spins is the number of nearest neighbours and is of order unity. The number of $He^3$ nuclei is the number in volume $b^3$ than can interact which is $\sim b^3(\mathrm{d}n/\mathrm{d}\varepsilon)kT$. However the requirement that there is also an empty state of energy $\sim\Delta$ away reduces this by the factor $\sim b^3(\mathrm{d}n/\mathrm{d}\varepsilon)\Delta$.

## 2. CMN-SOLID HE$^3$

It would be very interesting to have experimental data on the magnetic

† This gives $K \sim 10^{-13} T^{-2}$ W cm deg$^{-1}$ ($T$ in deg K) which can be compared with the phonon conductivity $K_p \sim \lambda T^3$ W cm deg$^{-1}$(49) where $\lambda$ is the phonon mean free path and is comparable to the "diameter" of the sample. So $K_s/K_p \sim 10^{-13} T^{-5}/\lambda$ and in a powder of 50 μ size they should be comparable at around 7 mk and $K_s$ should dominate at temperatures much below. However the data in reference 41 indicate that the conductivity in CMN powder (55 μ) is still due to phonons at 3 mk so (III.5) may be somewhat of an overestimate.

conductance between CMN and solid $He^3$. A theoretical discussion has been given by Guyer[(45)] (see also reference 50) who pointed out that the splittings of the nuclear spin levels of a $He^3$ atom in the solid due to spin fluctuations of its neighbours are very similar in magnitude to those in CMN. Energy can thus be exchanged rather effectively by spin flips between $Ce^{3+}$ and $He^3$. We expect the transmission coefficient for this process to be $\alpha_s \sim (g_n\mu_n/\mu_e)^2$ so that

$$h_K = \alpha_s h_i \sim \tfrac{1}{12}\, n_a \hbar^{-1} \Delta^3 (g_n\mu_n/\mu_e)^2/k_B T^2$$

which is equal to Guyer's expression apart from numerical factors. Guyer calculated this to be

$$h_K \sim 10^{-10} T^{-2} W \mathrm{cm}^{-2}\,\mathrm{deg}^{-1}$$

indicating that the magnetic conductance to the solid is always greater than that to the liquid for $T < 1$K and is $\sim$100 times larger at 10mK. However we note that it is not clear that overall the process will be more efficient because of the difficulty of getting the energy away from the surface into the phonon system of solid $He^3$: $T_2$ is slow as is the effective $T_1$.

## 3. ORDERED MAGNETIC SYSTEMS

In an ordered material in which the spins are on a regular lattice, individual spin flips are no longer possible and energy is transported through the system by magnons. Hence we expect the Kapitza conductance to depend on the rate at which energy can be transported from the magnon system to the $He^3$ quasiparticles. Leggett and Vuorio[(44)] report preliminary calculations of this process which show that the conductance should increase with temperature. They do not specify a temperature dependence but note that it should be less rapid than the $T^3$ dependence of phonon conductance. This is supported by the recent work of Mills and Beal-Monod[(51)] who show that $h_K \propto T^{3/2}$ for a regular lattice of spins which are ordered ferromagnetically. They also show that the temperature dependence increases when the spins are spatially disordered but that it is still less than $T^3$. We see then that theory predicts that the magnetic conductance should rise to a maximum at $T \sim T_c$ and then fall. We note that the temperature dependence is such that it will always become larger than the phonon conductance at low enough temperatures.

It seems likely that examples of magnetic conductance at $T < T_c$ are present in the work of Bishop, Mota and Wheatley[(52)] and of Avenel *et al.*[(53)] Bishop *et al.* made measurements on Pt wires in the temperature range 3–8mK and found that $h_K \propto T^2$. The conductance was substantially reduced by $He^4$ plating (by a factor of up to 7) indicating that it was probably due to a magnetic process. The similarity in the conductance obtained for two samples of different purity led the authors to suggest it could be an intrinsic effect and they speculated that it might be due to coupling between the $He^3$ and the conduction electrons. However Mills and Beal-Monod[46] showed that for dipolar and exchange interactions such a conductance should vary as $T^3 \ln T$ and $T^3$ respectively and also should be many orders of magnitude less than that observed. In their second paper[51] they show that a dilute alloy of magnetic impurities experiencing random internal fields such as Pt(Fe) should have a conductance $h_K \propto T^2$ as observed.

The work of Avenel *et al.*[53] was on Fe(30 ppm) in Pd, Gd(0·2%) in Au and Mn (3 ppm) in Cu. The measurements were made in the range 3–50 mK in which the Gd in Au sample was known to be largely ordered. The conductances varied approximately as $T^3$ at the highest temperatures but rose quite rapidly above the $T^3$ line as the temperature fell with $h_K(\text{Au}) \propto T$ and $h_K(\text{Pd})$ $T^{1 \cdot 4}$. However these data refer to the total conductance–magnetic plus phonon—and in a recent analysis of the data[51] in which the phonon conductance was subtracted $h_K(\text{Au})$ had a maximum at $\sim 10$ mK and tended towards a $T^2$ dependence at the lowest temperatures and that there was also a suggestion of a maximum in the Pd(Fe) conductance. An attempt was made to plate the surface of the Pd sample with $He^4$ but no effect was found and the authors suggest that the $He^4$ became trapped in a heat exchanger.

## IV. Conclusion

It is possible to believe that the main problem of phonon conductance is largely solved and that the solution lies in the non-conservation of $k_{\parallel}$ assumed in acoustic theory: the uncertainty in $k_{\parallel}$ introduced by scattering in the surface allows very efficient transfer through the surface modes. However even if this optimism is justified, and it is not shared by all workers in the field, much detail remains to be cleared up. We need to establish experimentally that the energy really does go into phonon modes in the helium and not into single particle excitations and we need to determine whether the very strong scattering assumed in this approach really does exist. Further theoretical work is desirable perhaps aimed towards obtaining approximate analytical solutions for $\bar{\alpha}_1$ since the present macroscopic expressions have so far been solved only by computer solution. It would also be interesting to place on a more quantitative footing the arguments relating to the role of Rayleigh modes and to insensitivity to the properties of helium advanced here. There has also been no work to date to explain why for example the transmission coefficient to solid neon should not also be enhanced; in fact Guo and Maris[2] showed that it is somewhat smaller than that to liquid helium. It would be agreeable if one could increase the density and velocity of sound of liquid helium independently until they became equal to those of neon. At some point $\bar{\alpha}_1$ would have to become dependent on the properties of the helium and there would need to be a maximum in $\bar{\alpha}_1$ before it fell to the value for neon. It seems likely that the fall will be found to have more to do with the velocity than the density. The velocity determines the solid angle from which one gets "extra" transmission and also the depth of the Rayleigh mode. The shallower the mode (i.e. the smaller $c_2$) the bigger the scattering effect of surface roughness. Finally there are many aspects of phonon conductance which are peripheral to the main problem but which contain a lot of interesting physics. Some of these were mentioned briefly in Section II.

It would seem too that the physics of magnetic conductance is broadly understood both above and below the ordering temperature although again there is detail to fill in. It would be of interest too to have measurements to solid $He^3$ and also to the superfluid phases of $He^3$.

## REFERENCES

1. P. L. Kapitza. *J. Phys. (Moscow)* **4**, 181 (1941).
2. The first experiments to provide information on the polarization dependence of $\bar{\alpha}_1$ were by C. H. Guo and H. J. Maris, *Phys. Rev. Letts.* **29**, 855 (1972); and T. J. B. Swanenburg and J. Wolter, *Phys. Rev. Letts.* **31**, 693 (1973). We also mention the work of C. H. Anderson and E. S. Sabisky, *In* "Physical Acoustics" (W. P. Mason and R. N. Thurston, Eds.) Academic Press, New York (1971), Vol. 8, p. 1, which was the first to give information using monochromatic phonons.
3. I. M. Khalatnikov. *Zh. Eksper. Teor. Fiz.* **22**, 687 (1952); R. M. Mazo, PhD thesis, Yale University (1955); W. A. Little, *Can. J. Phys.* **37**, 334 (1959); I. L. Bekarevich and I. M. Khalatnikov, *Zh. Eksper. Teor. Fiz.* **39**, 1699 (1960) (*Sov. Phys. JETP* **12**, 1187 (1961)). This last calculation which is for $He^3$ is not strictly an application of acoustic theory but it leads essentially to the acoustic result.
4. L. J. Challis. *J. Phys. C.* **7**, 481 (1974).
5. J. D. N. Cheeke, B. Hebral, J. Richard, and R. R. Turkington, *Phys. Letters* **46A**, 81 (1973).
6. F. Wagner, F. J. Kollarits, and M. Yaqub. *Phys. Rev. Letts.* **32**, 1117 (1974).
7. A. C. Anderson, J. I. Connolly, and J. C. Wheatley. *Phys. Rev.* **135**, 910 (1964).
8. R. A. Sherlock, A. F. G. Wyatt, N. G. Mills, and N. A. Lockerbie. *Phys. Rev. Letts.* **29**, 1299 (1972) and R. A. Sherlock, N. G. Mills, and A. F. G. Wyatt, to be published.
9. T. Ishiguro, and T. A. Fjeldly. *Phys. Letters* **45A**, 127 (1973).
10. F. W. Sheard, R. M. Bowley, and G. A. Toombs. *Phys. Rev.* **A8**, 3135 (1973). See also H. Haug. *Phys. Letters* **45A**, 170 (1973); R. M. Bowley, F. W. Sheard, T. J. Sluckin, and G. A. Toombs. *In* "Proc. Symposium on Microwave Acoustics, Lancaster 1974", to be published.
11. H. Kinder and W. Dietsche, to be published in *Phys. Rev. Letts.*
12. L. J. Challis, K. Dransfelt, and J. Wilks. *Proc. Roy Soc.* **A260**, 31 (1961).

13. Kuang Wey-Yen, *Zh. Eksper. Teor. Fiz.* **42**, 921 (1962) (*Sov. Phys. JETP* **15**, 635 (1962)). It would appear that there is an increase in $\bar{\alpha}_1$ when the helium becomes solid. J. T. Folinsbee, and A. C. Anderson. *Phys. Rev. Letts.* **31**, 1580 (1973).
14. N. S. Snyder, N. B. S. Technical note 385 (1969), (see also *Cryogenics* **10**, 89 (1970)). This review paper includes a very useful collection of experimental data.
15. W. L. Johnson and A. C. Anderson. *Phys. Letters* **37A**, 101 (1971); J. D. Siegwarth and R. Radebaugh. *In* "Proc. 13th Int. Conf. Low Temperature Physics 1972", R. H. Kropschot and K. D. Timmerhaus, Eds.), Plenum Press (in press).
16. E. A. Jones and J. C. A. van der Sluijs. *Cryogenics* **13**, 535 (1973).
17. C. F. Mate and S. P. Sawyer. *In* "Proc. 11th Int. Conf. Low Temperature Physics St Andrews 1968", (J. F. Allen, D. M. Finlayson, and D. M. McCall, Eds.) (St Andrews University Press, 1968), p. 579. These data are very similar for both phases of $He^3$ but differ significantly for $He^4$. A. C. Anderson and W. L. Johnson. *J. Low Temp. Phys.* **7**, 1 (1972) ($He^3$); C. J. Guo and H. J. Maris, reference 2 ($He^4$) ; A. R. Long, R. A. Sherlock and A. F. G. Wyatt. *J. Low Temp. Phys.* **15**, 523 (1974) ($He^4$); H. Kinder and W. Dietsche reference 11 ($He^3$ and $He^4$) to be published. There have also been measurements to thick films of $He^4$ (G. L. Pollack reference 18).
18. e.g. J. D. N. Cheeke. *J. Phys. Paris, Colloque Suppl.* **C3**, 129 (1970). The theoretical position and in particular the work of Little, Andreev and Bloch is discussed by L. J. Challis and J. D. N. Cheeke, *Proc. Roy Soc.* **A304**, 479 (1968), by G. L. Pollack, *Rev. Mod. Phys.* **1**, 48 (1969) in his review of the Kapitza problem and recently by Peterson and Anderson, reference 30.
19. M. F. Whelan and D. V. Osborne. *J. Low Temp. Phys.* **8**, 449 (1972).
20. J. D. N. Cheeke, B. Hebral, J. Richard and R. R. Turkington. *Phys. Rev. Letts.* **32**, 658 (1974).
21. L. J. Opsal and G. L. Pollack. *Phys. Rev.* **A9**, 2227 (1974). This paper shows that a reasonable quantitative description can be given of $\bar{\alpha}_1$ if both attentuation in the solid and dense layer effects are included. The existence of the maximum in $\bar{\alpha}_1$ at 1K not found in reference 12 is thought by Pollack (private communication) to be due to differences in the density of the helium in the region closest to the solid assumed in the two calculations.
22. M. Vuorio. *J. Phys. C* **5**, 1216 (1972); see also J. C. A. van der Sluijs, E. A. Jones and A. E. Alnaimi. *Cryogenics* **14**, 95 (1974).
23. L. J. Challis and G. A. Toombs. *In* "Proc. 12th Int. Conf. Low Temperature Physics Kyoto 1970", (E. Kanda, ed.), Keigaku Publishing Co., Tokoyo (1971) p. 851; G. A. Toombs and L. J. Challis. *J. Phys. C* **4**, 1085 (1971).
24. J. E. Lennard-Jones and C. Strachan. *Proc. Roy Soc.* **A150**, 442 (1935); C. Strachan, *Proc. Roy. Soc.* **A150**, 456 (1935); A. F. Devonshire. *Proc. Roy. Soc.* **A158**, 269 (1937).
25. R. G. Johnson and W. A. Little. *Phys. Rev.* **130**, 569, (1963).
26. G. A. Toombs, F. W. Sheard and M. J. Rice. *Solid State Comm.* **10**, 601 (1972).
27. M. J. Rice and G. A. Toombs. *Phys. Rev.* **A5**, 2259 (1972).
28. A. R. Long. *J. Low Temp. Phys.* (in press).
29. H. Haug and K. Weiss. *Phys. Letters* **40A**, 19 (1972).
30. R. E. Peterson and A. C. Anderson. *Phys. Letters* **40A**, 317 (1972).
31. R. E. Peterson and A. C. Anderson. *J. Low Temp. Phys.* **11**, 639 (1973).
32. L. J. Challis and J. D. N. Cheeke, reference 18.
33. F. W. Sheard and G. A. Toombs. *J. Phys.* **C5**, L166 (1972); G. A. Toombs and

R. M. Bowley, *J. Phys.* **C6**, L406 (1973); F. W. Sheard, R. M. Bowley, and G. A. Toombs, reference 10; H. Haug, reference 10.
34. T. J. Sluckin, G. A. Toombs and F. W. Sheard. *Solid State Comm.* **14**, 203 (1974).
35. M. J. Rice, F. W. Sheard and G. A. Toombs. To be published.
36. I. M. Khalatnikov and I. N. Ademenko. *Zh Eksper. Teor. Fiz.* **63**, 745 (1972) (*Sov. Phys. JETP* **36**, 391 (1973)); M. Vuorio. *J. Low Temp. Phys.* **10**, 781 (1973).
37. W. M. Saslow. *J. Low Temp. Phys.* **11**, 255 (1973).
38. W. R. Abel, A. C. Anderson, W. C. Black and J. C. Wheatley. *Phys. Rev. Letts.* **16**, 273 (1966). A similar effect has since been seen in $Na_3\ Ce(C_7H_3NO_4)_3$ $15H_2O$ (CDP). J. H. Bishop and A. C. Mota. *Phys. Letters* **43A**, 511 (1973).
39. W. R. Abel, A. C. Anderson, W. C. Black and J. C. Wheatley. *Phys. Rev.* **147**, 111 (1966).
40. W. C. Black, A. C. Mota, J. C. Wheatley, J. H. Bishop and P. M. Brewster. *J. Low Temp. Physics* **4**, 391 (1971).
41. J. H. Bishop, D. W. Cutter, A. C. Mota and J. C. Wheatley. *J. Low Temp. Physics* **10**, 379, (1973).
42. J. C. Wheatley. *Phys. Rev.* **165**, 304 (1968).
43. O. Evenson, D. F. Brewer and A. L. Thomson. *In* "Proc. 11th Int. Conf. Low Temperature Physics, St Andrews 1968" (J. F. Allen, D. M. Finlayson, D. M. McCall, Eds.) St Andrews University Press (1968), p. 125.
44. A. J. Leggett and M. Vuorio. *J. Low Temp. Physics* **3**, 359 (1970).
45. R. A. Guyer. *J. Low Temp. Physics* **10**, 157 (1973).
46. D. L. Mills and M. T. Beal-Monod. *Phys. Rev.* **A10**, 343 (1974).
47. D. S. Betts. Private communication.
48. A. G. Redfield. *IBM J. Res. Develop.* **1**, 19 (1957). An analogy between Redfield's theory of nuclear spin-conduction electron relaxation time in metals and the present problem has been made in reference 41.
49. J. E. Robichaux and A. C. Anderson. *Phys. Rev.* **B2**, 5035 (1970); J. P. Harrison and J. P. Pendrys. *Phys. Rev.* **B7**, 3902 (1973).
50. A. J. Leggett, remarks at European Physical Society conference on Thermal Contact and Thermometry, Albe 1972.
51. D. L. Mills and M. T. Beal-Monod (to be published). I am very grateful to Prof. J. C. Wheatley for his notes on this paper which I have not yet seen.
52. J. H. Bishop, A. C. Mota and J. C. Wheatley. *In* "Proc. 13th Int. Conf. Low Temperature Physics, Boulder 1972" (R. H. Kropschot and K. D. Timmerhaus, Eds.) Plenum Press (in press).
53. O. Avenel, M. P. Berglund, R. G. Gylling, N. E. Phillips, A. Vetleseter, and M. Vuorio. *Phys. Rev. Letts.* **32**, 76 (1973).

# AUTHOR INDEX

Numbers in parentheses are reference numbers and are included to assist in locating references in the text where the authors' names are not mentioned. Numbers in italics indicate the pages on which references are listed in full.

## C

## D



## E



## F

## G



## H

### I



### J



### K

## N



## O



## P

## R



## S

## T

## XY



## Z

# SUBJECT INDEX

## F



## G



## H



## J



## K



## L



## M

## N



## O



## P



## Q



## R



## S